Finite Mathematics

Finite Mathematics

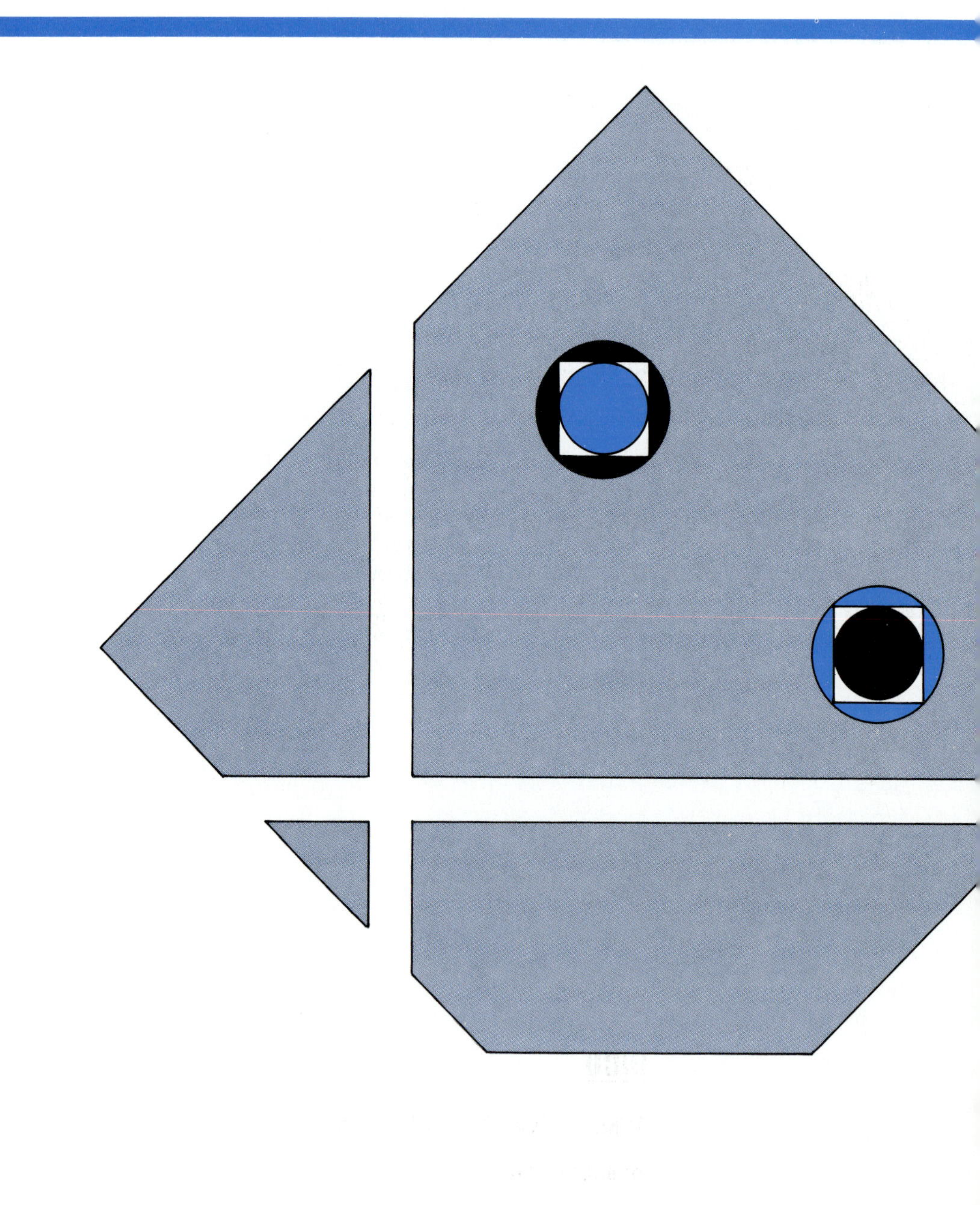

Howard L. Rolf
Mathematics Department,
Baylor University

Gareth Williams
Mathematics Department,
Stetson University

wcb

WM. C. BROWN PUBLISHERS

DUBUQUE, IOWA

Library of Congress Catalog Card Number: 87-24112

ISBN 0-697-06769-6

Printed in the United States of America.
10 9 8 7 6 5 4 3 2 1

To Anita Ward Rolf
Donna Williams

Contents

0

Review Topics 1

Functions and Lines 21

Linear Systems 65

3

Linear Programming 151

4

Linear Programming—Simplex Method 177

5

Sets and Counting 237

Probability 291

Statistics 385

Mathematics of Finance 479

Game Theory 525

Appendix A 550

Preface

Mathematics as we know it came into existence through an evolutionary process that continues today. Occasionally a mathematical idea will fade away as it is replaced by a better idea. Old ideas become modified, or new and significant concepts are born and take their place. Many disciplines have found that mathematical concepts are useful in understanding and applying the ideas of that discipline. Some mathematical concepts were developed in an attempt to solve problems in a particular discipline. This book deals with some of the topics that are especially useful in business, the life sciences, and the social sciences.

Prerequisites

This book assumes at least one year of high school algebra. Chapter 0 provides a brief review for those who may need to refresh their memory.

Features

We have concentrated on writing that is clear, friendly, and considerate of the student. The text contains over 300 examples to illustrate and develop the concepts. Over 1900 exercises provide ample means for the student to apply mathematical concepts to problems. They also illustrate applications of the concepts.

The exercises are graded into three levels: Level I for routine problems, Level II for elementary word problems and somewhat more challenging problems, and Level III for the more difficult problems.

As a pedagogical aid to the students, examples are cross-referenced with exercises. Students can go directly to an exercise that is related to the example in order to test their understanding of the example. Selected exercises refer to an example that illustrates the concept needed for the exercise. Warnings of mathematical pitfalls and reminders of concepts learned earlier appear throughout the text. They are flagged with a bold **Warning** or **Reminder**.

Sequence of Topics

The book was written to provide flexibility in the choice and order of topics. Some topics must, of necessity, be covered in sequence. The chapters that are prerequisites are shown in the following diagram.

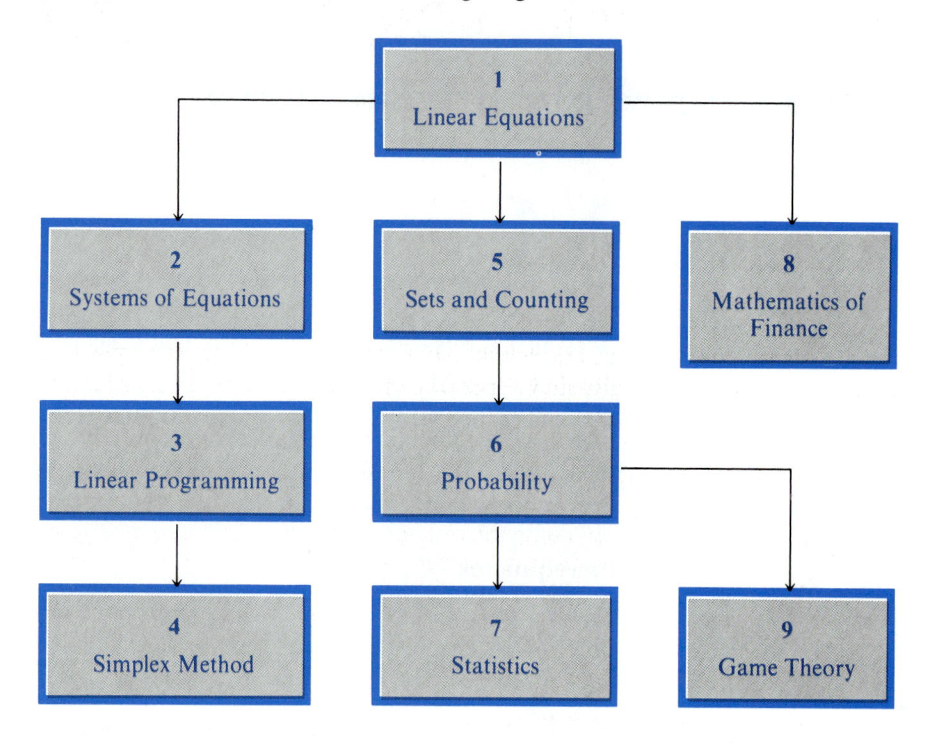

Acknowledgments

A number of people contributed to the writing of this book. We are most grateful to the following reviewers who provided constructive criticism, helped to clarify ideas and to improve the presentation of the concepts:

Dr. David Baugham
Department of Mathematics
College of DuPage
22nd & Lambert Rd.
Glen Ellyn, 60137

Professor Bill Bompart
Department of Mathematics
Augusta College
Augusta, GA 30910

Professor Donald Cathcart
Department of Math Sciences
Salisbury State College
Salisbury, MD 21801

Professor Joseph B. Dennin, Jr.
Fairfield University
Department of Math and Computer Science
Fairfield, CT 06430-7524

Professor Gail Earles
St. Cloud State University
Math & Statistics Department
Engineering & Computer Center
Saint Cloud, MN 56301

Professor Garrett Etgen
University Houston
Department of Mathematics
University Park
Houston, TX 77004

Professor Paul O'Meara
University of Colorado
Department of Mathematics
Denver, CO 80202

Professor Martin Sternstein
Ithaca College
Department of Mathematics
Ithaca, NY 14850

Professor Louis A. Talman
2273 Cherry St.
Denver, CO 80207

Professor Arnold R. Vobach
University of Houston
Department of Math
4800 Calhoun Rd.
Houston, TX 77004

Professor Thomas J. Woods
Central Connecticut State University
1615 Stanley St.
New Britain, CT 06050

Professor Donald L. Zalewski
Northern Michigan University
Department of Math & Computer Science
Marquette, MI 49855

The monumental task of transforming terrible handwriting into a readable manuscript fell upon Dee Nieman, Karen LaPoint, and Joan Maxwell. Thanks for your patience and a job well done.

Finite Mathematics

This chapter contains basic algebra topics that are necessary for the materials in this book. You are encouraged to study these topics for which you need review and skip those topics with which you are familiar.

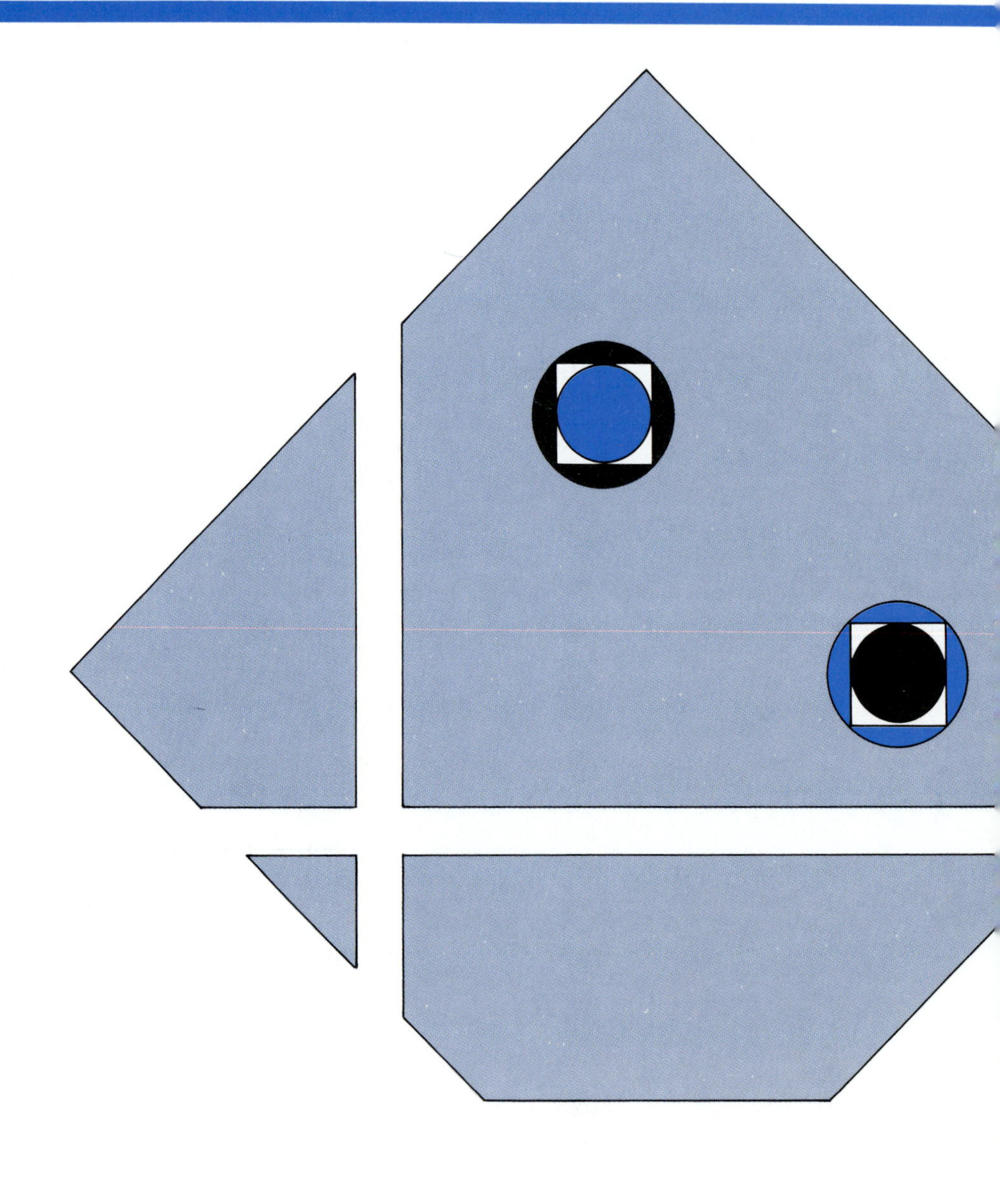

Review Topics

- Properties of Real Numbers
- Solving Linear Equations
- Coordinate Systems
- Linear Inequalities and Interval Notation

0–1 Properties of Real Numbers

The Real Number Line

The most basic numbers in our study of mathematics are the **natural** numbers;

$$1, 2, 3, 4, \ldots$$

These are, in fact, the numbers that we use to count. They can be represented graphically as points on a line, as in Figure 0–1.

Figure 0–1

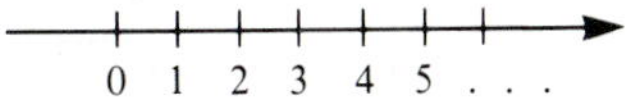

0 1 2 3 4 5 . . .

Begin with a point 0 on the line and a convenient unit scale. Starting at 0, mark off equal lengths to the right. These marks represent $1, 2, 3, \ldots$.

We can mark off lengths in the opposite direction from 0 as well to get the negative numbers $-1, -2, -3, \ldots$ (Figure 0–2). This collection of numbers

$$\ldots, -4, -3, -2, -1, 0, 1, 2, 3, 4, \ldots$$

is called the set of **integers**. We call the number represented by the symbol 0, "zero." It is neither positive nor negative.

Figure 0–2

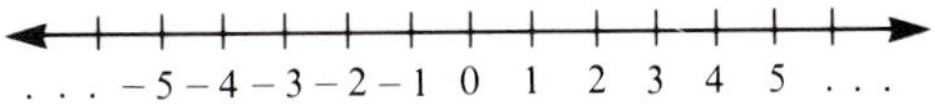

. . . −5 −4 −3 −2 −1 0 1 2 3 4 5 . . .

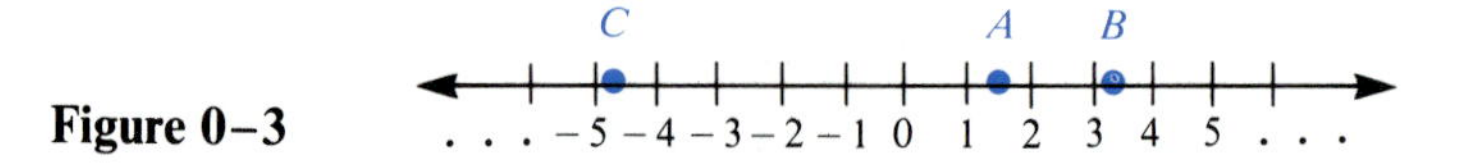

Figure 0–3

Other numbers, the **fractions**, can be represented with points between the integers. In Figure 0–3, point A, halfway between 1 and 2, represents $1\frac{1}{2}$; point B, one quarter of the way between 3 and 4, represents $3\frac{1}{4}$. Point C is three quarters of the way between -4 and -5; so it represents $-4\frac{3}{4}$. These numbers together with the integers are called **rational numbers**.

All these points can be expressed in terms of finite or infinite **decimals**. Point A is the point 1.5, B the point 3.25, and C the point -4.75. Every rational number can be expressed either as a decimal that terminates, such as A, B, and C above, or as a decimal that repeats infinitely. For example, $5\frac{1}{3}$ is a rational number that can be written in decimal form as 5.333...; the 3 repeats endlessly.

There are, however, certain numbers called **irrational** numbers that do not have any pattern of repetition in their decimal form. One such number is $\sqrt{2}$. Its decimal form is 1.414213....

The set of all rational and irrational numbers is called the set of **real numbers**. One way to visualize the set of real numbers is to think of each point on the line as a real number.

The Arithmetic of Real Numbers

There are four useful operations on the set of real numbers: addition, subtraction, multiplication, and division. Table 0–1 summarizes some rules that govern these operations.

Table 0–1

Rule		Example
Division by zero is not allowed.		5/0 and 0/0 have no meaning.
	Rules of Operations	
$a + b = b + a$	Numbers can be added in either order.	$3 + 7 = 7 + 3$
$ab = ba$	Numbers can be multiplied in either order.	$5 \times 8 = 8 \times 5$
$a(b + c) = ab + ac$	A common number can be factored from each term in a sum.	$3(4a + 5b) = 12a + 15b$

Table 0–1 *(continued)*

Rule		Example
	Rules of Signs	
$-a = (-1)a$	The negative of a number a is the product $(-1)a$.	$-12 = (-1)12$
$-(-a) = a$		$-(-7) = 7$
$(-a)b = a(-b) = -ab$	The product of a positive and a negative is negative.	$(-4)8 = -32$ $5(-3) = -15$
$(-a)(-b) = ab$	The product of two negative numbers is positive.	$(-2)(-7) = 14$
$(-a) + (-b) = -(a + b)$	The sum of two negative numbers is negative.	$(-2) + (-3) = -5$
$-(a - b) = (-a) + b$		$-(8 - 3) = -8 + 3$
$\frac{-a}{b} = \frac{a}{(-b)} = -\left(\frac{a}{b}\right)$ $(b \neq 0)$	Division using a positive and a negative number is negative.	$-\frac{18}{3} = -6$ $\frac{22}{(-11)} = -2$
	Arithmetic of Fractions	
$\frac{ac}{bc} = \frac{a}{b}$ $(b, c \neq 0)$	The value of a fraction is unchanged if both numerator and denominator are multiplied, or divided, by the same number.	$\frac{12}{8} = \frac{3}{2}$ $\frac{4}{3} = \frac{20}{15}$
$\frac{a}{d} + \frac{b}{d} = \frac{a + b}{d}$ $(d \neq 0)$	To add two fractions with the same denominators, add the numerators and keep the same denominator.	$\frac{2}{7} + \frac{3}{7} = \frac{5}{7}$
$\frac{a}{d} + \frac{b}{c} = \frac{ac}{dc} + \frac{bd}{dc} = \frac{ac + bd}{dc}$ $(c, d \neq 0)$	To add two fractions with different denominators, convert them to fractions with the same denominators by multiplying numerator and denominator of each by the denominator of the other.	$\frac{2}{5} + \frac{3}{4} = \frac{2(4)}{5(4)} + \frac{3(5)}{5(4)}$ $= \frac{2(4) + 3(5)}{5(4)}$ $= \frac{8 + 15}{20}$ $= \frac{23}{20}$

Table 0–1 *(continued)*

Rule		Example
	Arithmetic of Fractions	
$\frac{a}{d} \times \frac{b}{c} = \frac{ab}{dc}$ $(c, d \neq 0)$	To multiply two fractions, multiply their numerators and multiply their denominators.	$\frac{3}{8} \times \frac{2}{7} = \frac{6}{56} = \frac{3}{28}$
$\frac{a}{d} \div \frac{b}{c} = \frac{a}{d} \times \frac{c}{b} = \frac{ac}{db}$ $(b, c, d \neq 0)$	To divide by a fraction, invert the divisor and multiply.	$\frac{3}{4} \div \frac{2}{5} = \frac{3}{4} \times \frac{5}{2} = \frac{15}{8}$
$\frac{a/d}{b/c} = \frac{a}{d} \div \frac{b}{c} = \frac{a}{d} \times \frac{c}{b} = \frac{ac}{db}$ $(b, c, d \neq 0)$		$\frac{4/5}{8/9} = \frac{4}{5} \div \frac{8}{9} = \frac{4}{5} \times \frac{9}{8} = \frac{1}{5} \times \frac{9}{2} = \frac{9}{10}$

0–1 Exercises

Evaluate each of the following:

1. $(-1)13$
2. $(-1)(-7)$
3. $-(-23)$
4. $(-10)(-4)$
5. $(-5)(6)$
6. $(-2)(-4)$
7. $5(-7)$
8. $(-6) + (-11)$
9. $-(7 - 2)$
10. $(-10)/5$
11. $21/(-3)$
12. $(5 \times 3)/(7 \times 3)$
13. $(-4) + (-6)$
14. $(-3)(-2)$
15. $(-4)2$
16. $\frac{4}{9} + \frac{2}{9}$
17. $\frac{5}{3} + \frac{4}{3}$
18. $\frac{4}{11} - \frac{2}{11}$
19. $\frac{12}{5} - \frac{3}{5}$
20. $\frac{6}{10} - \frac{13}{10}$
21. $\frac{2}{3} + \frac{3}{4}$
22. $\frac{5}{8} - \frac{1}{3}$
23. $\frac{5}{6} - \frac{7}{4}$
24. $\frac{5}{12} - \frac{1}{6}$
25. $\frac{2}{5} + \frac{1}{4}$
26. $(-3) + 6$
27. $\frac{4}{7} - \frac{3}{5}$
28. $\frac{2}{3} \times \frac{4}{5}$
29. $(\frac{3}{4})/(\frac{9}{8})$
30. $\frac{3}{8} + \frac{2}{5}$
31. $(\frac{2}{7})/(\frac{4}{5})$
32. $(\frac{4}{3})(\frac{6}{7})$

33. $(\frac{1}{3})(\frac{1}{5})$

34. $6 \times (-3)$

35. $\frac{2}{5} \times \frac{4}{3}$

36. $(-\frac{2}{3})(\frac{1}{9})$

37. $(-\frac{3}{5})(-\frac{4}{7})$

38. $\frac{4}{5} \div \frac{2}{15}$

39. $\frac{3}{11} + \frac{1}{3}$

40. $(-\frac{4}{9})/(\frac{5}{2})$

41. $\frac{5}{7} \div \frac{15}{28}$

42. $(\frac{4}{9})/(\frac{16}{3})$

43. $(\frac{5}{8}) \div (\frac{1}{3})$

44. $(\frac{1}{2} - \frac{1}{3})(\frac{5}{7})$

45. $(\frac{3}{4} + \frac{1}{5}) \div (\frac{2}{9})$

46. $5(4a + 2b)$

47. $-2(3a + 11b)$

48. $2(a - 3b)$

49. $-5(2a + 10b)$

0–2 Solving Linear Equations

Numerous disciplines including science, technology, social sciences, business, manufacturing, and government find mathematical techniques essential in day-to-day operations. They depend heavily on mathematical equations that describe conditions or relationships between quantities.

In an equation such as

$$4x - 5 = 7$$

the symbol x, called a **variable**, represents an arbitrary, an unspecified, or an unknown number just as John Doe and Jane Doe often denote an arbitrary, unspecified, or unknown person.

The equation $4x - 5 = 7$ may be true or false depending on the choice of the number x. If we substitute the number 3 for x in

$$4x - 5 = 7$$

both sides become equal and we say that $x = 3$ is a **solution** of the equation. If 5 is substituted for x, then both sides are not equal so $x = 5$ is not a solution.

It may help a businessman to have an equation describing the relationship between sales and profits, for example. However, he may also need to find a solution to the equation in order to make some decision.

One basic procedure for solving an equation is to obtain a sequence of equivalent equations with the goal of isolating the variable on one side of the equation and the appropriate number on the other side.

The following two operations help to isolate the variable and find the solution.

1. The same number may be added to or subtracted from both sides of an equation.
2. Both sides of an equation may be multiplied or divided by a nonzero number.

Either of these operations yields another equation that is equivalent to the first, that is, a second equation having the same solution as the first.

Example 1 (*Compare Exercise 3*)
Solve the equation $3x + 4 = 19$.

Solution We begin to isolate x by subtracting 4 from both sides.

$$3x + 4 - 4 = 19 - 4$$
$$3x = 15$$

Next divide both sides by 3

$$\frac{3x}{3} = \frac{15}{3}$$

$x = 5$ is the solution.

We can check our answer by substituting $x = 5$ into the original equation.

$3(5) + 4 = 15 + 4 = 19$ so the solution checks.

Example 2 (*Compare Exercise 6*)
Solve $4x - 2 = 2x + 12$.

Solution

$4x - 2 = 2x + 12$	First, add 2 to both sides
$4x - 2 + 2 = 2x + 12 + 2$	
$4x = 2x + 14$	Next, subtract $2x$ from both sides
$4x - 2x = 2x + 14 - 2x$	
$2x = 14$	Now, divide both sides by 2
$x = 7$	

CHECK $4(7) - 2 = 28 - 2 = 26$ (left side) and $2(7) + 12 = 14 + 12 = 26$ (right side) so it checks.

Example 3 (*Compare Exercise 9*)
Solve $7x + 13 = 0$.

Solution

$7x + 13 = 0$	Subtract 13 from both sides
$7x = -13$	Divide both sides by 7
$x = -\frac{13}{7}$	

The above examples all use **linear equations**.

Definition

> A **linear equation** in **one variable** is an equation that can be written in the form
>
> $$ax + b = 0 \quad \text{where} \quad a \neq 0$$
>
> A **linear equation** in **two variables** is an equation that can be written in the form
>
> $$y = ax + b \quad \text{where} \quad a \neq 0$$

Example 4 (*Compare Exercise 13*)

Solve $\frac{3x-5}{2}+\frac{x+7}{3}=8$.

Solution We show two ways to solve this. First, use rules of fractions to combine the terms on the left side:

$$\frac{3x-5}{2}+\frac{x+7}{3}=8 \quad \text{Convert fractions to the same denominator}$$

$$\frac{3(3x-5)}{6}+\frac{2(x+7)}{6}=8 \quad \text{Now add the fractions}$$

$$\frac{3(3x-5)+2(x+7)}{6}=8$$

$$\frac{9x-15+2x+14}{6}=8$$

$$\frac{11x-1}{6}=8 \quad \text{Now multiply both sides by 6}$$

$$11x-1=48$$

$$11x=49$$

$$x=\frac{49}{11}$$

An alternate, and simpler, method is the following:

$$\frac{3x-5}{2}+\frac{x+7}{3}=8$$

Multiply through by 6.

$$3(3x-5)+2(x+7)=48$$

$$9x-15+2x+14=48$$

$$11x-1=48$$

$$11x=49$$

$$x=\frac{49}{11}$$

0–2 Exercises

Determine which of the following values of x are solutions to the equations in Exercises 1 and 2. Use $x = 1, 2, -3, 0, 4$, and -2.

1. $2x-4=-10$
2. $3x+1=x+5$

Solve the following equations:

3. (*See Example 1*)
$2x - 3 = 5$

4. $-4x + 2 = 6$

5. $4x - 3 = 5$

6. (*See Example 2*)
$7x + 2 = 3x + 4$

7. $5 - x = 8 + 3x$

8. $2x - 4 = -5x + 2$

9. (*See Example 3*)
$12x + 21 = 0$

10. $7x - 4 = 0$

11. $3(x - 5) + 4(2x + 1) = 9$

12. $6(4x + 5) + 7 = 2$

13. (*See Example 4*)
$\frac{2x + 3}{3} + \frac{5x - 1}{4} = 2$

14. $\frac{4x + 7}{6} + \frac{2 - 3x}{5} = 5$

15. $\frac{12x + 4}{2x + 7} = 4$

16. $\frac{x + 1}{x - 1} = \frac{3}{4}$

17. The U-Drive-It Rental charges $0.20 per mile plus $112 per week for car rental. The weekly rental fee for a car is represented by a linear equation

$$y = 0.20x + 112$$

where x is the number of miles driven and y is the weekly rental charge.

(a) Determine the rental fee if the car is driven 650 miles during the week.

(b) Determine the rental fee if the car is driven 1500 miles.

(c) The weekly rental fee is $302. How many miles were driven?

18. Joe Cool has a summer job of selling real estate in a subdivision development. He receives a base pay of $100 per week plus $50 for each lot sold. Thus, the equation

$$y = 50x + 100$$

represents his weekly income where x is the number of lots sold.

(a) What is his weekly income if he sells 7 lots?

(b) What is his weekly income if he sells 15 lots?

(c) If he receives $550 one week, how many lots did he sell?

19. A Boy Scout troop collects aluminum cans for a project. The recycling center weighs the cans in a container that weighs 8 pounds, so the Boy Scouts are paid according to the equation

$$y = 0.21(x - 8)$$

where x is the weight in pounds given by the scale and y is the payment in dollars.

(a) How much money do the Boy Scouts receive if the scale reads 42 pounds?

(b) How much do they receive if the scale reads 113 pounds?

(c) They received $11.13 for one weekend's collection. What was the reading on the scale?

20. The tuition and fees paid by students at a local junior college is given by the equation

$$y = 27x + 85$$

where x is the number of hours enrolled and y is the total tuition-fees cost in dollars.

(a) How much does a student pay who is enrolled in 13 hours?

(b) A student who pays $517 is enrolled in how many hours?

0–3 Coordinate Systems

We have all seen a map, a house plan, or a wiring diagram that shows information recorded on a flat surface. Each of these uses some notation unique to the subject to convey the desired information. In mathematics, we often use a flat surface called a **plane** to draw figures and locate points. A reference system in the plane helps to record and communicate information accurately. The standard mathematical reference system consists of a horizontal and a vertical line (called **axes**). These two perpendicular axes form a *Cartesian*, or **rectangular, coordinate system**. They intersect at a point called the **origin**.

We name the horizontal axis the **x-axis**, and we name the vertical axis the **y-axis**. The origin is labeled O.

Two numbers are used to describe the location of a point in the plane and they are recorded in the form (x, y). For example, $x = 3$ and $y = 2$ for the point (3, 2). The first number, 3, called the **x-coordinate** or **abscissa**, represents the horizontal distance from the y-axis to the point. The second number, 2, called the **y-coordinate** or **ordinate**, represents the vertical distance measured from the x-axis to the point. The point (3, 2) is shown as point P in Figure 0–4. Points located to the right of the y-axis have positive x-coordinates; those to the left have

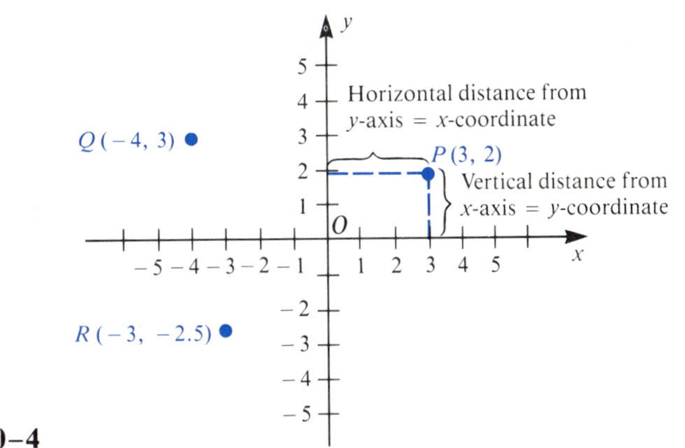

Figure 0–4

negative x-coordinates. The y-coordinate is positive for points located above the x-axis and negative for those located below.

Figure 0–4 shows other examples of points in this coordinate system: Q is the point $(-4, 3)$ and R is the point $(-3, -2.5)$. The origin O has coordinates $(0, 0)$.

Figure 0–5 shows the points $(-3, 2)$, $(-4, -2)$, $(1, 1)$, and $(1, -2)$ plotted on the Cartesian coordinate system.

The coordinate axes divide the plane into four parts called **quadrants**. The quadrants are labeled I, II, III, and IV as shown in Figure 0–6. Point A is in the first quadrant, B in the second quadrant, C in the third, and D in the fourth. Points A and E are in the same quadrant.

René Descartes (1596–1650), a French philosopher-mathematician, invented the Cartesian coordinate system. His invention of the coordinate system is one of the outstanding events in the history of mathematics because it combined algebra and geometry in a way that enables us to use algebra to solve geometry problems and to use geometry to clarify algebraic concepts.

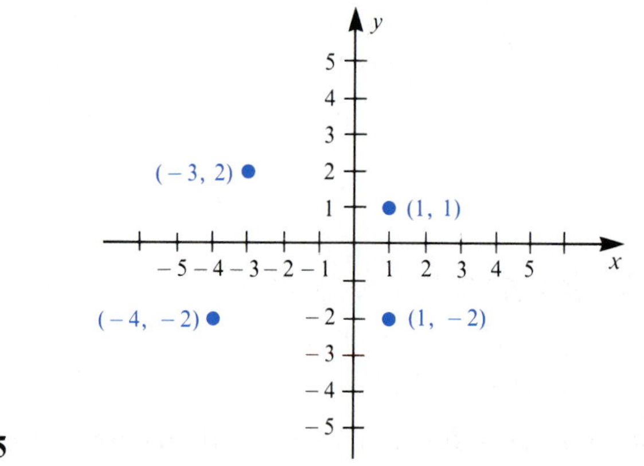

Figure 0–5

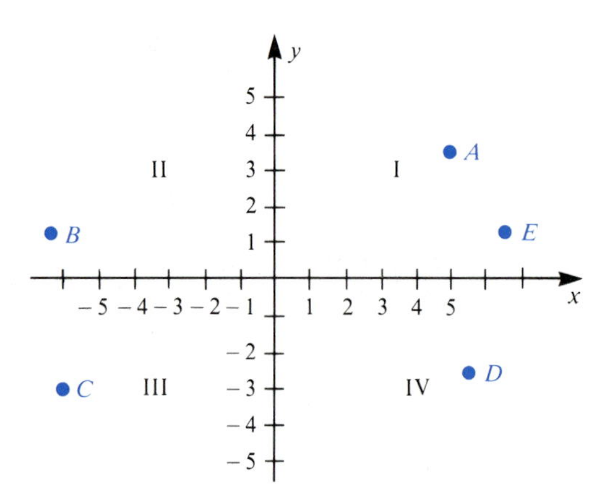

Figure 0–6

0–3 Exercises

1. The following are the coordinates of points in a rectangular Cartesian coordinate system. Plot these points.

$$(-5, 4), (-2, -3), (-2, 4), (1, 5), (2, -5)$$

2. What are the coordinates of the points P, Q, R, and S in the coordinate system in Figure 0–7?

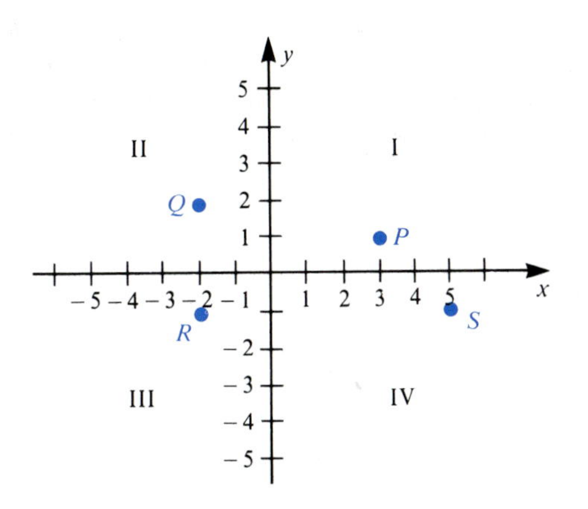

Figure 0–7

3. Locate the following points on a Cartesian coordinate system:

$$(-2, 5), (3, -2), (0, 4), (-2, 0), (\tfrac{7}{2}, 2),$$
$$(\tfrac{2}{3}, \tfrac{9}{4}), (-4, -2), (0, -5), (0, -2), (-6, -3)$$

4. Give the coordinates of A, B, C, D, E, and F in the coordinate system shown in Figure 0–8.

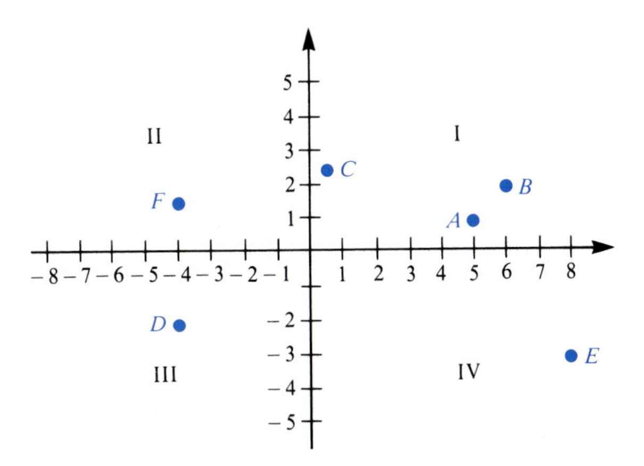

Figure 0–8

5. Note that all points in the first quadrant have positive x-coordinates and positive y-coordinates. What are the characteristics of the points in

   (a) the second quadrant?

   (b) the third quadrant?

   (c) the fourth quadrant?

6. For each case shown in Figure 0–9, find the property the points have in common.

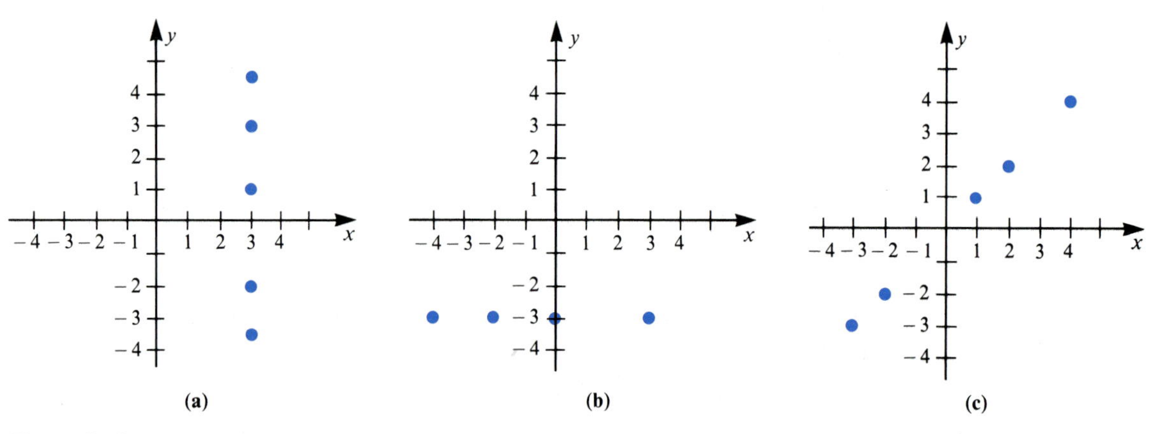

Figure 0–9

7. An old map gives these instructions to find a buried treasure:

   Start at giant oak tree. Go north 15 paces, then east 22 paces to a half buried rock. The key to the treasure chest is buried at the spot that is 17 paces west and 13 paces north of the rock.

   From the place where the key is buried, go 32 paces west and 16 paces south to the location of the buried treasure. Use a coordinate system to represent the location of the oak tree, the rock, the key, and the treasure.

0–4 Linear Inequalities and Interval Notation

Solving Inequalities
Interval Notation

We frequently use inequalities in our daily conversation. They may take the form "Which store has the lower price?" "Did you make a higher grade?" "Our team scored more points." "My expenses are greater than my income." Statements such as these basically state that one quantity is greater than another. Statements using the terms "greater than" or "less than" are called **inequalities**. Our goal is to solve inequalities. First, we give some terminology and notation.

The symbol < means "less than" and > means "greater than." Just remember that each of these symbols points to the smaller quantity. The notation $a > b$ and $b < a$ have exactly the same meaning. We may state the definition of $a > b$ in three ways. At times one may be more useful than the other, so choose the most appropriate one.

> If a and b are real numbers, the following statements have the same meaning.
>
> **(a)** $a > b$ means that a lies to the right of b on a number line.
>
> **(b)** $a > b$ means that there is a positive number p such that
> $a = b + p$.
>
> **(c)** $a > b$ means that $a - b$ is a positive number p.

The positive numbers lie to the right of zero on a number line and the negative numbers lie to the left.

Example 1 (*Compare Exercise 1*)
The numbers 5, 8, 17, −2, −3, and −15 are plotted on a number line in Figure 0–10. Notice the following.

1. **(a)** 17 lies to the right of 5.
   (b) $17 = 5 + 12$
   (c) $17 - 5 = 12$

   Each of these three statements is equivalent to saying that $17 > 5$.
2. $8 > -3$ because $8 - (-3) = 8 + 3 = 11$. (By part c)
3. $-2 > -15$ because $-2 = -15 + 13$. (By part b where $p = 13$)

Figure 0–10

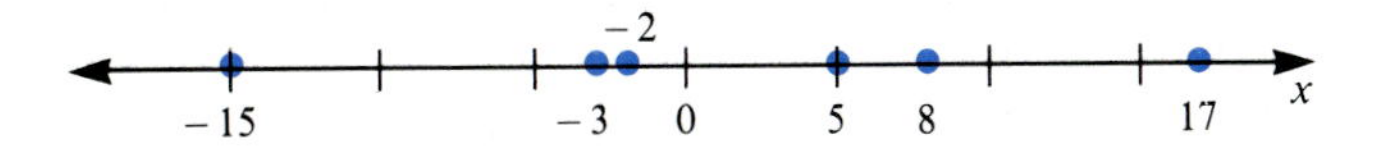

Solving Inequalities

By the **solution** of an inequality like

$$3x + 5 > 23$$

we mean the value, or values, of x that make the statement true. The method for solving inequalities is similar to that for solving equations. We want to operate on an inequality in a way that gives an equivalent inequality, but which enables us to determine the solution.

Here are some simple examples of useful properties of inequalities.

Example 2

1. Since $18 > 4$, $18 + 6 > 4 + 6$, that is, $24 > 10$. (6 added to both sides)
2. Since $23 > -1$, $23 - 7 > -1 - 7$, that is, $16 > -8$. (7 subtracted from both sides)

3. Since $6 > 2$, $4(6) > 4(2)$, that is, $24 > 8$. (Both sides multiplied by 4)
4. Since $10 > 3$, $-2(10) < -2(3)$, that is, $-20 < -6$. (Both sides multiplied by -2)

Warning! The inequality symbol reverses when we multiply each side by a negative number.

5. Since $-15 > -21$, $\frac{-15}{3} > \frac{-21}{3}$, that is, $-5 > -7$. (Divide both sides by 3)
6. Since $20 > 6$, $\frac{20}{-2} < \frac{6}{-2}$, that is, $-10 < -3$. (Divide both sides by -2)

Warning! The inequality symbol reverses when dividing each side by a negative number.

These examples illustrate basic properties that are useful in solving inequalities.

Properties of Inequalities

For real numbers a, b, and c, the following are true.

1. If $a > b$, then $a + c > b + c$.
2. If $a > b$, then $a - c > b - c$.
3. If $a > b$ and c is positive, then $ca > cb$.
4. If $a > b$ and c is negative, then $ca < cb$. Notice the change from $>$ to $<$.
5. If $a > b$ and c is positive, then $\frac{a}{c} > \frac{b}{c}$.
6. If $a > b$ and c is negative, then $\frac{a}{c} < \frac{b}{c}$. Notice the change from $>$ to $<$.

Similar properties hold if each inequality symbol is reversed.

We use these properties to solve an inequality; that is, to find the values of x that make the inequality true. In general, we proceed by finding equivalent inequalities that will eventually isolate x on one side of the inequality and the appropriate number on the other side.

Example 3 (*Compare Exercise 2*)
Solve the inequality $3x + 5 > 14$.

Solution Begin with the given inequality.

$3x + 5 > 14$ Next, subtract 5 from each side (Property 2)

$3x > 9$ Now divide each side by 3 (Property 5)

$x > 3$

Thus, all x greater than 3 make the inequality true. This solution can be graphed on a number line as shown in Figure 0–11. The empty circle indicates that $x = 3$ is omitted from the solution and the heavy line indicates the values of x included in the solution.

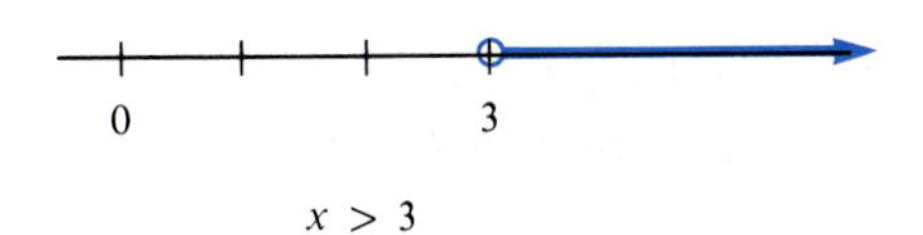

Figure 0–11 $x > 3$

Example 4 (*Compare Exercise 10*)
Solve the inequality $5x - 17 > 8x + 14$ and indicate the solution on a graph.

Solution Start with the given inequality.

$$5x - 17 > 8x + 14 \quad \text{Now add 17 to both sides (Property 1)}$$
$$5x > 8x + 31 \quad \text{Now subtract } 8x \text{ from both sides (Property 2)}$$
$$-3x > 31 \quad \text{Now divide both sides by } -3 \text{ (Property 6)}$$
$$x < -\frac{31}{3} \quad \text{This reverses the inequality symbol}$$

Thus, the solution consists of all x to the left of $-31/3$. See Figure 0–12.

We will also use the symbols $\geq$ (greater than or equal to) and $\leq$ (less than or equal to). All the properties of inequalities hold if $<$ is replaced with $\leq$ and $>$ is replaced with $\geq$. An inequality with $\geq$ simply includes the possibility of an equality as well as greater than.

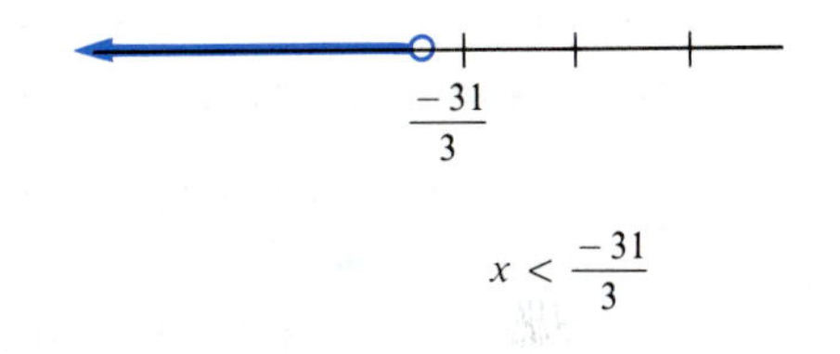

Figure 0–12 $x < \frac{-31}{3}$

Example 5 (*Compare Exercise 7*)
Solve and graph $2(x - 3) \leq 3(x + 5) - 7$.

Solution

$$2(x - 3) \leq 3(x + 5) - 7 \quad \text{First perform the indicated multiplications}$$
$$2x - 6 \leq 3x + 15 - 7$$
$$2x - 6 \leq 3x + 8 \quad \text{Now add 6 to both sides (Property 1)}$$
$$2x \leq 3x + 14 \quad \text{Subtract } 3x \text{ from both sides (Property 2)}$$
$$-x \leq 14 \quad \text{Multiply both sides by } -1 \text{ (Property 4)}$$
$$x \geq -14$$

Since the solution includes -14 and all numbers greater, the graph shows a solid circle at -14 (see Figure 0–13).

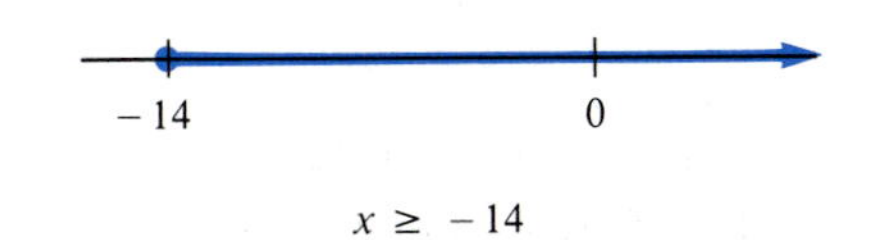

Figure 0–13 $x \geq -14$

The next example illustrates a problem that involves two inequalities.

Example 6 (*Compare Exercise 16*)
Solve and graph $3 < 2x + 5 \leq 13$.

Solution This inequality means both $3 < 2x + 5$ *and* $2x + 5 \leq 13$. Solve it in a manner similar to the preceding examples except that you try to isolate the x in the middle.

Begin with the given inequality.

$3 < 2x + 5 \leq 13$	Subtract 5 from all parts of the inequality
$-2 < 2x \leq 8$	Divide each part by 2
$-1 < x \leq 4$	The solution consists of all numbers between -1 and 4, including 4 but not including -1.

The graph of the solution (see Figure 0–14) shows an empty circle at -1 because -1 is not a part of the solution. It shows a solid circle at 4 because 4 is a part of the solution. The solid line between -1 and 4 indicates that all numbers between -1 and 4 are included in the solution.

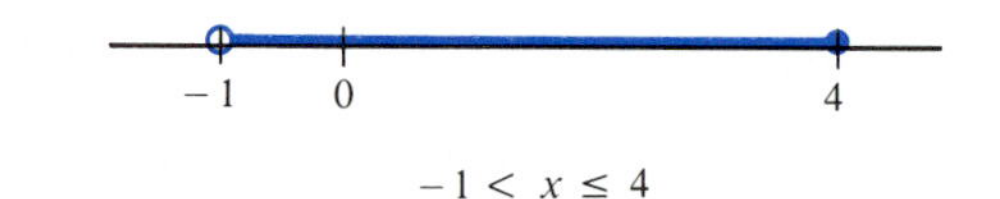

Figure 0–14 $-1 < x \leq 4$

Interval Notation

There is yet another notation for indicating the solution of an inequality. It is the **interval notation**. The portion of the number line that represents the solution of an inequality is identified by its endpoints; brackets or parentheses indicate whether or not the endpoint is included in the solution. A parenthesis indicates that the endpoint is not included and a bracket indicates that the endpoint is included. The notation $(-1, 4]$ indicates the set of all numbers between -1 and 4 with -1 excluded and 4 included in the set. The notation $(-1, 4)$ indicates that both -1 and 4 are excluded.

The notation $(-1, \infty)$ means all numbers greater than -1. The symbol ∞ denotes infinity and indicates that there is no upper bound to the interval.

Table 0–2 shows the variations of the interval notation.

Example 7 (*Compare Exercise 22*)
Solve $1 \leq 2(x - 5) + 3 < 5$.

Solution

$1 \leq 2(x - 5) + 3 < 5$	Multiply to remove parentheses
$1 \leq 2x - 10 + 3 < 5$	
$1 \leq 2x - 7 < 5$	Add 7 throughout
$8 \leq 2x < 12$	Divide through by 2
$4 \leq x < 6$	

The solution consists of all values of x in the interval $[4, 6)$ and the graph is shown in Figure 0–15.

Table 0–2.

Inequality notation		Interval notation		Graph of interval
General	Example	General	Example	
$a < x < b$	$-1 < x < 4$	(a, b)	$(-1, 4)$	
$a \le x < b$	$-1 \le x < 4$	$[a, b)$	$[-1, 4)$	
$a < x \le b$	$-1 < x \le 4$	$(a, b]$	$(-1, 4]$	
$a \le x \le b$	$-1 \le x \le 4$	$[a, b]$	$[-1, 4]$	
$x < b$	$x < 4$	$(-\infty, b)$	$(-\infty, 4)$	
$x \le b$	$x \le 4$	$(-\infty, b]$	$(-\infty, 4]$	
$a < x$	$-1 < x$	(a, ∞)	$(-1, \infty)$	
$a \le x$	$-1 \le x$	$[a, \infty)$	$[-1, \infty)$	

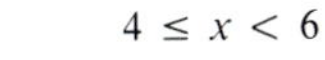

Figure 0–15 $4 \le x < 6$

Example 8 *(Compare Exercise 29)*
The total points on an exam given by Professor Passmore are determined by: 20 points plus 2.5 points for each correct answer. A total score in [70, 80) is a C. If Scott made a C on the exam, how many questions did he answer correctly?

Solution The score on an exam is given by $20 + 2.5x$, where x is the number of correct answers. Thus, the condition for a C is

$$70 \le 20 + 2.5x < 80$$

Solve for x to obtain the number of correct answers.

$$70 \le 20 + 2.5x < 80$$

$$50 \le 2.5x < 60$$

$$\frac{50}{2.5} \le x < \frac{60}{2.5}$$

$$20 \le x < 24$$

In this case only whole numbers make sense so, Scott got 20, 21, 22, or 23 correct answers.

0–4 Exercises

I.

1. (*See Example 1*) The following inequalities are of the form $a > b$. Verify the truth or falsity of each one by using the property $a > b$ means $a - b$ is a positive number.

(a) $9 > 3$ **(b)** $4 > 0$ **(c)** $-5 > 0$

(d) $-3 > -15$ **(e)** $\frac{5}{6} > \frac{2}{3}$

Solve the inequalities in Exercises 2 through 9. State your solution using inequalities.

2. (*See Example 3*) $2x + 5 > 17$

3. $3x - 5 < x + 4$

4. $12 > 1 - 5x$

5. $5x - 22 \leq 7x + 10$

6. $13x - 5 \leq 7 - 4x$

7. (*See Example 5*) $3(x + 4) < 2(x - 3) + 14$

8. $14 < 3x + 8 < 32$

9. $-9 \leq 3(x + 2) - 15 < 27$

Solve the inequalities in Exercises 10 through 17. Graph the solution.

10. (*See Example 4*) $3x + 2 \leq 4x - 3$

11. $6x + 5 < 5x - 4$

12. $3x + 2 < 2x - 3$

13. $78 < 6 - 3x$

14. $4(x - 2) > 5(2x + 1)$

15. $3(2x + 1) < -1(3x - 10)$

16. (*See Example 6*) $-16 < 3x + 5 < 22$

17. $124 > 5 - 2x \geq 68$

Solve the inequalities in Exercises 18 through 23. Give the solution in interval form.

18. $5x - 7 > 3$

19. $3x + 4 \leq 1$

20. $-3x + 4 < 2x - 6$

21. $-7x + 4 \geq 2x + 3$

22. (*See Example 7*) $-45 < 4x + 7 \leq -10$

23. $16 > 2x - 10 \geq 4$

II.

Solve the following inequalities.

24. $\frac{2x - 5}{3} < \frac{x + 7}{4}$

25. $\frac{6x + 5}{-2} \geq \frac{4x - 3}{5}$

26. $\frac{3}{4} < \frac{7x + 1}{6} < \frac{5}{2}$

27. $\frac{2}{3} < \frac{x + 5}{-4} \leq \frac{3}{2}$

III.

28. A sporting goods store runs a special on jogging shoes. The manager expects to make a profit if the number of shoes sold, x, satisfies $32x - 4230 > 2x + 480$. How many shoes must be sold in order to make a profit?

29. (*See Example* 8) Professor Tuff computes a grade on a test by $35 + 5x$, where x is the number of correct answers. A grade in [75, 90) is a B. If a student receives a B, how many correct answers were given?

30. On a final exam any grade in [85, 100] was an A. The professor gave 3 points for each correct answer and then adjusted the grades by adding 25 points. If a student made an A, how many correct answers were given?

IMPORTANT TERMS

0–1	**Natural numbers**	**Rational numbers**	**Irrational numbers**
	Real numbers		
0–2	**Variable**	**Solution**	**Linear equation**
0–3	**Cartesian coordinate system**	**Rectangular coordinate system**	**Origin**
	***x*-axis**	***y*-axis**	**Abscissa**
	Ordinate	**Quadrants**	
0–4	$>, <, \geq, \leq$	**Inequalities**	**Properties of inequalities**
	Interval notation		

1

Topics in this chapter can be applied to:

Fees and Expenses • Cost and Revenue Functions • Operating Costs • Depreciation • Break-Even Analysis • Simple Interest • Production Capacity

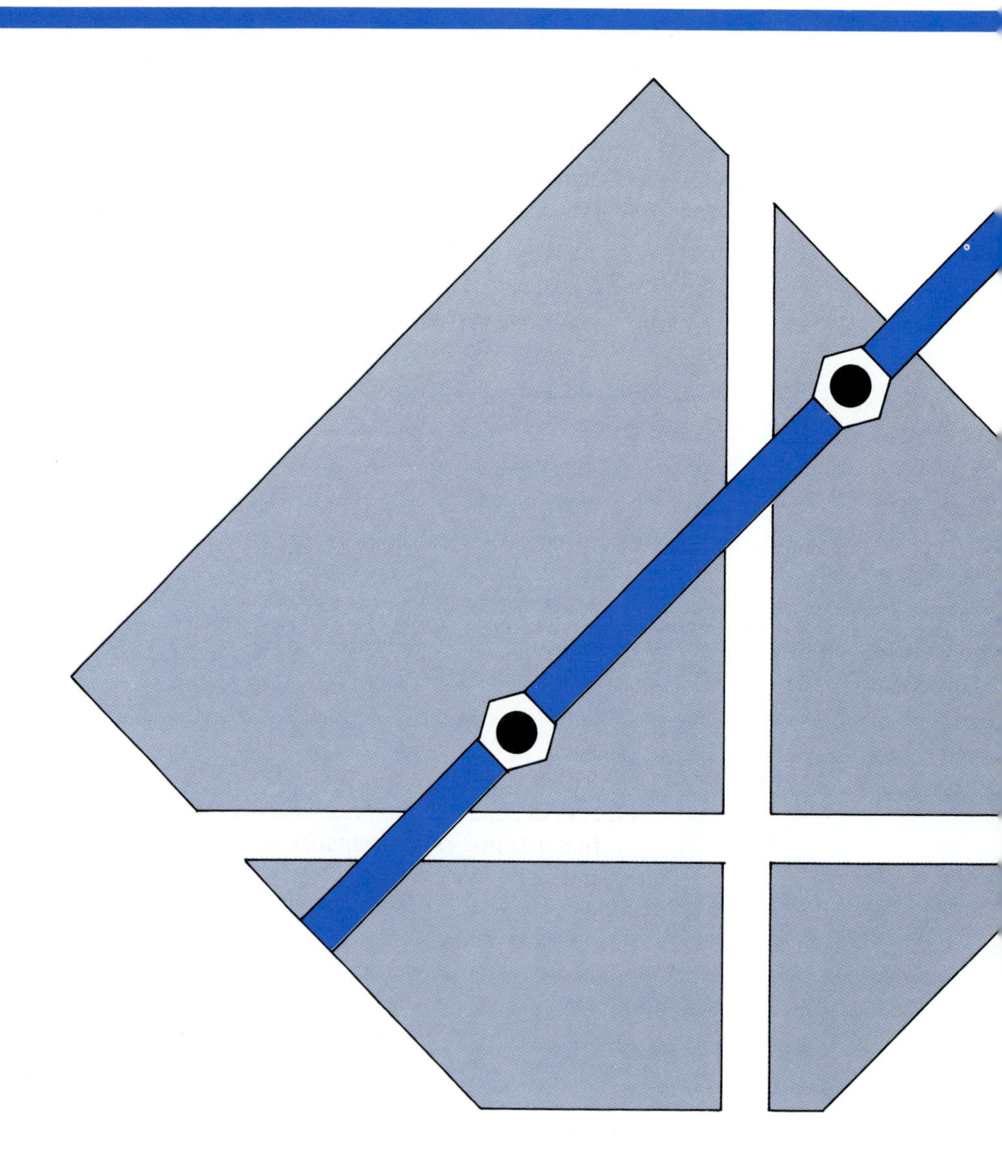

Functions and Lines

- Functions
- Graphs and Lines
- Applications of Linear Functions
- Linear Inequalities in Two Variables

1–1 Functions

Nearly all practical problems involve two or more quantities that are related in some fashion. The area of a square is related to the length of its sides; the amount withheld for FICA is related to an employee's salary; and the amount charged for sales tax depends on the price of an item.

In mathematics we formalize certain kinds of relationships between quantities and call them **functions**. Nearly always in this course, the quantities involved are measured by real numbers.

A function consists of three parts: a set of numbers called a **domain**, a set of numbers called the **range**, and a *rule* which assigns a **unique** number in the range to each number in the domain. The rule gives the relationship between the quantities and is often given in the form of an equation. We generally use the letter x to represent a number from the domain and y to represent a number from the range. We call x and y **variables**. The variable of the domain, x, is the **independent variable** and the variable of the range, y, is the **dependent variable**.

Emphasis: The rule for a function assigns exactly one number y for each number x in the domain. The number, y, that corresponds to a number x is called the **value** of the function at x.

Example 1 (*Compare Exercise 1*)
A consultant's fee is $300 for miscellaneous expenses plus $50 per hour. In this instance, the domain consists of positive numbers of hours, and the range consists of numbers of dollars (fees). The rule that determines the fee that corresponds to a certain number of consulting hours is given by the formula

$$y = 50x + 300$$

where x is the number of hours consulted and y is the total fee in dollars. Notice that the formula gives exactly one fee for each number of consulting hours.

Example 2 (*Compare Exercise 3*)
When a family goes to a concert, the amount paid depends on the number attending, that is, the total admission is a function of the number attending. The ticket office usually has a chart giving total admission so, for them, the rule is a chart something like this:

Number of Tickets	*Total Admission*
1	$ 6.50
2	$13.00
3	$19.50
4	$26.00
5	$32.50
6	$39.00

From this chart, the domain is the set {1, 2, 3, 4, 5, 6} and the range is {6.50, 13.00, 19.50, 26.00, 32.50, 39.00}.

Mathematicians have a standard notation for functions. For example, the equation $y = 50x + 300$ is often written as

$$f(x) = 50x + 300$$

$f(x)$ is read "f of x" indicating "f is a function of x." The $f(x)$ notation is especially useful to indicate a substitution of a number for x. $f(3)$ looks like 3 has been put in place of x in $f(x)$. This is the correct interpretation. $f(3)$ represents the **value** of the function when 3 is substituted for x in

$$f(x) = 50x + 300$$

that is,

$$\begin{aligned} f(3) &= 50(3) + 300 \\ &= 150 + 300 \\ &= 450 \end{aligned}$$

Example 3 (*Compare Exercise 5*)

$$\text{If } f(x) = -7x + 22, \text{ then}$$
$$f(2) = -7(2) + 22 = 8$$
$$f(-1) = -7(-1) + 22$$
$$= 7 + 22$$
$$= 29$$

$$\text{If } f(x) = 4x - 11, \text{ then}$$
$$f(5) = 4(5) - 11$$
$$= 20 - 11$$
$$= 9$$
$$f(0) = 4(0) - 11$$
$$= 0 - 11$$
$$= -11$$

$$\text{If } f(x) = x(4 - 2x), \text{ then}$$
$$f(6) = 6(4 - 2(6))$$
$$= 6(4 - 12)$$
$$= 6(-8)$$
$$= -48$$
$$f(a) = a(4 - 2a)$$

Note: In this example no mention was made of the domain and range of the functions. Normally we assume that the domain is the set of numbers that may be used in the expression and the range is the set of all numbers that may occur as a value of the function. As noted in preceding examples, the numbers in the domain and range may represent quantities that occur in units like hours, dollars, pounds, etc.

Example 4 (*Compare Exercise 9*)
Baked potatoes contain 26 calories per ounce (before adding a topping). Define a function as follows.

The domain of the function is the set of positive numbers representing weights in ounces.

The range is the set of positive numbers representing the number of calories.

The calorie function is

$$f(x) = 26x$$

where x represents the weight of the potato in ounces, and $f(x)$ is the corresponding number of calories.

(a) What is $f(3)$? $f(1.5)$? $f(2.4)$?
(b) If $f(x) = 143$ calories, how much did the potato weigh?

Solution **(a)**

$$f(3) = 26(3) = 78$$
$$f(1.5) = 26(1.5) = 39$$
$$f(2.4) = 62.4$$

(b) If $f(x) = 143$, then

$$143 = 26x$$

$$x = \frac{143}{26} = 5.5$$

So, if the potato contains 143 calories, it weighs 5.5 ounces.

You may wonder how to distinguish between the domain and range of a function. While there are no exact rules, you generally select the range as those quantities that depend on the other set of quantities. For example, in the calorie function most of us think of the number of calories depending on the potato selected, so the domain represents weights of potatoes.

Example 5 (*Compare Exercise 23*)
All lots in a new subdivision are 200 feet deep. The area of a lot is given by the function

$$f(x) = 200x$$

where x represents the width of the lot in feet and $f(x)$ the area in square feet.

(a) What is the area of a lot 150 feet wide?
(b) What is $f(275)$?
(c) A family wants an acre lot (44,000 square feet). How wide a lot should they buy?

Solution **(a)** If $x = 150$,

$$f(150) = 200(150) = 30{,}000$$

so the area is 30,000 square feet.

(b) $f(275) = 200(275) = 55{,}000$ square feet
(c) We are given $f(x) = 44{,}000$, so $44{,}000 = 200x$

$$x = \frac{44{,}000}{200} = 220$$

So, a lot containing 44,000 square feet is 220 feet wide.

Example 6 We use letters other than f and x to represent functions and variables. We might refer to the cost of producing x items as $C(x) = 5x + 540$; the price of x pounds of steak as $p(x) = 2.19x$; the area of a circle of radius r as $A(r) = \pi r^2$; or the distance an object falls in t seconds as $d(t) = 16t^2$.

1–1 Exercises

I.

1. (*See Example 1*) A tree service company charges $20 plus $15 per hour to trim trees so the rule relating their fee and hours worked is

$$y = 15x + 20$$

where x is the number of hours worked and y is their fee.

(a) Describe the domain of this function.

(b) Describe the range of this function.

2. An appliance repairman charges $30 plus $20 per hour for house calls.

(a) Write the rule that relates hours worked and his fee.

(b) Describe the domain of this function.

(c) Describe the range of this function.

3. (*See Example 2*) The price of movie tickets is given by the chart

Number of Tickets	*Total Admission*
1	$4.75
2	$9.50
3	$14.25
4	$19.00
5	$23.75
6	$28.50
7	$33.25

(a) What is the domain of the function defined by this chart?

(b) What is $f(3)$?

4. Tickets to a football game cost $14 each.

(a) Make a chart showing the total cost function for the purchase of 1, 2, 3, 4, 5, or 6 tickets.

(b) What is the domain of the function as shown in the chart?

5. (*See Example 3*) $f(x) = 4x - 3$. Determine

(a) $f(1)$

(b) $f(-2)$

(c) $f\left(\frac{1}{2}\right)$

(d) $f(a)$ (*See Example 4*)

6. $f(x) = x(2x - 1)$. Determine

(a) $f(3)$ **(b)** $f(-2)$
(c) $f(0)$ **(d)** $f(b)$

7. $f(x) = \dfrac{x + 1}{x - 1}$. Determine

(a) $f(5)$ **(b)** $f(-6)$
(c) $f(0)$ **(d)** $f(2c)$

8. $f(x) = -4x + 7$. Determine

(a) $f(a)$ **(b)** $f(y)$ **(c)** $f(a + 1)$ **(d)** $f(a + h)$
(e) $f(3a)$ **(f)** $f(2b + 1)$

II.

9. (*See Example 4*) The function for calories in french fried potatoes is given by

$$f(x) = 78x$$

where x represents the weight of french fried potatoes in ounces and $f(x)$ is the number of calories.

(a) Find $f(5)$; $f(2.5)$; $f(6.4)$
(b) If a serving of french fried potatoes contains 741 calories, what is the weight of the serving?

10. A small cheese pizza is cut into four pieces. The calcium content of pizza is given by the function

$$f(x) = 221x$$

where x is the number of pieces of pizza and $f(x)$ is the quantity of calcium in mg.

(a) How many mg of calcium is contained in one pizza?
(b) The recommended daily requirement for calcium is 750 mg. How many pieces should be eaten to intake that amount?

11. Swimming requires 9 calories per minute of energy so the function for calories used in swimming is given by

$$f(x) = 9x$$

where x is the number of minutes and $f(x)$ is the number of calories used.

(a) How many calories are used in swimming for one hour?
(b) A swimmer wants to use 750 calories. How long should she swim?

III.

Each of the following statements describes a function. Write an equation of the function.

12. The cost of grapes at the Corner Grocery is 49¢ per pound.
13. The cost of catering a hamburger cookout is \$25 service charge plus \$1.25 per hamburger.
14. The monthly income of a salesman is \$500 plus 5% of sales.
15. The sale price of all items in The Men's Clothing Store is 20% off the regular price.
16. The monthly sales of Pappa's Pizza is \$1200 plus \$3 for each dollar spent on advertising.
17. Mr. Parsons hauls sand and gravel. His hauling costs (per load) are overhead costs of \$12.00 per load and operating costs of \$0.60 per mile.
18. Becky has a lawn mowing service. She charges a base price of \$5 plus \$4 per hour.
19. Paloma University found that a good estimate of its operating budget is \$5,000,000 plus \$3,500 per student.
20. Peoples Bank collects a monthly service charge of \$2.00 plus \$0.10 per check.
21. An automobile dealer's invoice cost is .88 of the list price of an automobile.
22. A telephone company provides measured phone service. Their rate is \$7.60 per month plus \$0.05 per call.
23. (*See Example 5*) A street is 30 feet wide. The area of a section of the street is given by

$$f(x) = 30x$$

where x represents the length of that section of the street in feet and $f(x)$ is the area.

(a) What is the area of a section 450 feet long?

(b) What is $f(125)$?

(c) A 650-ft long section of the street is paved at a cost of \$0.40 per square foot. How much did the paving cost?

(d) A contractor paved 15,900 square feet of the street. How long was the section?

24. If $f(x) = (x + 2)(x - 1)$ and $g(x) = \dfrac{7x + 4}{x + 1}$, find $f(3) + g(2)$.

1–2 Graphs and Lines

Definition of a Graph

"A picture is worth a thousand words" may be an overworked phrase, but it does convey an important idea. You may even occasionally use the expression "Oh, I see!" when you really grasp a difficult concept. A graph shows a picture of a function, and can help you to understand the behavior of the function. Imagine how difficult it would be to convey all the information (and the drama!) in the following example without the aid of the graphs.

The 1975 Masters Golf Tournament in Augusta, Georgia, was one of the most dramatic events in the history of golf. In early rounds several players held or shared the lead, but, by the end, all eyes were focused on a battle among three of the finest players in the game: Jack Nicklaus, Tom Weiskopf, and Johnny Miller.

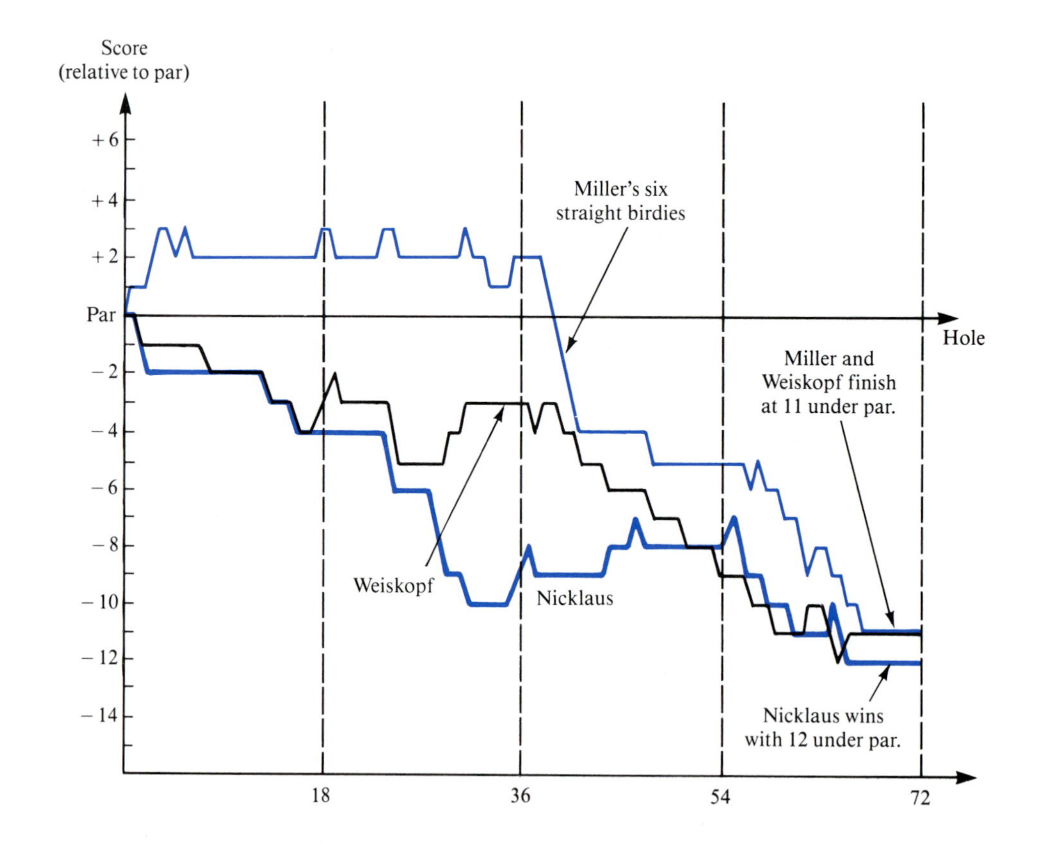

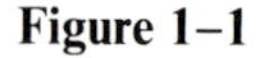
Figure 1–1

The preceding graphs (one for each player) tell the story. Follow Miller's plunge over the opening holes and his record-breaking six straight birdies later. Follow the slow, descending plateaus of Weiskopf's flawless game over much of the third day. Follow Nicklaus's early dominance and his seesaw battle with Weiskopf. Would Miller have won on a 73rd hole?

This example makes another important point. The graph seems to indicate that Miller was $2\frac{1}{2}$ over par after $17\frac{1}{2}$ holes; such an interpretation would be silly—golfers never talk about half of a hole. It is much harder to get information from a graph drawn with just dots. Remember those drawings you made when you were a child by "connecting-the-dots"? The picture made a lot more sense after you drew in the connecting lines. In Figure 1–1 we sacrificed some technical accuracy by "connecting-the-dots" but got a better picture of what happened by doing so. Professional users of mathematics do the same thing; an accountant may let $C(x)$ represent the cost of manufacturing x items, or a manager may let $E(x)$ represent an efficiency index when x people are involved in a large project. In reality, the domains of these functions involve only positive integers. But many methods of mathematics require the domain of the function to be an interval or intervals, rather than isolated points. These methods have proven so powerful in solving problems that people set up their functions using such domains. They may then have to use some common sense in interpreting their answer. If the manager finds that the most efficient number of people to assign to a project is 54.87, she will probably end up using either 54 or 55 people.

Definition

The **graph of a function** f is the set of points (x, y) in the plane that satisfy the equation $y = f(x)$. In other words, the graph of f is the collection of points of the form $(x, f(x))$.

As we develop different mathematical techniques throughout this text, we will use some concrete applications. This in turn will require some familiarity with the functions involved and some idea of the shape of their graphs. We start with the simpliest functions and graphs.

Linear Functions and Straight Lines

Definition

A function is called a **linear function** if its rule—its defining equation—can be written $f(x) = mx + b$. Such a function is called linear because its graph is a straight line.

Example 1 (*Compare Exercise 1*)
Draw the graph of $f(x) = 2x + 5$.

Solution The graph will be a straight line, and it takes just two points to determine a straight line. If we let $x = 1$, then we have $f(1) = 7$; if we let $x = 4$, then $f(4) = 13$. This means that the points $(1, 7)$ and $(4, 13)$ are on the graph of $f(x) = 2x + 5$.

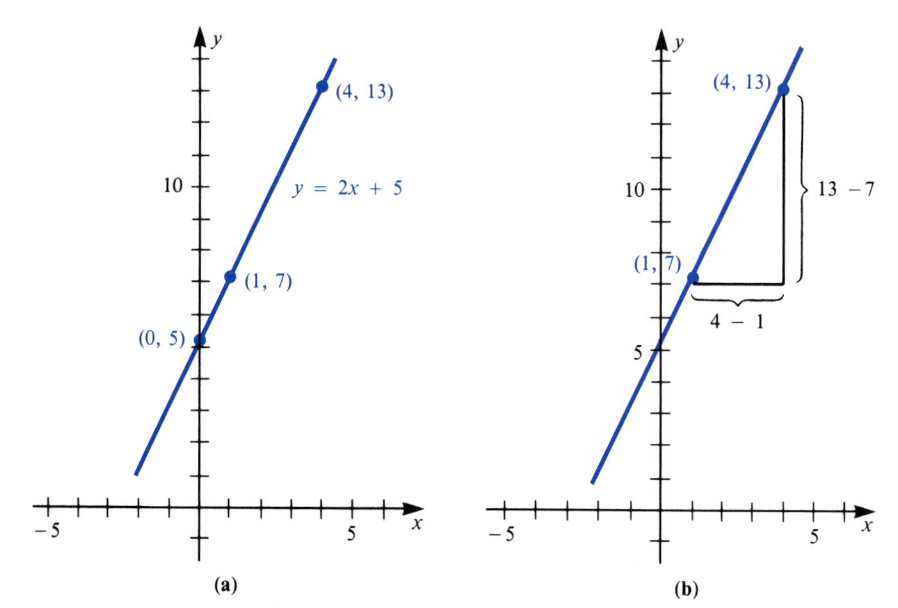

Figure 1–2

Since we also use y for $f(x)$, we could also say that these points are on the line $y=2x+5$. By plotting the points (1, 7) and (4, 13) and drawing the line through them, we obtain Figure 1–2(a). (We can use any pair of x values to get two points on the line.) It is usually a good idea to plot a third point to help catch any error. Because $f(0) = 5$, the point (0, 5) is also on the graph of f.

Slope and Intercept

When the equation of a line is written in the form $y = 2x + 5$ ($y = mx + b$ is the general form), the constants 2 and 5 (m and b in general) give key information about the line. The constant b (5 in this case) is the value of y when $x = 0$. Thus, $(0, b)$ (or (0, 5) in our example) is a point on the line. We refer to b as the **y-intercept** of the line; it gives the point where the line intercepts the y-axis. The constant m gives information about the direction, or slant, of a line. We call m the **slope** of the line. We will give more information about the slope later.

Example 2 (*Compare Exercise 7*)
What are the slope and y-intercept of each of the following lines?

(a) $y = 3x - 5$ **(b)** $y = -6x + 15$

Solution **(a)** For the line $y = 3x - 5$, the slope $m = 3$ and the y-intercept $b = -5$.
(b) For the line $y = -6x + 15$, the slope is -6 and the y-intercept is 15.

Warning! The coefficient of x in a general linear equation does not automatically give you the slope of the line. When the equation of the line is in the form $y = mx + b$, the coefficient of x is the slope of the line and the constant term is the y-intercept. If the equation is in another form, it is a good idea to change to this form to determine the slope and y-intercept.

Example 3 (*Compare Exercise 11*)
What are the slope and y-intercept of the line $3x + 2y - 4 = 0$?

Solution We must write the equation $3x + 2y - 4 = 0$ in the slope-intercept form, $y = mx + b$
Solve the equation

$$3x + 2y - 4 = 0$$

for y.

$$2y = -3x + 4$$
$$y = -\tfrac{3}{2}x + 2$$

Thus, the slope-intercept form is $y = -\frac{3}{2}x + 2$. Now we can say the slope is $-\frac{3}{2}$ and the y-intercept is 2.

The slope relates to the way the line slants in the following manner. Select two points on the line $y = 2x + 5$, such as (1, 7) and (4, 13). Compute the difference in the y-coordinates of the two points: $13 - 7 = 6$. (See Figure 1–2(b).) Now compute the difference in x-coordinates: $4 - 1 = 3$. The quotient $\frac{6}{3} = 2$ is m, the slope of the line $y = 2x + 5$. Following this procedure with any other two points on the line $y = 2x + 5$ will also yield the answer 2.

The following is the general formula showing how to compute the slope of a line.

Slope Formula

Choose two points P and Q on the line. Let (x_1, y_1) be the coordinates of P and (x_2, y_2) be the coordinates of Q. The **slope** of the line, m, is given by the equation

$$m = \frac{y_2 - y_1}{x_2 - x_1} \quad \text{where } x_2 \neq x_1$$

The slope is the difference in the y-coordinates divided by the difference in the x-coordinates.

Be sure you subtract the x and y coordinates in the same order. Figure 1–3 shows the geometric meaning of this quotient.

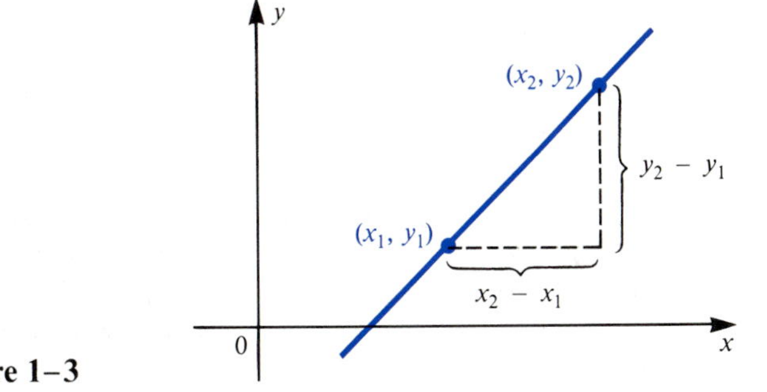

Figure 1–3

Example 4 (*Compare Exercise 15*)
Find the slope of the line through the points (1, 2) and (5, 6).

Solution We let $(x_1, y_1) = (1, 2)$ and $(x_2, y_2) = (5, 6)$. From the definition for m,

$$m = \frac{6 - 2}{5 - 1} = \frac{4}{4} = 1$$

Figure 1–4 shows the geometry.

Notice that it doesn't matter which point we call (x_1, y_1) and which one we call (x_2, y_2); it doesn't affect the computation of m. If we label the points differently in Example 4, the computation becomes

$$m = \frac{2 - 6}{1 - 5} = \frac{-4}{-4} = 1$$

The answer is the same.

Be sure to subtract the x- and y-coordinates in the same order.

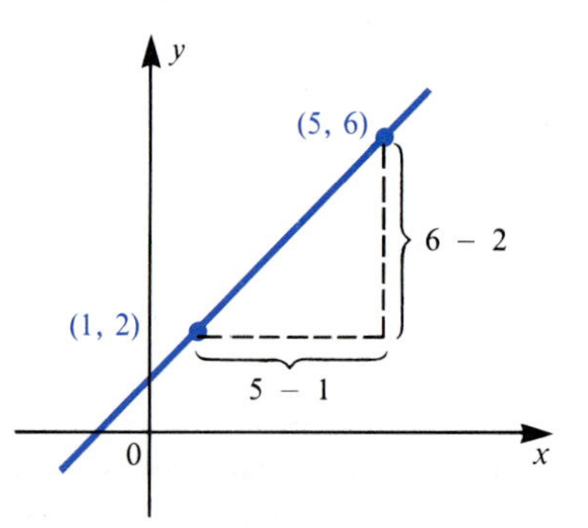

Figure 1–4

Example 5 Determine the slope of the line through the points (4, 5) and (6, 2).

Solution If we let (x_1, y_1) be (4, 5) and (x_2, y_2) be (6, 2), we get

$$m = \frac{y_2 - y_1}{x_2 - x_1} = \frac{2 - 5}{6 - 4} = \frac{-3}{2} = -\frac{3}{2}$$

The geometry is shown in Figure 1–5.

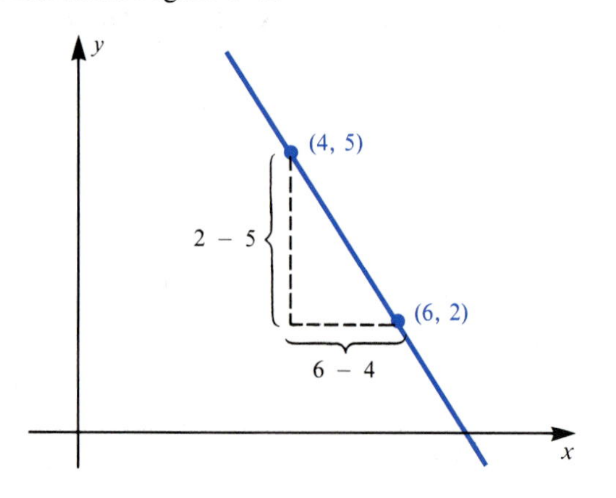

Figure 1–5

Example 6 (*Compare Exercise 19*)
Determine the slope of the line through the points (2, 5) and (6, 5).

Solution The slope of the line is

$$m = \frac{5-5}{6-2} = \frac{0}{4} = 0$$

Whenever $m = 0$, the equation $f(x) = 0x + b$ is written more simply as $f(x) = b$; f is called a constant function. The graph of a constant function is a line parallel to the x-axis; such a line has an equation of the form $y = b$ and is called a **horizontal line**. (See Figure 1–6.)

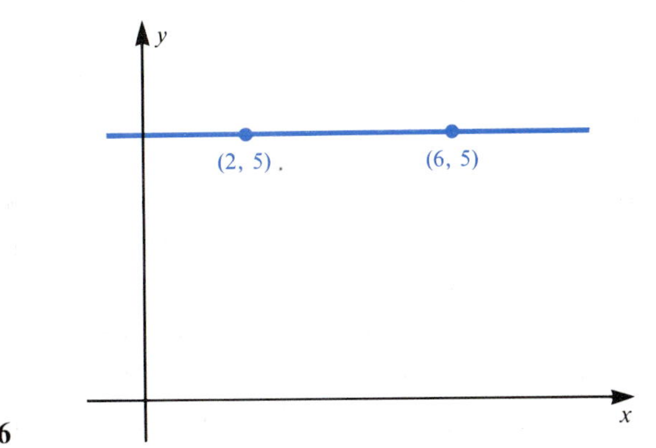

Figure 1–6

Example 7 (*Compare Exercise 23*)
Determine the equation of the line through the points (4, 1) and (4, 3).

Solution We can try to use the rule for computing the slope, but we obtain the quotient

$$\frac{3-1}{4-4} = \frac{2}{0}$$

which doesn't make sense because division by zero is not defined. The slope is not defined. When we plot the two points, however, we have no difficulty in drawing the line through them. See Figure 1–7.

The line, parallel to the y-axis, is called a **vertical line**. A point lies on this line when the first coordinate of the point is 4, so the equation of the line is $x = 4$.

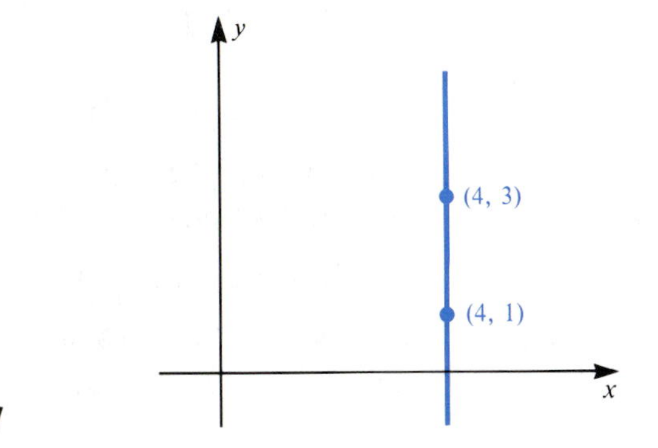

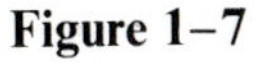
Figure 1–7

Whenever $x_2 = x_1$, you get a 0 in the denominator when computing the slope, so we say that the *slope does not exist* for such a line.

Be careful: A vertical line does *not* have a slope, but it does have an equation.

The slope of a line can be positive, negative, zero, or even not exist. These situations are depicted in Figure 1–8.

This figure shows the relationship between the slope and the slant of the line. If $m > 0$, the graph slants up as x moves to the right. If $m < 0$, the graph slants down as x moves to the right. If $m = 0$, the graph remains at the same height. If m does not exist, the line is vertical and the line is *not* the graph of a linear function. The equation of a vertical line cannot be written in form $f(x) = mx + b$ for there is no m.

We conclude this section by showing how to find equations of lines. Linear functions arise in many applied settings. When an application provides the appropriate two pieces of information, you can find an equation of the corresponding line. Two pieces of information that determine a line are

1. the slope and a point on the line, or
2. two points on the line.

We will show each form in a particular application and then give the general method of solving the problem.

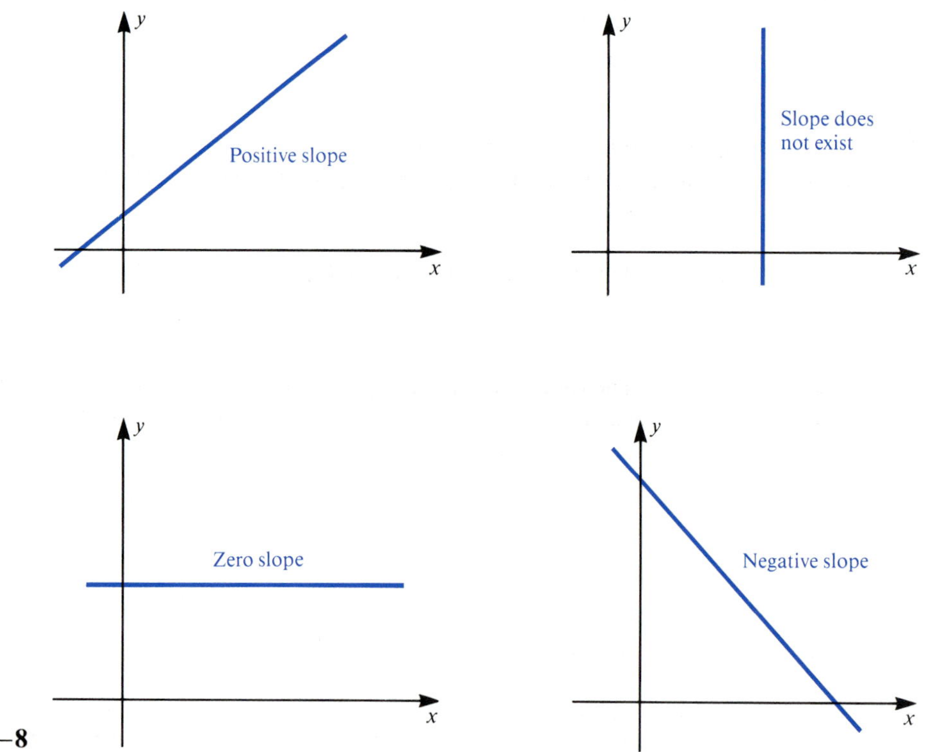

Figure 1–8

Example 8 **(point–slope given)**
(Compare Exercise 58)
A manufacturer of watches knows that it costs \$3.14 in materials to make each watch. Also, the manufacturer must pay \$7415 per week on other costs such as property taxes, payroll, and so on. Find the weekly costs of the company as a function of the number of watches produced.

Solution Let x be the number of watches produced each week, and C be the total weekly costs for the company in dollars. The amount \$7415 is called the **fixed cost** for the company; it does not depend on the number of watches produced. The other cost to the company is called the **variable cost**, and depends directly on the number of items produced. In this example the variable cost is $3.14x$. The **total cost** is the sum of the fixed cost and the variable cost. Thus, C and x are related by the equation $C = 3.14x + 7415$. C is a linear function of x with slope $m = 3.14$ and $(0, 7415)$ a point on the line.

Example 9 *(Compare Exercise 27)*
Determine an equation of the line with slope 3 and y-intercept 2.

Solution This example gives the same kind of information as Example 8. Note that the fixed cost corresponds to b in the equation $f(x) = mx + b$. The phrase "y-intercept 2" means that the point $(0, 2)$ is on the line. Since $m = 3$ and $b = 2$, the equation is

$$y = 3x + 2$$

Example 10 *(Compare Exercise 33)*
Determine an equation of the line that has slope -2 and passes through the point $(-3, 5)$.

Solution The value of $m = -2$, so the line has an equation of form

$$y = -2x + b$$

The solution will be complete when the value of b is found. Since the point $(-3, 5)$ lies on the line, $x = -3$ and $y = 5$ must be a solution to $y = -2x + b$. Just substitute those values into $y = -2x + b$ to obtain

$$5 = (-2)(-3) + b$$
$$5 = 6 + b$$
$$5 - 6 = b$$

Thus, $b = -1$ and the equation of the line is $y = -2x - 1$.

There is a formula, called the point-slope formula, that may be used to write an equation of the line in one step when the slope and point are given.

Point-Slope Formula

We will work with the same information in Example 9, $m = -2$ and the point $(-3, 5)$. Use the slope formula with $(-3, 5)$ as (x_1, y_1) and use an arbitrary point, (x, y), as (x_2, y_2). Because $m = -2$, we can write

$$-2 = \frac{y - 5}{x - (-3)} = \frac{y - 5}{x + 3}$$

Multiply both sides by $x + 3$ to obtain

$$-2(x + 3) = y - 5$$

The formula is usually written as

$$y - 5 = -2(x + 3)$$

Point-Slope Formula

If a line has slope m and passes through (x_1, y_1), an equation of the line is

$$y - y_1 = m(x - x_1)$$

Example 11 (*Compare Exercise 37*)
Find an equation of the line with slope 4 that passes through $(-1, 5)$.

Solution $x_1 = -1$ and $y_1 = 5$; $m = 4$, so the point-slope formula gives us

$$y - 5 = 4(x - (-1))$$

The arithmetic to write this equation in the form $y = mx + b$ is

$$y - 5 = 4x + 4$$
$$y = 4x + 4 + 5$$
$$y = 4x + 9$$

Example 12 **(2 points given)**
Businesses can deduct the cost of equipment in various ways; one method is called straight-line depreciation. For example, a piano teacher buys a new piano for \$5000 and can depreciate the value of the piano for the next 6 years, when it will be worth \$800. Find the value of the piano as a function of its age.

Solution Let t be the number of years after the piano was purchased and let v represent the value of the piano. When $t = 0$, $v = 5000$, so the graph contains the point $(0, 5000)$. When $t = 6$, $v = 800$ so the graph also contains the point $(6, 800)$. The graph is a straight line with slope

$$\frac{5000 - 800}{0 - 6} = \frac{4200}{-6} = -700$$

The negative slope means that the value of the piano is decreasing.

Now that we have slope we can use either of the two points to generate an equation of the line. Using the point $(0, 5000)$, we have $v - 5000 = -700(t - 0) = -700t$; $v = -700t + 5000$. This equation is valid for the application when $0 \le t \le 6$.

Just as we had a point-slope formula, we have a formula for finding an equation of a line using the coordinates of two points on the line.

Two-Point Formula

> If a line passes through the points (x_1, y_1) and (x_2, y_2), with $x_1 \neq x_2$, an equation of the line is
>
> $$y - y_1 = \frac{y_2 - y_1}{x_2 - x_1}(x - x_1)$$

Observe: This is basically the point-slope form of a line using

$$m = \frac{y_2 - y_1}{x_2 - x_1}$$

and (x_1, y_1) as the point on the line. You can think of this as a variation of the point-slope formula. First, use the two points to compute m, and then use one of the points and the point-slope formula.

Example 13 (*Compare Exercise 41*)
Determine an equation of the straight line through the points (1, 3) and (4, 7).

Solution Let (x_1, y_1) be the point (1, 3) and let (x_2, y_2) be the point (4, 7); then

$$m = \frac{7 - 3}{4 - 1} = \frac{4}{3}$$

$y - 3 = \frac{4}{3}(x - 1)$ is an equation of this line. We can simplify by multiplying both sides by 3, obtaining

$$3y - 9 = 4x - 4$$
$$3y = 4x + 5$$

We conclude this section with a discussion of parallel lines.

Parallel Lines

Definition

> Two lines are **parallel** if they have the same slope or if they are both vertical lines.

Example 14 (*Compare Exercise 47*)
Is the line through the points (1, 2) and (3, 3) parallel to the line through the points (−3, 2) and (5, 6)?

Solution Let L_1 be the line through (1, 2) and (3, 3), and let m_1 be its slope.

$$m_1 = \frac{3 - 2}{3 - 1} = \frac{1}{2}$$

Let L_2 be the line through $(-3, 2)$ and $(5, 6)$, and let m_2 be its slope.

$$m_2 = \frac{6 - 2}{5 - (-3)} = \frac{4}{8} = \frac{1}{2}$$

The slopes are identical; thus, the lines are parallel. (See Figure 1–9.)

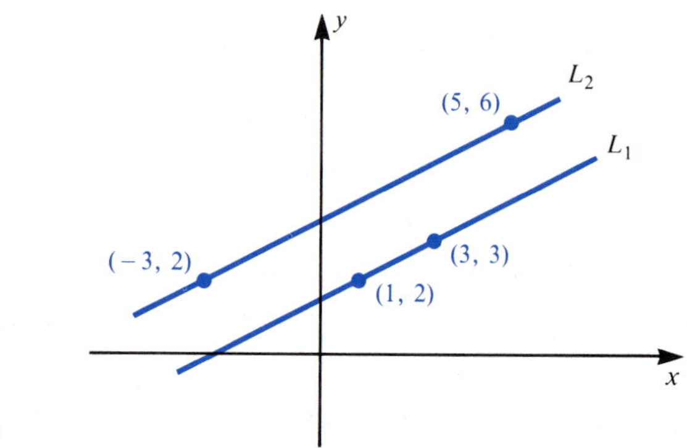

Figure 1–9

Example 15 Is the line through (5, 4) and (−1, 2) parallel to the line through (3, −2) and (4, 4)?

Solution The slope of the first line is

$$m_1 = \frac{4 - 2}{5 - (-1)} = \frac{2}{6} = \frac{1}{3}$$

The slope of the second line is

$$m_2 = \frac{-2 - 4}{3 - 4} = \frac{-6}{-1} = 6$$

Since the slopes are not equal, the lines are not parallel.

1–2 Exercises

I.

Draw the graphs of the lines in Exercises 1 through 6.

1. (*See Example 1*) $f(x) = 3x + 8$

2. $f(x) = 4x - 2$

3. $f(x) = x + 7$

4. $f(x) = -2x + 5$

5. $f(x) = -3x - 1$

6. $f(x) = \frac{2}{3}x + 4$

Find the slope and y-intercept for the lines in Exercises 7 through 10.

7. (*See Example 2*) $y = 7x + 22$

8. $y = 13x - 4$

9. $y = -\frac{2}{5}x + 6$

10. $y = -\frac{1}{4}x - \frac{1}{3}$

Find the slope and y-intercept of the lines in Exercises 11 through 14.

11. (*See Example 3*) $2x + 5y - 3 = 0$

12. $4x + y - 3 = 0$

13. $x - 3y + 6 = 0$

14. $5x - 2y = 7$

Determine the slopes of the straight lines through the parts of points in Exercises 15 through 22.

15. (*See Example 4*) (1, 1), (2, 3)

16. (−1, 0), (1, 2)

17. (−1, −1), (−1, −3)

18. (4, −1), (6, −1)

19. (*See Example 6*) (5, −2), (−3, −2)

20. (8, 3), (1, 3)

21. (−1, 0), (−4, 0)

22. (0, 0), (17, 0)

Find an equation of the line through the given points in Exercises 23 through 26.

23. (*See Example 7*) (3, 2), (3, 5)

24. (−4, 6), (−4, 9)

25. (10, 0), (10, 7)

26. (−6, −1), (−6, 13)

Find equations of the lines in Exercises 27 through 32 with the given slope and y-intercept.

27. (*See Example 9*) slope 4 and y-intercept 3

28. slope −2 and y-intercept 5

29. $m = -1, b = 6$

30. $-\frac{3}{4}, b = 7$

31. $m = \frac{1}{2}, b = 0$

32. $m = 3.5, b = -1.5$

Find equations of the lines in Exercises 33 through 36 with the given slope and passing through the given point.

33. (*See Example 10*) slope −4 and point (2, 1)

34. slope 6 and point (−1, −1)

35. slope $\frac{1}{2}$ and point (5, 4)

36. slope −1.5 and point (2.6, 5.2)

Use the point-slope formula to find the equations of the lines in Exercises 37 through 40 with given slope and point.

37. (*See Example 11*) slope 7 and point (1, 5)

38. slope −2 and point (3, 1)

39. slope $\frac{1}{5}$ and point (9, 6)

40. slope $-\frac{2}{3}$ and point (−1, 4)

In Exercises 41 through 46, determine the equations of the straight lines through the pairs of points, and sketch the lines.

41. (*See Example 13*) (−1, 0), (2, 1)

42. (3, 0), (1, −1)

43. (0, 0), (1, 2)

44. (1, 3), (1, 5)

45. (2, 4), (5, 4)

46. (0, 3), (7, 0)

47. (*See Example 14*) Is the line through the points (8, 2) and (3, −3) parallel to the line through (6, −1) and (16, 9)?

48. Is the line through (5, 4) and (1, -2) parallel to the line through (1, 2) and (6, 8)?

49. Is the line through (9, -1) and (2, 8) parallel to the line through (3, 5) and (10, -4)?

Determine if the pairs of lines in Exercises 50 through 53 are parallel.

50. $y = 6x + 22$
$y = 6x - 17$

51. $3x + 2y = 5$
$6x + 4y = 15$

52. $x - 2y = 3$
$2x + y = 1$

53. $3x - 5y = 4$
$-6x + 10y = -8$

54. Write the equation of the line through (-1, 5) that is parallel to $y = 3x + 4$.

55. Write the equation of the line through (2, 6) that is parallel to $3x + 2y = 17$.

56. Write the equation of the line with y-intercept 8 that is parallel to $5x + 7y = -2$.

57. Write the equation of the line through (-4, 5) that is parallel to the line through (6, 1) and (2, -3).

II.

58. (*See Example 8*) A manufacturer of lawnmowers estimates that the material for each lawnmower costs \$85.00. The manufacturer's fixed operating costs are \$4250 per week. Find the weekly costs of the company.

59. A hamburger place estimates that the materials for each hamburger cost \$0.67 and fixed daily operating expenses are \$480. Find the daily costs of the place.

60. (*See Example 12*) Don's Auto Parts bought a new delivery van for \$12,000. Don expects the van to be worth \$1500 in five years. Find the value of the van as a function of its age.

61. A wholesale company paid \$1340 for 500 items, and later bought 800 of the same item for \$1760. Assume the cost is a linear function of the number of items. Write the equation.

62. The calorie content of four large shrimp is 36 calories; that for eleven shrimp is 99 calories. Write the calorie content as a linear function of the number of shrimp.

63. A male college student requires 3000 calories per day to maintain his daily activities and to maintain a constant weight. He wants to gain weight. Each pound of body fat requires an additional 3500 calories. Write his daily calorie intake as a function of the number of pounds gained per day.

III.

64. (*See Example 8*) In a cost equation like $C(x) = 3x + 15$, 15 is the fixed cost and $3x$ is the variable cost. The coefficient of x, 3 in this case, is the **unit cost**.

Find the cost equation for the following:

(a) Fixed cost = \$745, unit cost = \$4.50

(b) Fixed cost = \$9248, unit cost = \$22.75
(c) Fixed cost = \$860, unit cost = \$2.47

65. The slope of a line is the amount of change in y when x increases by one unit. For $y = 3x + 7$, y increases by 3 when x increases by 1. For $y = -2x + 4$, y decreases by 2 when x increases by 1. (The negative value indicates a decrease in y.)

Find the change in y when x increases by 1:

(a) $y = 4x - 5$
(b) $y = -3x + 4$
(c) $y = \frac{2}{3}x + 7$
(d) $y = -\frac{1}{2}x + \frac{2}{5}$
(e) $3y + 2x - 4 = 0$
(f) $y = 17$

66. A female college student needs 2100 calories per day to maintain her present weight and daily activities. She wants to lose x pounds per day. Each pound of body fat is equivalent to 3500 calories. Write her daily calorie intake as a linear function of the number of pounds of weight lost per day.

67. An auto rental firm charges a fixed daily rate and a mileage charge. One customer rents a car for one day and drives it 125 miles. His bill is \$35.75. Another customer rents a car for one day and drives 265 miles. Her bill is \$51.65. Write the linear equation relating miles driven and total cost.

1–3 Applications of Linear Functions

Cost-Volume Functions
Revenue Function
Break-Even Analysis
Simple Interest
Straight-Line Depreciation

In practical problems the relationship between the variables can be quite complicated. For example, the variables and their relationship that affect the stock market still defy the best of analysts. However, many times a linear relationship can be used to provide reasonable and useful information for solving practical problems. This section contains several applications of the linear function.

Cost-Volume Function

A manufacturer of mopeds conducted a study of production costs and found that fixed costs averaged \$5600 per week and material costs averaged \$359 per moped. This information can be stated as

$$C = 359x + 5600$$

where x represents the number of mopeds produced, also called the **volume**, and C is the total cost of producing x mopeds. A linear function like this is used when

1. there are **fixed costs** such as rent, utilities, and salaries which are the same each week independent of the number of items produced;
2. there are **variable costs** which depend on the number of items produced, such as the cost of materials for the items, packaging, and shipping costs.

The moped example is a linear **cost-volume function** (often simply called the cost function). It is appropriate when the general form of the cost function C is given by

$$C(x) = ax + b$$

where

x is the number of items (volume),
b is the fixed cost in dollars,
ax is the variable cost in dollars,
a is the *unit* cost (the cost per item) in dollars,
C is the total cost in dollars of producing x items.

Notice the form of the cost function. It is essentially the slope-intercept form of a line where the slope is the unit cost and the intercept is the fixed cost.

Example 1 (*Compare Exercise 1*)
If the cost function of manufacturing mopeds is given by $C(x) = 359x + 5600$, then

(a) determine the cost of producing 700 mopeds per week.
(b) determine how many mopeds were produced if the production cost for one week was $200,178.

Solution **(a)** Substitute $x = 700$ into the cost equation to obtain

$$\begin{aligned} C(700) &= 359(700) + 5600 \\ &= 251{,}300 + 5600 \\ &= 256{,}900 \end{aligned}$$

So, the total cost is $256,900.

(b) This information gives us $C(x) = 200{,}178$, so we have

$$200{,}178 = 359x + 5600$$

We need to solve this for x.

$$\begin{aligned} 200{,}178 - 5600 &= 359x \\ 194{,}578 &= 359x \\ x &= \frac{194{,}578}{359} = 542 \end{aligned}$$

so 542 mopeds were produced.

Revenue Function

If a sporting goods store sells mopeds for \$798 each, the total income (**revenue**) in dollars from mopeds is 798 times the number of mopeds sold. This illustrates the general concept of a **revenue function**; it gives the total revenue obtained from the sale of x items. In the moped example, the revenue function is given by

$$R(x) = 798x$$

where x represents the number of mopeds sold, 798 is the selling price in dollars for each item, and $R(x)$ is the total revenue in dollars from x items.

Example 2 (*Compare Exercise 5*)
The sporting goods store has a sale on mopeds at \$725 each.

(a) Give the revenue function.
(b) The store sold 23 mopeds. What was the total revenue?
(c) One salesperson sold \$5075 worth of mopeds. How many did she sell?

Solution **(a)** The revenue function is given by

$$R(x) = 725x$$

(b) The revenue for 23 mopeds is obtained from the revenue function when $x = 23$:

$$R(23) = 725(23) = 16{,}675$$

The revenue in this case is \$16,675.

(c) This gives $R(x) = 5075$ so

$$725x = 5075$$

$$x = \frac{5075}{725} = 7$$

So, she sold 7 mopeds.

Break-Even Analysis

Break-even analysis answers a common management question: At what sales volume will we break even? When do revenues equal costs? Greater sales will induce a profit while lesser sales will show a loss.

The break-even point occurs when the cost equals the revenue, so the cost and revenue functions can be used to determine the break-even point. In function notation this is written as $C(x) = R(x)$.

Example 3 (*Compare Exercise 8*)
A department store pays \$99 each for tape decks. Their monthly fixed costs are \$1250. They sell the tape decks for \$189.95 each.

(a) What is the cost-volume function?
(b) What is the revenue function?
(c) What is the break-even point?

Solution Let x represent the number of tape decks sold.

(a) The cost function is given by

$$C(x) = 99x + 1250$$

(b) The revenue function is defined by

$$R(x) = 189.95x$$

(c) The break-even point occurs when cost equals revenue,

$$C(x) = R(x)$$

Writing out the function gives

$$99x + 1250 = 189.95x$$

The solution of this equation gives the break-even point.

$$\begin{aligned} 99x + 1250 &= 189.95x \\ 1250 &= 189.95x - 99x \\ 1250 &= 90.95x \\ x &= \frac{1250}{90.95} \\ &= 13.74 \end{aligned}$$

Because x represents the number of tape decks, we use the next integer, 14, as the number sold per month to break even. If more than 14 are sold, there will be a profit. If fewer than 14 are sold, there will be a loss.

Example 4 (*Compare Exercise 15*)
A temporary secretarial service has a fixed weekly cost of \$730. The wages and benefits of the secretaries amount to \$3.93 per hour. A firm who employs a secretary pays Temporary Service \$6.40 per hour. How many hours per week of secretarial service must Temporary Service place in order to break even?

Solution First, write the cost and revenue functions. The fixed cost is \$730 and the unit cost is \$3.93, so the cost function is given by

$$C(x) = 3.93x + 730$$

where x is the number of hours placed each week. The revenue function is given by

$$R(x) = 6.40x$$

Equating cost and revenue we have

$$3.93x + 730 = 6.40x$$

This equation reduces to

$$6.40x - 3.93x = 730$$
$$2.47x = 730$$
$$x = \frac{730}{2.47}$$
$$= 295.5$$

Temporary Service must place secretaries for a total of 296 hours per week (rounded up) in order to break even.

Simple Interest

Our modern economy depends on borrowed money. Very few families would own a house or a car without credit. Business depends on borrowed money for day-to-day operations and for long-term expansion. When we borrow money from a bank, we pay them for the use of it. Interest is the "rent" paid to use money for a period of time. The amount of interest is usually based on a specified percent of the money borrowed. If you borrow \$800 at a simple interest rate of 10% per year, the interest paid is

$$I = 800(.10)t$$

where I is the interest in dollars and t is the length of the loan in years. If \$800 is borrowed at 10% per year for three years, the amount of interest is found by computing

$$I = 800(.10)(3) = 240$$

The amount of interest paid is \$240.

In general, the interest is

$$I = Prt$$

where P is the amount borrowed (principal), r is the interest rate written in decimal form, and t is the length of the loan.

Example 5 (*Compare Exercise 16*)
Find the interest paid on a \$1500 loan if the interest rate is 8.5% and the loan is for 2.5 years.

Solution The decimal form of 8.5% is .085 so the formula for interest becomes

$$I = 1500(.085)(2.5)$$
$$= 318.75$$

The interest paid is \$318.75.

Notice that when P and r are fixed, I is a linear function of t.

Example 6 (*Compare Exercise 23*)
If Brady borrows \$2000 at a 12% interest rate to pay college expenses, the total amount due at the end of the loan is the original amount borrowed plus the interest. This amount due, A, for t years is

$$A = 2000 + 2000(.12)t$$

If the loan is for 2 years, the amount due, in dollars, is

$$\begin{aligned} \mathbf{A} &= 2000 + 2000(.12)(2) \\ &= 2000 + 480 = 2480 \end{aligned}$$

The general expression for the amount due at the end of t years is

$$A = P + Prt$$

Notice that when P and r are fixed, A is a linear function of t.

Straight-Line Depreciation

When a corporation buys a fleet of cars, they expect them to decline in value due to wear and tear. If they purchase new cars for \$11,500 each, they may expect them to be worth only \$2500 three years later. This decline in value is called **depreciation**. The value of an item after deducting depreciation is called its **book value**. In three years each car depreciated \$9000, and its book value at the end of three years was \$2500. For tax and accounting purposes a company will report depreciation and book value each year during the life of an item. The Internal Revenue Service allows several methods of depreciation. The simplest is **straight line depreciation**. This method assumes that the book value is a linear function of time, that is,

$$BV = mx + b$$

where BV is the book value and x is the number of years. For example, each car had a book value of \$11,500 when $x = 0$ ("brand new" occurs at zero years). When $x = 3$, its book value declined to \$2500. This information is equivalent to giving two points (0, 11500) and (3, 2500) on the straight line representing book value. (See Figure 1–10.)

We obtain the linear equation of the book value by finding the equation of a line through these two points. The slope of the line is

$$\begin{aligned} m = \frac{y_2 - y_1}{x_2 - x_1} &= \frac{2500 - 11{,}500}{3 - 0} \\ &= \frac{-9000}{3} = -3000 \end{aligned}$$

and the y-intercept is 11,500 so the equation is

$$BV = -3000x + 11{,}500$$

The book value at the end of two years is

$$BV = -3000(2) + 11{,}500 = -6000 + 11{,}500 = 5500$$

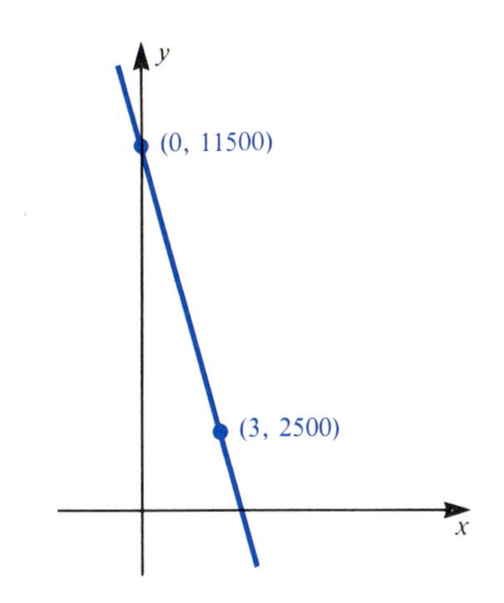

Figure 1–10

The negative value of the slope indicates that the book value is decreasing by \$3000 each year. This annual decrease is the **annual depreciation**.

Generally, a company will estimate the number of years of useful life of an item. The value of the item at the end of its useful life is called its **scrap value**. The values of x are restricted to $0 \le x \le n$, where n is the number of years of useful life.

Example 7 (*Compare Exercise 19*)

Acme Manufacturing Co. purchases a piece of equipment for \$28,300 and estimates its useful life as 8 years. At the end of its useful life, its scrap value is estimated at \$900.

(a) Find the linear equation expressing the relationship between book value and time.

(b) Find the annual depreciation.

(c) Find the book value for the first, fifth, and seventh year.

Solution **(a)** The line passes through the two points (0, 28300) and (8, 900). Thus, the slope is

$$m = \frac{(900 - 28{,}300)}{(8 - 0)} = -3425$$

and the y-intercept is 28,300 giving the equation

$$BV = -3425x + 28{,}300$$

(b) The annual depreciation is obtained from the slope and is \$3425.

(c) For year 1, $BV = -3425(1) + 28{,}300 = 24{,}875$.
For year 5, $BV = -3425(5) + 28{,}300 = 11{,}175$.
For year 7, $BV = -3425(7) + 28{,}300 = 4{,}325$.

Thus, after 7 years the book value of the equipment is \$4325.

1–3 Exercises

I.

1. (*See Example 1*) The weekly cost function of manufacturing x bicycles is given by

$$C(x) = 43x + 2300$$

   (a) Determine the cost of producing 180 bicycles per week.

   (b) One week the total production cost was $11,889. How many bicycles were produced that week?

2. A software company produces a home-accounting system. Their cost function for producing x systems per month is given by

$$C(x) = 16.25x + 28{,}300$$

   (a) Determine the cost of producing 2500 systems per month.

   (b) One month their production costs were $63,010. How many systems did they produce?

3. A company has determined that the relationship between cost and volume (the cost-volume formula) for a certain product is $C(x) = 3x + 400$.

   (a) Determine the fixed cost and the unit cost.

   (b) Find the total costs when the production is 600 units and 1000 units, respectively.

4. The cost-volume formula for a certain product is $y = 2.5x + 750$. Determine the fixed cost and the unit cost. Find the total costs when the production is

   (a) 100 units

   (b) 300 units

   (c) 650 units

5. (*See Example 2*) A store sells jogging shoes for $32 per pair.

   (a) Write the rule that gives the revenue function.

   (b) The store sold 78 pairs. What was the revenue?

   (c) One day the store sold $672 worth of jogging shoes. How many pairs did they sell?

6. Tony's Cassette Warehouse sells cassettes for $6.25 each.

   (a) Write the rule for the revenue function.

   (b) What is the revenue from selling 265 cassettes?

7. Cold Pizza sells frozen pizzas for $3.39 each.

   (a) Write the rule that gives the revenue function.

   (b) What is the revenue from selling 834 pizzas?

8. (*See Example 3*) A clothing store pays \$57 each for sports coats and has a fixed monthly cost of \$780. They sell the coats for \$79 each.

(a) What is the rule for the cost-volume function?

(b) What is the rule for the revenue function?

(c) What is the break-even point?

9. The monthly expenses of The Campus Copy Shop are given by the cost equation

$$C(x) = 3690 + 0.025x$$

where x is the number of pages copied in a month. The revenue function is

$$R(x) = .055x$$

Find the break-even point of The Campus Copy Shop.

10. The Academic T-Shirt Company did a cost study and found that it cost \$1400 to produce 600 "I Love Math" T-shirts. The total cost is \$1600 when the volume is 700 T-shirts. Determine the linear equation that describes the relationship of cost to volume.

(a) What is the fixed cost?

(b) What is the unit cost?

11. A Toy Co. estimates that total costs are \$1000 when their volume is 500 Fastback cars, and \$1200 when their volume is 900 cars.

(a) Determine the cost-volume formula.

(b) What are the fixed cost and the unit cost?

(c) What are the estimated total costs when the volume is 1200 units?

12. Find the cost equation if the fixed cost is \$700 and the cost-per-unit volume is \$2.50. What are the total costs when the volume produced is 400 units?

13. Find the cost-volume formula if the fixed cost is \$500 and the cost-per-unit volume is \$4. What are the total costs when the volume produced is 800 units?

14. A company has a cost function given by

$$C(x) = 22x + 870$$

and a revenue function of

$$R(x) = 37.50x$$

Find the break-even point.

15. (*See Example 4*) A Computer Shop sells computers. The shop has fixed costs of \$1500 per week. Their average cost per computer is \$649 and the average selling price is \$899.

(a) Write the rule for the cost function.

(b) Write the rule for the revenue function.

(c) Find the cost of selling 37 computers per week.

(d) Find the revenue from selling 37 computers.

(e) Find the break-even point.

16. (*See Example 5*) Compute the simple interest on $4800 for 1.5 years at 11% interest rate.

17. Compute the simple interest on $500 for 2 years at 7% interest rate.

18. Compute the simple interest on $950 for 1.75 years at 8% interest rate.

19. (*See Example 7*) A TV costs $425, has a scrap value of $25, and a useful life of 8 years. Find

(a) the linear equation relating book value and number of years.

(b) the annual depreciation.

(c) the book value for year 3.

20. A machine costs $1500 and has a useful life of 10 years. If it has a scrap value of $200, find

(a) the linear equation relating book value and number of years.

(b) the annual depreciation.

(c) the book value after 7 years.

21. An automobile costs $9750, has a useful life of 6 years, and a scrap value of $300. Find

(a) the linear equation relating book value and number of years.

(b) the annual depreciation.

(c) the book value for years 2 and 5.

II.

22. Jones borrows $3500 for 3 years at 8% simple interest. How much interest does Jones pay?

23. (*See Example 6*) Angie borrowed $750 for 2 years at 9% interest. What is the amount due at the end of two years?

24. Joe obtained a loan of $2500 for 4.5 years at 9% simple interest. What amount was due at the end of that time?

25. Alfred borrowed $1850 at 10.5% simple interest. The amount due at the end of the loan was $2529.88. What was the length of the loan?

26. A health club membership costs $35 per month.

(a) What is the monthly revenue function for the club?

(b) What is the monthly revenue if the club has 1238 members?

(c) The revenue increased $595 in one month. What was the increase in membership?

III.

27. A company's records showed that the daily fixed costs for one of their production lines was \$1850 and the total cost of one day's production of 320 items was \$3178. What is the cost-volume function?

28. How much should be invested at 8% simple interest in order to have \$2000 in 18 months?

29. A specialty store sells personalized telephones. Their weekly cost function is given by

$$C(x) = 28x + 650$$

and the break-even point is $x = 65$ phones per week. What is the revenue function?

30. The break-even point for a tanning salon is 260 memberships, which gives them \$3120 monthly revenue. If they sell only 200 memberships, they will lose \$330 per month.

(a) What is their revenue function?

(b) What is their cost function?

31. The profit function is revenue minus cost, that is,

$$P(x) = R(x) - C(x)$$

(a) The cost and revenue functions for Acme Manufacturing are given by

$$C(x) = 28x + 465$$
$$R(x) = 52x$$

(i) What is the profit function?

(ii) What is the profit from selling 25 items?

(b) The weekly expenses of selling x bicycles in The Bike Shop are given by the cost function

$$C(x) = 1200 + 130x$$

and revenue is given by

$$R(x) = 210x$$

(i) What is the profit function?

(ii) Find the profit from selling 18 bicycles in a week.

(c) Another bicycle shop has monthly fixed costs of \$5200 and unit costs of \$145. They sell their bicycles for \$225 each.

(i) Write the profit function.

(ii) What is the profit from selling 75 bicycles per month?

1–4 Linear Inequalities in Two Variables

The recommended minimum daily requirement of vitamin B_6 for adults is 2.0 mg. A deficiency of vitamin B_6 in men may increase their cholesterol level and lead to a thickening and degeneration in the walls of their arteries. Many meats, bread, and vegetables contain no vitamin B_6. Fruits usually contain this vitamin. For example, one small banana contains 0.45 mg, and one ounce of grapes contains 0.02 mg of vitamin B_6. What quantities of bananas and grapes should you consume in order to exceed the minimum requirements?

Mathematically, this question is equivalent to asking for solutions of the inequality

$$.45x + .02y > 2.0$$

where x = the number of bananas eaten and y = the number of ounces of grapes eaten.

Let's see how we solve inequalities like this. If a point is selected, say (3, 5), and substituted into the inequality, we obtain

$$\begin{aligned} .45(3) + .02(5) &= 1.35 + .10 \\ &= 1.45 \end{aligned}$$

which is not greater than 2.0. Since (3, 5) makes $.45x + .02y > 2.0$ false, (3, 5) is *not* a solution. However, the point (4, 12) is a solution because

$$\begin{aligned} .45(4) + .02(12) &= 1.80 + .24 \\ &= 2.04 \end{aligned}$$

Thus, $x = 4$ and $y = 12$ makes the inequality true. Actually, an infinity of points make the statement true.

If an arbitrary point is selected and its coordinates are substituted into

$$.45x + .02y$$

we can expect one of three different outcomes:

$$\begin{aligned} .45x + .02y &= 2.0 \\ .45x + .02y &< 2.0 \\ .45x + .02y &> 2.0 \end{aligned}$$

You recognize the first of these, $.45x + .02y = 2.0$, as the equation of a straight line. Any point that makes this statement true lies on that line. This line holds the key to finding the solution to the original inequality. A line divides the plane into two parts. The areas on either side of the line are called **half planes** (Figure 1–11).

All the points that satisfy

$$.45x + .02y < 2.0$$

lie on one side of the line $.45x + .02y = 2.0$, and all the points that satisfy

$$.45x + .02y > 2.0$$

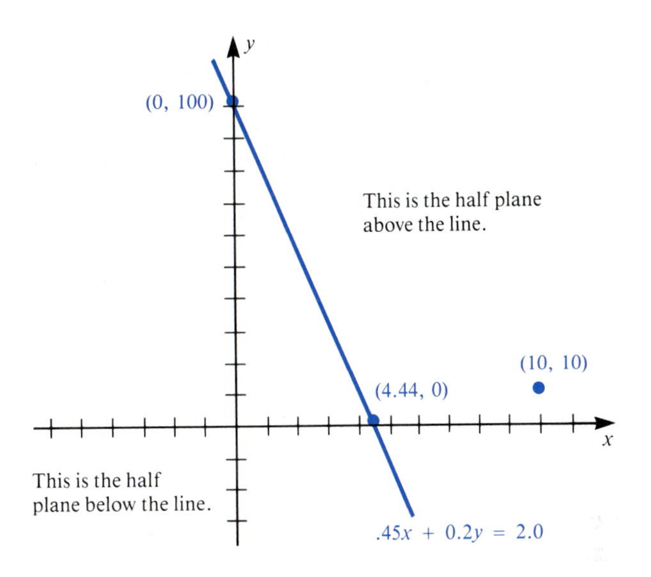

Figure 1–11 The line $.45x + .02y = 2.0$ divides the plane into two half planes.

lie on the other side. We find the solution when we determine which half plane satisfies $.45x + .02y > 2.0$.

There is a simple way to decide. Just pick a point and try it. For example, the point (10, 10) is in the half plane above the line. Since $.45(10) + .02(10) = 4.5 + .2$, the point (10, 10) satisfies the inequality $.45x + .02y > 2.0$. Consequently, *all* points above the line satisfy the same inequality. The graph of the inequality is the half plane above the line. Indicate this by shading that half plane (Figure 1–12). Since the points on the line do not satisfy the inequality, use a dotted line for $.45x + .02y = 2.0$.

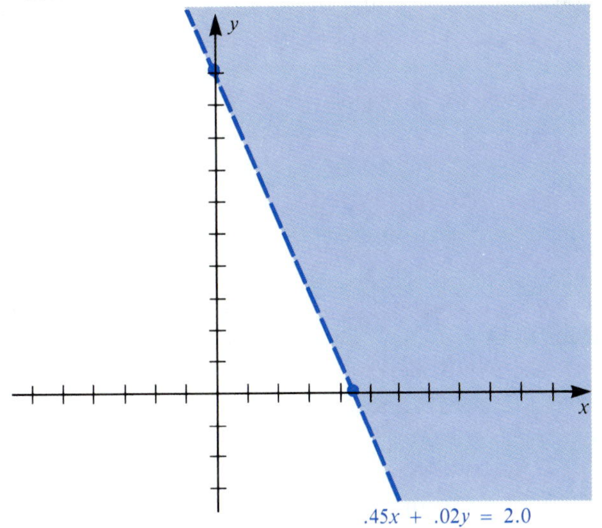

Figure 1–12 The graph of $.45x + .02y > 2.0$

Example 1 (*Compare Exercise 1*)
Solve $2x + 3y < 12$.

Solution: First, replace the inequality by equals and obtain

$$2x + 3y = 12$$

This line divides the plane into two half planes (see Figure 1–13), the part where

$$2x + 3y > 12$$

and the part where

$$2x + 3y < 12$$

To decide which part satisfies

$$2x + 3y < 12$$

select a point and try it. For example, the point (0, 0) is in the half plane below the line. Since $2(0) + 3(0) = 0 < 12$, the point (0,0) satisfies the inequality $2x + 3y < 12$. Consequently *all* points below the line satisfy the same inequality. The graph of the inequality is the half plane below the line. Indicate this by shading that half plane (Figure 1–14). Since the points on the line do not satisfy the inequality, use a dotted line for $2x + 3y = 12$.

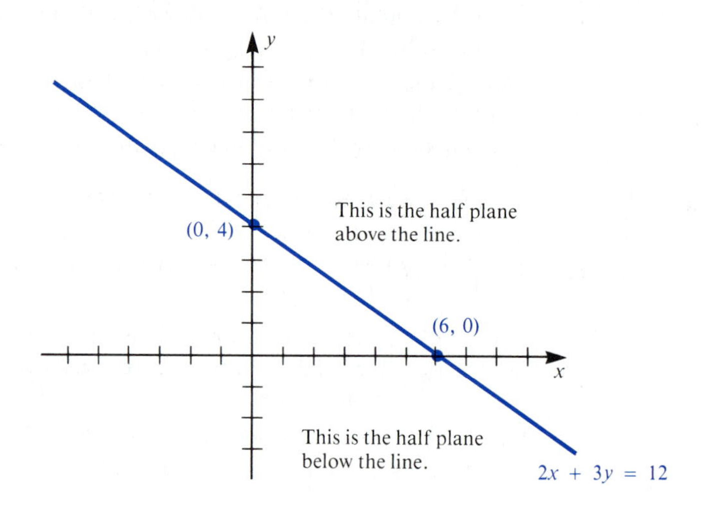

Figure 1–13 The line $2x + 3y = 12$ divides the plane into two half planes.

If you select the point (5, 6), a point above the line, you find that

$$2(5) + 3(6) = 28 > 12$$

so the half plane above the line is not the correct one. This is another way to conclude that the correct part is the half plane below the line.

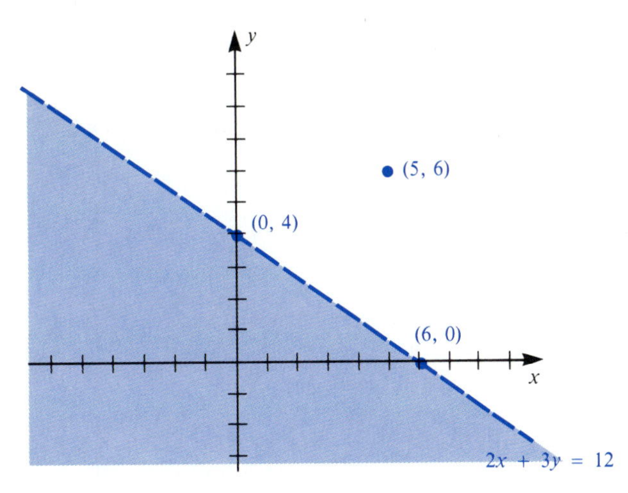

Figure 1–13 The graph of $2x + 3y < 12$

Procedure for Graphing a Linear Inequality

Graph the inequality of the form $ax + by < c$.
The procedure also applies if the inequality symbols are $\leq$, $>$, or $\geq$.

1. Graph the line $ax + by = c$. This line is the boundary between two half planes.
2. Select a point not on the line from one half plane. The point (0, 0) is usually a good choice when it is not on the line.
3. Substitute the coordinates of the point for x and y in the inequality.
   - **(a)** If the selected point satisfies the inequality, then shade the half plane where the point lies. These points are on the graph.
   - **(b)** If the selected point does not satisfy the inequality, shade the half plane opposite the point.
   - **(c)** If the inequality symbol is $<$ or $>$, use a dotted line for the graph of $ax + by = c$. This indicates that the points on the line are *not* a part of the graph.
   - **(d)** If the inequality symbol is $\leq$ or $\geq$, use a solid line for the graph of $ax + by = c$. This indicates that the line is a part of the graph.

Example 2 (*Compare Exercise 5*)
Graph the solution to $5x + 2y \leq 10$.

Solution: First, graph the line $5x + 2y = 10$. Use a solid line because the points on the line are part of the solution. Second, select a point in one of the half planes. The point (0, 0) is below the line and $5(0) + 2(0) \leq 10$ is true. Therefore, the points below the line are in the solution. Shade the half plane below the line. (See Figure 1–15.)

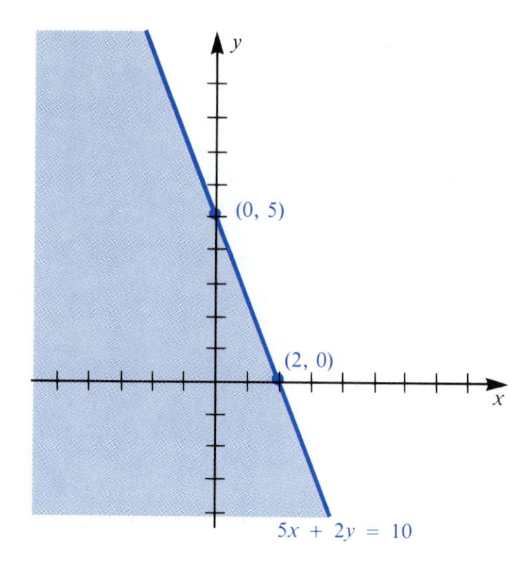

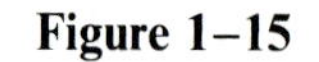

Figure 1–15 Graph of $5x + 2y \leq 10$

Example 3 (*Compare Exercise 9*)
Graph $4x - 3y > 24$.

Solution: Graph the line $4x - 3y = 24$. Use a dotted line since the points on the line are not a part of the solution. The point (1, 2) is above the line and $4(1) - 3(2) > 24$ is false, so the half plane below the line is the desired graph. (See Figure 1–16.)

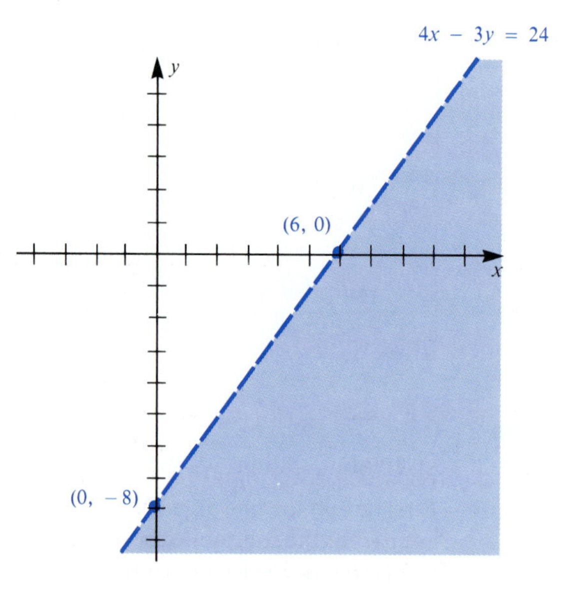

Figure 1–16 The graph of $4x - 3y > 24$

Example 4 (*Compare Exercise 15*)
An automobile assembly plant has an assembly line that produces the Hatchback Special and the Sportster. Each Hatchback requires 2.5 hours of assembly-line time, and each Sportster requires 3.5 hours. The assembly line has a maximum operating time of 140 hours per week. Graph the number of each car that can be produced in one week.

Solution: Let x be the number of Hatchback Specials and y the number of Sportsters produced per week. The total amount of assembly-line time required is

$$2.5x + 3.5y$$

Since the total assembly-line time is restricted to 140 hours, we need to graph the solution of the inequality

$$2.5x + 3.5y \leq 140$$

Graph the line $2.5x + 3.5y = 140$. Use a solid line. Since the point (0, 0) is a solution to the system, it is in the half plane of the graph. Figure 1–17 shows the graph of the inequality.

However, Figure 1–18 is a more appropriate graph since it is not practical to include negative values for the number of autos produced. Notice that the points (20, 20), (10, 30), and (40, 10) all lie in the region of solutions. This tells us that the combination 20 Hatchbacks and 20 Sportsters could be produced, or 10 Hatchbacks and 30 Sportsters, or 40 Hatchbacks and 10 Sportsters.

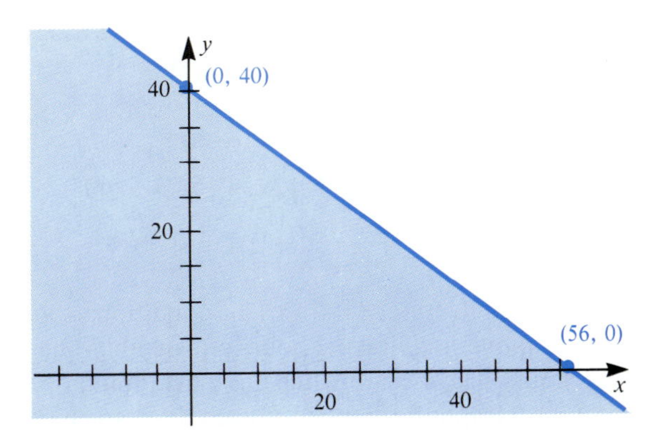

Figure 1–17

This example illustrates how methods of graphing an inequality are used to find the region that satisfies an inequality, but the context dictates that common sense be used in interpreting the answer. For example, the point (10.3, 5.2) is in the region of solutions, but it doesn't make sense to talk about producing 10.3 Hatchbacks and 5.2 Sportsters. Once the region of solutions is found, only points with whole number coordinates are reasonable.

If the problem had found the number of acres a farmer should plant to wheat and the number to corn, fractional values would be appropriate.

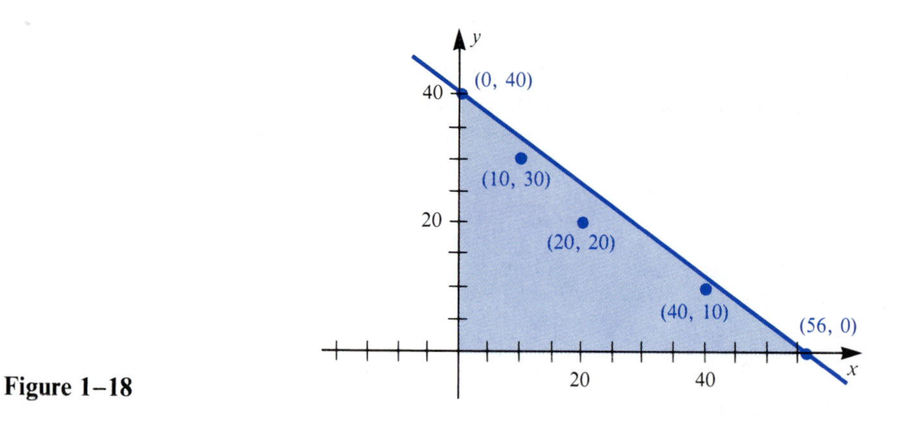

Figure 1–18

1–4 Exercises

I.

Graph the solution for the inequalities in Exercises 1 through 14.

1. (*See Example 1*) $5x + 4y < 20$
2. $5x + 3y < 15$
3. $6x + 5y < 30$
4. $2x - 10y < 30$
5. (*See Example 2*) $6x + 8y \leq 24$
6. $15x + 55y \leq 1650$
7. $3x - 7y \leq 21$
8. $4x - 5y \leq 20$
9. (*See Example 3*) $9x - 6y > 30$
10. $4x + 3y > 12$
11. $4x - 3y > 12$
12. $9x - 2y < 18$
13. $-2x - 5y > 10$
14. $-12x + 15y < -60$

II.

15. (*See Example 4*) A manufacturer produces air conditioners and commercial fans. Each air conditioner requires 3.2 hours of assembly-line time, and each fan requires 1.8 hours. The assembly line has a maximum operating time of 144 hours per week. Graph the number of each that can be produced in one week.

16. A furniture manufacturer makes desks and tables. There are 350 hours of labor available each week in the finishing room. Each table requires 2.5 hours in the finishing room, and each desk requires 4 hours. Graph the number of each that can be produced in one week.

17. A service club agrees to donate at least 500 hours of community service. A full member is to give 4 hours and a pledge gives 6 hours.

    (a) Write the inequality that expresses this information.

    (b) Graph the inequality.

18. Acme Manufacturing has two production lines. Line A can produce 200 gadgets per hour, and line B can produce 350 widgets per hour. Because of warehouse limitations, the total number of gadgets and widgets produced must not exceed 75,000.

(a) Express this information with an inequality.

(b) Graph the inequality.

19. An advertising agency has an advertising budget of $75,000 to advertise a client's product. They need to decide how to schedule the advertising between TV and newspaper ads. Each TV spot costs $900, and each Sunday newspaper ad costs $830. How many of each can they schedule?

20. An inventor wants to invest in tax exempt bonds with a 7.5% return and taxable bonds with a 9% return so that the total return will be at least $8100. How much of each can be purchased?

IMPORTANT TERMS

1–1	**Function**	**Domain**	**Range**
	Rule of a function	**Independent variable**	**Dependent variable**
1–2	**Graph**	**Linear function**	**Slope**
	***y*-intercept**	**Point-slope formula**	**Two-point formula**
	Fixed cost	**Variable cost**	**Parallel lines**
1–3	**Cost-volume function**	**Fixed costs**	**Variable costs**
	Unit costs	**Revenue function**	**Break-even analysis**
	Break-even point	**Simple interest**	**Depreciation**
	Book value	**Straight-line depreciation**	**Annual depreciation**
1–4	**Linear inequality**	**Half plane**	**Solution of a linear inequality**

REVIEW EXERCISES

1. What are the domains of the following functions?

(a) $f(x) = 1.45x$
where x is the number of burgers and $f(x)$ is the total cost in dollars.

(b) $f(x) = 9x$
where x is the weight, in ounces, of a steak and $f(x)$ is the amount of fat, in grams, in the steak.

2. If $f(x) = 8x - 4$ find

(a) $f(2)$ **(b)** $f(-3)$ **(c)** $f(\frac{1}{2})$ **(d)** $f(c)$

3. If $f(x) = \frac{(x + 2)}{(x - 1)}$ and $g(x) = 5x + 3$, find $f(2) + g(3)$.

4. If $f(x) = (x + 5)(2x - 1)$, find

(a) $f(1)$ **(b)** $f(0)$ **(c)** $f(-5)$ **(d)** $f(a - 5)$

5. Apples cost \$1.20 per pound so the price of a bag of apples is

$$f(x) = 1.20x$$

where x is the weight in pounds and $f(x)$ is the purchase price in dollars.

(a) What is $f(3.5)$?
(b) A bag of apples cost \$3.30. How much did it weigh?

6. Tuition and fees charges at a university are given by

$$f(x) = 135x + 450$$

where x is the number of semester hours enrolled and $f(x)$ is the total cost of tuition and fees.

(a) Find $f(15)$
(b) If a student's bill for tuition and fees were \$2205, in how many semester hours was he enrolled?

7. Write an equation of the function described by the following statements.

(a) All the shoes on this table are \$29.95 per pair.
(b) A catering service charges \$40 plus \$1.25 per person to cater a reception.

8. Sketch the graph of

(a) $f(x) = 2x - 5$
(b) $6x + 10y = 30$

9. Find the slope and y-intercept for the following lines.

(a) $y = -2x + 3$
(b) $y = \frac{2}{3}x - 4$
(c) $4y = 5x + 6$
(d) $6x + 7y + 5 = 0$

10. Find the slope of the line through the following pairs of points

(a) $(2, 7)$ and $(-3, 4)$
(b) $(6, 8)$ and $(-11, 8)$
(c) $(4, 2)$ and $(4, 6)$

11. Find the equation of the following lines.

(a) with slope $-\frac{3}{4}$ and y-intercept 5.
(b) with slope 8 and y-intercept -3
(c) with slope -2 and passing through $(5, -1)$

(d) with slope 0 and passing through (11, 6)
(e) passing through (5, 3) and $(-1, 4)$
(f) passing through $(-2, 5)$ and $(-2, -2)$
(g) passing through (2, 7) and parallel to $4x - 3y = 22$.

12. Determine if the following pairs of lines are parallel.

(a) $7x - 4y = 12$ and $-21x + 12y = 17$
(b) $3x + 2y = 13$ and $2x - 3y = 28$

13. Is the line through (5, 19) and $(-2, 7)$ parallel to the line through (11, 3) and $(-1, -5)$?
14. Is the line through (4, 0) and $(7, -2)$ parallel to the line through (7, 4) and (10, 2)?
15. A manufacturer has fixed costs of $12,800 per month and a unit cost of $36 per item produced. What is the cost function?
16. The weekly cost function of a manufacturer is

$$C(x) = 83x + 960$$

(a) What are the weekly fixed costs?
(b) What is the unit cost?

17. The cost function of producing x bags of Hi-Gro fertilizer per week is

$$C(x) = 3.60x + 2850$$

(a) What is the cost of producing 580 bags per week?
(b) If the production costs for one week amounted to $5208, how many bags were produced?

18. A shoe shop has a special sale where all jogging shoes are $28.50 per pair. Write the revenue function for jogging shoes.
19. A T-shirt shop pays $6.50 each for T-shirts. Their weekly fixed expenses are $675. They sell the T-shirts for $11.00 each.

(a) What is their revenue function?
(b) What is their cost function?
(c) What is their break-even point?

20. What is the simple interest on a $4500 loan at 9% interest rate for 18 months?
21. What is the amount due at the end of two years for a $1500 loan at 7% interest rate?
22. A student borrowed $800 for 9 months. The interest charge was $51. What was the interest rate?
23. A company purchases a piece of equipment for $17,500. The useful life is 8 years and the scrap value at the end of 8 years is $900.

(a) Find the equation relating book value and its age using straight-line depreciation.
(b) What is the annual depreciation?
(c) What is the book value for the fifth year?

24. The function for the book value of a truck is

$$f(x) = -2300x + 16{,}500$$

where x is its age and $f(x)$ its book value.

(a) What did the truck cost?

(b) If its useful life is 7 years, what is its scrap value?

25. Graph the solution to the following inequalities.

(a) $5x + 7y < 70$

(b) $2x - 3y > 18$

(c) $x + 9y \leq 21$

26. An assembly plant has two production lines. Line A can produce 65 items per hour, and Line B can produce 105 per hour. The loading dock can ship a maximum of 1500 items per day.

(a) Express the information with an inequality.

(b) Graph the inequality.

2

Topics in this chapter can be applied to:

Diet and Calorie Intake • Supply and Demand Analysis • Competitive Markets • Production Scheduling • Investments • Cost Control • Diet and Nutrition • Billing • Energy Consumption • Manufacturing • Coding and Decoding • Medical Diets • Industrial Interdependencies • Input-Output Models

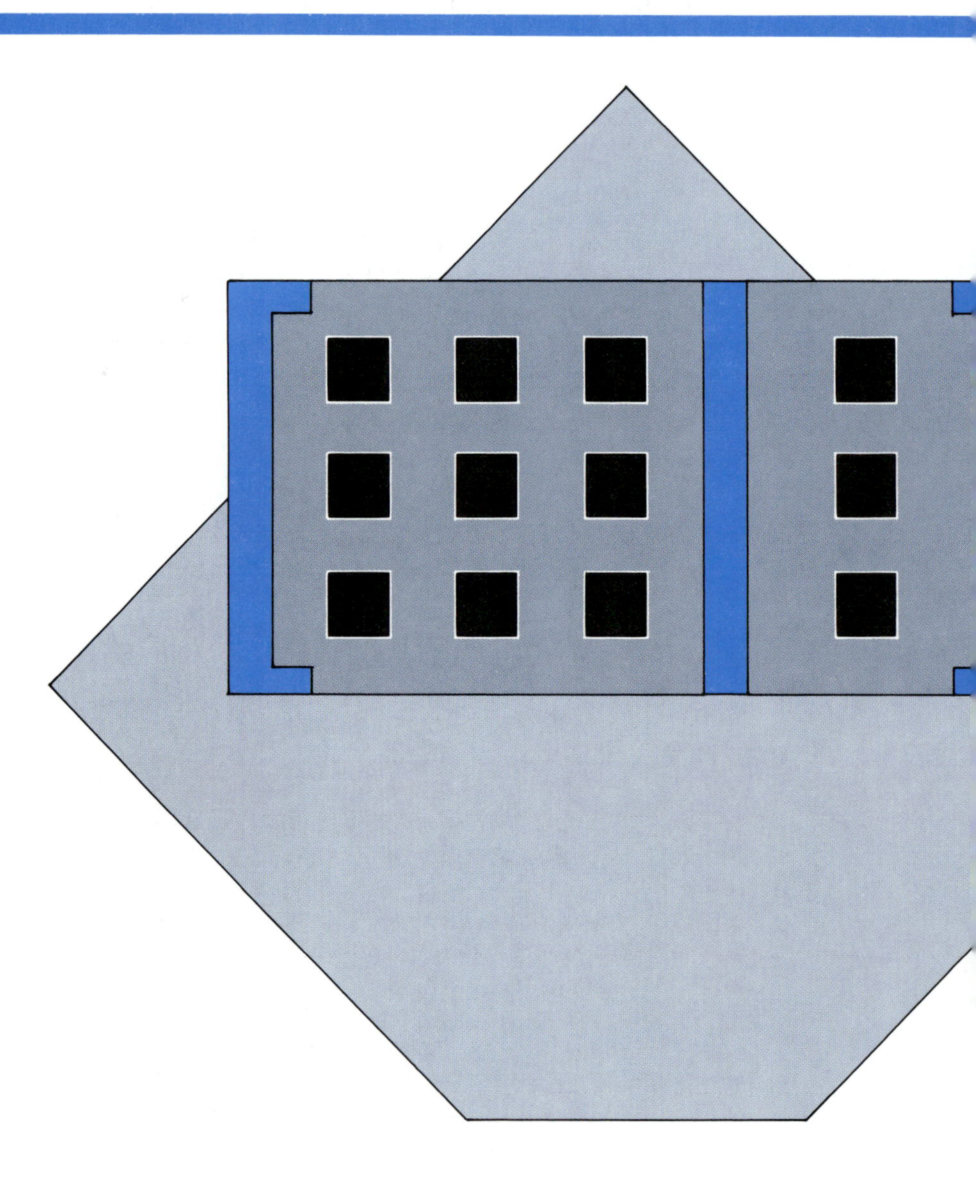

Linear Systems

- Systems of Two Equations
- Systems of Three Equations
- Gauss-Jordan Method
- Matrix Operations
- Multiplication of Matrices
- The Inverse of a Matrix
- Leontif Input-Output Model in Economics

2–1 Systems of Two Equations

Solution by Graphing
Substitution Method
Method of Elimination
Inconsistent Systems
Systems That Have Many Solutions
Supply and Demand Analysis

In Chapter 1 you learned some applications of linear equations. While many problems in mathematics, business, engineering, and sciences can be solved using a linear equation, many problems involve more than one linear equation. We use a simple example to illustrate the concept.

The Paperback Bookstore stocks two best sellers, one fiction and one nonfiction. They buy a total of 100 books. The fiction books cost \$3 each, and the nonfiction cost \$2 each. The total cost was \$245. How many of each did they buy?

The above information can be stated mathematically in the following manner.

Let x represent the number of fiction books and let y represent the number of nonfiction books purchased.

$$x + y = 100 \text{ is the total number of books purchased}$$
$$3x + 2y = 245 \text{ is the total cost}$$

We want to find the number of each kind of book that makes *both* equations true. The pair of equations is called a **system of equations**. A pair of numbers, one a value of x and the other a value of y, that makes *both* equations true is called a **solution to the system of equations**.

Solution by Graphing

We use the following simple system to illustrate the geometric meaning of solving a system of linear equations.

Example 1 Solve the system

$$2x - y = 3$$
$$x + 2y = 4$$

Solution Geometrically, each of these equations represents a line. When the lines are graphed with the same coordinate axes, they intersect at the point (2, 1) (Figure 2–1). The values $x = 2$ and $y = 1$ satisfy both equations (check to be sure). Thus, they form a solution to the system.

Emphasis. The pair of numbers $x = 2$ and $y = 1$ form *one* solution.

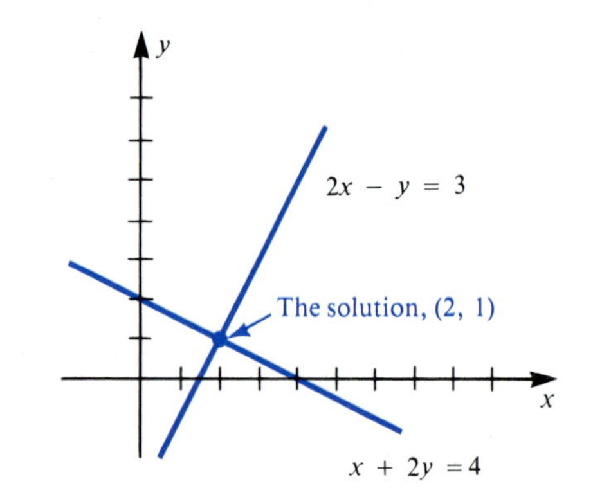

Figure 2–1

We can estimate the solution to a system by graphing the lines and noting the intersection of the lines. However, the graph must be accurate, and even so the precision of the solution may be in doubt. Precise solutions may be obtained through algebraic techniques. We will use two algebraic techniques, the **substitution** method and the **elimination** method.

Substitution Method

The basic approach of the substitution method is

1. Solve for a variable in one of the equations.
2. Substitute for that variable in the other equation.

Example 2 (*Compare Exercise 1*)
Solve the system

$$2x - y = 3$$
$$x + 2y = 4$$

by substitution.

Solution In this case it is easy to solve for x in the second equation because its coefficient is 1. We obtain

$$x = 4 - 2y$$

Substitute this expression for x in the first equation, $2x - y = 3$.

$$\begin{aligned} 2(4 - 2y) - y &= 3 \\ 8 - 4y - y &= 3 \\ 8 - 5y &= 3 \\ 8 &= 3 + 5y \\ 5 &= 5y \\ y &= 1 \end{aligned}$$

Now substitute 1 for y in $x = 4 - 2y$ to obtain $x = 4 - 2 = 2$. You could also substitute in $2x - y = 3$. Thus, the solution to the system is (2, 1). You may also solve for y in the first equation and substitute it in the second equation. This will yield the same solution (try it). It is good idea to check your solution in both equations.

Example 3 (*Compare Exercise 13*)
Solve the system

$$2x - 3y = -18$$
$$4x + 5y = 8$$

Solution Solve for x in one equation. From the first equation we obtain

$$\begin{aligned} 2x &= -18 + 3y \\ x &= \frac{1}{2}(-18 + 3y) \\ &= -9 + \frac{3}{2}y \end{aligned}$$

Substitute the expression for x in the second equation.

$$\begin{aligned} 4\left(-9 + \frac{3}{2}y\right) + 5y &= 8 \\ -36 + 6y + 5y &= 8 \\ -36 + 11y &= 8 \\ 11y &= 44 \\ y &= 4 \end{aligned}$$

Now substitute $y = 4$ into one of the equations. We use

$$x = -9 + \left(\frac{3}{2}\right)y \text{ to obtain}$$

$$\begin{aligned} x &= -9 + \frac{3}{2}(4) \\ &= -9 + 6 \\ &= -3 \end{aligned}$$

The pair $x = -3$, $y = 4$ gives the solution to the system.

Method of Elimination

The method of elimination finds the solution by systematically modifying the system to simpler systems. It does so in a manner that gives a system with the same solutions as the original system. We call this an **equivalent** system. The elimination method is especially useful because it can be used with systems with several variables and equations. The elimination method modifies a system by eliminating a variable from some equation in order to produce a simpler system. As you study the examples, observe that the operations used to transform a system of equations into a simpler, yet equivalent, system are the following:

To transform one system of linear equations into an equivalent system use one or more of the following:

1. Interchange two equations.
2. Multiply or divide one or more equations by a nonzero constant.
3. Multiply one equation by a constant and add it to or subtract it from another equation.

We illustrate this method with the system

$$\begin{aligned} 3x - y &= 3 \\ x + 2y &= 8 \end{aligned}$$

The arithmetic is a little easier if we eliminate a variable that has 1 as a coefficient. It is usually more convenient if we use the top equation to eliminate x, so interchange the two equations to get

$$\begin{aligned} x + 2y &= 8 \\ 3x - y &= 3 \end{aligned}$$

Now eliminate x from the bottom equation as follows.

$-3x - 6y = -24$ multiply the top equation by -3 because it gives $-3x$, the negative of the x term in the bottom equation

$3x - y = 3$ Now add it to the bottom equation (new second equation)

$-7y = -21$

Since the new equation came from equations in the system, it is true whenever the system is true. It replaces the equation $3x - y = 3$ to give the system

$$\begin{aligned} x + 2y &= 8 \\ -7y &= -21 \end{aligned}$$

Notice that the bottom equation has been modified so that the variable x has been **eliminated** from it. Simplify the bottom equation further by dividing by -7.

$$\begin{aligned} x + 2y &= 8 \\ y &= 3 \end{aligned}$$

This system has the same solution as the original system, but it has the advantage of giving the value of y at the common solution, namely 3. Now substitute 3 for y in the top equation to obtain

$$x + 2(3) = 8$$

(Actually, you can substitute y into either of the original equations.) This simplifies to $x = 2$ so the solution to the system is (2, 3).

Example 4 (*Compare Exercise 17*)
Solve this system by elimination.

$$\begin{aligned} 2x - 3y &= -19 \\ 5x + 7y &= 25 \end{aligned}$$

Solution We want to eliminate x and find the value of y for the common solution. We modify the system of equations as follows:

$-10x + 15y = 95$	Multiply the top equation by -5
$10x + 14y = 50$	Multiply the bottom equation by 2
$29y = 145$	Now add
$y = 5$	Divide through by 29 to find y

Replace the bottom equation to obtain the modified, and equivalent, system

$$\begin{aligned} 2x - 3y &= -19 \\ y &= 5 \end{aligned}$$

Now substitute this value of y into one of the equations. We use the first to obtain

$$\begin{aligned} 2x - 3(5) &= -19 \\ 2x &= -4 \\ x &= -2 \end{aligned}$$

The solution to the original system is $(-2, 5)$.

Observe: Each of the original equations was multiplied by a number so the resulting equations have the same coefficients of x, except for the sign. Then it was easy to eliminate x from the second equation by adding.

It is a good practice to check your solutions because errors in arithmetic sometimes occur. To check, substitute your solution into *all* of the equations in the original system. If one or more of the equations is not satisfied by your solution, an error has occurred.

Example 5 (*Compare Exercise 49*)
A woman has been ordered to control her diet carefully. She selects milk and bagels for breakfast. How much of each should she serve in order to consume 700 calories and 28 g of protein. Each cup of milk contains 170 calories and 8 g of protein. Each bagel contains 138 calories and 4 g of protein.

Solution Let x be the number of cups of milk and y the number of bagels. Then the total number of calories is

$$170x + 138y$$

and the total protein is

$$8x + 4y$$

so we need to solve the system

$$\begin{aligned} 8x + 4y &= 28 \\ 170x + 138y &= 700 \end{aligned}$$

Divide the top equation by 4 and the bottom equation by 2 to simplify somewhat.

$$\begin{aligned} 2x + y &= 7 \\ 85x + 69y &= 350 \end{aligned}$$

Next, multiply the top equation by -69 and add it to the bottom equation in order to eliminate y.

$$\begin{aligned} -138x - 69y &= -483 \\ 85x + 69y &= \quad 350 \\ \hline -53x \qquad &= -133 \end{aligned}$$

$$x = \frac{133}{53} = 2.509 \quad \text{(rounded)}$$

Now substitute and solve for y

$$\begin{aligned} 2(2.509) + y &= 7 \\ 5.018 + y &= 7 \\ y &= 1.982 \end{aligned}$$

It is reasonable to round these answers to 2.5 cups of milk and 2 bagels.

Inconsistent Systems

Each system in the preceding examples has exactly one solution. Do not expect this always to be the case. If the equations represent two parallel lines, they have

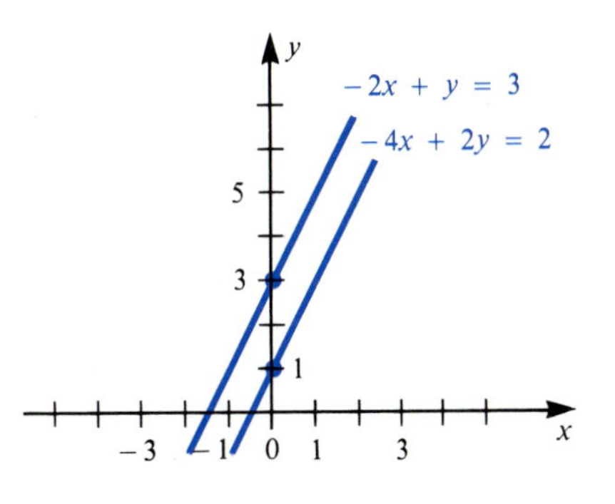

Figure 2–2 Parallel lines

no points in common and a solution to the system does not exist. The graph of the two lines

$$-2x + y = 3$$
$$-4x + 2y = 2$$

is shown in Figure 2–2. Each line has slope 2, and they do not intersect so no solution exists for the system.

Example 6 (*Compare Exercise 29*)
What happens when we try to solve the system

$$-2x + y = 3$$
$$-4x + 2y = 2$$

To eliminate x from the second equation, multiply the first by -2 and add it to the second. The new system is

$$-2x + y = 3$$
$$0y = -4$$
$$0 = -4$$

The last equation gives an inconsistency, $0 = -4$. When an attempt to solve a system leads to an inconsistency, the system has no solution. This will happen when the system represents two different parallel lines.

Systems That Have Many Solutions

Suppose you graph the two lines

$$9x - 3y = 6$$
$$6x - 2y = 4$$

You will find their graphs are identical so they coincide. Furthermore, both have slope 3 with y-intercept-2 so, when you put them in the slope-intercept form, $y = mx + b$, you will find that the slopes and intercepts are the same so their graphs are identical. Every point on this line is a solution to the system (see Figure 2–3).

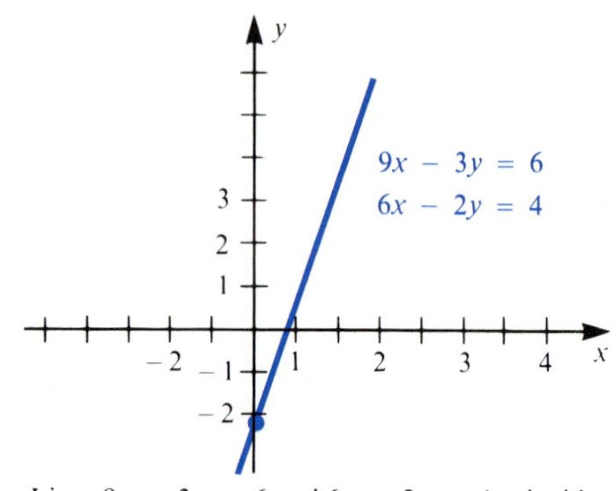

Figure 2–3 Lines $9x - 3y = 6$ and $6x - 2y = 4$ coincide.

Example 7 (*Compare Exercise 31*)
Let's see what happens when we try to solve the system

$$9x - 3y = 6$$
$$6x - 2y = 4$$

We eliminate x as follows.

$$\begin{aligned} -18x + 6y &= -12 \\ 18x - 6y &= 12 \\ \hline 0 &= 0 \end{aligned}$$

Multiply the first equation by -2
Multiply the second equation by 3
Add the two equations

The system is transformed into the system

$$9x - 3y = 6$$
$$0 = 0$$

The system reduces to a single equation; any solution to this equation gives a solution to the system. There are an infinite number of solutions to this equation which may be represented by solving for y:

$$y = 3x - 2 \quad \text{where } x \text{ may be any number.}$$

For example, solutions are obtained when

$$x = 3 \text{ and } y = 7$$
$$x = -5 \text{ and } y = -17$$
$$x = 0 \text{ and } y = -2$$

We can also describe the infinity of solutions in **parametric form**. The letter k, called a **parameter**, represents an arbitrary value of x. When substituting $x = k$ into one of the equations, we get the corresponding value of y,

$$y = 3k - 2$$

The point $(k, 3k - 2)$ represents a solution to the system when any real number is used for k. We also say the system is a **dependent** system.

Supply and Demand Analysis

Department stores are well aware that they can sell large quantities of goods if they advertise a reduction in price. The lower the price, the more they sell. Retailers understand this relationship between the price of a commodity and the consumer **demand** (the amount consumers buy). They also know that there may be more than one relationship between price and demand depending on circumstances. In times of shortages a different psychology takes effect, and prices tend to *increase* when demand *increases*. We will analyze one market situation, the **perfectly competitive market** of supply and demand.

In the competitive situation, a decrease in price can cause an increase in demand (when a store has a sale). This suggests that demand is a function of price. On other occasions, a store may lower prices because an item is in great demand and they expect to increase profits by a greater volume. This suggests that price is a function of demand. Since the cause and effect relationship between price and demand can go either way—a change in price causes a change in demand; or a change in demand can cause a change in price—we need to choose the way the demand equation is written. The analysis is easier if we write the demand equation (and the supply equation in the next paragraph) so that the price is a function of demand (or supply).

Example 8 (*Compare Exercise 41*)

The Bike Shop held an annual sale. The consumer price and demand relationship for the Ten-Speed Special was

$$y = -2x + 179$$

where x is the number of bikes in demand at the price y. The negative slope, -2, indicates that as demand increases, the price decreases; and as demand decreases, the price increases. This relationship between price and demand is a linear function. Its graph illustrates the decrease in price with an increase in demand (see Figure 2–4). When demand increases from 10 to 40, prices drop from \$159 to \$99.

The Bike Shop cannot lower prices indefinitely because the supplier wants to make a profit also. In a competitive situation a price increase gives the supplier incentive to produce more. When prices fall, the supplier tends to produce less. The quantity produced by the supplier is called **supply**. Suppose Bike Manufacturing produces the Ten-Speed Special and the relationship between supply and price is given by the linear function

$$y = 1.5x + 53$$

The graph of this equation illustrates that an increase in price leads to a higher supply (see Figure 2–5). When the price increases from \$83 to \$128 the supply increases from 20 to 50 units.

Supply and demand are two sides of a perfectly competitive market. They interact to determine the price of a commodity. The price of a commodity settles down in the market to one at which the amount willingly supplied and the amount willingly demanded are equal. This price is called the **equilibrium price**. The equilibrium price may be determined by solving a system of equations. In our example the system is

$$y = -2x + 179$$
$$y = 1.5x + 53$$

Solve the system to obtain the equilibrium solution by either substitution or the elimination method. The solution is $x = 36$ and $y = 107$. (Be sure you can find this solution.) The equilibrium price is $107 when the supply and demand are 36 bikes. See Figure 2–6.

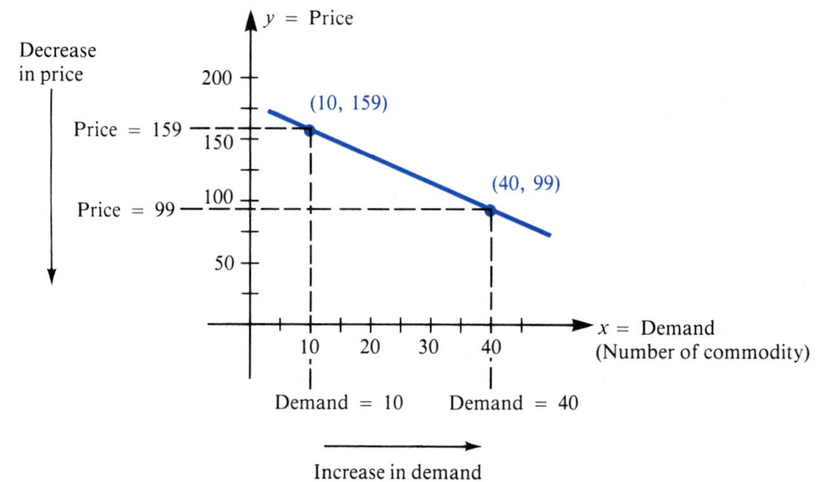

Figure 2–4 The price-demand function $y = -2x + 179$

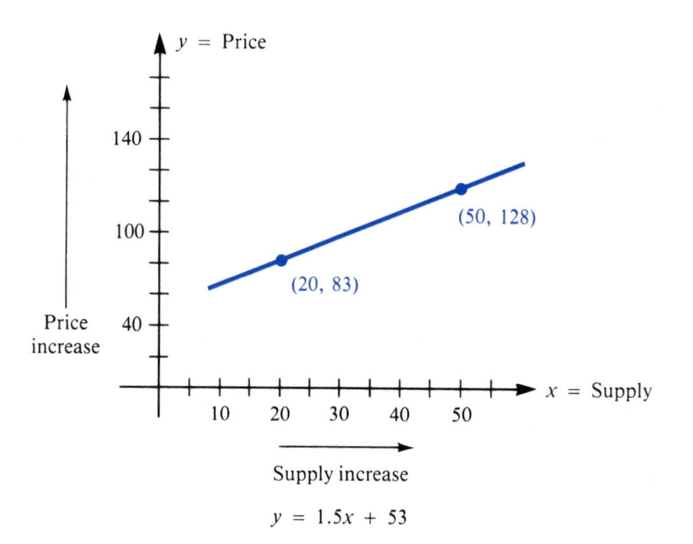

Figure 2–5 $y = 1.5x + 53$

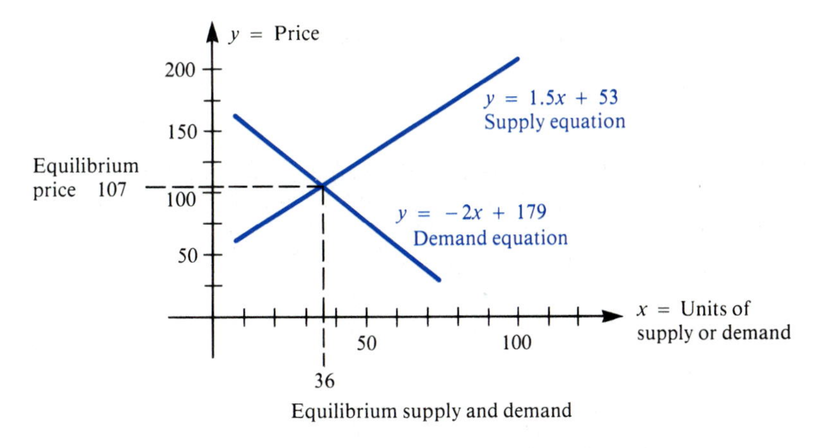

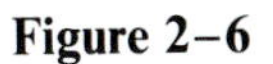

Figure 2–6 Equilibrium solution of supply and demand equations.

2–1 Exercises

I.

Solve the systems of equations in Exercises 1 through 16 by substitution.

1. (*See Example 2*)
$4x - y = 5$
$x + 2y = 8$

2. $2x + y = 7$
$x - 3y = 7$

3. $5x - y = -15$
$x + y = -3$

4. $2x - y = -1$
$2x + y = -3$

5. $y = 5x$
$6x - 2y = 12$

6. $x = 5 - 2y$
$3x - y = 15$

7. $7x - y = 32$
$2x + 3y = 19$

8. $x - 2y = -16$
$3x - 4y = -34$

9. $5x + 2y = 14$
$x - 3y = 30$

10. $4x - y = 13$
$3x - 5y = 31$

11. $22x + y = 81$
$8x - 3y = 16$

12. $3x - y = -47$
$x + 2y = 17$

13. (*See Example 3*)
$6x - 3y = 9$
$9x - 15y = 31$

14. $0.06x - y + 1.4 = 0$
$0.07x - 0.04y - 0.62 = 0$

15. $y = 3x - 5$
$12x - 4y - 20 = 0$

16. $5x - y = 8$
$7.5x - 1.5y = 10$

Solve the systems of equations in Exercises 17 through 28 by elimination.

17. (*See Example 4*)
$3x - 4y = 22$
$2x + 5y = 7$

18. $2x + y = 13$
$3x + 5y = 16$

19. $6x - y = 18$
$2x + y = 2$

20. $2x - 3y = -14$
$3x + 4y = -4$

21. $-2x + y = 7$
$6x + 12y = 24$

22. $7x - 2y = 14$
$-21x + 6y = -42$

23. $2x + y = -9$
$4x + 3y = 1$

24. $2x + 3y = 237$
$6x - 2y = 370$

25. $7x + 3y = -1.5$
$2x - 5y = -30.3$

26. $11x - 2y = 387$
$3x + 6y = -189$

27. $2x - 3y = -0.27$
$5x - 2y = 0.04$

28. $49x - 27y + 47 = 0$
$14x - 45y + 30 = 0$

The systems of equations in Exercises 29 through 34 do not have unique solutions. Determine if each system has no solution or an infinite number of solutions.

29. (*See Example 6*)
$6x - 9y = 8$
$10x - 15y = -20$

30. $8x + 10y = 18$
$20x + 25y = 45$

31. (*See Example 7*)
$8x + 10y = 2$
$12x + 15y = 3$

32. $3x - 7y = 20$
$-6x + 14y = 15$

33. $x - 6y = 4$
$5x - 30y = 20$

34. $4x + 14y = 15$
$6x + 21y = 26$

Determine the equilibrium solutions in Exercises 35 through 40. The demand equation is given first and the supply equation second.

35. (*See Example 7*)
$y = -3x + 15$
$y = 2x - 5$

36. $y = -8x + 200$
$y = 3x - 20$

37. $y = -4x + 130$
$y = x - 20$

38. $y = -6x + 68$
$y = 3x - 4$

39. $y = -5x + 83$
$y = 4x - 52$

40. $y = -8x + 2000$
$y = 6x - 800$

II.

41. (*See Example 8*) The demand equation for portable TV sets is $y = -2.5x + 148$ where x is the demand quantity and y is the price. The supply equation is $y = 1.7x + 43$ where x is the supply quantity and y is the price. Find the equilibrium solution.

42. The research department of a major corporation keeps careful records on its product. It finds that the demand equation is $y = -18x + 970$ where x is the demand and y is the price. The supply equation is $y = 15x + 640$. Find the equilibrium solution.

III.

43. A child has \$14.35 worth of nickels and dimes in her piggy bank. There is a total of 165 coins. How many of each does she have?

44. A child has 34 nickels and dimes. Their total value is \$2.35. How many of each does he have?

45. The sum of two numbers is 14. If one is multiplied by 3 and the other by 4, the sum of these two products is 48. Find the numbers.

46. A woman invested \$5000 in securities. Part of the money was invested at 8% and part at 9%. The total annual income was \$415. How much was invested at each rate?

47. The Beef Pit sells two kinds of sandwiches, chopped beef and smoked sausage. One day they sold 115 sandwiches for a total of \$138.90. The chopped beef cost \$1.30 each and the smoked sausage cost \$1.10 each. Find the number of each kind sold.

48. A developer has two size lots in his development; one sells for \$2500 and the other for \$3000. One month he sold 22 lots for a total of \$61,500. How many of each did he sell?

49. (*See Example 5*) An orange contains 50 mg of calcium and 0.5 mg of iron. An apple contains 8 mg of calcium and 0.4 mg of iron. How many of each is required to obtain 151 mg of calcium and 2.55 mg of iron?

50. A plant has two production lines, I and II. Line I can produce 5 tons of regular dog food per hour and 3 tons of premium per hour. Line II can produce 3 tons of regular dog food per hour and 6 tons of premium. How many hours of production should be scheduled in order to produce 360 tons of premium and 460 tons of regular dog food?

51. A businessman has \$50,000 to invest in a tax-free fund and in a money market fund. The tax-free fund pays 7.4% and the money market pays 8.8%. How much should he invest in each in order to get a return of \$4071 per year?

52. A grocer blends two types of coffee that sell for \$3.20 and \$3.60 per pound to obtain 100 pounds of coffee that sells for \$3.50. How much of each kind should he use?

2–2 Systems of Three Equations

Elimination Method

Some applications of linear equations have more than two variables and may involve more than two equations. Let's make up an example with nice numbers to illustrate the procedure for solving such a system.

Example 1 (*Compare Exercise 1*)
A toy store set up some bargain tables in the shopping mall. Three children, Al, Bob, and Cindy, bought some items. Based on the following information determine how many items each child purchased.

(a) The total number of items purchased was 5.

(b) Al paid \$1 for each item he purchased, Bob paid \$2 each, and Cindy paid \$3 each. The total spent by all three was \$10.

(c) The toy store gave balloons when items were purchased. Al got two balloons for each item he purchased; Bob and Cindy got one balloon for each item. The children received a total of 6 balloons.

Solution Let's state this information in mathematical form and determine the number of items each child purchased. Let

x = number of items purchased by Al

y = number of items purchased by Bob

z = number of items purchased by Cindy

The given information may be written as follows

$$\begin{aligned} x + y + z &= 5 && \text{(Total number of items)} \\ x + 2y + 3z &= 10 && \text{(Total value of purchases)} \\ 2x + y + z &= 6 && \text{(Total number of balloons)} \end{aligned}$$

The solution to this system of three equations in three variables gives the number of items sold to each child. Before we solve this system, let's look at some basic ideas.

A set of values for x, y, z that satisfies all three equations is called a **solution** to the system. ***Be sure*** you understand that a solution consists of three numbers, one each for x, y, and z.

You know that a linear equation in two variables represents a line in two-dimensional space. However, a linear equation in three variables does not represent a line, it represents a plane in three-dimensional space. A solution to a system of three equations in three variables corresponds to a point that lies in all three planes. Figure 2–7 illustrates some possible ways the planes might intersect. If the three planes have one point in common, the solution will be unique. If the planes have no points in common, there will be no solution to the system. In case the planes have many points in common, the system will have many solutions.

We need not stop with a system of three variables. There are applications that require larger systems with more variables. While larger systems are more difficult to interpret geometrically and are more tedious to solve, they also can have a unique solution, no solution, or many solutions.

The method of elimination used to solve systems of two equations can be adapted to solve larger systems. Let's rewrite the system of equations from our example and show its method of solution.

$$\begin{aligned} x_1 + x_2 + x_3 &= 5 \\ x_1 + 2x_2 + 3x_3 &= 10 \\ 2x_1 + x_2 + x_3 &= 6 \end{aligned}$$

Notice that we use x_1, x_2, and x_3 instead of x, y, and z for the variables. This notation is better for a system with several variables because we won't run out of letters for a larger number of variables.

First, eliminate x_1 from the second equation by subtracting the first equation from the second equation.

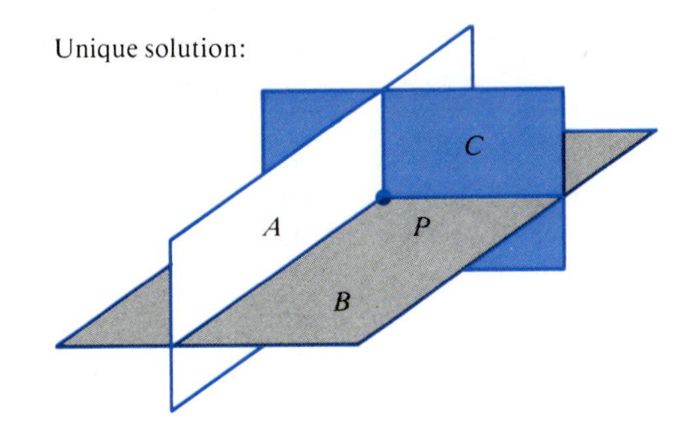

Three planes A, B, C intersect at a single point P;
P corresponds to a unique solution.

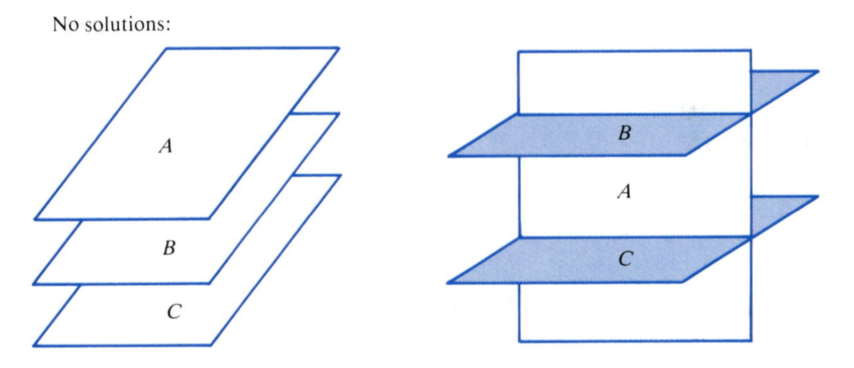

Planes A, B, C have no point of common intersection, no solution.

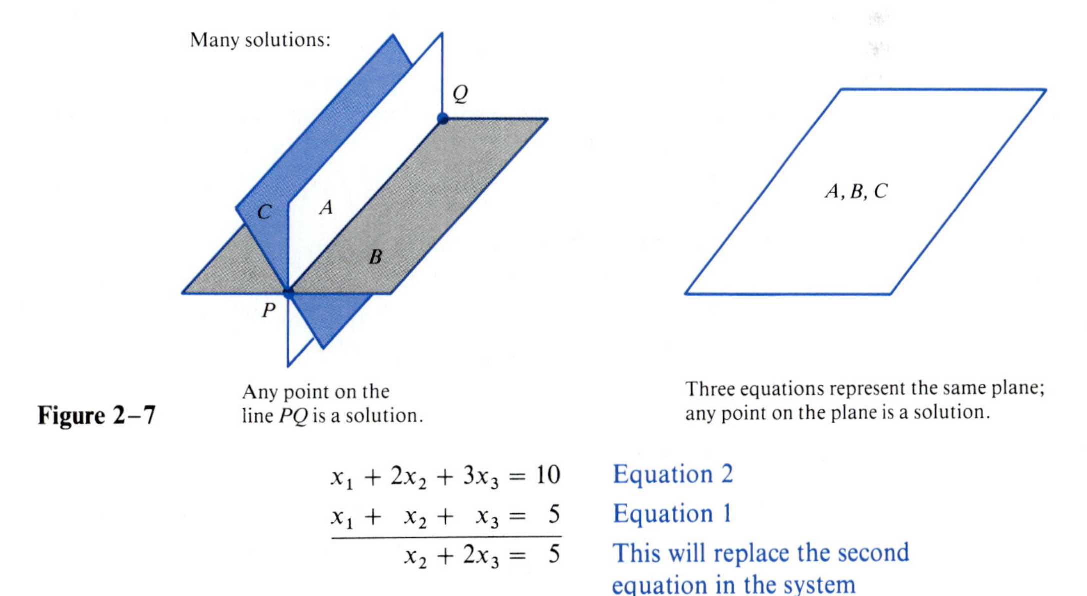

Figure 2–7

Any point on the line PQ is a solution.

Three equations represent the same plane; any point on the plane is a solution.

$$\begin{array}{rl} x_1 + 2x_2 + 3x_3 = 10 & \text{Equation 2} \\ x_1 + x_2 + x_3 = 5 & \text{Equation 1} \\ \hline x_2 + 2x_3 = 5 & \text{This will replace the second equation in the system} \end{array}$$

Notice: The coefficients of the variables determine the steps taken throughout.

Now eliminate x_1 from the third equation by multiplying the first equation by -2 and subtracting from the third.

$$\begin{array}{rl} 2x_1 + x_2 + x_3 = 6 & \text{Equation 3} \\ \underline{-2x_1 - 2x_2 - 2x_3 = -10} & -2 \times \text{Equation 1} \\ - x_2 - x_3 = -4 & \text{This will replace the third equation in the system} \end{array}$$

This gives the new, and simpler, equivalent system

$$\begin{aligned} x_1 + x_2 + x_3 &= 5 \\ x_2 + 2x_3 &= 5 \\ -x_2 - x_3 &= -4 \end{aligned}$$

Observe that the variable x_1 has been eliminated from both the second and third equations. These two equations form a system of two equations in the two variables x_2 and x_3.

In order to describe the operations performed in a concise manner we will use the notation Eq 2 − Eq 1 to mean that the first equation is subtracted from the second. −2Eq 1 + Eq 3 means to multiply the first equation by −2 and add it to the third. We continue the process by eliminating more variables from the new system.

Now we can eliminate the variable x_2 from the first and third equations by using the second equation.

$$\begin{array}{rll} x_1 + x_2 + x_3 = 5 & \text{Eq 1} - \text{Eq 2 yields} & x_1 - x_3 = 0 \\ x_2 + 2x_3 = 5 & \text{remains} & x_2 + 2x_3 = 5 \\ -x_2 - x_3 = -4 & \text{Eq 3} + \text{Eq 2 yields} & x_3 = 1 \end{array}$$

The system on the right is equivalent to the original system, and we can complete the solution by eliminating x_3 from the first and second equation of this latest system.

$$\begin{array}{rll} x_1 - x_3 = 0 & \text{Eq 1} + \text{Eq 3 yields} & x_1 = 1 \\ x_2 + 2x_3 = 5 & \text{Eq 2} - 2\text{Eq 3 yields} & x_2 = 3 \\ x_3 = 1 & \text{remains} & x_3 = 1 \end{array}$$

The solution to this system and thus the solution to the original system is $x_1 = 1$, $x_2 = 3$, $x_3 = 1$, which is also written (1, 3, 1). This solution tells us that Al bought 1 item, Bob 3 items, and Cindy 1 item.

Matrices

The method of elimination may be used to solve larger systems, but it is tedious and errors are easily made. Let's look at a method that reduces the work and gives a more efficient approach to solving a system. It also is more adaptable to a computer. The method we will learn is called the **Gauss-Jordan Method** and uses **matrices**. This method "ignores" the variables in the equations.

Basically, it is the elimination method made more concise by matrix notation. Since the coefficients of the variables determine the multipliers used in the elimination method, some labor can be saved if we can avoid writing the variables and concentrate on the coefficients. Matrices enable us to keep track of the variables without writing them. The formal definition of a matrix is the following.

Definition

A **matrix** is a rectangular array of numbers. The numbers in the array are called the **elements** of the matrix. The array is enclosed with brackets.

An array composed of a single row of numbers is called a **row** matrix.

An array composed of a single column of numbers is called a **column** matrix.

Here are some examples of matrices.

$$\begin{bmatrix} 1 & 2 & 3 \\ 0 & -1 & 1 \end{bmatrix} \quad \begin{bmatrix} 2 & 3 \\ 1 & 1 \\ 4 & 1 \end{bmatrix} \quad \begin{bmatrix} 1 & 2 & 3 \\ 4 & 5 & 6 \\ 0 & 1 & 2 \end{bmatrix}$$

The location of each element in a matrix is described by the row and column in which it lies. Count the rows from the top of the matrix and the columns from the left.

	Col. 1	*Col. 2*	*Col. 3*	*Col. 4*
Row 1	4	6	-3	2
Row 2	1	5	9	-2
Row 3	7	8	3	4

The element 9 is in row 2 and column 3 of the matrix. This location is called the (2, 3) location with the row indicated first and the column second. The element 7 is in the (3, 1) location and -3 is in the (1, 3) location.

Example 2 (*Compare Exercise 6*)
For the matrix

$$\begin{bmatrix} 2 & -6 & -5 & -1 & 0 \\ 1 & 7 & 6 & -4 & 4 \\ 9 & 5 & -8 & 3 & -2 \end{bmatrix}$$

find

(a) the (1, 1) element.

(b) the (2, 5) element.

(c) the (3, 3) element.

(d) the location of -4.

(e) the location of 0.

Solution

(a) 2 is the (1, 1) element.

(b) 4 is the (2, 5) element.

(c) The (3, 3) element is -8.

(d) -4 is in the (2, 4) location.

(e) 0 is in the (1, 5) location.

Let's pause to point out that you encounter matrices more often than you might realize. When you go to a fast food place, the cashier may record your order on a device that resembles Figure 2–8. When you order three Big Burgers, the cashier presses the Big Burger location, the (1, 2) location, to record the number ordered. Your completed order forms a matrix, with some entries zero, which the computer uses to compute your bill. Both the cashier and the company computer understand that a number in a location counts the number of the item represented by the location.

The rows and columns of a matrix often represent categories. A gradebook shows a rectangular array of numbers. Each column represents a test and each row represents a student.

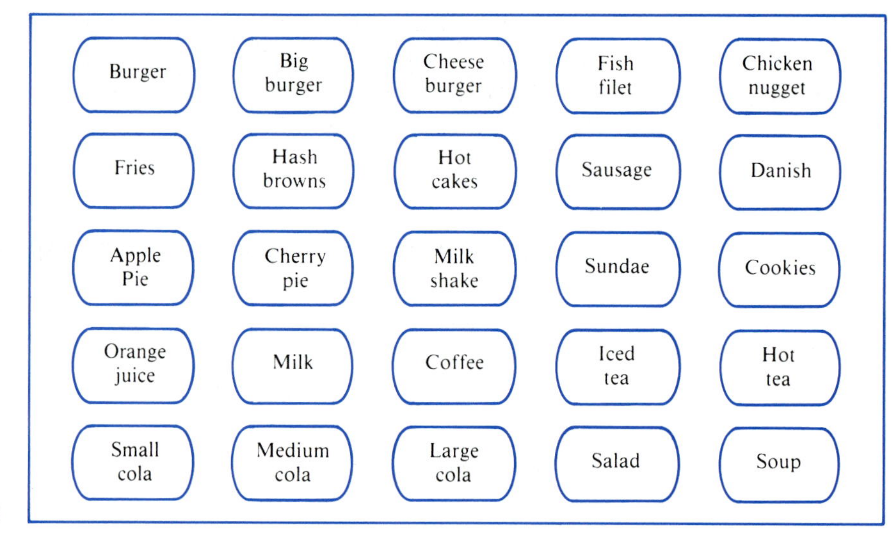

Figure 2–8

Matrices and Systems of Equations

Let us use the following system to see how matrices relate to systems of linear equations.

$$\begin{aligned} x_1 + x_2 + x_3 &= 3 \\ 2x_1 + 3x_2 + x_3 &= 5 \\ x_1 - x_2 - 2x_3 &= -5 \end{aligned}$$

One matrix may be formed by using only the coefficients of the system. This gives the **coefficient matrix**

$$\begin{bmatrix} 1 & 1 & 1 \\ 2 & 3 & 1 \\ 1 & -1 & -2 \end{bmatrix}$$

Column 1 lists the coefficients of x_1, column 2 lists the coefficients of x_2, and column 3 lists the coefficients of x_3. Notice that they are listed in the order first, second, and third equations, so the first row represents the left side of the first equation, and so on.

A matrix that also includes the numbers on the right-hand sides of the equations is called the **augmented matrix** of the system

$$\begin{bmatrix} 1 & 1 & 1 & 3 \\ 2 & 3 & 1 & 5 \\ 1 & -1 & -2 & -5 \end{bmatrix}$$

The augmented matrix gives complete information about the system of equations, provided we agree that each row represents an equation and each column, except the last, consists of the coefficients of one variable. Conversely, any matrix can be thought of as an augmented matrix of a system of linear equations.

Example 3 (*Compare Exercise 8*)
Write the coefficient matrix and the augmented matrix of the system

$$\begin{aligned} 5x_1 - 7x_2 + 2x_3 &= 17 \\ -x_1 + 3x_2 + 8x_3 &= 12 \\ 6x_1 + 9x_2 - 4x_3 &= -23 \end{aligned}$$

Solution The coefficient matrix is

$$\begin{bmatrix} 5 & -7 & 2 \\ -1 & 3 & 8 \\ 6 & 9 & -4 \end{bmatrix}$$

and the augmented matrix is

$$\left[\begin{array}{rrr|r} 5 & -7 & 2 & 17 \\ -1 & 3 & 8 & 12 \\ 6 & 9 & -4 & -23 \end{array}\right]$$

Example 4 (*Compare Exercise 14*)
Write the system of linear equations represented by the augmented matrix

$$\left[\begin{array}{cccc|c} 3 & 7 & 2 & -3 & 8 \\ 4 & 0 & -5 & 7 & -2 \end{array}\right]$$

Solution The system is

$$\begin{aligned} 3x_1 + 7x_2 + 2x_3 - 3x_4 &= 8 \\ 4x_1 \qquad\quad - 5x_3 + 7x_4 &= -2 \end{aligned}$$

We can solve a system of linear equations by using its augmented matrix. Since each row in the matrix represents an equation, we perform the same operations on rows of the matrix as we do on the equations in the system. These operations are as follows:

Row Operations

1. Interchange two rows.
2. Multiply or divide a row by a nonzero constant.
3. Multiply a row by a constant and add it to or subtract it from another row.

Two augmented matrices are **equivalent** if one is obtained from the other by using row operations.

Gauss-Jordan Method

In order to illustrate the matrix technique, we solve a system of linear equations using the augmented matrix and row operations. This technique is basically a simplification of the elimination method. You will soon find that even this "simplified" method can be tedious and it is easy to make arithmetic errors. However, a method such as this is widely used by computers to solve systems of equations. While the examples and exercises use equations that yield relatively simple answers, actual applications don't always have nice, neat, unique solutions.

Example 5 Solve the system of equations.

$$\begin{aligned} x_1 + x_2 + x_3 &= 5 \\ x_1 + 2x_2 + 3x_3 &= 10 \\ 2x_1 + x_2 + x_3 &= 6 \end{aligned}$$

Solution This system was solved earlier in Example 1. It is used so you can concentrate on the procedure. We show the solution of this sytem by both the elimination method and the method using matrices. They are shown in parallel so you can see the relationship of the two methods.

Sequence of Equivalent Systems of Equations	**Corresponding Equivalent Augmented Matrices**
Original system:	*Original augmented matrix:*
$\begin{aligned} x_1 + x_2 + x_3 &= 5 \\ x_1 + 2x_2 + 3x_3 &= 10 \\ 2x_1 + x_2 + x_3 &= 6 \end{aligned}$	$\left[\begin{array}{ccc\|c} 1 & 1 & 1 & 5 \\ 1 & 2 & 3 & 10 \\ 2 & 1 & 1 & 6 \end{array}\right]$
Eliminate x_1 from the second equation by multiplying the first equation by -1 and adding to the second. Eliminate x_1 from the third equation by multiplying the first equation by -2 and adding to the third.	Get 0 in the second row, first column by multiplying the first row by -1 and adding to the second. Get 0 in the third row, first column by multiplying the first row by -2 and adding to the third.
$\begin{aligned} x_1 + x_2 + x_3 &= 5 \\ x_2 + 2x_3 &= 5 \\ -x_2 - x_3 &= -4 \end{aligned}$	$\left[\begin{array}{ccc\|c} 1 & 1 & 1 & 5 \\ 0 & 1 & 2 & 5 \\ 0 & -1 & -1 & -4 \end{array}\right]$
Eliminate x_2 from the first equation by multiplying the second equation by -1 and adding to the first. Eliminate x_2 from the third equation by adding the second equation to the third.	Get 0 in the first row, second column by multiplying second row by -1 and adding to the first. Get 0 in the third row, second column by adding the second row to the third.
$\begin{aligned} x_1 - x_3 &= 0 \\ x_2 + 2x_3 &= 5 \\ x_3 &= 1 \end{aligned}$	$\left[\begin{array}{ccc\|c} 1 & 0 & -1 & 0 \\ 0 & 1 & 2 & 5 \\ 0 & 0 & 1 & 1 \end{array}\right]$
Eliminate x_3 from the first equation by adding the third equation to the first. Eliminate x_3 from the second equation by multiplying the third equation by -2 and adding to the second.	Get 0 in the first row, third column by adding the third row to the first. Get 0 in the second row, third column by multiplying the third row by -2 and adding to the second.
$\begin{aligned} x_1 &= 1 \\ x_2 &= 3 \\ x_3 &= 1 \end{aligned}$	$\left[\begin{array}{ccc\|c} 1 & 0 & 0 & 1 \\ 0 & 1 & 0 & 3 \\ 0 & 0 & 1 & 1 \end{array}\right]$
	Read the solution from this augmented matrix. The first row gives $x_1 = 1$, the second row gives $x_2 = 3$, and the third row gives $x_3 = 1$.

This technique of using row operations to reduce an augmented matrix to a simple matrix is called the **Gauss-Jordan Method**. The form of the final matrix is such that the solution to the original system can easily be read from the matrix. Notice that the final matrix in the example above was

$$\left[\begin{array}{ccc|c} 1 & 0 & 0 & 1 \\ 0 & 1 & 0 & 3 \\ 0 & 0 & 1 & 1 \end{array}\right]$$

For the moment ignore the last column of the matrix. The remaining columns have zeros everywhere except in the (1, 1), (2, 2), and (3, 3) locations. These are called the **diagonal** locations. The Gauss-Jordan method attempts to reduce the augmented matrix until there are 1's in the diagonal locations and 0's elsewhere (except in the last column).

We now look at another example and focus attention on the procedure for arriving at this desired diagonal form. In this section we will focus our attention on augmented matrices that can be reduced to this diagonal form. The cases where the diagonal form is not possible are studied in the next section.

A new notation is introduced in the next example to reduce the writing involved. When we are reducing a matrix and you see

$$\tfrac{1}{4}\text{R1 gives } [1 \quad 2 \quad -3 \quad 11] \to \text{R1}$$

this means that row 1 of the current matrix is divided by 4 and gives the new row $[1 \quad 2 \quad -3 \quad 11]$ which is placed in row 1 of the next matrix. The notation $-2\text{R2} + \text{R3} \to \text{R3}$ means that row 2 of the current matrix is multiplied by -2 and added to row 3. The result becomes row 3 of the next matrix.

Example 6 (*Compare Exercise 26*)

Solve this system of equations by reducing the augmented matrix to the diagonal form.

$$\begin{aligned} 4x_1 + 8x_2 - 12x_3 &= 44 \\ 3x_1 + 6x_2 - 8x_3 &= 32 \\ -2x_1 - x_2 \qquad\quad &= -7 \end{aligned}$$

The augmented matrix of this system is

$$\left[\begin{array}{ccc|c} 4 & 8 & -12 & 44 \\ 3 & 6 & -8 & 32 \\ -2 & -1 & 0 & -7 \end{array}\right]$$

We now use row operations to find the solution to the system.

	Matrix	This Operation on Present Matrix	Put in New Row
Need 1 here	$\left[\begin{array}{rrr\|r} 4 & 8 & -12 & 44 \\ 3 & 6 & -8 & 32 \\ -2 & -1 & 0 & -7 \end{array}\right]$	$\left(\frac{1}{4}\right)$R1 gives [1 2 −3 11]	→ R1
Need 0 here	$\left[\begin{array}{rrr\|r} 1 & 2 & -3 & 11 \\ 3 & 6 & -8 & 32 \\ -2 & -1 & 0 & -7 \end{array}\right]$	−3R1 + R2 gives [0 0 1 −1] 2R1 + R3 gives [0 3 −6 15]	→ R2 → R3
Need 1 here	$\left[\begin{array}{rrr\|r} 1 & 2 & -3 & 11 \\ 0 & 0 & 1 & -1 \\ 0 & 3 & -6 & 15 \end{array}\right]$	Interchange row 2 and row 3	
	$\left[\begin{array}{rrr\|r} 1 & 2 & -3 & 11 \\ 0 & 3 & -6 & 15 \\ 0 & 0 & 1 & -1 \end{array}\right]$	$\left(\frac{1}{3}\right)$R2 gives [0 1 −2 5]	→ R2
Need 0 here	$\left[\begin{array}{rrr\|r} 1 & 2 & -3 & 11 \\ 0 & 1 & -2 & 5 \\ 0 & 0 & 1 & -1 \end{array}\right]$	−2R2 + R1 gives [1 0 1 1]	→ R1
Need 0 here	$\left[\begin{array}{rrr\|r} 1 & 0 & 1 & 1 \\ 0 & 1 & -2 & 5 \\ 0 & 0 & 1 & -1 \end{array}\right]$	−R3 + R1 gives [1 0 0 2] 2R3 + R2 gives [0 1 0 3]	→ R1 → R2

$$\left[\begin{array}{rrr|r} 1 & 0 & 0 & 2 \\ 0 & 1 & 0 & 3 \\ 0 & 0 & 1 & -1 \end{array}\right]$$

This matrix is in a diagonal form and represents the system

$$\begin{aligned} x_1 \quad\quad &= 2 \\ x_2 \quad &= 3 \\ x_3 &= -1 \end{aligned}$$

so the solution is (2, −3, 1).

The next example uses a more abbreviated notation in the solution.

Example 7 (*Compare Exercise 30*)
Solve the system

$$\begin{aligned} x_1 - x_2 + x_3 + 2x_4 &= 1 \\ 2x_1 - x_2 \quad\quad + 3x_4 &= 0 \\ -x_1 + x_2 + x_3 + x_4 &= -1 \\ x_2 \quad\quad + x_4 &= 1 \end{aligned}$$

Solution The augmented matrix of this system is

$$\left[\begin{array}{cccc|c} 1 & -1 & 1 & 2 & 1 \\ 2 & -1 & 0 & 3 & 0 \\ -1 & 1 & 1 & 1 & -1 \\ 0 & 1 & 0 & 1 & 1 \end{array}\right]$$

Performing the indicated row operations produces the following sequence of equivalent augmented matrices.

$$\left[\begin{array}{cccc|c} 1 & -1 & 1 & 2 & 1 \\ 2 & -1 & 0 & 3 & 0 \\ -1 & 1 & 1 & 1 & -1 \\ 0 & 1 & 0 & 1 & 1 \end{array}\right] \quad \begin{array}{r} \\ -2R1 + R2 \to R2 \\ R1 + R3 \to R3 \\ \\ \end{array}$$

$$\left[\begin{array}{cccc|c} 1 & -1 & 1 & 2 & 1 \\ 0 & 1 & -2 & -1 & -2 \\ 0 & 0 & 2 & 3 & 0 \\ 0 & 1 & 0 & 1 & 1 \end{array}\right] \quad \begin{array}{r} R2 + R1 \to R1 \\ \\ \\ -R2 + R4 \to R4 \end{array}$$

$$\left[\begin{array}{cccc|c} 1 & 0 & -1 & 1 & -1 \\ 0 & 1 & -2 & -1 & -2 \\ 0 & 0 & 2 & 3 & 0 \\ 0 & 0 & 2 & 2 & 3 \end{array}\right] \quad \left(\frac{1}{2}\right)R3 \to R3$$

$$\left[\begin{array}{cccc|c} 1 & 0 & -1 & 1 & -1 \\ 0 & 1 & -2 & -1 & -2 \\ 0 & 0 & 1 & \frac{3}{2} & 0 \\ 0 & 0 & 2 & 2 & 3 \end{array}\right] \quad \begin{array}{r} R3 + R1 \to R1 \\ 2R3 + R2 \to R2 \\ \\ -2R3 + R4 \to R4 \end{array}$$

$$\left[\begin{array}{cccc|c} 1 & 0 & 0 & \frac{5}{2} & -1 \\ 0 & 1 & 0 & 2 & -2 \\ 0 & 0 & 1 & \frac{3}{2} & 0 \\ 0 & 0 & 0 & -1 & 3 \end{array}\right] \quad \begin{array}{r} \left(\frac{5}{2}\right)R4 + R1 \to R1 \\ 2R4 + R2 \to R2 \\ \left(\frac{3}{2}\right)R4 + R3 \to R3 \\ -R4 \to R4 \end{array}$$

$$\left[\begin{array}{cccc|c} 1 & 0 & 0 & 0 & \frac{13}{2} \\ 0 & 1 & 0 & 0 & 4 \\ 0 & 0 & 1 & 0 & \frac{9}{2} \\ 0 & 0 & 0 & 1 & -3 \end{array}\right]$$

This matrix is in the reduced diagonal form so we can read the solution to the system:

$$x_1 = \frac{13}{2}, \quad x_2 = 4, \quad x_3 = \frac{9}{2}, \quad x_4 = -3$$

The examples in this section all have unique solutions. In such cases, the last column of the diagonal form gives the unique solution.

2–2 Exercises

I.

Use the elimination method to solve the systems in Exercises 1 through 5.

1. (*See Example 1*)

$$\begin{aligned} x + y - z &= -1 \\ x - y + z &= 5 \\ x - y - z &= 1 \end{aligned}$$

2.

$$\begin{aligned} x + y + 2z &= 12 \\ 2x + 3y - z &= -2 \\ -3x + 4y + z &= -8 \end{aligned}$$

3.

$$\begin{aligned} x + 4y - 2z &= 21 \\ 3x - 6y - 3z &= -18 \\ 2x + 4y + z &= 37 \end{aligned}$$

4.

$$\begin{aligned} x + y + z &= 2 \\ 4x + 3y + 2z &= -4 \\ -2x + y - z &= 2 \end{aligned}$$

5.

$$\begin{aligned} 2x + 4y - 6z &= -2 \\ 4x - 3y + z &= 11 \\ 3x + 2y - 2z &= 7 \end{aligned}$$

6. (*See Example 2*) For the matrix

$$\begin{bmatrix} 1 & 2 & 3 & 5 \\ -6 & 7 & 0 & 9 \\ 7 & 2 & 1 & 8 \end{bmatrix}$$

(a) give the (1, 1), (2, 2), (3, 3), and (3, 4) elements.

(b) what is the location of 9?

7. For the matrix

$$\begin{bmatrix} 5 & 4 & 6 & -9 & 3 & -5 \\ 7 & 8 & 0 & 4 & 2 & 8 \\ 9 & -5 & 6 & 6 & -7 & 3 \end{bmatrix}$$

(a) give the (1, 1), (2, 2), (3, 4), (2, 5), and (1, 6) elements.

(b) what are the locations that contain -5?

Determine the matrix of coefficients and the augmented matrix for each of the systems of linear equations in Exercises 8 through 13.

8. (*See Example 3*)

$$\begin{aligned} x_1 - x_2 &= 1 \\ 2x_1 + x_2 &= 3 \end{aligned}$$

9.

$$\begin{aligned} x_1 - x_2 + x_3 &= 4 \\ x_1 + 3x_2 + 4x_3 &= 2 \\ 2x_1 - 2x_2 + x_3 &= 5 \end{aligned}$$

10.

$$\begin{aligned} x_2 - x_3 &= 5 \\ x_1 + 2x_2 - 4x_3 &= 2 \\ 2x_1 - x_2 &= 4 \end{aligned}$$

11.

$$\begin{aligned} 6x_1 + 4x_2 - x_3 &= 0 \\ -x_1 - x_2 - x_3 &= 7 \\ 5x_1 &= 15 \end{aligned}$$

12. $$\begin{aligned} 2x_1 + x_2 - x_3 + x_4 &= -6 \\ 3x_1 - x_2 + 2x_3 &= 7 \\ 8x_1 + x_2 - 9x_3 + x_4 &= 8 \\ -x_1 - 3x_2 + 8x_3 + x_4 &= 9 \end{aligned}$$

13. $$\begin{aligned} 7x_1 - x_2 + x_3 + 2x_4 &= 4 \\ x_2 - 8x_3 - x_4 &= -1 \\ x_1 + 7x_3 + x_4 &= -2 \\ x_1 + x_2 + x_3 + x_4 &= 6 \end{aligned}$$

Any matrix can be interpreted as an augmented matrix defining a system of linear equations. Write out the systems defined by the matrices in Exercises 14 through 18.

(*See Example 4*)

14. $\begin{bmatrix} 2 & 1 & 5 \\ 3 & -2 & 3 \end{bmatrix}$

15. $\begin{bmatrix} 1 & 3 & 2 & 0 \\ 2 & 4 & -3 & 2 \\ -1 & 6 & 1 & 4 \end{bmatrix}$

16. $\begin{bmatrix} 3 & 5 & -4 & 1 \\ -1 & 2 & 2 & 2 \\ 2 & 3 & 0 & 3 \end{bmatrix}$

17. $\begin{bmatrix} 0 & 2 & 1 & 3 \\ 1 & 3 & 0 & 5 \\ 4 & 2 & 3 & 8 \end{bmatrix}$

18. $\begin{bmatrix} 4 & 3 & 0 & 8 & 5 \\ 5 & 6 & 5 & 9 & 1 \\ 0 & 8 & 4 & -2 & 0 \\ -1 & -2 & 7 & 4 & 2 \end{bmatrix}$

19. What are the diagonal elements of

$$\begin{bmatrix} -3 & 1 & 5 & 4 \\ 0 & 2 & 3 & -7 \\ 9 & 8 & -6 & 11 \\ 4 & 7 & -5 & 13 \end{bmatrix}$$

In Exercises 20 through 25 a row operation is indicated to the right of the matrix. Determine the matrix that results from performing the operation.

20. $\begin{bmatrix} 2 & 4 & -4 & 2 \\ 1 & 2 & -1 & 3 \\ -1 & 0 & 4 & 2 \end{bmatrix}$ $\left(\frac{1}{2}\right)R1 \to R1$

21. $\begin{bmatrix} 0 & -1 & 3 & 4 \\ 2 & 1 & 5 & 2 \\ -3 & 1 & -2 & 6 \end{bmatrix}$ $R2 + R3 \to R3$

22. $\begin{bmatrix} 1 & 2 & 3 & -1 \\ 0 & 1 & 2 & 1 \\ 0 & -4 & 5 & -3 \end{bmatrix}$ $-2R2 + R1 \to R1$, $4R2 + R3 \to R3$

23. $\begin{bmatrix} 1 & 0 & 4 & 4 \\ 0 & 1 & -3 & 2 \\ 0 & 0 & 1 & 3 \end{bmatrix}$ $-4R3 + R1 \to R1$, $3R3 + R2 \to R2$

24. $\begin{bmatrix} 1 & -2 & 0 & 3 \\ -2 & -1 & 3 & 4 \\ 3 & 1 & 2 & 6 \end{bmatrix}$ $2R1 + R2 \to R2$, $-3R1 + R3 \to R3$

25. $\begin{bmatrix} 1 & 0 & 2 & 3 \\ 0 & 1 & 4 & 1 \\ 0 & 0 & -2 & 6 \end{bmatrix}$ $\left(-\frac{1}{2}\right)R3 \to R3$

Solve the systems of equations in Exercises 26 through 31 by reducing the augmented matrix.

26. (*See Example 6*)
$$\begin{aligned} x_1 + 2x_2 - x_3 &= 3 \\ x_1 + 3x_2 - x_3 &= 4 \\ x_1 - x_2 + x_3 &= 4 \end{aligned}$$

27. $$\begin{aligned} x_1 + x_2 + 2x_3 &= 5 \\ 3x_1 + 2x_2 &= 4 \\ 2x_1 - x_3 &= 2 \end{aligned}$$

28. $$\begin{aligned} 2x_1 + 4x_2 + 2x_3 &= 6 \\ 2x_1 + x_2 + x_3 &= 16 \\ x_1 + x_2 + 2x_3 &= 9 \end{aligned}$$

29. $$\begin{aligned} x_1 + 2x_3 &= 5 \\ x_2 - 30x_3 &= -17 \\ x_1 - 2x_2 + 4x_3 &= 10 \end{aligned}$$

30. (*See Example 7*)

$$\begin{aligned} x_1 + x_2 + x_3 + x_4 &= 4 \\ x_1 + 2x_2 - x_3 - x_4 &= 7 \\ 2x_1 - x_2 - x_3 - x_4 &= 8 \\ x_1 - x_2 + 2x_3 - 2x_4 &= -7 \end{aligned}$$

31.

$$\begin{aligned} x_1 + x_2 + 2x_3 + x_4 &= 3 \\ x_1 + 2x_2 + x_3 + x_4 &= 2 \\ x_1 + x_2 + x_3 + 2x_4 &= 1 \\ 2x_1 + x_2 + x_3 + x_4 &= 4 \end{aligned}$$

The systems of linear equations in Exercises 32 through 47 all have the same number of equations as variables, and all have unique solutions. Solve each system using the method of Gauss-Jordan elimination.

32.

$$\begin{aligned} x_1 - 2x_2 &= -8 \\ 2x_1 - 3x_2 &= -11 \end{aligned}$$

33.

$$\begin{aligned} 2x_1 + 2x_2 &= 4 \\ 3x_1 + 2x_2 &= 3 \end{aligned}$$

34.

$$\begin{aligned} x_1 \qquad + x_3 &= 3 \\ 2x_2 - 2x_3 &= -4 \\ -x_2 - 2x_3 &= 5 \end{aligned}$$

35.

$$\begin{aligned} x_1 + x_2 + 3x_3 &= 6 \\ x_1 + 2x_2 + 4x_3 &= 9 \\ 2x_1 + x_2 + 6x_3 &= 11 \end{aligned}$$

36.

$$\begin{aligned} 2x_2 + 4x_3 &= 8 \\ 2x_1 + 2x_2 \qquad &= 6 \\ x_1 + x_2 + x_3 &= 5 \end{aligned}$$

37.

$$\begin{aligned} x_1 - 2x_2 - 4x_3 &= -9 \\ -x_1 + 5x_2 + 10x_3 &= 21 \\ 2x_1 - 3x_2 - 5x_3 &= -13 \end{aligned}$$

38.

$$\begin{aligned} x_1 + 2x_2 + 3x_3 &= 14 \\ 2x_1 + 5x_2 + 8x_3 &= 36 \\ -x_1 - x_2 \qquad &= -4 \end{aligned}$$

39.

$$\begin{aligned} x_1 - x_2 - x_3 &= -1 \\ -2x_1 + 6x_2 + 10x_3 &= 14 \\ 2x_1 + x_2 + 6x_3 &= 9 \end{aligned}$$

40.

$$\begin{aligned} 2x_1 + 2x_2 - 4x_3 &= 14 \\ 3x_1 + x_2 + x_3 &= 8 \\ 2x_1 - x_2 + 2x_3 &= -1 \end{aligned}$$

41.

$$\begin{aligned} x_1 - 2x_2 - 6x_3 &= -17 \\ 2x_1 - 6x_2 - 16x_3 &= -46 \\ x_1 + 2x_2 - x_3 &= -5 \end{aligned}$$

42.

$$\begin{aligned} \frac{3}{2}x_1 \qquad + 3x_3 &= 15 \\ -x_1 + 7x_2 - 9x_3 &= -45 \\ 2x_1 \qquad + 5x_3 &= 22 \end{aligned}$$

43.

$$\begin{aligned} -3x_1 - 6x_2 - 15x_3 &= -3 \\ x_1 + \frac{3}{2}x_2 + \frac{9}{2}x_3 &= \frac{1}{2} \\ -2x_1 - \frac{7}{2}x_2 - \frac{17}{2}x_3 &= -2 \end{aligned}$$

II.

44.

$$\begin{aligned} 3x_1 + 6x_2 \qquad - 3x_4 &= 3 \\ x_1 + 3x_2 - x_3 - 4x_4 &= -12 \\ -x_1 - x_2 + x_3 + 2x_4 &= 8 \\ 2x_1 + 3x_2 \qquad\qquad &= 8 \end{aligned}$$

45.

$$\begin{aligned} x_1 + 2x_2 + 2x_3 + 5x_4 &= 11 \\ 2x_1 + 4x_2 + 2x_3 + 8x_4 &= 14 \\ x_1 + 3x_2 + 4x_3 + 8x_4 &= 19 \\ -x_1 - x_2 + x_3 \qquad &= 2 \end{aligned}$$

46.

$$\begin{aligned} x_1 + x_2 + 2x_3 + 6x_4 &= 11 \\ 2x_1 + 3x_2 + 6x_3 + 19x_4 &= 36 \\ 3x_2 + 4x_3 + 15x_4 &= 28 \\ x_1 - x_2 - x_3 - 6x_4 &= -12 \end{aligned}$$

47.

$$\begin{aligned} x_2 + 2x_3 + 6x_4 &= 21 \\ -x_1 + x_2 + x_3 + 5x_4 &= 12 \\ x_1 - x_2 - x_3 - 4x_4 &= -9 \\ 3x_1 - 2x_2 \qquad - 6x_4 &= -4 \end{aligned}$$

III.

48. An electronics firm has three production facilities: I, II, and III. Each one produces radios, stereos, and TV sets. Their production capacities are:

Plant I: 10 radios, 12 stereos, and 6 TV sets per hour.

Plant II: 7 radios, 10 stereos, and 8 TV sets per hour.

Plant III: 5 radios, 4 stereos, and 13 TV sets per hour.

(a) The firm receives an order for 1365 radios, 1530 stereos, and 1890 TV sets. How many hours should each plant be scheduled in order to produce these amounts?

(b) How many hours should each plant be scheduled to fill an order of 1095 radios, 1230 stereos, and 1490 TV sets?

49. A woman invested $40,000 in three stocks. The first year stock A paid 6% dividends and increased 3% in value; stock B paid 7% dividends and increased 4% in value; stock C paid 8% dividends and increased 2% in value. If the total dividends were $2730 and the total increase in value was $1080, how much was invested in each stock?

50. As a part of a promotional campaign the T-Shirt Company packaged thousands of cartons of T-shirts for their retail outlets. Each carton contained small, medium, and large sizes. Three types of cartons were packed according to quantities as shown in the table. The entries in the table give the number (in dozens) of each size T-shirt in the carton.

		Carton		
		A	*B*	*C*
Size	*Small*	2	5	4
	Medium	5	8	6
	Large	3	2	10

The promotion was a flop, and their warehouse was full of the packed cartons. When they received an order from T-Shirt Orient for 45 dozen small, 78 dozen medium, and 52 dozen large, they wanted to fill the order with the packed cartons in order to save repacking costs. How many of each carton should they send to fill the order?

51. Curt has one hour to spend at the athletic club where he will jog, play handball, and ride a bicycle. Jogging uses 13 calories per minute, handball 11, and cycling 7. He jogs twice as long as he rides the bicycle. How long should he participate in each of these activities in order to use 660 calories?

52. A study showed that for young women a breakfast containing approximately 21 g protein, 46 g carbohydrate, 20 g fat, and 450 calories is a nutritious breakfast that prevents a hungry feeling before lunch. The table shows the content of four breakfast foods. How much of each should be served in order to obtain the desired amounts of protein, carbohydrates, fat, and calories?

	Protein	*Carbohydrates*	*Fat*	*Calories*
1 cup orange juice	2	24	0	110
1 scrambled egg	7	1	8	110
1 slice bread	2	10	6	100
1 cup skim milk	9	13	0	85

53. Concert Special, Inc. has a package deal for three concerts, the Rock Stars, the Smooth Sounds, and the Baroque Band. The customer must buy a ticket to all three concerts in order to get the special prices. This chart shows the way ticket prices are divided among the musical groups.

Ticket Prices for

	High School Students	*College Students*	*Adults*
Rock Stars	$4	$6	$8
Smooth Sounds	$4	$5	$9
Baroque Band	$2	$7	$7

The total ticket sales were: $14,980 for the Rock Stars, $14,430 for the Smooth Sounds, and $14,450 for the Baroque Band. Determine the number of tickets sold to high school students, to college students, and to adults.

2–3 Gauss-Jordan Method

Reduced Echelon Form
Summary

The previous section dealt with the Gauss-Jordan method of solving a system of equations. Those systems had the same number of equations as variables and had unique solutions. Generally, a system may have a unique solution, no solution, or many solutions, and the number of equations can be different from the number of variables. Basically, solving these systems starts with the augmented matrix of the system of equations. Then a sequence of row operations eventually gives a simpler matrix which yields the solution directly. In the last section the matrices reduced to a diagonal form. Sometimes an augmented matrix cannot be reduced to such a diagonal form. However, a **reduced echelon form** is always possible.

Reduced Echelon Form

We give the general definition of reduced echelon form. The diagonal forms of the preceding section also conform to this definition.

Definition

A matrix is in **reduced echelon form** if all the following are true.

1. All rows consisting entirely of zeros are grouped at the bottom of the matrix.
2. The first nonzero number in a row is 1. This element is called the *leading* 1.
3. The leading 1 of a row is to the right of the leading 1 of the previous row.
4. All elements directly above and below a leading 1 are zeros.

The following matrices are all in reduced echelon form. Check the conditions in the definition to make sure you understand why.

$$\begin{bmatrix} 1 & 0 & 0 & 2 \\ 0 & 1 & 0 & 4 \\ 0 & 0 & 1 & -2 \end{bmatrix} \quad \begin{bmatrix} 1 & 4 & 0 & 0 \\ 0 & 0 & 1 & 0 \\ 0 & 0 & 0 & 1 \end{bmatrix} \quad \begin{bmatrix} 1 & 2 & 0 & 4 \\ 0 & 0 & 1 & 3 \\ 0 & 0 & 0 & 0 \end{bmatrix}$$

$$\begin{bmatrix} 1 & 4 & 1 & 0 & 4 & 0 & 2 \\ 0 & 0 & 0 & 1 & 3 & 0 & -2 \\ 0 & 0 & 0 & 0 & 0 & 1 & 3 \end{bmatrix}$$

$$\begin{bmatrix} 1 & 0 & 5 & 0 & 4 & 0 & 5 \\ 0 & 1 & 2 & 0 & -3 & 0 & 0 \\ 0 & 0 & 0 & 1 & 2 & 0 & -2 \\ 0 & 0 & 0 & 0 & 0 & 1 & 3 \end{bmatrix}$$

The following matrices are not in reduced echelon form:

$$\begin{bmatrix} 1 & 2 & 0 & 4 & 0 \\ 0 & 0 & 0 & 0 & 0 \\ 0 & 0 & 1 & 3 & 0 \\ 0 & 0 & 0 & 0 & 1 \end{bmatrix}$$

There is a row consisting of zeros that is not at the bottom of the matrix.

$$\begin{bmatrix} 1 & 2 & 0 & 3 & 0 \\ 0 & 0 & 3 & 4 & 0 \\ 0 & 0 & 0 & 0 & 1 \end{bmatrix}$$

The first nonzero element in row 2 is not 1.

$$\begin{bmatrix} 1 & 0 & 0 & 2 \\ 0 & 0 & 1 & 4 \\ 0 & 1 & 0 & 3 \end{bmatrix}$$

The leading 1 in row 3 is not to the right of the leading 1 in row 2.

$$\begin{bmatrix} 1 & 2 & 0 & 4 \\ 0 & 1 & 0 & -3 \\ 0 & 0 & 1 & 2 \\ 0 & 0 & 0 & 0 \end{bmatrix}$$

The element directly above the leading 1 in row 2 is not 0.

We now work through the details of modifying a matrix until we obtain the reduced echelon form. As we work through it, notice how we use row operations to obtain the leading 1 in row 1, row 2, and so on, and then get zeros in the rest of a column with a leading 1.

Example 1 (*Compare Exercise 11*)
Find the reduced echelon form of the matrix

$$\begin{bmatrix} 0 & 1 & -3 & 2 \\ 2 & 4 & 6 & -4 \\ 3 & 5 & 2 & 2 \end{bmatrix}$$

Solution

	Matrices	Row Operations	Comments
Need 1 here	$\begin{bmatrix} \textcircled{0} & 1 & -3 & 2 \\ 2 & 4 & 6 & -4 \\ 3 & 5 & 2 & 2 \end{bmatrix}$	$R2 \leftrightarrow R1$	Interchange row 1 and row 2 to get nonzero number at top of column 1.
Need 1 here	$\begin{bmatrix} \textcircled{2} & 4 & 6 & -4 \\ 0 & 1 & -3 & 2 \\ 3 & 5 & 2 & 2 \end{bmatrix}$	$\left(\frac{1}{2}\right)R1 \to R1$	Divide row 1 by 2 to get a 1.
Need 0 here	$\begin{bmatrix} 1 & 2 & 3 & -2 \\ 0 & 1 & -3 & 2 \\ \textcircled{3} & 5 & 2 & 2 \end{bmatrix}$	$-3R1 + R3 \to R3$	Get zeros in rest of column 1.
Leading 1 here	$\begin{bmatrix} 1 & 2 & 3 & -2 \\ 0 & \textcircled{1} & -3 & 2 \\ 0 & -1 & -7 & 8 \end{bmatrix}$		Now get leading 1 in row 2. No changes necessary this time.
Need 0 here	$\begin{bmatrix} 1 & \textcircled{2} & 3 & -2 \\ 0 & 1 & -3 & 2 \\ 0 & \textcircled{-1} & -7 & 8 \end{bmatrix}$	$-2R2 + R1 \to R1$ $R2 + R3 \to R3$	Zero entries above and below leading 1 of row 2.
Need leading 1 here	$\begin{bmatrix} 1 & 0 & 9 & -6 \\ 0 & 1 & -3 & 2 \\ 0 & 0 & \textcircled{-10} & 10 \end{bmatrix}$	$\left(-\frac{1}{10}\right)R3 \to R3$	Get a leading 1 in the next row.
Need 0 here	$\begin{bmatrix} 1 & 0 & \textcircled{9} & -6 \\ 0 & 1 & \textcircled{-3} & 2 \\ 0 & 0 & 1 & -1 \end{bmatrix}$	$-9R3 + R1 \to R1$ $3R3 + R2 \to R2$	Zero entries above. leading 1 in row 3.
	$\begin{bmatrix} 1 & 0 & 0 & 3 \\ 0 & 1 & 0 & -1 \\ 0 & 0 & 1 & -1 \end{bmatrix}$		This is the reduced echelon form.

Example 2 (*Compare Exercise 19*)
Find the reduced echelon form of this matrix.

$$\begin{bmatrix} 0 & 0 & 2 & -2 & 2 \\ 3 & 3 & -3 & 9 & 12 \\ 4 & 4 & -2 & 11 & 12 \end{bmatrix}$$

Solution Again we show much of the detailed row operations.

	Matrices	Row Operations	Comments
Need 1 here (circled 0, row 1 column 1)	$\begin{bmatrix} \textcircled{0} & 0 & 2 & -2 & 2 \\ 3 & 3 & -3 & 9 & 12 \\ 4 & 4 & -2 & 11 & 12 \end{bmatrix}$	$R1 \leftrightarrow R2$	
Need 1 here (circled 3, row 1 column 1)	$\begin{bmatrix} \textcircled{3} & 3 & -3 & 9 & 12 \\ 0 & 0 & 2 & -2 & 2 \\ 4 & 4 & -2 & 11 & 12 \end{bmatrix}$	$\left(\frac{1}{3}\right)R1 \rightarrow R1$	
Need 0 here (circled 4, row 3 column 1)	$\begin{bmatrix} 1 & 1 & -1 & 3 & 4 \\ 0 & 0 & 2 & -2 & 2 \\ \textcircled{4} & 4 & -2 & 11 & 12 \end{bmatrix}$	$-4R1 + R3 \rightarrow R3$	
Leading 1 row 2 (circled 1, row 2 column 3)	$\begin{bmatrix} 1 & 1 & -1 & 3 & 4 \\ 0 & 0 & \textcircled{1} & -1 & 1 \\ 0 & 0 & 2 & -1 & -4 \end{bmatrix}$		The leading 1 of row 2 must come from row 2 or below. Since all entries in column 2 are zero in row 2 and 3, go to column 3 for leading 1.
Need 0 here (circled −1 in row 1 and 2 in row 3, column 3)	$\begin{bmatrix} 1 & 1 & \textcircled{-1} & 3 & 4 \\ 0 & 0 & 1 & -1 & 1 \\ 0 & 0 & \textcircled{2} & -1 & -4 \end{bmatrix}$	$R2 + R1 \rightarrow R1$ $-2R2 + R3 \rightarrow R3$	
Need 0 here (circled 2 in row 1 and −1 in row 2, column 4)	$\begin{bmatrix} 1 & 1 & 0 & \textcircled{2} & 5 \\ 0 & 0 & 1 & \textcircled{-1} & 1 \\ 0 & 0 & 0 & 1 & -6 \end{bmatrix}$	$-2R3 + R1 \rightarrow R1$ $R3 + R2 \rightarrow R2$	
	$\begin{bmatrix} 1 & 1 & 0 & 0 & 17 \\ 0 & 0 & 1 & 0 & -5 \\ 0 & 0 & 0 & 1 & -6 \end{bmatrix}$		This is reduced echelon form.

We now solve various systems of equations to illustrate the method of Gauss-Jordan elimination.

Example 3 (*Compare Exercise 35*)
Solve, if possible, the system

$$\begin{aligned} 2x_1 - 4x_2 + 12x_3 - 10x_4 &= 58 \\ -x_1 + 2x_2 - 3x_3 + 2x_4 &= -14 \\ 2x_1 - 4x_2 + 9x_3 - 6x_4 &= 44 \end{aligned}$$

Solution We start with the augmented matrix and convert it to reduced echelon form.

$$\left[\begin{array}{rrrr|r} 2 & -4 & 12 & -10 & 58 \\ -1 & 2 & -3 & 2 & -14 \\ 2 & -4 & 9 & -6 & 44 \end{array}\right] \quad \left(\frac{1}{2}\right)R1 \rightarrow R1$$

$$\left[\begin{array}{rrrr|r} 1 & -2 & 6 & -5 & 29 \\ -1 & 2 & -3 & 2 & -14 \\ 2 & -4 & 9 & -6 & 44 \end{array}\right] \quad \begin{array}{r} R1 + R2 \rightarrow R2 \\ -2R1 + R3 \rightarrow R3 \end{array}$$

$$\left[\begin{array}{rrrr|r} 1 & -2 & 6 & -5 & 2 \\ 0 & 0 & 3 & -3 & 15 \\ 0 & 0 & -3 & 4 & -14 \end{array}\right] \quad \left(\frac{1}{3}\right)R2 \rightarrow R2$$

$$\left[\begin{array}{rrrr|r} 1 & -2 & 6 & -5 & 29 \\ 0 & 0 & 1 & -1 & 5 \\ 0 & 0 & -3 & 4 & -14 \end{array}\right] \quad \begin{array}{r} -6R2 + R1 \rightarrow R1 \\ \\ 3R2 + R3 \rightarrow R3 \end{array}$$

$$\left[\begin{array}{rrrr|r} 1 & -2 & 0 & 1 & -1 \\ 0 & 0 & 1 & -1 & 5 \\ 0 & 0 & 0 & 1 & 1 \end{array}\right] \quad \begin{array}{r} -R3 + R1 \rightarrow R1 \\ R3 + R2 \rightarrow R2 \end{array}$$

$$\left[\begin{array}{rrrr|r} 1 & -2 & 0 & 0 & -2 \\ 0 & 0 & 1 & 0 & 6 \\ 0 & 0 & 0 & 1 & 1 \end{array}\right]$$

This matrix is the reduced echelon form of the augmented matrix. It represents the system of equations

$$\begin{aligned} x_1 - 2x_2 \qquad &= -2 \\ x_3 \quad &= \ \ 6 \\ x_4 &= \ \ 1 \end{aligned}$$

When the reduced echelon form gives an equation with more than one variable like $x_1 - 2x_2 = -2$, the system has many solutions. Many sets of x_1, x_2, x_3, and x_4 satisfy these equations. It is customary to solve the first equation for x_1 in terms of x_2. Doing this, we get $x_1 = 2x_2 - 2$, $x_3 = 6$, and $x_4 = 1$ as the solution.

Specific solutions are found by assigning values to x_2. For example, $x_2 = 3$ gives $x_1 = 4$, $x_3 = 6$, and $x_4 = 1$. In general, we assign the arbitrary value k to x_2

and solve for x_1. The **arbitrary solution** can then be expressed as $x_1 = 2k - 2$, $x_2 = k$, $x_3 = 6$, and $x_4 = 1$. As k ranges over the set of real numbers, we get all the solutions. In such a case, k is called a **parameter**. For example, when $k = 1$, we get $x_1 = 0$, $x_2 = 1$, $x_3 = 6$, and $x_4 = 1$ as a solution. When $k = -2$, we get $x_1 = -6$, $x_2 = -2$, $x_3 = 6$, and $x_4 = 1$ as a solution.

The reduction of an augmented matrix can be tedious. However, this method reduces the solution of a system of equations to a routine. This routine can be carried out by a computer. When dozens of variables are involved, a computer is the only practical way to solve such a system. We want you to be able to perform this routine, so we have two more examples to help you.

Example 4 (*Compare Exercise 43*)
Solve, if possible, the system

$$\begin{aligned} x_1 + 2x_2 - x_3 + 3x_4 + x_5 &= 2 \\ 2x_1 + 4x_2 - 2x_3 + 6x_4 + 3x_5 &= 6 \\ -x_1 - 2x_2 + x_3 - x_4 + 3x_5 &= 4 \end{aligned}$$

We get the augmented matrix and start reducing to echelon form.

$$\left[\begin{array}{rrrrr|r} 1 & 2 & -1 & 3 & 1 & 2 \\ 2 & 4 & -2 & 6 & 3 & 6 \\ -1 & -2 & 1 & -1 & 3 & 4 \end{array}\right] \quad \begin{array}{r} -2R1 + R2 \to R2 \\ R1 + R3 \to R3 \end{array}$$

$$\left[\begin{array}{rrrrr|r} 1 & 2 & -1 & 3 & 1 & 2 \\ 0 & 0 & 0 & 0 & 1 & 2 \\ 0 & 0 & 0 & 2 & 4 & 6 \end{array}\right] \quad R2 \leftrightarrow R3 \text{ (Exchange R2 and R3)}$$

$$\left[\begin{array}{rrrrr|r} 1 & 2 & -1 & 3 & 1 & 2 \\ 0 & 0 & 0 & 2 & 4 & 6 \\ 0 & 0 & 0 & 0 & 1 & 2 \end{array}\right] \quad \left(\frac{1}{2}\right)R2 \to R2$$

$$\left[\begin{array}{rrrrr|r} 1 & 2 & -1 & 3 & 1 & 2 \\ 0 & 0 & 0 & 1 & 2 & 3 \\ 0 & 0 & 0 & 0 & 1 & 2 \end{array}\right] \quad -3R2 + R1 \to R1$$

$$\left[\begin{array}{rrrrr|r} 1 & 2 & -1 & 0 & -5 & -7 \\ 0 & 0 & 0 & 1 & 2 & 3 \\ 0 & 0 & 0 & 0 & 1 & 2 \end{array}\right] \quad \begin{array}{r} 5R3 + R1 \to R1 \\ -2R3 + R2 \to R2 \end{array}$$

$$\left[\begin{array}{rrrrr|r} 1 & 2 & -1 & 0 & 0 & 3 \\ 0 & 0 & 0 & 1 & 0 & -1 \\ 0 & 0 & 0 & 0 & 1 & 2 \end{array}\right]$$

This matrix is the reduced echelon form of the augmented matrix. The corresponding system of equations is

$$\begin{aligned} x_1 + 2x_2 - x_3 \quad &= \ \ 3 \\ x_4 \quad &= -1 \\ x_5 &= \ \ 2 \end{aligned}$$

Solving for x_1 in terms of the remaining variables, we get

$$x_1 = -2x_2 + x_3 + 3, x_4 = -1, x_5 = 2$$

The variable x_1 is expressed in terms of the two variables x_2 and x_3, so we make arbitrary choices of each. Let us assign the arbitrary values k to x_2 and r to x_3. The arbitrary solution then is

$$x_1 = -2k + r + 3, x_2 = k, x_3 = r, x_4 = -1, x_5 = 2$$

We get specific solutions by letting k and r take on different values.

It is possible for a system to have no solution. We illustrate this in the following example.

Example 5 (*Compare Exercise 38*)
This example has no solutions. We attempt to solve the system

$$\begin{aligned} x_1 - \ \ x_2 + 2x_3 &= \ \ 3 \\ 2x_1 - 2x_2 + 5x_3 &= \ \ 4 \\ x_1 + 2x_2 - \ \ x_3 &= -3 \\ 2x_2 + 2x_3 &= \ \ 1 \end{aligned}$$

Solution We get the augmented matrix

$$\left[\begin{array}{rrr|r} 1 & -1 & 2 & 3 \\ 2 & -2 & 5 & 4 \\ 1 & 2 & -1 & -3 \\ 0 & 2 & 2 & 1 \end{array}\right]$$

Now proceed to use row operations to reduce the matrix.

$$\left[\begin{array}{rrr|r} 1 & -1 & 2 & 3 \\ 2 & -2 & 5 & 4 \\ 1 & 2 & -1 & -3 \\ 0 & 2 & 2 & 1 \end{array}\right] \quad \begin{array}{r} -2\text{R}1 + \text{R}2 \to \text{R}2 \\ -\text{R}1 + \text{R}3 \to \text{R}3 \end{array}$$

$$\left[\begin{array}{rrr|r} 1 & -1 & 2 & 3 \\ 0 & 0 & 1 & -2 \\ 0 & 3 & -3 & -6 \\ 0 & 2 & 2 & 1 \end{array}\right] \quad \left(\frac{1}{3}\right)\text{R3 and interchange R2 and R3}$$

$$\left[\begin{array}{rrr|r} 1 & -1 & 2 & 3 \\ 0 & 1 & -1 & -2 \\ 0 & 0 & 1 & -2 \\ 0 & 2 & 2 & 1 \end{array}\right] \begin{array}{l} R2 + R1 \to R1 \\ \\ \\ -2R2 + R4 \to R4 \end{array}$$

$$\left[\begin{array}{rrr|r} 1 & 0 & 1 & 1 \\ 0 & 1 & -1 & -2 \\ 0 & 0 & 1 & -2 \\ 0 & 0 & 4 & 5 \end{array}\right] \begin{array}{r} -R3 + R1 \to R1 \\ R3 + R2 \to R2 \\ \\ -4R3 + R4 \to R4 \end{array}$$

$$\left[\begin{array}{rrr|r} 1 & 0 & 0 & 3 \\ 0 & 1 & 0 & -4 \\ 0 & 0 & 1 & -2 \\ 0 & 0 & 0 & 13 \end{array}\right] \left(\frac{1}{13}\right) R4 \to R4$$

$$\left[\begin{array}{rrr|r} 1 & 0 & 0 & 3 \\ 0 & 1 & 0 & -4 \\ 0 & 0 & 1 & -2 \\ 0 & 0 & 0 & 1 \end{array}\right]$$

This matrix is not in reduced echelon form; zeros still have to be created above the 1 in the last row. However, in such a situation, when the last nonzero row is of the form $[0 \quad 0 \quad \cdots \quad 0 \quad 1]$ there is no need to proceed further. The system has no solutions. To see this, let us write down the system that corresponds to the above matrix. We get

$$\begin{aligned} x_1 \qquad\qquad &= 3 \\ x_2 \qquad &= -4 \\ x_3 &= -2 \\ 0 &= 1 \end{aligned}$$

Because this last equation is false, this system cannot be satisfied for any values of x_1, x_2, and x_3; therefore, the original system has no solutions.

In general, a system with fewer equations than variables has many solutions.

Example 6 (*Compare Exercise 37*)
Solve the system

$$\begin{aligned} x_1 + 2x_2 - x_3 &= -3 \\ 4x_1 + 3x_2 + x_3 &= 13 \end{aligned}$$

Solution: Set up the augmented matrix and solve

$$\left[\begin{array}{rrr|r} 1 & 2 & -1 & -3 \\ 4 & 3 & 1 & 13 \end{array}\right] \begin{array}{l} \\ -4R1 + R2 \to R2 \end{array}$$

$$\left[\begin{array}{rrr|r} 1 & 2 & -1 & -3 \\ 0 & -5 & 5 & 25 \end{array}\right] -\frac{1}{5} R2 \to R2$$

$$\left[\begin{array}{ccc|c} 1 & 2 & -1 & -3 \\ 0 & 1 & -1 & -5 \end{array}\right] \quad -2R2 + R1 \rightarrow R1$$

$$\left[\begin{array}{ccc|c} 1 & 0 & 1 & 7 \\ 0 & 1 & -1 & -5 \end{array}\right]$$

This matrix is in reduced echelon form and represents the equation

$$x_1 = 7 - x_3$$
$$x_2 = -5 + x_3$$

Since x_3 can be arbitrarily chosen, this system has many solutions.

Each nonzero row in the reduced echelon matrix gives the value of one variable—either as a number or expressed in terms of another variable. When there are fewer equations than variables, there are not enough rows in the matrix to give a row for each variable. This means that you can solve only for part of the variables and they will be expressed in terms of the remaining variables (parameters). Examples 3, 4, and 6 illustrate the relationship between the number of variables solved in terms of parameters. Notice the following:

In Example 3, with three equations and four variables, three of the variables were solved in terms of one parameter.

In Example 4, with three equations and five variables, three of the variables were solved in terms of two parameters.

In Example 6, with two equations and three variables, two variables were solved in terms of one parameter.

The relationship is:

If there are k equations with n variables, $n > k$, then k of the variables can be solved in terms of $n - k$ parameters. The system has many solutions. In case a row reduces to all 0's in the reduced echelon form of the augmented matrix, the number of equations is reduced by one.

Summary

The nonzero rows of the reduced echelon matrix give the needed information about the solutions to a system of equations. Three situations are possible.

1. No solution.
   At least one row has all 0's in the coefficient portion of the matrix (to the left of the vertical line) and a nonzero entry to the right of the vertical line.

$$\left[\begin{array}{ccc|c} 1 & 0 & 0 & 3 \\ 0 & 1 & 0 & 2 \\ 0 & 0 & 0 & 5 \end{array}\right] \quad \text{No solution}$$

   Two more possibilities arise when a solution exists.

2. The solution is unique.
   The number of nonzero rows equals the number of variables in the system.

$$\left[\begin{array}{cc|c} 1 & 0 & 5 \\ 0 & 1 & -2 \end{array}\right] \quad \left[\begin{array}{ccc|c} 1 & 0 & 0 & -4 \\ 0 & 1 & 0 & 3 \\ 0 & 0 & 1 & 2 \\ 0 & 0 & 0 & 0 \end{array}\right]$$ Unique solution

3. Infinite number of solutions.
The number of nonzero rows is less than the number of variables in the system.

$$\left[\begin{array}{cccc|c} 1 & 0 & 0 & 2 & 3 \\ 0 & 1 & 0 & 5 & 2 \\ 0 & 0 & 1 & 3 & 4 \end{array}\right] \quad \left[\begin{array}{ccc|c} 1 & 0 & 1 & 2 \\ 0 & 1 & 2 & 4 \\ 0 & 0 & 0 & 0 \end{array}\right]$$ Infinite number of solutions

2–3 Exercises

I.

State whether or not the matrices in Exercises 1 through 10 are in reduced echelon form. If a matrix is not in reduced echelon form, explain why it is not.

1. $\left[\begin{array}{cccc|c} 1 & 0 & 0 & 0 & 0 \\ 0 & 0 & 1 & 2 & 3 \\ 0 & 0 & 0 & 0 & 0 \end{array}\right]$

2. $\left[\begin{array}{ccc|c} 1 & 0 & 0 & 3 \\ 0 & 1 & 0 & 4 \\ 0 & 0 & 2 & 1 \end{array}\right]$

3. $\left[\begin{array}{cccc|c} 1 & 0 & 0 & 3 & 2 \\ 0 & 2 & 0 & 6 & 1 \\ 0 & 0 & 1 & 2 & 3 \end{array}\right]$

4. $\left[\begin{array}{ccccc|c} 1 & 6 & 0 & 0 & 2 & -1 \\ 0 & 0 & 1 & 0 & 4 & 3 \\ 0 & 0 & 0 & 1 & 3 & 1 \end{array}\right]$

5. $\left[\begin{array}{ccc|c} 1 & 0 & 0 & 2 \\ 0 & 0 & 1 & 3 \\ 0 & 1 & 0 & 4 \end{array}\right]$

6. $\left[\begin{array}{cccc|c} 1 & 2 & 0 & 0 & 4 \\ 0 & 0 & 1 & 0 & 6 \\ 0 & 0 & 0 & 1 & 5 \end{array}\right]$

7. $\left[\begin{array}{ccccc} 1 & 0 & 4 & 2 & 6 \\ 0 & 1 & 2 & 3 & 4 \\ 0 & 0 & 0 & 1 & 2 \\ 0 & 0 & 0 & 0 & 1 \end{array}\right]$

8. $\left[\begin{array}{ccccc} 1 & 5 & 4 & 2 & 1 \\ 0 & 0 & 1 & 5 & 3 \\ 0 & 0 & 0 & 0 & 1 \\ 0 & 0 & 0 & 0 & 0 \end{array}\right]$

9. $\left[\begin{array}{cccc|c} 1 & 5 & 0 & 0 & 0 \\ 0 & 0 & 2 & 0 & 0 \\ 0 & 0 & 0 & 1 & 0 \\ 0 & 0 & 0 & 0 & 1 \end{array}\right]$

10. $\left[\begin{array}{cccccc|c} 1 & 2 & 0 & 2 & 0 & 2 & 0 \\ 0 & 0 & 1 & 3 & 0 & 4 & 0 \\ 0 & 0 & 0 & 0 & 1 & 3 & 0 \\ 0 & 0 & 0 & 0 & 0 & 0 & 1 \end{array}\right]$

Interpret each of the matrices in Exercises 11 through 18 as being a matrix in the sequence that leads to the reduced echelon form of a given system of linear equations. Find the next matrix in the sequence in each case.

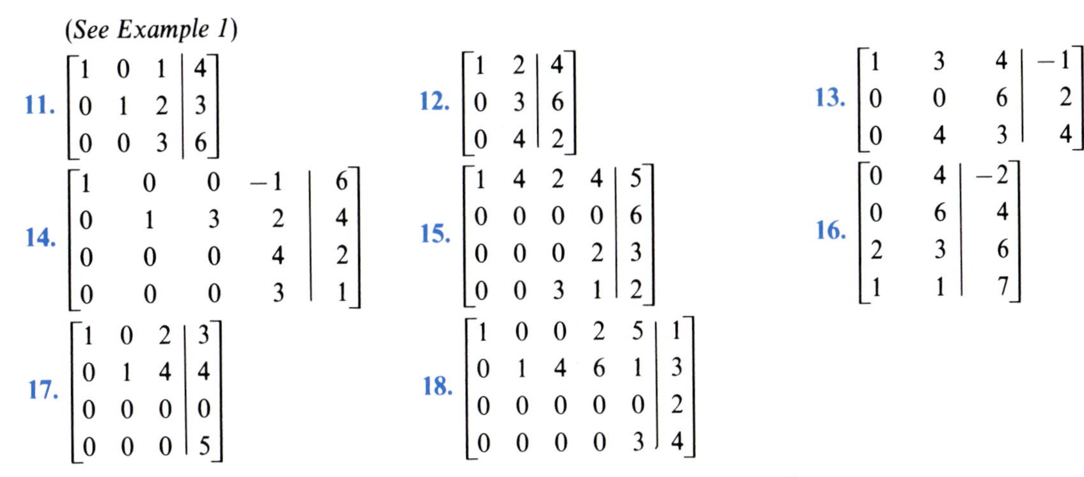

(*See Example 1*)

11. $\left[\begin{array}{ccc|c} 1 & 0 & 1 & 4 \\ 0 & 1 & 2 & 3 \\ 0 & 0 & 3 & 6 \end{array}\right]$

12. $\left[\begin{array}{cc|c} 1 & 2 & 4 \\ 0 & 3 & 6 \\ 0 & 4 & 2 \end{array}\right]$

13. $\left[\begin{array}{ccc|c} 1 & 3 & 4 & -1 \\ 0 & 0 & 6 & 2 \\ 0 & 4 & 3 & 4 \end{array}\right]$

14. $\left[\begin{array}{cccc|c} 1 & 0 & 0 & -1 & 6 \\ 0 & 1 & 3 & 2 & 4 \\ 0 & 0 & 0 & 4 & 2 \\ 0 & 0 & 0 & 3 & 1 \end{array}\right]$

15. $\left[\begin{array}{cccc|c} 1 & 4 & 2 & 4 & 5 \\ 0 & 0 & 0 & 0 & 6 \\ 0 & 0 & 0 & 2 & 3 \\ 0 & 0 & 3 & 1 & 2 \end{array}\right]$

16. $\left[\begin{array}{cc|c} 0 & 4 & -2 \\ 0 & 6 & 4 \\ 2 & 3 & 6 \\ 1 & 1 & 7 \end{array}\right]$

17. $\left[\begin{array}{ccc|c} 1 & 0 & 2 & 3 \\ 0 & 1 & 4 & 4 \\ 0 & 0 & 0 & 0 \\ 0 & 0 & 0 & 5 \end{array}\right]$

18. $\left[\begin{array}{ccccc|c} 1 & 0 & 0 & 2 & 5 & 1 \\ 0 & 1 & 4 & 6 & 1 & 3 \\ 0 & 0 & 0 & 0 & 0 & 2 \\ 0 & 0 & 0 & 0 & 3 & 4 \end{array}\right]$

In Exercises 19 through 22, reduce each matrix to its reduced echelon form.

(*See Example 2*)

19. $\left[\begin{array}{ccc|c} 0 & -1 & 2 & 1 \\ 1 & -2 & 1 & 4 \\ 2 & 1 & 0 & 1 \end{array}\right]$

20. $\left[\begin{array}{ccc|c} 1 & 3 & 5 & 4 \\ -3 & -4 & 0 & 3 \\ 0 & 2 & 1 & 1 \end{array}\right]$

21. $\left[\begin{array}{ccc|c} 2 & 4 & 8 & -4 \\ 3 & 5 & -1 & 6 \\ 1 & -1 & -9 & 10 \end{array}\right]$

22. $\left[\begin{array}{ccc|c} 3 & -2 & 5 & 1 \\ -6 & 4 & -10 & -2 \\ 12 & -8 & 20 & 4 \end{array}\right]$

Each of the matrices in Exercises 23 through 31 is in reduced echelon form. Write down the system of equations that corresponds to each matrix. Solve each system, if possible.

23. $\left[\begin{array}{cccc|c} 1 & 0 & 0 & 0 & 2 \\ 0 & 1 & 2 & 0 & 3 \\ 0 & 0 & 0 & 1 & -1 \end{array}\right]$

24. $\left[\begin{array}{cccc|c} 1 & 2 & 0 & 0 & 2 \\ 0 & 0 & 1 & 0 & 3 \\ 0 & 0 & 0 & 1 & 4 \end{array}\right]$

25. $\left[\begin{array}{ccccc|c} 1 & 0 & 2 & 0 & 2 & -1 \\ 0 & 1 & 4 & 0 & 3 & 2 \\ 0 & 0 & 0 & 1 & 4 & 1 \end{array}\right]$

26. $\left[\begin{array}{cccc|c} 1 & 0 & 2 & 0 & 0 \\ 0 & 1 & 4 & 0 & 0 \\ 0 & 0 & 0 & 1 & 0 \\ 0 & 0 & 0 & 0 & 1 \end{array}\right]$

27. $\left[\begin{array}{cccc|c} 1 & 0 & 2 & 0 & 2 \\ 0 & 1 & -3 & 0 & 1 \\ 0 & 0 & 0 & 1 & 3 \\ 0 & 0 & 0 & 0 & 0 \end{array}\right]$

28. $\left[\begin{array}{cccccc|c} 1 & 3 & 4 & 0 & 0 & 0 & 1 \\ 0 & 0 & 0 & 1 & 0 & 0 & 2 \\ 0 & 0 & 0 & 0 & 1 & 0 & 4 \\ 0 & 0 & 0 & 0 & 0 & 1 & 3 \end{array}\right]$

29. $\left[\begin{array}{cccccc|c} 1 & 2 & 0 & 1 & 0 & 0 & 4 \\ 0 & 0 & 1 & 3 & 0 & 0 & -1 \\ 0 & 0 & 0 & 0 & 1 & 0 & 3 \\ 0 & 0 & 0 & 0 & 0 & 1 & 2 \end{array}\right]$

30. $\left[\begin{array}{ccc|c} 1 & 0 & 0 & 0 \\ 0 & 1 & 0 & 0 \\ 0 & 0 & 1 & 0 \\ 0 & 0 & 0 & 1 \end{array}\right]$

31. $\left[\begin{array}{cccccc|c} 1 & 2 & 0 & 4 & 0 & 1 & 7 \\ 0 & 0 & 1 & 3 & 0 & -1 & 3 \\ 0 & 0 & 0 & 0 & 1 & 4 & 2 \end{array}\right]$

Solve each system of equations in Exercises 32 through 45 (if possible).

32. $\begin{aligned} x_1 - x_2 + x_3 &= 3 \\ -2x_1 + 3x_2 + x_3 &= -8 \\ 4x_1 - 2x_2 + 10x_3 &= 10 \end{aligned}$

33. $\begin{aligned} x_2 + 2x_3 &= 7 \\ x_1 - 2x_2 - 6x_3 &= -18 \\ x_1 - x_2 - 2x_3 &= -5 \\ 2x_1 - 5x_2 - 15x_3 &= -46 \end{aligned}$

II.

34. $\begin{aligned} 2x_1 - 4x_2 - 14x_3 &= 6 \\ x_1 - x_2 - 5x_3 &= 4 \\ 2x_1 - 4x_2 - 17x_3 &= 9 \\ -x_1 + 3x_2 + 10x_3 &= -3 \\ 2x_2 + 2x_3 &= 4 \end{aligned}$

35. (*See Example 3*)
$\begin{aligned} x_1 + x_2 + x_3 - x_4 &= -3 \\ 2x_1 + 3x_2 + x_3 - 5x_4 &= -9 \\ x_1 + 3x_2 - x_3 - 6x_4 &= 7 \end{aligned}$

36. $\begin{aligned} x_1 + 2x_2 - x_3 - x_4 &= 0 \\ x_1 + 2x_2 + x_4 &= 4 \\ -x_1 - 2x_2 + 2x_3 + 4x_4 &= 5 \\ -x_1 - x_2 - x_3 &= 1 \end{aligned}$

37. (*See Example 6*)
$\begin{aligned} x_1 + 2x_2 - x_3 &= -13 \\ 2x_1 + 5x_2 + 3x_3 &= -3 \end{aligned}$

38. (*See Example 5*)
$\begin{aligned} 2x_1 - 4x_2 + 16x_3 - 14x_4 &= 12 \\ -x_1 + 5x_2 - 17x_3 + 19x_4 &= -2 \\ x_1 - 3x_2 + 11x_3 - 11x_4 &= 4 \end{aligned}$

39. $\begin{aligned} x_1 + x_2 - 3x_3 + x_4 &= 4 \\ -2x_1 - 2x_2 + 6x_3 - 2x_4 &= 3 \end{aligned}$

40. $\begin{aligned} x_1 - x_2 + 2x_3 &= 7 \\ 3x_1 - 4x_2 + 18x_3 - 13x_4 &= 17 \\ 2x_1 - 2x_2 + 2x_3 - 4x_4 &= 12 \\ -x_1 + x_2 - x_3 + 2x_4 &= -4 \\ -3x_1 + x_2 - 8x_3 - 10x_4 &= -29 \end{aligned}$

41. $\begin{aligned} x_1 + x_2 - x_3 - x_4 &= -1 \\ 3x_1 - 2x_2 - 4x_3 + 2x_4 &= 1 \\ 4x_1 - x_2 - 5x_3 + x_4 &= 5 \end{aligned}$

42. $\begin{aligned} x_1 + 6x_2 - x_3 - 4x_4 &= 0 \\ -2x_1 - 12x_2 + 5x_3 + 17x_4 &= 0 \\ 3x_1 + 18x_2 - x_3 - 6x_4 &= 0 \end{aligned}$

43. (*See Example 4*)
$\begin{aligned} x_1 - 2x_2 + x_3 + x_4 - 2x_5 &= -9 \\ 5x_1 + x_2 - 6x_3 - 2x_4 + x_5 &= 21 \end{aligned}$

44. $\begin{aligned} 4x_1 + 8x_2 - 12x_3 &= 28 \\ -x_1 - 2x_2 + 3x_3 &= -7 \\ 2x_1 + 4x_2 - 8x_3 &= 16 \\ -3x_1 - 6x_2 + 9x_3 &= -21 \end{aligned}$

45. $\begin{aligned} x_1 + 3x_2 + 6x_3 - 2x_4 &= -7 \\ -2x_1 - 5x_2 - 10x_3 + 3x_4 &= 10 \\ x_1 + 2x_2 + 4x_3 &= 0 \\ x_2 + 2x_3 - 3x_4 &= -10 \end{aligned}$

III.

46. $\begin{aligned} x_1 + 2x_2 - 2x_3 + 7x_4 + 7x_5 &= 0 \\ -x_1 - 2x_2 + 2x_3 - 9x_4 - 11x_5 &= 0 \\ x_3 - 2x_4 - x_5 &= 0 \\ 2x_1 + 4x_2 - 3x_3 + 12x_4 + 13x_5 &= 0 \end{aligned}$

47.
$$\begin{aligned} x_1 + 2x_2 - 3x_3 + 2x_4 + 5x_5 - x_6 &= 0 \\ -2x_1 - 4x_2 + 6x_3 - x_4 - 4x_5 + 5x_6 &= 0 \\ 3x_1 + 6x_2 - 9x_3 + 5x_4 + 13x_5 - 4x_6 &= 0 \end{aligned}$$

48.
$$\begin{aligned} x_1 + x_2 + 3x_3 + 6x_4 + 24x_5 + 48x_6 &= 112 \\ 2x_1 + 2x_2 + 6x_3 + 13x_4 + 52x_5 + 104x_6 &= 241 \\ x_3 + 2x_4 + 8x_5 + 16x_6 &= 37 \\ -x_1 - x_2 - 4x_3 - 7x_4 - 29x_5 - 58x_6 &= -136 \\ x_1 + 2x_2 + 5x_3 + 11x_4 + 42x_5 + 84x_6 &= 197 \end{aligned}$$

49.
$$\begin{aligned} x_1 + 2x_2 + 3x_3 - x_5 - 2x_6 + x_7 &= 2 \\ 2x_1 + 4x_2 + 6x_3 - x_5 - 3x_6 + 3x_7 &= 7 \\ x_1 + 2x_2 + 3x_3 + x_4 + x_6 + x_7 &= 3 \\ -3x_1 - 6x_2 - 9x_3 + 2x_4 + 7x_5 + 14x_6 + 2x_7 &= 5 \end{aligned}$$

50. A contractor builds houses, duplexes, and apartment units. He has financial backing to build 250 units. He makes a profit of $4500 on each house, $4000 on each duplex, and $3000 on each apartment unit. Each house requires 10 man-months of labor, each duplex requires 12 man-months, and each apartment requires 6 man-months. How many of each should he build if he has 2050 man-months of labor available and wishes to make a total profit of $875,000?

51. An investor bought $45,000 in stocks, bonds, and money market funds. The total invested in bonds and money market funds was twice the amount invested in stocks. The return on the stocks, bonds, and money markets was 10%, 7%, and 7.5%, respectively. The total return was $3660. How much was purchased of each?

2–4 Matrix Operations

Equal Matrices
Addition of Matrices
Scalar Multiplication

Matrices come in various shapes and sizes and can be classified by the number of rows and columns they have. For example,

$$\begin{bmatrix} 1 & 2 & 5 \\ 3 & 4 & 6 \end{bmatrix}$$

is a 2×3 matrix, for it has two rows and three columns.

$$\begin{bmatrix} 1 & 2 & 3 & 4 \\ 0 & -1 & 7 & 6 \\ 4 & 4 & 3 & 2 \end{bmatrix}$$

is a 3×4 matrix, having three rows and four columns. The convention used in describing the size of a matrix states the number of rows first, followed by the number of columns.

A matrix with the same number of rows as columns is said to be a **square matrix**. Two matrices are of the **same kind** if they have the same number of rows and the same number of columns. For example,

$$\begin{bmatrix} 1 & 3 & 2 \\ -1 & 3 & 4 \end{bmatrix} \text{ and } \begin{bmatrix} 2 & -1 & 7 \\ 6 & 3 & 4 \end{bmatrix}$$

are matrices of the same kind; they are both 2×3 matrices.

Equal Matrices

Two matrices of the same kind are said to be *equal* if and only if their corresponding components are equal. If the matrices are not the same kind, they are not equal.

Example 1 (*Compare Exercise 13*)

$$\begin{bmatrix} 3 & 7 \\ 5-1 & 4 \times 4 \end{bmatrix} = \begin{bmatrix} 6/2 & 7 \\ 4 & 16 \end{bmatrix}$$

because corresponding components are equal.

$$\begin{bmatrix} 1 & 2 & 5 \\ 3 & 6 & 4 \end{bmatrix} \neq \begin{bmatrix} 1 & 2 & 5 \\ 3 & -1 & 4 \end{bmatrix}$$

because the entries in row 2, column 2 are different; that is, the (2, 2) entries are not equal.

Addition of Matrices

A businessperson has two stores. He is interested in the daily sales of the regular size and the giant economy size of laundry soap. He can use matrices to record this information. These two matrices show sales for two days.

	Store			Store	
	1	*2*		*1*	*2*
Reg.	8	12	*Reg.*	6	5
Giant	9	7	*Giant*	11	4
	Day 1			Day 2	

The position in the matrix identifies the store and package size. For example, store 2 sold 7 giant sizes and 12 regular on day 1, and so on.

The total sales, by store and package size, can be obtained by **addition of matrices**. Doesn't it seem reasonable that the sum of the two sales matrices is

$$\begin{bmatrix} 8 & 12 \\ 9 & 7 \end{bmatrix} + \begin{bmatrix} 6 & 5 \\ 11 & 4 \end{bmatrix} = \begin{bmatrix} 14 & 17 \\ 20 & 11 \end{bmatrix}$$

because we just add the sales in each individual category to get the total in that category? This procedure applies generally to the addition of matrices.

Warning! The two matrices must be of the same kind.

Definition

> The **sum** of two matrices of the same kind is obtained by adding corresponding elements. If two matrices are not of the same kind, they cannot be added; we say that their sum does not exist. Subtraction is performed on matrices of the same kind by subtracting corresponding elements.

Example 2 (*Compare Exercise 19*)
For the following matrices:

$$A = \begin{bmatrix} -1 & 2 & 3 \\ 0 & 1 & 4 \end{bmatrix} \quad B = \begin{bmatrix} 3 & -2 & 6 \\ 7 & -1 & 2 \end{bmatrix} \quad C = \begin{bmatrix} -1 & 2 \\ 0 & 1 \end{bmatrix}$$

Determine the sums $A + B$ and $A + C$ if possible.

Solution

$$A + B = \begin{bmatrix} -1 & 2 & 3 \\ 0 & 1 & 4 \end{bmatrix} + \begin{bmatrix} 3 & -2 & 6 \\ 7 & -1 & 2 \end{bmatrix}$$

$$= \begin{bmatrix} -1+3 & 2-2 & 3+6 \\ 0+7 & 1-1 & 4+2 \end{bmatrix} = \begin{bmatrix} 2 & 0 & 9 \\ 7 & 0 & 6 \end{bmatrix}$$

Neither the sum $A + C$ nor $B + C$ exists since A and C, and B and C, are not matrices of the same kind. (Try adding these matrices using the rule.)

Let us extend our definition to enable us to add more than just two matrices. For example, define the sum of three matrices as

$$\begin{bmatrix} 1 & 2 \\ 0 & -1 \end{bmatrix} + \begin{bmatrix} 3 & 4 \\ 2 & 1 \end{bmatrix} + \begin{bmatrix} 5 & 2 \\ -1 & 0 \end{bmatrix} = \begin{bmatrix} 1+3+5 & 2+4+2 \\ 0+2-1 & -1+1+0 \end{bmatrix}$$

$$= \begin{bmatrix} 9 & 8 \\ 1 & 0 \end{bmatrix}$$

We add a string of matrices that are of the same kind by adding corresponding elements. The following example illustrates a use of this rule.

Example 3 (*Compare Exercise 35*)
A clinic has three doctors, each with his or her own speciality. Patients attending the clinic may see more than one doctor. The accounts are drawn up monthly and handled systematically through the use of matrices. We illustrate this accounting with four patients. It can easily be extended to accommodate any number of patients.

$$
\begin{array}{cc|ccc}
 & & & \textbf{Doctor} & \\
 & & I & II & III \\
\hline
 & A & 10 & 25 & 0 \\
\textbf{Patient} & B & 0 & 20 & 40 \\
 & C & 15 & 0 & 25 \\
 & D & 10 & 10 & 0
\end{array}
$$

The entries in the above matrix are in dollars and represent the bill from each doctor to each patient for a certain month.

For a given quarter, three such matrices represent monthly bills. To determine the quarterly bill to each patient from each doctor, add the three matrices. Assume that the monthly bills are as folllows:

First Month (I, II, III) + **Second Month** (I, II, III) + **Third Month** (I, II, III) = **Quarter** (I, II, III)

$$
\begin{matrix} A \\ B \\ C \\ D \end{matrix}
\begin{bmatrix} 10 & 25 & 0 \\ 0 & 20 & 40 \\ 15 & 0 & 25 \\ 10 & 10 & 0 \end{bmatrix}
+
\begin{bmatrix} 0 & 0 & 0 \\ 10 & 100 & 0 \\ 10 & 0 & 0 \\ 20 & 0 & 0 \end{bmatrix}
+
\begin{bmatrix} 20 & 0 & 0 \\ 0 & 0 & 0 \\ 0 & 0 & 20 \\ 15 & 15 & 0 \end{bmatrix}
=
\begin{bmatrix} 30 & 25 & 0 \\ 10 & 120 & 40 \\ 25 & 0 & 45 \\ 45 & 25 & 0 \end{bmatrix}
$$

Thus, A's bill from doctor I during the quarter would total \$30; B's quarterly bill from doctor III would total \$40; and so on.

Although this analysis and others like it can be carried out without the use of matrices, the handling of large quantities of data is often most efficiently done on computers using matrix techniques.

Scalar Multiplication

Let us now turn our attention to multiplying matrices by numbers. We are interested in defining the following type of operation:

$$3\begin{bmatrix} 1 & 2 \\ 4 & 1 \end{bmatrix}$$

The product is

$$3\begin{bmatrix} 1 & 2 \\ 4 & 1 \end{bmatrix} = \begin{bmatrix} 3 & 6 \\ 12 & 3 \end{bmatrix}$$

Note that the net effect is that every element in the matrix is multiplied by 3. Let us use this example as a prototype for a definition of **scalar multiplication**. (**Scalar** is a term used by mathematicians for **number**.)

Definition

> The operation of multiplying a matrix by a number is called **scalar multiplication**. It is carried out by multiplying every element in the matrix by the scalar.

Example 4 (*Compare Exercise 27*)

$$-2\begin{bmatrix} 1 & 2 \\ 3 & 0 \\ 1 & -4 \end{bmatrix} = \begin{bmatrix} -2 & -4 \\ -6 & 0 \\ -2 & 8 \end{bmatrix}$$

Example 5 (*Compare Exercise 40*)
The following matrix represents total trade figures between certain countries for the five-year period. The numbers are values in millions of U.S. dollars. We use this matrix to illustrate a manner in which scalar multiplication of a matrix can arise.

From: \ **To:**	*Canada*	*E.E.C.*	*Japan*	*U.S.A.*
Canada	0	12,600	4,890	63,690
E.E.C.	9,230	0	8,280	56,510
Japan	4,020	13,230	0	37,170
U.S.A.	54,750	59,690	25,190	0

The annual average over this period is $\frac{1}{5}$ of the five-year figure. We can get the annual figures by scalar multiplying the above matrix by $\frac{1}{5}$:

$\left(\frac{1}{5}\right)\mathbf{A} =$

From: \ **To:**	*Canada*	*E.E.C.*	*Japan*	*U.S.A.*
Canada	0	2,520	978	12,738
E.E.C.	1,846	0	1,656	11,302
Japan	804	2,646	0	7,434
U.S.A.	10,950	11,938	5,038	0

Example 6 (*Compare Exercise 43*)
A class of ten students had five tests during the quarter. A perfect score on each of the tests is 50. The scores are listed in Table 2–1. We can express these scores as

column matrices

$$\begin{bmatrix}40\\20\\40\\25\\35\\50\\22\\35\\28\\40\end{bmatrix}\begin{bmatrix}45\\15\\35\\40\\35\\46\\24\\27\\31\\35\end{bmatrix}\begin{bmatrix}30\\30\\25\\45\\38\\45\\30\\20\\25\\36\end{bmatrix}\begin{bmatrix}48\\25\\45\\40\\37\\48\\32\\41\\27\\32\end{bmatrix}\begin{bmatrix}42\\10\\46\\38\\39\\47\\29\\30\\31\\38\end{bmatrix}$$

To obtain each person's average, we use matrix addition to add the matrices and then scalar multiplication to multiply by $\frac{1}{5}$ (dividing by the number of tests). We get

$$\frac{1}{5}\begin{bmatrix}205\\100\\191\\188\\184\\236\\137\\153\\142\\181\end{bmatrix}=\begin{bmatrix}41.0\\20.0\\38.2\\37.6\\36.8\\47.2\\27.4\\30.6\\28.4\\36.2\end{bmatrix}$$

Column matrix giving each person's average score

Table 2–1

	Test 1	Test 2	Test 3	Test 4	Test 5
Anderson	40	45	30	48	42
Boggs	20	15	30	25	10
Chittar	40	35	25	45	46
Diessner	25	40	45	40	38
Farnam	35	35	38	37	39
Gill	50	46	45	48	47
Homes	22	24	30	32	29
Johnson	35	27	20	41	30
Schomer	28	31	25	27	31
Wong	40	35	36	32	38

Row matrices are also useful; a person's complete set of scores corresponds to a row matrix. For example,

$$\begin{bmatrix} 25 & 40 & 45 & 40 & 38 \end{bmatrix}$$

is a row matrix giving Diessner's scores.

This approach to analyzing test scores has the advantage of lending itself to implementation on the computer. A computer program can be written that will perform the desired matrix additions and scalar multiplications.

2–4 Exercises

I.

What kind of matrix is found in Exercises 1 through 12?

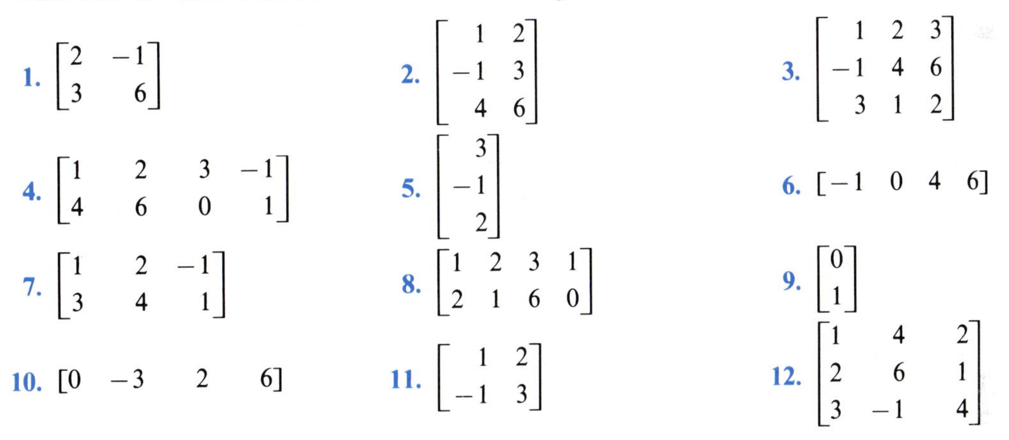

1. $\begin{bmatrix} 2 & -1 \\ 3 & 6 \end{bmatrix}$

2. $\begin{bmatrix} 1 & 2 \\ -1 & 3 \\ 4 & 6 \end{bmatrix}$

3. $\begin{bmatrix} 1 & 2 & 3 \\ -1 & 4 & 6 \\ 3 & 1 & 2 \end{bmatrix}$

4. $\begin{bmatrix} 1 & 2 & 3 & -1 \\ 4 & 6 & 0 & 1 \end{bmatrix}$

5. $\begin{bmatrix} 3 \\ -1 \\ 2 \end{bmatrix}$

6. $\begin{bmatrix} -1 & 0 & 4 & 6 \end{bmatrix}$

7. $\begin{bmatrix} 1 & 2 & -1 \\ 3 & 4 & 1 \end{bmatrix}$

8. $\begin{bmatrix} 1 & 2 & 3 & 1 \\ 2 & 1 & 6 & 0 \end{bmatrix}$

9. $\begin{bmatrix} 0 \\ 1 \end{bmatrix}$

10. $\begin{bmatrix} 0 & -3 & 2 & 6 \end{bmatrix}$

11. $\begin{bmatrix} 1 & 2 \\ -1 & 3 \end{bmatrix}$

12. $\begin{bmatrix} 1 & 4 & 2 \\ 2 & 6 & 1 \\ 3 & -1 & 4 \end{bmatrix}$

In Exercises 13 through 18, determine which of the pairs of matrices are equal.

(See Example 1)

13. $\begin{bmatrix} 2 & 1 & 3 \\ 4 & 0 & 2 \end{bmatrix}\begin{bmatrix} 4 & 0 & 2 \\ 2 & 1 & 3 \end{bmatrix}$

14. $\begin{bmatrix} \frac{3}{4} & 16 \\ 8 & \frac{1}{2} \end{bmatrix}\begin{bmatrix} .75 & 16 \\ 8 & .5 \end{bmatrix}$

15. $\begin{bmatrix} 5+2 & 5-2 \\ \frac{5}{2} & \frac{2}{5} \end{bmatrix}\begin{bmatrix} 7 & 3 \\ 2.5 & .4 \end{bmatrix}$

16. $\begin{bmatrix} 2 & 1 & 3 & 6 \\ 5 & 9 & 4 & 1 \end{bmatrix}\begin{bmatrix} 2 & 3 & 6 \\ 5 & 4 & 1 \end{bmatrix}$

17. $\begin{bmatrix} -1 & 4 & 1 \\ 3 & 2 & 0 \end{bmatrix}\begin{bmatrix} -1 & 3 \\ 4 & 2 \\ 1 & 0 \end{bmatrix}$

18. $\begin{bmatrix} 1 & 0 \\ 0 & 2 \end{bmatrix}\begin{bmatrix} 2 & 0 \\ 0 & 1 \end{bmatrix}$

If possible, add the matrices in Exercises 19 through 26. We say that the sum does not exist if the matrices cannot be added.

(See Example 2)

19. $\begin{bmatrix} 1 & -1 & 3 \\ 2 & 4 & 1 \end{bmatrix} + \begin{bmatrix} 2 & 4 & -1 \\ 5 & 0 & 2 \end{bmatrix}$

20. $\begin{bmatrix} 0 & 1 \\ 2 & 3 \\ -1 & 4 \end{bmatrix} + \begin{bmatrix} 3 & -1 \\ 2 & 0 \\ 4 & -6 \end{bmatrix}$

21. $\begin{bmatrix} 1 \\ 2 \end{bmatrix} + \begin{bmatrix} 3 \\ -1 \end{bmatrix} + \begin{bmatrix} 1 \\ 4 \end{bmatrix}$

22. $\begin{bmatrix} 1 & 2 \\ 3 & -1 \end{bmatrix} + \begin{bmatrix} 1 & .4 \\ 2 & 6 \\ 0 & 3 \end{bmatrix}$

23. $[1 \quad -1] + [2 \quad 3 \quad 0]$

24. $\begin{bmatrix} 1 & 2 & 4 \\ 0 & 1 & -3 \end{bmatrix} + \begin{bmatrix} 2 & -2 & -3 \\ -1 & 0 & -7 \end{bmatrix}$

25. $\begin{bmatrix} 1 & 4 & 2 \\ 3 & 0 & 1 \\ 6 & -1 & 5 \end{bmatrix} + \begin{bmatrix} 5 & 1 & 6 \\ 2 & 1 & 2 \\ -1 & -2 & -1 \end{bmatrix}$

26. $\begin{bmatrix} 1 & 0 & 1 & 0 \\ 0 & 1 & 0 & 1 \\ 2 & 1 & 2 & 1 \\ 1 & 2 & 1 & 2 \end{bmatrix} + \begin{bmatrix} 3 & 4 & 3 & 4 \\ 2 & 5 & 2 & 5 \\ 3 & 0 & 3 & 0 \\ 2 & 2 & 3 & 3 \end{bmatrix}$

Perform the scalar multiplications in Exercises 27 through 32.

(*See Example 4*)

27. $2\begin{bmatrix} 1 & 2 \\ 3 & 4 \end{bmatrix}$

28. $3\begin{bmatrix} -1 & 2 \\ 4 & 1 \\ 2 & 0 \end{bmatrix}$

29. $4[1 \quad -1 \quad 2 \quad -3]$

30. $-3\begin{bmatrix} 3 & 1 \\ -2 & 4 \end{bmatrix}$

31. $-0.5\begin{bmatrix} -1 \\ 2 \\ .3 \end{bmatrix}$

32. $0\begin{bmatrix} 1 & 2 & -1 \\ 3 & 1 & 4 \end{bmatrix}$

II.

33. If

$$A = \begin{bmatrix} 1 & 2 \\ 3 & 0 \end{bmatrix}, \quad B = \begin{bmatrix} -1 & 2 \\ 1 & 1 \end{bmatrix}, \quad \text{and} \quad C = \begin{bmatrix} 0 & 1 \\ 1 & 4 \end{bmatrix},$$

determine the matrices

(a) $2A$, $3B$, and $-2C$.

(b) $A + B$, $B + A$, $A + C$, and $B + C$

(c) $A + 2B$, $3A + C$, and $2A + B - C$

34. If

$$A = \begin{bmatrix} -1 & 4 & 5 \\ 0 & 1 & 2 \end{bmatrix}, \quad B = \begin{bmatrix} 1 & 1 & 0 \\ 2 & 3 & 1 \end{bmatrix}, \quad \text{and} \quad C = \begin{bmatrix} 4 & 5 & 0 \\ 2 & -1 & -1 \end{bmatrix},$$

determine

(a) $A - B$, $B + 3C$, and $2B + C$.

(b) $4A$, $-B$, and $3C$.

(c) $A + 3B$, $2A - B + C$, and $A + 2B - 2C$.

35. (*See Example 3*) A distributor furnishes PC computers, printers, and floppy disks to three customers. He summarizes monthly sales in a matrix.

	Customer								
	I	*II*	*III*	*I*	*II*	*III*	*I*	*II*	*III*
PC	5	7	8	8	6	5	10	4	7
Print	6	4	5	10	9	4	3	9	2
Disk	45	52	35	52	60	42	54	39	28
		June			**July**			**August**	

Find the three month total by item and by customer.

Find the value of x that makes the pairs of matrices equal in Exercises 36 through 39.

36. $\begin{bmatrix} 3 & x \\ 2 & 1 \end{bmatrix} = \begin{bmatrix} 3 & 9 \\ 2 & 1 \end{bmatrix}$

37. $\begin{bmatrix} 5x & 7 \\ 2 & 4 \end{bmatrix} = \begin{bmatrix} 15 & 7 \\ 2 & 4 \end{bmatrix}$

38. $\begin{bmatrix} 2x+3 & -2 \\ 6 & 1 \end{bmatrix} = \begin{bmatrix} 3x-1 & -2 \\ 6 & 1 \end{bmatrix}$

39. $\begin{bmatrix} 17 & 6x+4 \\ 94 & -39 \end{bmatrix} = \begin{bmatrix} 17 & 14x-13 \\ 94 & -39 \end{bmatrix}$

40. (*See Example 5*) A firm has three plants that all make small, regular, and giant sizes of detergent. The annual report shows the total production (in thousands), broken down by plant and size, in the following matrix.

	Plant		
	A	*B*	*C*
Small	65	110	80
Regular	90	135	60
Giant	75	112	84

Find the average monthly production by plant and size.

41. A manufacturer has two plants A and B which produce the pollutants sulfur dioxide, nitric oxide, and particulate matter in the amounts shown in the matrix. The amounts are in kilograms and represent the average per month.

	Sulphur Dioxide	*Nitric Oxide*	*Particulate Matter*
Plant A	230	90	140
Plant B	260	115	166

Find the annual total by plant and pollutant.

42. A cafeteria manager estimates that the amount of food needed to serve one person is

meat	4 oz.
peas	2 oz.
rice	3 oz.
bread	1 slice
milk	1 cup

Use matrix arithmetic to find the amount needed to serve 114 people.

43. (*See Example 6*) The test scores for five students are given below.

	Test 1	*Test 2*	*Test 3*
A	90	88	91
B	62	69	73
C	76	78	72
D	82	80	84
E	74	76	77

Use matrix arithmetic to find each student's average.

44. The total sales of regular and giant size detergent at two stores is given for three months.

	1	*2*	*1*	*2*	*1*	*2*
Regular	85	46	80	61	50	42
Giant	77	93	93	47	61	38
	March		**April**		**May**	

$$\begin{bmatrix} 85 & 46 \\ 77 & 93 \end{bmatrix} \quad \begin{bmatrix} 80 & 61 \\ 93 & 47 \end{bmatrix} \quad \begin{bmatrix} 50 & 42 \\ 61 & 38 \end{bmatrix}$$

Use matrix arithmetic to find the average monthly sales by store and size.

III.

45. The following matrix A represents energy production statistics for certain countries for the period 1980 through 1983. The units are in millions of metric tons of coal-equivalent energy.

	1980	*1981*	*1982*	*1983*
Canada	206.03	223.25	257.25	285.35
U.S.A.	2,062.53	2,029.17	2,065.22	2,052.26
U.K.	163.73	176.16	159.71	175.00
Japan	58.82	49.47	44.92	39.73

$$\begin{bmatrix} 206.03 & 223.25 & 257.25 & 285.35 \\ 2{,}062.53 & 2{,}029.17 & 2{,}065.22 & 2{,}052.26 \\ 163.73 & 176.16 & 159.71 & 175.00 \\ 58.82 & 49.47 & 44.92 & 39.73 \end{bmatrix} = A$$

The following matrix B gives energy consumption for the same countries:

	1980	*1981*	*1982*	*1983*
Canada	201.07	209.68	235.01	248.61
U.S.A.	2,269.49	2,317.04	2,426.07	2,516.44
U.K.	300.20	309.21	301.83	323.91
Japan	332.37	342.02	344.55	390.20

$= B$

Compute the matrix $A - B$. Give an interpretation of the elements of $A - B$. What is the significance of a negative element?

2–5 Multiplication of Matrices

Dot Product
Matrix Multiplication

You have learned to add matrices and multiply a matrix by a number. It is natural to ask if one can multiply two matrices together and whether it helps solve problems. Mathematicians have devised a way of multiplying two matrices. It may seem rather complicated, but it has many useful applications. For example, you will learn how to use matrix multiplication to solve a problem like the following.

A manufacturer makes tables and chairs. The time, in hours, required to assemble and finish the items is given by the matrix

	Chair	*Table*
Assemble	2	3
Finish	2.5	4.75

The total assembly and finishing time required to produce 950 chairs and 635 tables can be obtained by an appropriate matrix multiplication.

Before we show you how to multiply matrices, here's an overview.

1. We have two matrices A and B, and we will find their product AB. Call their product C ($AB = C$).
2. C is a matrix. The problem is to find the entries of C.
3. Each entry in C will depend on a row from matrix A and a column from matrix B.

We call the entry in row i and column j the (i, j) entry of a matrix. The (i, j) entry in C is a number obtained using all entries of row i in A and using all entries of column j in B. For example, the (2, 3) entry in C depends on row 2 of A and column 3 of B.

Dot Product

We use the two matrices

$$A = \begin{bmatrix} 1 & 2 \\ -1 & 3 \end{bmatrix} \quad \text{and} \quad B = \begin{bmatrix} -2 & 4 \\ 5 & -3 \end{bmatrix}$$

to illustrate matrix multiplication. We use the first row of A, $\mathrm{R1} = [1 \quad 2]$ and the first column of B

$$\mathrm{C1} = \begin{bmatrix} -2 \\ 5 \end{bmatrix}$$

to find the (1, 1) entry of the product. To do so, we need to find what is called the **dot product**, $\mathrm{R1} \cdot \mathrm{C1}$, of the row and column. It is

$$\mathrm{R1} \cdot \mathrm{C1} = [1 \quad 2] \cdot \begin{bmatrix} -2 \\ 5 \end{bmatrix} = 1(-2) + (2)(5) = -2 + 10 = 8.$$

Notice the following:

1. The dot product of a row and column is a single number.
2. The product is obtained by multiplying the first numbers from both the row and column, then the second numbers from both, etc., and then adding the results.

There are three other dot products possible using a row from A and a column from B. They are:

$$\mathrm{R1} \cdot \mathrm{C2} = [1 \quad 2] \cdot \begin{bmatrix} 4 \\ -3 \end{bmatrix} = 1(4) + 2(-3) = -2$$

$$\mathrm{R2} \cdot \mathrm{C1} = [-1 \quad 3] \cdot \begin{bmatrix} -2 \\ 5 \end{bmatrix} = (-1)(-2) + 3(5) = 2 + 15 = 17$$

$$\mathrm{R2} \cdot \mathrm{C2} = [-1 \quad 3] \cdot \begin{bmatrix} 4 \\ -3 \end{bmatrix} = (-1)(4) + 3(-3) = -13$$

The general form of the dot product of a row and column is

$$[a_1\ a_2 \cdots a_n] \cdot \begin{bmatrix} b_1 \\ b_2 \\ \vdots \\ b_n \end{bmatrix} = a_1 b_1 + a_2 b_2 + \cdots + a_n b_n$$

Note: The dot product is defined only when the row and column matrices have the same number of entries.

Example 1 (*Compare Exercise 7*)
Let the row matrix [.85 1.75 1.60] represent the prices of a loaf of bread, a six pack of soft drinks, and a package of granola bars, in that order. Let

$$\begin{bmatrix} 5 \\ 3 \\ 4 \end{bmatrix}$$

represent the quantity of bread (5), soft drinks (3), and granola bars (4) purchased in that order. Then,

$$[.85 \quad 1.75 \quad 1.60] \cdot \begin{bmatrix} 5 \\ 3 \\ 4 \end{bmatrix} = .85(5) + 1.75(3) + 1.60(4)$$

$$= 4.25 + 5.25 + 6.40 = 15.90$$

gives the total cost of the purchase.

Recall that we said that the entries in C depend on a row of A and a column of B. The entries are dot products; the (1, 2) entry in C is R1 · C2. For the given matrices A and B, the product AB is

$$\begin{bmatrix} [1 \quad 2] \cdot \begin{bmatrix} -2 \\ 5 \end{bmatrix} & [1 \quad 2] \cdot \begin{bmatrix} 4 \\ -3 \end{bmatrix} \\ [-1 \quad 3] \cdot \begin{bmatrix} -2 \\ 5 \end{bmatrix} & [-1 \quad 3] \cdot \begin{bmatrix} 4 \\ -3 \end{bmatrix} \end{bmatrix}$$

$$= \begin{bmatrix} 1(-2) + 2(5) & 1(4) + 2(-3) \\ (-1)(-2) + 3(5) & (-1)(4) + 3(-3) \end{bmatrix}$$

$$= \begin{bmatrix} 8 & -2 \\ 17 & -13 \end{bmatrix}$$

Example 2 (*Compare Exercise 11*)
Find the product AB of

$$A = \begin{bmatrix} 1 & 3 & 2 \\ -1 & 0 & 4 \end{bmatrix} \quad B = \begin{bmatrix} 7 & 5 \\ -2 & 6 \\ -3 & -4 \end{bmatrix}$$

Solution

$$AB = \begin{bmatrix} \text{R1} \cdot \text{C1} & \text{R1} \cdot \text{C2} \\ \text{R2} \cdot \text{C1} & \text{R2} \cdot \text{C2} \end{bmatrix}$$

$$= \begin{bmatrix} 1(7) + 3(-2) + 2(-3) & 1(5) + 3(6) + 2(-4) \\ (-1)(7) + 0(-2) + 4(-3) & -1(5) + 0(6) + 4(-4) \end{bmatrix}$$

$$= \begin{bmatrix} -5 & 15 \\ -19 & -21 \end{bmatrix}$$

Example 3 (*Compare Exercise 22*)
Multiply the matrices

$$A = \begin{bmatrix} 1 & 3 \\ 5 & 4 \end{bmatrix} \quad B = \begin{bmatrix} -1 & 6 \\ 2 & 7 \\ 0 & 8 \end{bmatrix}$$

Solution The product AB is not possible because a row-column dot product can occur only when the row of A and column of B have the same number of entries.

This example illustrates that two matrices may or may not have a product. There must be the same number of columns in the first matrix as the number of rows in the second in order for multiplication to be possible.

Matrix Multiplication

Multiplication of Matrices

Given matrices A and B, to find $AB = C$:

1. Check the number of columns of A and the number of rows of B. If they are equal, the product is possible. If they are not equal, no product is possible.
2. Form all possible dot products using a row from A and a column from B. The dot product of row i with column j gives the entry for the (i, j) position in C.

We now return to the problem at the beginning of the section.

Example 4 (*Compare Exercise 47*)
The time, in hours, required to assemble and finish a table and a chair is given by the matrix

	Chair	*Table*
Assemble	2	3
Finish	2.5	4.75

How long will it take to assemble and finish 950 chairs and 635 tables?

Solution Matrix multiplication gives the answer when we let

$$\begin{bmatrix} 950 \\ 635 \end{bmatrix}$$

be the column matrix that specifies the number of chairs and tables produced.
Multiply the matrices.

$$\begin{bmatrix} 2 & 3 \\ 2.5 & 4.75 \end{bmatrix} \begin{bmatrix} 950 \\ 635 \end{bmatrix} = \begin{bmatrix} 2(950) + 3(635) \\ 2.5(950) + 4.75(635) \end{bmatrix} = \begin{bmatrix} 3805 \\ 5391.25 \end{bmatrix}$$

The rows of the result correspond to the rows in the first matrix; the first row in each represents assembly time and the second represents finishing time. In the final matrix, 3805 is the *total* number of hours of assembly and 5391.25 is the *total* number of hours for finishing.

Example 5 Find AB and BA.

$$A = \begin{bmatrix} 1 & 3 \\ 5 & -2 \end{bmatrix} \quad B = \begin{bmatrix} 2 & 1 \\ 3 & -4 \end{bmatrix}$$

Solution

$$AB = \begin{bmatrix} 1 & 3 \\ 5 & -2 \end{bmatrix}\begin{bmatrix} 2 & 1 \\ 3 & -4 \end{bmatrix} = \begin{bmatrix} 11 & -11 \\ 4 & 13 \end{bmatrix}$$

$$BA = \begin{bmatrix} 2 & 1 \\ 3 & -4 \end{bmatrix}\begin{bmatrix} 1 & 3 \\ 5 & -2 \end{bmatrix} = \begin{bmatrix} 7 & 4 \\ -17 & 17 \end{bmatrix}$$

This example shows that AB and BA are not always equal. In fact, sometimes one of them may exist and the other not. The following example illustrates this.

Example 6 (*Compare Exercise 33*)
Find AB and BA, if possible.

$$A = \begin{bmatrix} 1 & 2 & 3 \\ -4 & 0 & -2 \\ 1 & 1 & 1 \end{bmatrix} \quad B = \begin{bmatrix} 5 & -2 \\ 1 & 4 \\ 2 & 3 \end{bmatrix}$$

Solution

$$AB = \begin{bmatrix} 1 & 2 & 3 \\ -4 & 0 & -2 \\ 1 & 1 & 1 \end{bmatrix}\begin{bmatrix} 5 & -2 \\ 1 & 4 \\ 2 & 3 \end{bmatrix} = \begin{bmatrix} 13 & 15 \\ -24 & 2 \\ 8 & 5 \end{bmatrix}$$

$$BA = \begin{bmatrix} 5 & -2 \\ 1 & 4 \\ 2 & 3 \end{bmatrix}\begin{bmatrix} 1 & 2 & 3 \\ -4 & 0 & -2 \\ 1 & 1 & 1 \end{bmatrix} = 5(1) + (-2)(-4) + ?(1)$$

Since we have no entry in row 1 to multiply by the bottom entry 1 of column 1, we cannot complete the computation for the product. BA does not exist.

2–5 Exercises

I.

Find the dot products in Exercises 1 through 6.

1. $[1 \quad 3] \cdot \begin{bmatrix} 2 \\ 4 \end{bmatrix}$

2. $[-2 \quad 5] \cdot \begin{bmatrix} 4 \\ 1 \end{bmatrix}$

3. $[6 \quad 5] \cdot \begin{bmatrix} 2 \\ 0 \end{bmatrix}$

4. $[3 \quad -1 \quad 2] \cdot \begin{bmatrix} -2 \\ 5 \\ 3 \end{bmatrix}$

5. $[1 \quad 0 \quad 1] \cdot \begin{bmatrix} 6 \\ 7 \\ 8 \end{bmatrix}$

6. $[2 \quad 1 \quad 3 \quad -2] \cdot \begin{bmatrix} 5 \\ 2 \\ -1 \\ 3 \end{bmatrix}$

7. (*See Example 1*) The price matrix of bread, milk, and cheese is, in that order, $[.90 \quad 1.85 \quad .65]$. The quantity of each purchased, in the same order, is given by the column matrix

$$\begin{bmatrix} 2 \\ 1 \\ 4 \end{bmatrix}$$

Find the total bill for the purchases.

8. Find the total bill for the purchase of hamburgers, fries, and drink where the price and quantities are listed in that order in the matrices.

$$\text{price matrix} = [1.65 \quad .80 \quad .59] \quad \text{quantity matrix} = \begin{bmatrix} 10 \\ 15 \\ 8 \end{bmatrix}$$

Find the products in Exercises 9 through 12.

9. $\begin{bmatrix} 3 & 1 \\ 2 & 4 \end{bmatrix}\begin{bmatrix} -2 & 3 \\ 1 & 2 \end{bmatrix}$

(*See Example 2*)

10. $\begin{bmatrix} -6 & 2 \\ 0 & 4 \end{bmatrix}\begin{bmatrix} 1 & -1 \\ 3 & 5 \end{bmatrix}$

11. $\begin{bmatrix} 2 & 1 & 4 \\ 3 & -1 & 5 \end{bmatrix}\begin{bmatrix} 3 & -2 \\ 0 & 2 \\ 6 & 1 \end{bmatrix}$

12. $\begin{bmatrix} 1 & 2 & 3 \\ 5 & -1 & 2 \end{bmatrix}\begin{bmatrix} 1 & 0 \\ 3 & 2 \\ 4 & 5 \end{bmatrix}$

Compute the products in Exercises 13 through 20, if they exist.

13. $\begin{bmatrix} 1 & 2 \\ -1 & 3 \end{bmatrix}\begin{bmatrix} 2 & 4 \\ 6 & 1 \end{bmatrix}$

14. $\begin{bmatrix} 2 & 1 \\ 3 & 6 \end{bmatrix}\begin{bmatrix} 0 & -1 \\ 2 & 4 \end{bmatrix}$

15. $\begin{bmatrix} 4 & 1 \\ 2 & 3 \end{bmatrix}\begin{bmatrix} 1 \\ 6 \end{bmatrix}$

16. $\begin{bmatrix} 1 & 2 & 3 \\ -1 & 4 & 6 \end{bmatrix}\begin{bmatrix} 0 & 1 \\ 2 & 3 \end{bmatrix}$

17. $\begin{bmatrix} 1 & 2 \\ 3 & -1 \\ 4 & 6 \end{bmatrix}\begin{bmatrix} 1 & 2 \\ 3 & 1 \end{bmatrix}$

18. $\begin{bmatrix} 1 & 2 & -1 \\ 0 & 3 & 1 \\ 2 & 4 & 6 \end{bmatrix}\begin{bmatrix} 1 & 2 \\ 2 & 2 \\ 3 & -1 \end{bmatrix}$

19. $\begin{bmatrix} -1 & 2 & 0 \\ 6 & 3 & 1 \\ 4 & 1 & 3 \end{bmatrix}\begin{bmatrix} 0 & 2 & 1 \\ -6 & 3 & -1 \\ 4 & 1 & 2 \end{bmatrix}$

20. $[1 \quad 2 \quad 3]\begin{bmatrix} 1 & 3 \\ 1 & 1 \\ 2 & 1 \end{bmatrix}$

Find the products in Exercises 21 through 28, if possible.

(*See Example 3*)

21. $\begin{bmatrix} 0 & 1 \\ 2 & 3 \end{bmatrix}\begin{bmatrix} 1 & -1 \\ 2 & 3 \end{bmatrix}$

22. $[1 \quad 1 \quad 2]\begin{bmatrix} 1 \\ 3 \end{bmatrix}$

23. $\begin{bmatrix} 1 & 2 & 3 \\ -1 & 4 & 6 \end{bmatrix}\begin{bmatrix} 0 & 1 \\ 2 & 3 \end{bmatrix}$

24. $\begin{bmatrix} 1 & 3 & 1 \\ 2 & 4 & 2 \end{bmatrix}\begin{bmatrix} 1 & 0 & 5 \\ 3 & -1 & 1 \\ 1 & 4 & 3 \end{bmatrix}$

25. $\begin{bmatrix} 1 & 2 \\ 3 & 1 \end{bmatrix}\begin{bmatrix} 4 & 1 \\ 5 & 2 \\ 6 & 3 \end{bmatrix}$

26. $\begin{bmatrix} 1 & 2 & 3 \end{bmatrix}\begin{bmatrix} 1 & 4 \\ 2 & 6 \end{bmatrix}$

27. $\begin{bmatrix} 1 & 2 & 3 \\ -1 & 4 & 6 \\ 2 & 1 & 3 \end{bmatrix}\begin{bmatrix} 1 \\ 2 \end{bmatrix}$

28. $\begin{bmatrix} 1 & 3 \\ 2 & 2 \end{bmatrix}\begin{bmatrix} 6 & 5 & -2 & -1 \\ 3 & 7 & 3 & 1 \end{bmatrix}$

Find AB and BA in Exercises 29 through 32.

29. $A = \begin{bmatrix} 1 & 2 \\ 0 & 1 \end{bmatrix}$ $B = \begin{bmatrix} 3 & 4 \\ -1 & 2 \end{bmatrix}$

30. $A = \begin{bmatrix} 1 & 2 & 3 \\ 4 & 0 & -2 \end{bmatrix}$ $B = \begin{bmatrix} 3 & 5 \\ 1 & -4 \\ 3 & 1 \end{bmatrix}$

31. $A = \begin{bmatrix} 1 & 3 \\ -1 & 2 \end{bmatrix}$ $B = \begin{bmatrix} 0 & 1 \\ -1 & 3 \end{bmatrix}$

32. $A = \begin{bmatrix} 1 & 3 & 4 \end{bmatrix}$ $B = \begin{bmatrix} 2 \\ 5 \\ 1 \end{bmatrix}$

In Exercises 33 through 36 find AB and BA, if possible.

(*See Example 6*)

33. $A = \begin{bmatrix} 1 & 2 \\ -3 & 1 \end{bmatrix}$ $B = \begin{bmatrix} 4 & 2 & 3 \\ 1 & 0 & 5 \end{bmatrix}$

34. $A = \begin{bmatrix} 1 & 2 & 3 \\ 4 & 1 & 2 \\ 3 & -1 & 0 \end{bmatrix}$ $B = \begin{bmatrix} 1 \\ 2 \\ 3 \end{bmatrix}$

35. $A = \begin{bmatrix} 1 & 3 & 5 \end{bmatrix}$ $B = \begin{bmatrix} -1 & 4 \\ 6 & 3 \\ 2 & 5 \end{bmatrix}$

36. $A = \begin{bmatrix} 2 & 4 & 6 \end{bmatrix}$ $B = \begin{bmatrix} 5 \\ 3 \end{bmatrix}$

II.

Perform the indicated matrix operations in Exercises 37 through 46.

37. $\begin{bmatrix} 1 & 2 & 3 \end{bmatrix}\begin{bmatrix} 1 & 2 \\ 0 & 1 \\ 4 & 3 \end{bmatrix}\begin{bmatrix} 1 & 3 \\ 2 & 4 \end{bmatrix}$

38. $\begin{bmatrix} 1 & 2 \\ 0 & 3 \end{bmatrix}\begin{bmatrix} 3 & 0 & -2 \\ 4 & 1 & 1 \end{bmatrix}\begin{bmatrix} 1 \\ 2 \\ 3 \end{bmatrix}$

39. $\begin{bmatrix} 0 & 1 & 2 & 3 \\ 4 & 1 & 2 & 5 \\ -1 & 2 & 3 & 1 \\ 1 & 2 & 9 & 7 \end{bmatrix}\begin{bmatrix} 1 & 2 & 3 & 2 & 6 \\ 2 & 4 & 1 & -9 & 2 \\ 3 & 6 & 0 & 7 & 3 \\ -1 & 7 & \frac{1}{2} & 8 & 4 \end{bmatrix}$

40. $\begin{bmatrix} 1 & 2 \\ 3 & 1 \end{bmatrix}\begin{bmatrix} 3 & 4 \\ -1 & 5 \end{bmatrix} + \begin{bmatrix} 0 & 2 \\ 1 & 1 \end{bmatrix}$

41. $\begin{bmatrix} -1 & 3 & 1 \\ 2 & 1 & 3 \end{bmatrix}\begin{bmatrix} 1 & 3 \\ 1 & 1 \\ 2 & 1 \end{bmatrix} + \begin{bmatrix} 1 & 2 \\ 3 & 1 \end{bmatrix}\begin{bmatrix} -1 & 2 \\ 4 & 3 \end{bmatrix}$

42. $\begin{bmatrix} 1 & 2 & 3 \end{bmatrix}\begin{bmatrix} 1 & 1 \\ 2 & 1 \\ -1 & 2 \end{bmatrix} - \begin{bmatrix} 1 & 2 \end{bmatrix}\begin{bmatrix} 1 & 4 \\ 3 & 1 \end{bmatrix}$

43. $\begin{bmatrix} 1 & 2 \\ 3 & 1 \end{bmatrix}\begin{bmatrix} 1 & 0 \\ 0 & 1 \end{bmatrix}$

44. $\begin{bmatrix} 1 & 0 \\ 0 & 1 \end{bmatrix}\begin{bmatrix} 3 & -1 \\ 2 & 6 \end{bmatrix}$

45. $\begin{bmatrix} 1 & 2 & 3 \\ 4 & -1 & 0 \\ 2 & 1 & 7 \end{bmatrix}\begin{bmatrix} 1 & 0 & 0 \\ 0 & 1 & 0 \\ 0 & 0 & 1 \end{bmatrix}$

46. $\begin{bmatrix} 1 & 2 & 3 & 4 \\ 0 & -1 & 4 & 2 \\ 3 & 1 & 2 & 0 \\ 1 & 2 & 3 & 1 \end{bmatrix}\begin{bmatrix} 1 & 0 & 0 & 0 \\ 0 & 1 & 0 & 0 \\ 0 & 0 & 1 & 0 \\ 0 & 0 & 0 & 1 \end{bmatrix}$

47. (*See Example 4*)
An electronics firm makes stereos and TV sets. The matrix below shows the time required for assembly and checking.

	Stereo	*TV*
Assembly	4	5.5
Check	1	2

Use matrices to determine the total assembly time and total checking time for 300 stereos and 450 TV sets.

48. A plant has two production lines I and II. Both produce ten-speed and three-speed bicycles. The number produced per hour is given by the matrix.

	Line I	*Line II*
Three-Speed	10	15
Ten-Speed	12	20

Find the number of each type bicycle that is produced if Line I operates 60 hours and Line II 48 hours.

49. Data from three supermarkets are summarized in this matrix.

	Store 1	*Store 2*	*Store 3*
Sugar (per lb)	64¢	62¢	68¢
Peaches (per can)	63¢	68¢	71¢
Chicken (per lb)	59¢	58¢	57¢
Bread (per loaf)	80¢	82¢	78¢

What is the total grocery bill at each store if the following purchase is made at each store: five pounds of sugar, three cans of peaches, three pounds of chicken, and two loaves of bread.

50. The following matrix gives the vitamin content of a typical breakfast in conveniently chosen units:

Vitamin

	A	B_1	B_2	C
Orange Juice	500	.2	0	129
Oatmeal	0	.2	0	0
Milk	1,560	.32	1.7	6
Biscuit	0	0	0	0
Butter	460	0	0	0

If you have 1 unit of orange juice, 1 unit of oatmeal, $\frac{1}{4}$ unit of milk, 2 units of biscuit, and 2 units of butter, find the matrix that tells how much of each type vitamin you have consumed.

51. Use the vitamin content from Exercise 50. Two breakfast menus are summarized in the matrix

Menus

	I	*II*
Orange Juice	.5	0
Oatmeal	1.5	1.0
Milk	.5	1.0
Biscuit	1.0	3.0
Butter	1.0	2.0

Find the matrix that tells the amount of each vitamin consumed in each diet.

2–6 The Inverse of a Matrix

Identity Matrix

You are familiar with the number fact

$$1a = a1 = a$$

where a is any real number. We call 1 the **identity** for multiplication.

We have no similar property for multiplication of matrices; there is no matrix I such that $AI = IA = A$ for all matrices A. However, there is such a matrix for certain sets of matrices. For example, if

$$A = \begin{bmatrix} 4 & 3 \\ 7 & 2 \end{bmatrix} \quad \text{and} \quad I = \begin{bmatrix} 1 & 0 \\ 0 & 1 \end{bmatrix}, \quad \text{then}$$

$$AI = \begin{bmatrix} 4 & 3 \\ 7 & 2 \end{bmatrix}\begin{bmatrix} 1 & 0 \\ 0 & 1 \end{bmatrix} = \begin{bmatrix} 4 & 3 \\ 7 & 2 \end{bmatrix} = A \quad \text{and}$$

$$IA = \begin{bmatrix} 1 & 0 \\ 0 & 1 \end{bmatrix}\begin{bmatrix} 4 & 3 \\ 7 & 2 \end{bmatrix} = \begin{bmatrix} 4 & 3 \\ 7 & 2 \end{bmatrix} = A$$

Furthermore, for any 2×2 matrix, A, the matrix I has the property that $AI = A$ and $IA = A$. This can be justified by using a 2×2 matrix with arbitrary entries

$$A = \begin{bmatrix} a & b \\ c & d \end{bmatrix}$$

$$\text{Now } AI = \begin{bmatrix} a & b \\ c & d \end{bmatrix}\begin{bmatrix} 1 & 0 \\ 0 & 1 \end{bmatrix}\begin{bmatrix} a \times 1 + b \times 0 & a \times 0 + b \times 1 \\ c \times 1 + d \times 0 & c \times 0 + d \times 1 \end{bmatrix}$$

$$= \begin{bmatrix} a & b \\ c & d \end{bmatrix} = A.$$

You should now multiply IA to verify that it is indeed A. Thus,

$$\begin{bmatrix} 1 & 0 \\ 0 & 1 \end{bmatrix}$$

is the identity for all 2×2 matrices. If we try to multiply the 3×3 matrix

$$A = \begin{bmatrix} 1 & 2 & 3 \\ 5 & 7 & 12 \\ 8 & 4 & -2 \end{bmatrix} \quad \text{by} \quad \begin{bmatrix} 1 & 0 \\ 0 & 1 \end{bmatrix}$$

we find we are unable to multiply at all because A has 3 columns and I has only two rows. So,

$$\begin{bmatrix} 1 & 0 \\ 0 & 1 \end{bmatrix}$$

is not the identity for 3×3 matrices. However, the matrix

$$I_3 = \begin{bmatrix} 1 & 0 & 0 \\ 0 & 1 & 0 \\ 0 & 0 & 1 \end{bmatrix}$$

Example 1 is an identity matrix for the set of all 3×3 matrices.

$$\begin{bmatrix} a & b & c \\ d & e & f \\ g & h & i \end{bmatrix}\begin{bmatrix} 1 & 0 & 0 \\ 0 & 1 & 0 \\ 0 & 0 & 1 \end{bmatrix} = \begin{bmatrix} a & b & c \\ d & e & f \\ g & h & i \end{bmatrix}$$

$$\begin{bmatrix} 1 & 0 & 0 \\ 0 & 1 & 0 \\ 0 & 0 & 1 \end{bmatrix}\begin{bmatrix} a & b & c \\ d & e & f \\ g & h & i \end{bmatrix} = \begin{bmatrix} a & b & c \\ d & e & f \\ g & h & i \end{bmatrix}$$

In general, if we let I_n be the $n \times n$ matrix with 1's on the main diagonal and 0's elsewhere, it is the identity matrix for the class of all $n \times n$ matrices. (The main diagonal runs from the upper left to lower right corner.)

Inverse of a Square Matrix

We can extend another number fact to matrices. The simple multiplication facts

$$2 \times \frac{1}{2} = 1$$

$$\frac{3}{4} \times \frac{4}{3} = 1$$

$$1.25 \times .8 = 1$$

have a common property. Each of the numbers 2, $\frac{3}{4}$, and 1.25 can be multiplied by another number to obtain 1. In general, for any real number a, except zero, there is a number b such that $a \times b = 1$. We call b the *inverse* of a. The standard notation for the inverse of a is a^{-1}.

Example 2 (*Compare Exercise 1*)

$$3^{-1} = \frac{1}{3};\ 2^{-1} = .5;\ \left(\frac{5}{8}\right)^{-1} = \frac{8}{5};\ .4^{-1} = 2.5;\ 625^{-1} = .0016.$$

In terms of matrix multiplication, there is a similar property. For example,

$$\begin{bmatrix} 1 & 1 \\ 1 & 2 \end{bmatrix}\begin{bmatrix} 2 & -1 \\ -1 & 1 \end{bmatrix} = \begin{bmatrix} 1 & 0 \\ 0 & 1 \end{bmatrix}$$

We can restate this equation as $AA^{-1} = I$ where

$$A = \begin{bmatrix} 1 & 1 \\ 1 & 2 \end{bmatrix} \quad \text{and} \quad A^{-1} = \begin{bmatrix} 2 & -1 \\ -1 & 1 \end{bmatrix}$$

We say that A^{-1} is the inverse of the matrix A.

In general, a matrix A has an inverse if there is a matrix A^{-1} that fulfills the conditions that $AA^{-1} = A^{-1}A = I$. Not all matrices have inverses. In fact, a matrix must be square in order to have an inverse, and some square matrices have no inverse. We now come to the problem of deciding if a square matrix has an inverse. If it does, how do we find it? Let's approach this problem with a simple 2×2 example.

Example 3 (*Compare Exercise 2*)
If we have the square matrix

$$A = \begin{bmatrix} 2 & 1 \\ 3 & 2 \end{bmatrix}$$

find, if possible, its inverse.

Solution We want to find a 2×2 matrix A^{-1} such that $AA^{-1} = I$. Since we don't know the entries in A^{-1}, let's enter variables a, b, c, and d and attempt to find their values.

$$\text{Write } A^{-1} = \begin{bmatrix} a & c \\ b & d \end{bmatrix}$$

The condition $AA^{-1} = I$ can now be written

$$\begin{bmatrix} 2 & 1 \\ 3 & 2 \end{bmatrix}\begin{bmatrix} a & c \\ b & d \end{bmatrix} = \begin{bmatrix} 1 & 0 \\ 0 & 1 \end{bmatrix}$$

We want to find values of a, b, c, and d so that the product on the left equals the identity matrix on the right. First, form the product AA^{-1}. We get

$$\overset{AA^{-1}}{\begin{bmatrix} (2a+b) & (2c+d) \\ (3a+2b) & (3c+2d) \end{bmatrix}} = \overset{I}{\begin{bmatrix} 1 & 0 \\ 0 & 1 \end{bmatrix}}$$

Recall that two matrices are equal only when they have equal entries in corresponding positions. So, the matrix equality reduces to the equations

$$\begin{aligned} 2a + b &= 1 \\ 3a + 2b &= 0 \end{aligned} \quad \textit{and} \quad \begin{aligned} 2c + d &= 0 \\ 3c + 2d &= 1 \end{aligned}$$

Notice that we have one system of two equations with variables a and b.

1. $2a + b = 1$ with augmented matrix $\left[\begin{array}{cc|c} 2 & 1 & 1 \\ 3 & 2 & 0 \end{array}\right]$
$3a + 2b = 0$
and a system with variables c and d:

2. $2c + d = 0$ with augmented matrix $\left[\begin{array}{cc|c} 2 & 1 & 0 \\ 3 & 2 & 1 \end{array}\right]$
$3c + 2d = 1$.

The solution to system 1 gives $a = 2$, $b = -3$. The solution to system 2 gives $c = -1$, $d = 2$ so the inverse of

$$A = \begin{bmatrix} 2 & 1 \\ 3 & 2 \end{bmatrix} \quad \text{is} \quad A^{-1} = \begin{bmatrix} 2 & -1 \\ -3 & 2 \end{bmatrix}$$

We check our results by computing AA^{-1} and $A^{-1}A$.

$$\begin{bmatrix} 2 & 1 \\ 3 & 2 \end{bmatrix}\begin{bmatrix} 2 & -1 \\ -3 & 2 \end{bmatrix} = \begin{bmatrix} 1 & 0 \\ 0 & 1 \end{bmatrix}$$

$$\begin{bmatrix} 2 & -1 \\ -3 & 2 \end{bmatrix}\begin{bmatrix} 2 & 1 \\ 3 & 2 \end{bmatrix} = \begin{bmatrix} 1 & 0 \\ 0 & 1 \end{bmatrix}$$

It checks.

Look at the two systems we just solved. The two systems have precisely the same coefficients; they differ only in the constant terms. The left-hand portions of the augmented matrices are exactly the same. In fact, they are the matrix A. Combine the two augmented matrices into one using the common coefficient portion on the left, and list both columns from the right sides. This gives the matrix

$$\left[\begin{array}{cc|cc} 2 & 1 & 1 & 0 \\ 3 & 2 & 0 & 1 \end{array}\right]$$

Now proceed in the same way you do to solve a system of equations with an augmented matrix, that is, use row operations to reduce the left hand portion to the identity. This gives the following sequence.

$$\left[\begin{array}{cc|cc} 2 & 1 & 1 & 0 \\ 3 & 2 & 0 & 1 \end{array}\right] \quad \left(\frac{1}{2}\right)\text{R1} \to \text{R1}$$

$$\left[\begin{array}{cc|cc} 1 & \frac{1}{2} & \frac{1}{2} & 0 \\ 3 & 2 & 0 & 1 \end{array}\right] \quad -3\text{R1} + \text{R2} \to \text{R2}$$

$$\left[\begin{array}{cc|cc} 1 & \frac{1}{2} & \frac{1}{2} & 0 \\ 0 & \frac{1}{2} & -\frac{3}{2} & 1 \end{array}\right] \quad -\text{R2} + \text{R1} \to \text{R1}$$

$$\left[\begin{array}{cc|cc} 1 & 0 & 2 & -1 \\ 0 & \frac{1}{2} & -\frac{3}{2} & 1 \end{array}\right] \quad 2\text{R2} \to \text{R2}$$

$$\left[\begin{array}{cc|cc} 1 & 0 & 2 & -1 \\ 0 & 1 & -3 & 2 \end{array}\right]$$

The final matrix has the identity matrix formed by the first two columns and A^{-1} formed by the last two columns. This is no accident; one may find the inverse of a square matrix in this manner.

Method to find the inverse of a square matrix

1. To find the inverse of a matrix A, form an augmented matrix $[A \mid I]$ by writing down the matrix A and then writing the identity matrix to the right of A.
2. Perform a sequence of row operations that reduces this matrix to reduced echelon form.
3. If the "A portion" of the reduced echelon form is the identity matrix, then the matrix found in the "I portion" is A^{-1}.
4. If the reduced echelon form produces a row in the A portion that is all zeros, then A has no inverse.

Example 4 (*Compare Exercise 8*)
Determine the inverse of the matrix

$$A = \begin{bmatrix} 1 & 2 & 0 \\ 2 & 1 & -1 \\ 3 & 1 & 1 \end{bmatrix}$$

Solution Adjoin I to A and we get

$$[A \mid I] = \left[\begin{array}{ccc|ccc} 1 & 2 & 0 & 1 & 0 & 0 \\ 2 & 1 & -1 & 0 & 1 & 0 \\ 3 & 1 & 1 & 0 & 0 & 1 \end{array}\right]$$

Now get zeros in column 1.

$$\left[\begin{array}{ccc|ccc} 1 & 2 & 0 & 1 & 0 & 0 \\ 0 & -3 & -1 & -2 & 1 & 0 \\ 0 & -5 & 1 & -3 & 0 & 1 \end{array}\right]$$

Next divide row 2 by -3 to get a leading 1.

$$\left[\begin{array}{ccc|ccc} 1 & 2 & 0 & 1 & 0 & 0 \\ 0 & 1 & \frac{1}{3} & \frac{2}{3} & -\frac{1}{3} & 0 \\ 0 & -5 & 1 & -3 & 0 & 1 \end{array}\right]$$

Now zero the rest of column 2.

$$\left[\begin{array}{ccc|ccc} 1 & 0 & -\frac{2}{3} & -\frac{1}{3} & \frac{2}{3} & 0 \\ 0 & 1 & \frac{1}{3} & \frac{2}{3} & -\frac{1}{3} & 0 \\ 0 & 0 & \frac{8}{3} & \frac{1}{3} & -\frac{5}{3} & 1 \end{array}\right]$$

Multiply row 3 by $\frac{3}{8}$ to get a leading 1.

$$\left[\begin{array}{ccc|ccc} 1 & 0 & -\frac{2}{3} & -\frac{1}{3} & \frac{2}{3} & 0 \\ 0 & 1 & \frac{1}{3} & \frac{2}{3} & -\frac{1}{3} & 0 \\ 0 & 0 & 1 & \frac{1}{8} & -\frac{5}{8} & \frac{3}{8} \end{array}\right]$$

Zero the rest of column 3.

$$\left[\begin{array}{ccc|ccc} 1 & 0 & 0 & -\frac{1}{4} & \frac{1}{4} & \frac{1}{4} \\ 0 & 1 & 0 & \frac{5}{8} & -\frac{1}{8} & -\frac{1}{8} \\ 0 & 0 & 1 & \frac{1}{8} & -\frac{5}{8} & \frac{3}{8} \end{array}\right]$$

The left portion reduced to the identity matrix, so the right portion contains A^{-1}.

$$A^{-1} = \begin{bmatrix} -\frac{1}{4} & \frac{1}{4} & \frac{1}{4} \\ \frac{5}{8} & -\frac{1}{8} & -\frac{1}{8} \\ \frac{1}{8} & -\frac{5}{8} & \frac{3}{8} \end{bmatrix}$$

Example 5 (*Compare Exercise 12*)
Find the inverse of

$$A = \begin{bmatrix} 1 & 3 \\ 3 & 9 \end{bmatrix}$$

Solution Adjoin I to A to obtain

$$\left[\begin{array}{cc|cc} 1 & 3 & 1 & 0 \\ 3 & 9 & 0 & 1 \end{array}\right]$$

Now reduce to echelon form using row operations.

$$\left[\begin{array}{cc|cc} 1 & 3 & 1 & 0 \\ 3 & 9 & 0 & 1 \end{array}\right] \quad -3R1 + R2 \to R2$$

$$\left[\begin{array}{cc|cc} 1 & 3 & 1 & 0 \\ 0 & 0 & -3 & 1 \end{array}\right]$$

Since the A portion of the augmented matrix has a row of 0's, A has no inverse.

Matrix Equations

We can write systems of equations using matrices and solve some systems using matrix inverses.

The matrix equation

$$\begin{bmatrix} 5 & 3 & -4 & 12 \\ 8 & -21 & 7 & -19 \\ 2 & 1 & -15 & 1 \end{bmatrix} \begin{bmatrix} x_1 \\ x_2 \\ x_3 \\ x_4 \end{bmatrix} = \begin{bmatrix} 7 \\ 16 \\ -22 \end{bmatrix}$$

becomes the following when the multiplication on the left is performed.

$$\begin{bmatrix} 5x_1 + 3x_2 - 4x_3 + 12x_4 \\ 8x_1 - 21x_2 + 7x_3 - 19x_4 \\ 2x_1 + x_2 - 15x_3 + x_4 \end{bmatrix} = \begin{bmatrix} 7 \\ 16 \\ -22 \end{bmatrix}$$

These matrices are equal only when corresponding components are equal; that is,

$$\begin{aligned} 5x_1 + 3x_2 - 4x_3 + 12x_4 &= 7 \\ 8x_1 - 21x_2 + 7x_3 - 19x_4 &= 16 \\ 2x_1 + x_2 - 15x_3 + x_4 &= -22 \end{aligned}$$

In general, we can write a system of equations in the compact matrix form

$$AX = B$$

where A is a matrix formed from the coefficients of the variables, X is a column matrix formed by listing the variables, and B is the column matrix formed from the constants in the system.

Example 6 (*Compare Exercise 21*)
For this system of equations

$$\begin{aligned} 2x_1 + 3x_2 - x_3 &= 4 \\ x_1 - x_2 + x_3 &= 1 \\ 3x_1 + x_2 - x_3 &= 2 \end{aligned}$$

The matrix of coefficients of this system is

$$\begin{bmatrix} 2 & 3 & -1 \\ 1 & -1 & 1 \\ 3 & 1 & -1 \end{bmatrix}$$

The augmented matrix is

$$\left[\begin{array}{ccc|c} 2 & 3 & -1 & 4 \\ 1 & -1 & 1 & 1 \\ 3 & 1 & -1 & 2 \end{array}\right]$$

and the system can be written in $AX = B$ form as

$$\begin{bmatrix} 2 & 3 & -1 \\ 1 & -1 & 1 \\ 3 & 1 & -1 \end{bmatrix} \begin{bmatrix} x_1 \\ x_2 \\ x_3 \end{bmatrix} = \begin{bmatrix} 4 \\ 1 \\ 2 \end{bmatrix}$$

Using A^{-1} to Solve a System

Now we can illustrate the use of the inverse in solving a system of equations when the matrix of coefficients has an inverse. There are times when it helps to be able to solve a system using the inverse matrix. One such situation occurs when a number of systems need to be solved, but all have the same coefficients. The constant terms change, but the coefficients don't. Here is a simple example.

A doctor treats patients who need adequate calcium and iron in their diet. The doctor has found that two foods, A and B, provide these. Each unit of food A has 0.5 mg iron and 25 mg calcium. Each unit of food B has 0.3 mg iron and 7 mg calcium. Let x = number of units of food A eaten by the patient; let y = number of units of food B eaten by the patient. Then $.5x + .3y$ gives the total mg of iron consumed by the patient and $25x + 7y$ gives the total mg of calcium. Suppose the doctor wants patient Jones to get 60 mg calcium and 6 mg iron. The amount of each food to be consumed is the solution to

$$\begin{aligned} .5x + .3y &= 6 \\ 25x + 7y &= 60 \end{aligned}$$

If patient Smith requires 80 mg calcium and 7 mg iron, the food required is found in the solution of the system

$$\begin{aligned} .5x + .3y &= 7 \\ 25x + 7y &= 80 \end{aligned}$$

These two systems have the same coefficients; they differ only in the constant terms.

The inverse of the coefficient matrix

$$A = \begin{bmatrix} .5 & .3 \\ 25 & 7 \end{bmatrix}$$

may be used to avoid going through the Gauss-Jordan elimination process with each patient.

Here's how A^{-1} may be used to solve a system. Let $AX = B$ be a system for which A actually has an inverse. When both sides of $AX = B$ are multiplied by A^{-1}, the equation reduces to

$$\begin{aligned} A^{-1}AX &= A^{-1}B \\ IX &= A^{-1}B \\ X &= A^{-1}B \end{aligned}$$

The product $A^{-1}B$ gives the solution. The solution to such a system exists, and it is unique.

Example 7 Solve the following system of equations using an inverse matrix:

$$\begin{aligned} x_1 + 2x_2 \quad &= 4 \\ 2x_1 + x_2 - x_3 &= 2 \\ 3x_1 + x_2 + x_3 &= -2 \end{aligned}$$

Solution This system can be written in matrix form as

$$\begin{bmatrix} 1 & 2 & 0 \\ 2 & 1 & -1 \\ 3 & 1 & 1 \end{bmatrix}\begin{bmatrix} x_1 \\ x_2 \\ x_3 \end{bmatrix} = \begin{bmatrix} 4 \\ 2 \\ -2 \end{bmatrix}$$

Hence,

$$\begin{bmatrix} x_1 \\ x_2 \\ x_3 \end{bmatrix} = \begin{bmatrix} 1 & 2 & 0 \\ 2 & 1 & -1 \\ 3 & 1 & 1 \end{bmatrix}^{-1}\begin{bmatrix} 4 \\ 2 \\ -2 \end{bmatrix}$$

The inverse was found in Example 4 so we use it to get

$$\begin{bmatrix} x_1 \\ x_2 \\ x_3 \end{bmatrix} = \begin{bmatrix} -\frac{1}{4} & \frac{1}{4} & \frac{1}{4} \\ \frac{5}{8} & -\frac{1}{8} & -\frac{1}{8} \\ \frac{1}{8} & -\frac{5}{8} & \frac{3}{8} \end{bmatrix}\begin{bmatrix} 4 \\ 2 \\ -2 \end{bmatrix} = \begin{bmatrix} -1 \\ \frac{5}{2} \\ -\frac{3}{2} \end{bmatrix}$$

Thus, the unique solution is

$$x_1 = -1,\ x_2 = \frac{5}{2},\ x_3 = -\frac{3}{2}$$

Example 8 (*Compare Exercise 34*)
Solve the systems

$$AX = B$$

where

$$A = \begin{bmatrix} 1 & 2 \\ 4 & 3 \end{bmatrix} \quad X = \begin{bmatrix} x \\ y \end{bmatrix}$$

Using

$$B = \begin{bmatrix} 6 \\ 3 \end{bmatrix}, \begin{bmatrix} 10 \\ 15 \end{bmatrix}, \text{ and } \begin{bmatrix} 2 \\ 11 \end{bmatrix}$$

Solution First find A^{-1}. Adjoin the identity matrix to A.

$$\left[\begin{array}{cc|cc} 1 & 2 & 1 & 0 \\ 4 & 3 & 0 & 1 \end{array}\right]$$

This reduces to

$$\left[\begin{array}{cc|cc} 1 & 0 & -\frac{3}{5} & \frac{2}{5} \\ 0 & 1 & \frac{4}{5} & -\frac{1}{5} \end{array}\right]$$

so the inverse of A is

$$\begin{bmatrix} -\frac{3}{5} & \frac{2}{5} \\ \frac{4}{5} & -\frac{1}{5} \end{bmatrix}$$

For $B = \begin{bmatrix} 6 \\ 3 \end{bmatrix}$ the solution is

$$= \begin{bmatrix} x \\ y \end{bmatrix} = \begin{bmatrix} -\frac{3}{5} & \frac{2}{5} \\ \frac{4}{5} & -\frac{1}{5} \end{bmatrix} \begin{bmatrix} 6 \\ 3 \end{bmatrix} = \begin{bmatrix} -\frac{12}{5} \\ \frac{21}{5} \end{bmatrix}$$

so $x = -\frac{12}{5}$, $y = \frac{21}{5}$ is the solution.

For $B = \begin{bmatrix} 10 \\ 15 \end{bmatrix}$

$$\begin{bmatrix} x \\ y \end{bmatrix} = \begin{bmatrix} -\frac{3}{5} & \frac{2}{5} \\ \frac{4}{5} & -\frac{1}{5} \end{bmatrix} \begin{bmatrix} 10 \\ 15 \end{bmatrix} = \begin{bmatrix} 0 \\ 5 \end{bmatrix}$$

For $B = \begin{bmatrix} 2 \\ 11 \end{bmatrix}$

$$\begin{bmatrix} x \\ y \end{bmatrix} = \begin{bmatrix} -\frac{3}{5} & \frac{2}{5} \\ \frac{4}{5} & -\frac{1}{5} \end{bmatrix} \begin{bmatrix} 2 \\ 11 \end{bmatrix} = \begin{bmatrix} \frac{16}{5} \\ -\frac{3}{5} \end{bmatrix}$$

Example 9 (*Compare Exercise 40*)
Let's return to the earlier example where a doctor prescribed foods containing calcium and iron.

Let x = the number of units of food A
y = the number of units of food B

where A contains .5 mg iron and 25 mg calcium and B contains .3 mg iron and 7 mg calcium per unit.

(a) Find the amount of each food for patient Jones who needs 1.3 mg iron and 49 mg calcium.

(b) Find the amount of each food for patient Smith who needs 2.6 mg iron and 106 mg calcium.

Solution **(a)** We need the solution to

$$\begin{aligned} .5x + .3y &= 1.3 \\ 25x + 7y &= 49 \end{aligned}$$

In matrix form this is

$$\begin{bmatrix} .5 & .3 \\ 25 & 7 \end{bmatrix}\begin{bmatrix} x \\ y \end{bmatrix} = \begin{bmatrix} 1.3 \\ 49 \end{bmatrix}$$

The inverse of

$$\begin{bmatrix} .5 & .3 \\ 25 & 7 \end{bmatrix} \text{ is } \begin{bmatrix} -1.75 & .075 \\ 6.25 & -.125 \end{bmatrix}$$

The solution to the system is

$$\begin{bmatrix} x \\ y \end{bmatrix} = \begin{bmatrix} -1.75 & .075 \\ 6.25 & -.125 \end{bmatrix}\begin{bmatrix} 1.3 \\ 49 \end{bmatrix} = \begin{bmatrix} -1.75(1.3) + .075(49) \\ 6.25(1.3) - .125(49) \end{bmatrix}$$

$$= \begin{bmatrix} 1.4 \\ 2.0 \end{bmatrix}$$

so 1.4 units of food A and 2.0 units of food B are required.

(b) In this case the solution is

$$\begin{bmatrix} x \\ y \end{bmatrix} = \begin{bmatrix} -1.75 & .075 \\ 6.25 & -.125 \end{bmatrix}\begin{bmatrix} 2.6 \\ 106.0 \end{bmatrix} = \begin{bmatrix} 3.4 \\ 3.0 \end{bmatrix}$$

Coding Theory

The following example illustrates how matrices having inverses are used in coding theory. Governments use sophisticated methods of coding and decoding messages. One type of code used that is extremely difficult to break makes use of a large matrix to encode a message. The receiver of the message decodes it using the inverse of the matrix. We illustrate the method with a 3×3 matrix.

Let the message be

PREPARE TO ATTACK

and the matrix be

$$\begin{bmatrix} -3 & -3 & -4 \\ 0 & 1 & 1 \\ 4 & 3 & 4 \end{bmatrix}$$

We assign a number to each letter of the alphabet. For convenience, let us associate each letter with its position in the alphabet. A is associated with 1, B with 2, etc. Let a space between words be denoted by the number 27. Thus, the message becomes

P	R	E	P	A	R	E	*	T	O	*	A	T	T	A	C	K
16	18	5	16	1	18	5	27	20	15	27	1	20	20	1	3	11

Since we are going to use a 3×3 matrix to encode, we break the enumerated message up into a sequence of 3×1 matrices as follows:

$$\begin{bmatrix} 16 \\ 18 \\ 5 \end{bmatrix} \begin{bmatrix} 16 \\ 1 \\ 18 \end{bmatrix} \begin{bmatrix} 5 \\ 27 \\ 20 \end{bmatrix} \begin{bmatrix} 15 \\ 27 \\ 1 \end{bmatrix} \begin{bmatrix} 20 \\ 20 \\ 1 \end{bmatrix} \begin{bmatrix} 3 \\ 11 \\ 27 \end{bmatrix}$$

Observe that we needed to add a space at the end of the message in order to complete the last matrix. We now put the message into code by multiplying each of the above column matrices by the encoding matrix:

$$\begin{bmatrix} -3 & -3 & -4 \\ 0 & 1 & 1 \\ 4 & 3 & 4 \end{bmatrix} \begin{bmatrix} 16 \\ 18 \\ 5 \end{bmatrix} = \begin{bmatrix} -122 \\ 23 \\ 138 \end{bmatrix} \begin{bmatrix} -3 & -3 & -4 \\ 0 & 1 & 1 \\ 4 & 3 & 4 \end{bmatrix} \begin{bmatrix} 16 \\ 1 \\ 18 \end{bmatrix} = \begin{bmatrix} -123 \\ 19 \\ 139 \end{bmatrix}$$

and so on. The column matrices obtained are

$$\begin{bmatrix} -122 \\ 23 \\ 138 \end{bmatrix} \begin{bmatrix} -123 \\ 19 \\ 139 \end{bmatrix} \begin{bmatrix} -176 \\ 47 \\ 181 \end{bmatrix} \begin{bmatrix} -130 \\ 28 \\ 145 \end{bmatrix} \begin{bmatrix} -124 \\ 21 \\ 144 \end{bmatrix} \begin{bmatrix} -150 \\ 38 \\ 153 \end{bmatrix}$$

The coded message is transmitted in the following linear form:

$$-122, 23, 138, -123, 19, 139, -176, 47, 181, -130,$$
$$28, 145, -124, 21, 144, -150, 38, 153.$$

Notice that the letter A is coded as 19, 145, and 144 in this message. For someone who does not know how the message was coded, this increases the difficulty of deciphering the code.

To decode the message, the receiver writes this string as a sequence of column matrices and repeats the technique using the inverse of the original matrix. The inverse in this case is

$$\begin{bmatrix} 1 & 0 & 1 \\ 4 & 4 & 3 \\ -4 & -3 & -3 \end{bmatrix}$$

Thus, for example, using the first three numbers in the coded message,

$$\begin{bmatrix} 1 & 0 & 1 \\ 4 & 4 & 3 \\ -4 & -3 & -3 \end{bmatrix} \begin{bmatrix} -122 \\ 23 \\ 138 \end{bmatrix} = \begin{bmatrix} 16 \\ 18 \\ 5 \end{bmatrix}$$

leading to 16, 18, 5, and PRE, the first three letters in the original message.

2–6 Exercises

I.

1. (*See Example 2*)

Find 25^{-1}, $\left(\frac{2}{3}\right)^{-1}$, $(-5)^{-1}$, $.75^{-1}$, 11^{-1}.

Find the inverse of the matrices in Exercises 2 through 11.

(*See Example 3*)

2. $\begin{bmatrix} 3 & 1 \\ 4 & 1 \end{bmatrix}$ **3.** $\begin{bmatrix} 1 & 2 \\ 3 & 5 \end{bmatrix}$ **4.** $\begin{bmatrix} 9 & 11 \\ 1 & 5 \end{bmatrix}$ **5.** $\begin{bmatrix} 3 & 2 \\ 4 & 3 \end{bmatrix}$

(*See Example 4*)

6. $\begin{bmatrix} 3 & 5 \\ 2 & 4 \end{bmatrix}$ **7.** $\begin{bmatrix} 3 & 3 \\ 6 & 5 \end{bmatrix}$ **8.** $\begin{bmatrix} 1 & 2 & 1 \\ 2 & -1 & 3 \\ 2 & 2 & 1 \end{bmatrix}$ **9.** $\begin{bmatrix} 1 & 2 & 3 \\ 0 & 1 & 2 \\ 4 & 5 & 3 \end{bmatrix}$

10. $\begin{bmatrix} 2 & 0 & 4 \\ -1 & 3 & 1 \\ 0 & 1 & 2 \end{bmatrix}$ **11.** $\begin{bmatrix} 0 & 3 & 3 \\ 1 & 2 & 3 \\ 1 & 4 & 6 \end{bmatrix}$

Find the inverse, if possible, of the following matrices.

(*See Example 5*)

12. $\begin{bmatrix} 4 & -2 \\ -2 & 1 \end{bmatrix}$ **13.** $\begin{bmatrix} 1 & 0 & 1 \\ 1 & -1 & 2 \\ 3 & -1 & 4 \end{bmatrix}$

Determine the inverses (if they exist) of the matrices in Exercises 14 through 20 using the Gauss-Jordan elimination method.

14. $\begin{bmatrix} 1 & 2 & -1 \\ 3 & -1 & 0 \\ 2 & -3 & 1 \end{bmatrix}$ **15.** $\begin{bmatrix} 1 & 2 & 1 \\ 1 & -3 & 2 \\ 2 & -1 & 3 \end{bmatrix}$ **16.** $\begin{bmatrix} 1 & 0 \\ 2 & 1 \end{bmatrix}$ **17.** $\begin{bmatrix} 2 & 1 \\ 4 & 3 \end{bmatrix}$

18. $\begin{bmatrix} 0 & 2 \\ -\frac{1}{3} & \frac{1}{3} \end{bmatrix}$ **19.** $\begin{bmatrix} 1 & 2 & 3 \\ 2 & -1 & 4 \\ 0 & -1 & 1 \end{bmatrix}$ **20.** $\begin{bmatrix} 1 & 2 & -1 \\ 2 & 4 & -3 \\ 1 & -2 & 0 \end{bmatrix}$

For each of the systems of equations in Exercises 21 through 24, write

(a) the augmented matrix
(b) the coefficient matrix
(c) the system in the form $AX = B$

21. (*See Example 6*)

$$\begin{aligned} 3x_1 + 4x_2 - 5x_3 &= 4 \\ 2x_1 - x_2 + 3x_3 &= -1 \\ x_1 + x_2 - x_3 &= 2 \end{aligned}$$

22.

$$\begin{aligned} x_1 - 4x_2 + 3x_3 &= 6 \\ x_1 + x_3 &= 2 \\ 2x_1 + 5x_2 - 6x_3 &= 1 \end{aligned}$$

23.

$$\begin{aligned} 4x + 5y &= 2 \\ 3x - 2y &= 7 \end{aligned}$$

24.

$$\begin{aligned} 7x_1 + 9x_2 - 5x_3 + x_4 &= 14 \\ 3x_1 + 5x_2 + 6x_3 - 8x_4 &= 23 \\ -2x_1 + x_2 + 17x_4 &= 12 \end{aligned}$$

II.

Express each of the systems in Exercises 25 through 27 as a single matrix equation, $AX = B$.

25.

$$\begin{aligned} x_1 + 3x_2 &= 5 \\ 2x_1 - x_2 &= 6 \end{aligned}$$

26.

$$\begin{aligned} 2x_1 - 3x_2 + x_3 &= 4 \\ 4x_1 - x_2 + 2x_3 &= -1 \\ x_1 + x_2 - x_3 &= 2 \end{aligned}$$

27.

$$\begin{aligned} x_1 + 2x_2 - 3x_3 + 4x_4 &= 0 \\ x_1 + x_2 + x_4 &= 5 \\ 3x_1 + 2x_2 + x_3 + 2x_4 &= 4 \end{aligned}$$

Find the inverse of the matrices in Exercises 28 through 29.

28.

$$\begin{bmatrix} -3 & -1 & 1 & -2 \\ -1 & 3 & 2 & 1 \\ 1 & 2 & 3 & -1 \\ -2 & 1 & -1 & -3 \end{bmatrix}$$

29.

$$\begin{bmatrix} 1 & 1 & 0 & 0 \\ 0 & 1 & 1 & 0 \\ 1 & 0 & 0 & 1 \\ 0 & 0 & 1 & 1 \end{bmatrix}$$

Solve the systems of equations in Exercises 30 through 33 by determining the inverse of the matrix of coefficients and then using matrix multiplication.

30.

$$\begin{aligned} x_1 + 3x_2 &= 5 \\ 2x_1 + x_2 &= 10 \end{aligned}$$

31.

$$\begin{aligned} x_1 + 2x_2 - x_3 &= 2 \\ x_1 + x_2 + 2x_3 &= 0 \\ x_1 - x_2 - x_3 &= 1 \end{aligned}$$

32.

$$\begin{aligned} x_1 - x_2 &= 1 \\ x_1 + x_2 + 2x_3 &= 2 \\ x_1 + 2x_2 + x_3 &= 0 \end{aligned}$$

33.

$$\begin{aligned} x_1 + x_2 + 2x_3 + x_4 &= 5 \\ 2x_1 + 2x_3 + x_4 &= 6 \\ x_2 + 3x_3 - x_4 &= 1 \\ 3x_1 + 2x_2 + 2x_4 &= 7 \end{aligned}$$

Using the matrix inverse method, solve the system of equations in Exercises 34 through 39 for each of the B matrices.

34. (*See Example 8*)

$$\begin{aligned} x + 2y - z &= b_1 \\ x + y + 2z &= b_2 \\ x - y - z &= b_3 \end{aligned} \qquad \begin{bmatrix} b_1 \\ b_2 \\ b_3 \end{bmatrix} = \begin{bmatrix} 1 \\ 2 \\ 3 \end{bmatrix}, \begin{bmatrix} 0 \\ 1 \\ 4 \end{bmatrix}, \begin{bmatrix} 5 \\ 2 \\ 3 \end{bmatrix}$$

35.

$$\begin{aligned} x_1 + 2x_2 &= b_1 \\ 3x_1 + 5x_2 &= b_2 \end{aligned} \qquad \begin{bmatrix} b_1 \\ b_2 \end{bmatrix} = \begin{bmatrix} 3 \\ 8 \end{bmatrix}, \begin{bmatrix} 4 \\ 9 \end{bmatrix}, \begin{bmatrix} 3 \\ 7 \end{bmatrix}$$

36.

$$\begin{aligned} x_1 + x_2 &= b_1 \\ 2x_1 + 3x_2 &= b_2 \end{aligned} \qquad \begin{bmatrix} b_1 \\ b_2 \end{bmatrix} = \begin{bmatrix} 0 \\ 1 \end{bmatrix}, \begin{bmatrix} 5 \\ 13 \end{bmatrix}, \begin{bmatrix} 1 \\ 2 \end{bmatrix}$$

37. $\begin{aligned} x_1 - 2x_2 + 3x_3 &= b_1 \\ x_1 - x_2 + 2x_3 &= b_2 \\ 2x_1 - 3x_2 + 6x_3 &= b_3 \end{aligned}$ $\quad \begin{bmatrix} b_1 \\ b_2 \\ b_3 \end{bmatrix} = \begin{bmatrix} 6 \\ 5 \\ 14 \end{bmatrix}, \begin{bmatrix} -5 \\ -3 \\ -8 \end{bmatrix}, \begin{bmatrix} 4 \\ 3 \\ 9 \end{bmatrix}$

38. $\begin{aligned} x_1 + 2x_2 - x_3 &= b_1 \\ -x_1 - x_2 + x_3 &= b_2 \\ 3x_1 + 7x_2 - x_3 &= b_3 \end{aligned}$ $\quad \begin{bmatrix} b_1 \\ b_2 \\ b_3 \end{bmatrix} = \begin{bmatrix} -1 \\ 1 \\ -1 \end{bmatrix}, \begin{bmatrix} 6 \\ -4 \\ 18 \end{bmatrix}, \begin{bmatrix} 0 \\ -2 \\ -4 \end{bmatrix}$

39. $\begin{aligned} x_1 + x_2 + 5x_3 &= b_1 \\ x_1 + 2x_2 + 8x_3 &= b_2 \\ 2x_1 + 4x_2 + 15x_3 &= b_3 \end{aligned}$ $\quad \begin{bmatrix} b_1 \\ b_2 \\ b_3 \end{bmatrix} = \begin{bmatrix} 2 \\ 5 \\ 10 \end{bmatrix}, \begin{bmatrix} 3 \\ 2 \\ 4 \end{bmatrix}$

40. (*See Example 9*) A doctor advises her patients to eat two foods for vitamins A and C. The contents per unit of food are given as follows:

	Food	
	A	*B*
Vitamin C (mg)	32	24
Vitamin A (iu)	900	425

$$\begin{bmatrix} 32 & 24 \\ 900 & 425 \end{bmatrix} = M$$

$$\text{It turns out that } M^{-1} = \begin{bmatrix} -.053125 & .003 \\ .1125 & -.004 \end{bmatrix}$$

Let x = number of units of food A
y = number of units of food B
b_1 = desired intake of vitamin C
b_2 = desired intake of vitamin A

(a) Show that the $MX = B$ describes the relationship between units of food consumed and desired intake of vitamins.

(b) If a patient eats 3.2 units of food A and 2.5 units of food B, what is the vitamin C and vitamin A intake?

(c) If a patient eats 1.5 units of food A and 3.0 units of B, what is the vitamin A and vitamin C intake?

(d) The doctor wants a patient to intake 107.2 mg of vitamin C and 2315 iu of vitamin A. How many units of each food should be eaten?

(e) The doctor wants a patient to intake 104 mg vitamin C and 2575 iu vitamin A. How many units of each food should be eaten?

41. Encode the message RETREAT using the matrix

$$\begin{bmatrix} 4 & -3 \\ 3 & -2 \end{bmatrix}$$

42. Encode the message ATTACK AT DAWN using the matrix

$$\begin{bmatrix} 1 & 2 & 1 \\ 2 & 3 & 1 \\ -2 & 0 & 1 \end{bmatrix}$$

43. Decode the message 49, 38, -5, -3, -61, -39, which was encoded using the matrix of Exercise 41.

44. Decode the message 71, 100, -1, 28, 43, -5, 84, 122, -11, 77, 112, -13, 61, 89, -8, 71, 104, -13, which was encoded using the matrix of Exercises 42.

2–7 Leontief Input-Output Model in Economics

Input-Output Matrix
Output Matrix, Consumer Demand Matrix, Internal Demand Matrix

The Leontief input-output model is used to analyze the interdependence of industries in an economic situation. Wassily Leontief received the Nobel Prize in Economics in 1973 for his work in this area. The practical applications have proliferated; it has become a standard tool for investigating economic structures ranging from cities and corporations to states and countries. In practice, a large number of variables are required to describe an economic situation. Thus, the problems are quite complicated, so they can best be handled with computers using matrix techniques. We will consider one of the simplest forms of the problem.

Suppose we imagine a simple economy with just two industries: electricity and steel. These industries exist to produce electricity and steel for the consumers. However, both production processes themselves use electricity and steel. The electricity industry uses steel in the generating equipment and uses electricity to light the plant and to heat and cool the buildings. The steel industry uses electricity to run some of their equipment, and that equipment in turn contains steel components.

An **input-output matrix** describes the interdependence of the industries. It is convenient to express the quantity of each product in terms of its dollar value. We are interested in the quantities of each product that are used internally by the processes themselves and the quantities needed by the consumers (demand).

Example 1 *(Compare Exercise 1)*
Suppose the electricity industry uses \$0.15 worth of its own electricity and \$0.05 worth of steel for each \$1 worth of electricity produced. Furthermore, suppose the steel industry uses \$0.40 worth of electricity and \$0.10 worth of its own steel

for each \$1 of steel produced. Summarize this information with the input-output matrix

	Electricity	*Steel*	
Electricity	.15	.40	$= A$
Steel	.05	.10	

$$\begin{array}{l} \textit{Electricity} \\ \textit{Steel} \end{array} \begin{bmatrix} .15 & .40 \\ .05 & .10 \end{bmatrix} = A$$

There is one row for each industry. The row labeled electricity gives the quantity of electricity (\$0.15) needed to produce \$1 worth of electricity and the quantity of electricity (\$0.40) needed to produce \$1 worth of steel. The second row shows the quantity of steel (\$0.05) needed to produce \$1 worth of electricity and quantity of steel (\$0.10) needed to produce \$1 worth of steel.

You may be wondering why this information is presented in matrix form. It makes it easier to answer questions such as

1. "The production capacity of industry is \$9 million worth of electricity and \$7 million worth of steel. How much of each is consumed internally by the production processes?"
2. "The consumers want \$6 million worth of electricity and \$8 million worth of steel for their use. How much of each should be produced in order to satisfy their demands and to provide for that consumed internally?"

Before we learn how to answer these two questions, let's make some observations that will help set up the problems.

Let x = the dollar value of electricity produced and let y = the dollar value of steel produced. These include that used internally and that available to the consumers. Then the total amounts consumed internally are

$$\text{Electricity consumed internally} = .15x + .40y$$
$$\text{Steel consumed internally} = .05x + .10y$$

Notice that this can be expressed in matrix form as

$$\begin{bmatrix} \text{Electricity consumed internally} \\ \text{Steel consumed internally} \end{bmatrix} = \begin{bmatrix} .15 & .40 \\ .05 & .10 \end{bmatrix}\begin{bmatrix} x \\ y \end{bmatrix} = A\begin{bmatrix} x \\ y \end{bmatrix}$$

So if production capacities are \$9 million worth of electricity ($x = 9$) and \$7 million worth of steel ($y = 7$), the amount consumed internally is

$$\begin{bmatrix} .15 & .40 \\ .05 & .10 \end{bmatrix}\begin{bmatrix} 9 \\ 7 \end{bmatrix} = \begin{bmatrix} 4.15 \\ 1.15 \end{bmatrix}$$

\$4.15 million and \$1.15 million worth of electricity and steel.

Another fact relates the amount of electricity and steel produced to that available to the consumer:

$$[\text{Amount Produced}] = \begin{bmatrix} \text{Amount consumed} \\ \text{internally} \end{bmatrix} + \begin{bmatrix} \text{Amount available} \\ \text{to consumer} \end{bmatrix}$$

We call these matrices Output, Internal Demand, and Consumer Demand matrices, respectively.

In the case where \$9 million and \$7 million worth of electricity were produced with \$4.15 and \$1.15 million consumed internally,

$$\underset{\textbf{Output}}{\begin{bmatrix} 9 \\ 7 \end{bmatrix}} = \underset{\textbf{Internal Demand}}{\begin{bmatrix} 4.15 \\ 1.15 \end{bmatrix}} + \underset{\textbf{Consumer Demand}}{\begin{bmatrix} \text{Electricity available to consumer} \\ \text{Steel available to consumer} \end{bmatrix}}$$

we get \$4.85 million and \$5.85 million worth of electricity and steel available to the consumers.

If we call the output matrix X, the consumer demand matrix D, and the input-output matrix A, then the internal demand matrix is AX and

$$X = AX + D$$

expresses the relation between output, internal demand, and consumer demand.

The question, "What total output is necessary to supply consumers with \$6 million worth of electricity and \$8 million worth of steel?" asks the output when consumer demand is given. Using the same input-output matrix, we want to find x and y (output) so that

$$\underset{X}{\begin{bmatrix} x \\ y \end{bmatrix}} = \underset{A}{\begin{bmatrix} .15 & .40 \\ .05 & .10 \end{bmatrix}} \underset{X}{\begin{bmatrix} x \\ y \end{bmatrix}} + \underset{D}{\begin{bmatrix} 6 \\ 8 \end{bmatrix}}$$

Notice that the variables x and y appear in two matrices.

Let's use the general form $X = AX + D$ and apply some matrix algebra to find the matrix X.

You need to solve for the matrix X in

$$X = AX + D$$

This is equivalent to solving for X in the following:

$$X - AX = D$$
$$IX - AX = D$$
$$(I - A)X = D$$
$$X = (I - A)^{-1}D$$

The last equation is the most helpful. To find the total production X that meets the final demand D and also provides the quantities needed to carry out the internal production processes, find the inverse of the matrix $I - A$ and multiply it by the matrix D. Using

$$A = \begin{bmatrix} .15 & .40 \\ .05 & .10 \end{bmatrix}$$

$$I - A = \begin{bmatrix} .85 & -.40 \\ -.05 & .90 \end{bmatrix}$$

and

$$(I - A)^{-1} = \begin{bmatrix} 1.208 & .537 \\ .0671 & 1.141 \end{bmatrix}$$

where the entries in $(I - A)^{-1}$ are rounded. For the demand matrix

$$D = \begin{bmatrix} 6 \\ 8 \end{bmatrix}$$

$$X = \begin{bmatrix} 1.208 & .537 \\ .0671 & 1.141 \end{bmatrix}\begin{bmatrix} 6 \\ 8 \end{bmatrix} = \begin{bmatrix} 11.544 \\ 9.531 \end{bmatrix}$$

so \$11.544 million worth of electricity and \$9.531 million worth of steel must be produced in order to provide \$6 million worth of electricity and \$8 million worth of steel to the consumers and to provide for the electricity and steel used internally in production.

Example 2 (*Compare Exercise 8*)
An input-output matrix for electricity and steel is

$$A = \begin{bmatrix} .25 & .20 \\ .50 & .20 \end{bmatrix}$$

(a) If the production capacity of electricity is \$15 million and the production capacity for steel is \$20 million, how much of each is consumed internally for capacity production?

(b) How much electricity and steel must be produced in order to have \$5 million worth of electricity and \$8 million worth of steel available for consumer use?

Solution **(a)** We are given

$$A = \begin{bmatrix} .25 & .20 \\ .50 & .20 \end{bmatrix} \quad \text{and} \quad X = \begin{bmatrix} 15 \\ 20 \end{bmatrix}$$

We want to find AX.

$$AX = \begin{bmatrix} .25 & .20 \\ .50 & .20 \end{bmatrix}\begin{bmatrix} 15 \\ 20 \end{bmatrix} = \begin{bmatrix} 7.75 \\ 11.50 \end{bmatrix}$$

so \$7.75 million worth of electricity and \$11.50 million worth of steel are consumed internally.

(b) We are given

$$A = \begin{bmatrix} .25 & .20 \\ .50 & .20 \end{bmatrix} \quad D = \begin{bmatrix} 5 \\ 8 \end{bmatrix}$$

and we need to solve for X in $(I - A)^{-1}D = X$.

$$I - A = \begin{bmatrix} 1 & 0 \\ 0 & 1 \end{bmatrix} - \begin{bmatrix} .25 & .20 \\ .50 & .20 \end{bmatrix} = \begin{bmatrix} .75 & -.20 \\ -.50 & .80 \end{bmatrix}$$

The Gauss-Jordan method gives

$$(I - A)^{-1} = \begin{bmatrix} 1.60 & .40 \\ 1.00 & 1.50 \end{bmatrix} \quad \text{(Check it)}$$

Then the output is

$$(I - A)^{-1}D = \begin{bmatrix} 1.60 & .40 \\ 1.00 & 1.50 \end{bmatrix}\begin{bmatrix} 5 \\ 8 \end{bmatrix} = \begin{bmatrix} 11.2 \\ 17.0 \end{bmatrix}$$

The two industries must produce \$11.2 million and \$17.0 million in order to have \$5 million and \$8 million worth of their products available to the consumer.

Example 3 (*Compare Exercise 9*)
An economy consists of three industries having the following input-output matrix A. Compute the output levels required of the industries to meet the demands of the other industries and of the consumers for each of the three given demand levels.

$$A = \begin{bmatrix} \frac{1}{5} & \frac{1}{5} & \frac{3}{10} \\ \frac{1}{2} & \frac{1}{2} & 0 \\ 0 & 0 & \frac{1}{5} \end{bmatrix} \quad D = \begin{bmatrix} 9 \\ 12 \\ 16 \end{bmatrix}, \begin{bmatrix} 6 \\ 9 \\ 8 \end{bmatrix}, \begin{bmatrix} 12 \\ 18 \\ 32 \end{bmatrix}$$

Solution The units of D are millions of dollars. We wish to compute the X's that correspond to the various D's. We need to find X as follows for each D.

$$X = (I - A)^{-1}D$$

For our matrix A,

$$I - A = \begin{bmatrix} 1 & 0 & 0 \\ 0 & 1 & 0 \\ 0 & 0 & 1 \end{bmatrix} - \begin{bmatrix} \frac{1}{5} & \frac{1}{5} & \frac{3}{10} \\ \frac{1}{2} & \frac{1}{2} & 0 \\ 0 & 0 & \frac{1}{5} \end{bmatrix} = \begin{bmatrix} \frac{4}{5} & -\frac{1}{5} & -\frac{3}{10} \\ -\frac{1}{2} & \frac{1}{2} & 0 \\ 0 & 0 & \frac{4}{5} \end{bmatrix}$$

$(I - A)^{-1}$ is computed using Gauss-Jordan elimination:

$$(I - A)^{-1} = \begin{bmatrix} \frac{5}{3} & \frac{2}{3} & \frac{5}{8} \\ \frac{5}{3} & \frac{8}{3} & \frac{5}{8} \\ 0 & 0 & \frac{5}{4} \end{bmatrix}$$

We can efficiently compute $X = (I - A)^{-1}D$ for each of the three values of D by forming a matrix having the various values of D as columns:

$$X = \underbrace{\begin{bmatrix} \frac{5}{3} & \frac{2}{3} & \frac{5}{8} \\ \frac{5}{3} & \frac{8}{3} & \frac{5}{8} \\ 0 & 0 & \frac{5}{4} \end{bmatrix}}_{(I-A)^{-1}} \underbrace{\begin{bmatrix} 9 & 6 & 12 \\ 12 & 9 & 18 \\ 16 & 8 & 32 \end{bmatrix}}_{\text{various values of } D} = \underbrace{\begin{bmatrix} 33 & 21 & 52 \\ 57 & 39 & 88 \\ 20 & 10 & 40 \end{bmatrix}}_{\text{corresponding outputs}}$$

The output levels of the three industries to meet the demands

$$\begin{bmatrix} 9 \\ 12 \\ 16 \end{bmatrix}, \quad \begin{bmatrix} 6 \\ 9 \\ 8 \end{bmatrix}, \quad \text{and} \quad \begin{bmatrix} 12 \\ 18 \\ 32 \end{bmatrix}$$

are

$$\begin{bmatrix} 33 \\ 57 \\ 20 \end{bmatrix}, \quad \begin{bmatrix} 21 \\ 39 \\ 10 \end{bmatrix}, \quad \text{and} \quad \begin{bmatrix} 52 \\ 88 \\ 40 \end{bmatrix}$$

respectively, the units being millions of dollars.

Today the concept of a world economy has become a tangible reality. In 1973, the United Nations commissioned an input-output model of the world economy. This model was developed with special financial support from the Netherlands. The aim of the model was to transform the vast collection of economic facts that describe the world economy into an organized system from which economic projections could and have been made.

In the model, the world is divided into 15 distinct geographic regions, each one described by an individual input-output matrix. The regions are then linked by a larger matrix which is used in an input-output model. Overall, more than 200 variables enter into the model, and naturally the computations are done on a computer. By feeding in projected values for certain variables, researchers use the model in a variety of ways to create scenarios of future world economic possibilities.

2–7 Exercises

1. (*See Example 1*) The following input-output matrix defines the interdependency of five industries. Each entry gives the dollar value of the row industry's output required to produce one dollar's worth of output of the column industry.

	1	2	3	4	5
1. Auto	0.03	0.15	0.05	0.05	0.10
2. Steel	0.40	0.20	0.10	0.10	0.10
3. Electricity	0.10	0.25	0.20	0.10	0.20
4. Coal	0.10	0.20	0.30	0.15	0.10
5. Chemical	0.05	0.10	0.05	0.02	0.05

Determine

(a) the amount of electricity consumed in producing \$1 worth of steel.

(b) the amount of steel consumed in producing \$1 worth in the auto industry.

(c) the industry that requires the largest amount of coal per \$1 output.

(d) the industry that requires the largest amount of electricity per \$1 output.

(e) on which industry the auto industry is most dependent.

Exercises 2 through 5 give the input-output matrix and the output of some industries. Determine the amount consumed internally by the production processes.

(*See Example 2*)

2. $A = \begin{bmatrix} .15 & .08 \\ .30 & .20 \end{bmatrix} \quad X = \begin{bmatrix} 8 \\ 12 \end{bmatrix}$

3. $A = \begin{bmatrix} .10 & .20 \\ .25 & .15 \end{bmatrix} \quad X = \begin{bmatrix} 20 \\ 15 \end{bmatrix}$

4. $A = \begin{bmatrix} .06 & .12 & .09 \\ .15 & .05 & .10 \\ .08 & .04 & .02 \end{bmatrix} \quad X = \begin{bmatrix} 8 \\ 14 \\ 10 \end{bmatrix}$

5. $A = \begin{bmatrix} .03 & 0 & .02 & .06 \\ .08 & .02 & 0 & .05 \\ .07 & .10 & .01 & .04 \\ .05 & .04 & .02 & .06 \end{bmatrix} \quad X = \begin{bmatrix} 10 \\ 30 \\ 20 \\ 40 \end{bmatrix}$

Compute $(I - A)^{-1}$ for the matrices in Exercises 6 and 7.

6. $A = \begin{bmatrix} .2 & .3 \\ .2 & .3 \end{bmatrix}$

7. $A = \begin{bmatrix} .32 & .16 \\ .22 & .36 \end{bmatrix}$

8. (*See Example 2*) Find the output required to meet the consumer demand and internal demand for the following input-output matrix and consumer demand matrix.

$$A = \begin{bmatrix} .24 & .08 \\ .12 & .04 \end{bmatrix} \quad D = \begin{bmatrix} 15 \\ 12 \end{bmatrix}$$

The economies in Exercises 9 through 13 are either two or three industries. Determine the output levels required of each industry to meet the demands of the other industries and of the consumer. The units are millions of dollars.

9. (*See Example 3*)

$$A = \begin{bmatrix} 0.20 & 0.60 \\ 0.40 & 0.10 \end{bmatrix} \quad D = \begin{bmatrix} 24 \\ 12 \end{bmatrix}, \begin{bmatrix} 8 \\ 6 \end{bmatrix}, \text{ and } \begin{bmatrix} 0 \\ 12 \end{bmatrix}$$

10. $$A = \begin{bmatrix} 0.10 & 0.40 \\ 0.30 & 0.20 \end{bmatrix} \quad D = \begin{bmatrix} 6 \\ 12 \end{bmatrix}, \begin{bmatrix} 18 \\ 6 \end{bmatrix}, \text{ and } \begin{bmatrix} 24 \\ 12 \end{bmatrix}$$

11. $$A = \begin{bmatrix} 0.30 & 0.60 \\ 0.35 & 0.10 \end{bmatrix} \quad D = \begin{bmatrix} 42 \\ 84 \end{bmatrix}, \begin{bmatrix} 0 \\ 10 \end{bmatrix}, \begin{bmatrix} 14 \\ 7 \end{bmatrix}, \text{ and } \begin{bmatrix} 42 \\ 42 \end{bmatrix}$$

12. $$A = \begin{bmatrix} 0.20 & 0.20 & 0.10 \\ 0 & 0.40 & 0.20 \\ 0 & 0.20 & 0.60 \end{bmatrix} \quad D = \begin{bmatrix} 4 \\ 8 \\ 8 \end{bmatrix}, \begin{bmatrix} 0 \\ 8 \\ 16 \end{bmatrix}, \text{ and } \begin{bmatrix} 8 \\ 24 \\ 8 \end{bmatrix}$$

13. $$A = \begin{bmatrix} 0.20 & 0.20 & 0 \\ 0.40 & 0.40 & 0.60 \\ 0.40 & 0.10 & 0.40 \end{bmatrix} \quad D = \begin{bmatrix} 36 \\ 72 \\ 36 \end{bmatrix}, \begin{bmatrix} 36 \\ 0 \\ 18 \end{bmatrix}, \begin{bmatrix} 3 \\ 0 \\ 0 \end{bmatrix}, \text{ and } \begin{bmatrix} 0 \\ 18 \\ 18 \end{bmatrix}$$

The economies in Exercises 14 through 16 are either two or three industries. The output level of each industry is given. Determine the amounts consumed internally and the amounts available for the consumer from each industry.

14. $$A = \begin{bmatrix} 0.20 & 0.40 \\ 0.50 & 0.10 \end{bmatrix} \quad X = \begin{bmatrix} 8 \\ 10 \end{bmatrix}$$

15. $$A = \begin{bmatrix} 0.10 & 0.20 & 0.30 \\ 0 & 0.10 & 0.40 \\ 0.50 & 0.40 & 0.20 \end{bmatrix} \quad X = \begin{bmatrix} 10 \\ 10 \\ 20 \end{bmatrix}$$

16. $$A = \begin{bmatrix} 0.10 & 0.10 & 0.20 \\ 0.20 & 0.10 & 0.30 \\ 0.40 & 0.30 & 0.15 \end{bmatrix} \quad X = \begin{bmatrix} 6 \\ 4 \\ 5 \end{bmatrix}$$

IMPORTANT TERMS

2–1	**System of equations**	**Solution of a system**	**Elimination method**
	Inconsistent system	**Many solutions to a system**	**Parametric form of a solution**
	Dependent system	**Supply and demand**	**Equilibrium price**
2–2	**Matrix**	**Row matrix**	**Column matrix**
	Coefficient matrix	**Augmented matrix**	**Row operations**
	Equivalent augmented matrices	**Gauss-Jordan Method**	**Diagonal locations**
2–3	**Reduced echelon form**	**No solution**	**Unique solution**
	Many solutions		
2–4	**Square matrix**	**Equal matrices**	**Addition of matrices**
	Scalar multiplication		
2–5	**Dot product**	**Matrix multiplication**	

2–6	**Identity matrix**	**Inverse matrix**	**Matrix equations**
2–7	**Input-output model**	**Input-output matrix**	**Output matrix**
	Internal demand matrix	**Consumer demand matrix**	

REVIEW EXERCISES

Solve the systems in Exercises 1 through 2 by substitution.

1. $$\begin{aligned} 3x - 2y &= 5 \\ 2x + 4y &= 9 \end{aligned}$$

2. $$\begin{aligned} x + 5y &= 2 \\ 3x - 7y &= 12 \end{aligned}$$

Solve the systems in Exercises 3 through 6 by elimination.

3. $$\begin{aligned} 5x - y &= 34 \\ 2x + 3y &= 0 \end{aligned}$$

4. $$\begin{aligned} x + 3y - 2z &= -15 \\ 4x - 3y + 5z &= 50 \\ 3x + 2y - 2z &= -4 \end{aligned}$$

5. $$\begin{aligned} x - 2y + 3z &= 3 \\ 4x + 7y - 6z &= 6 \\ -2x + 4y + 12z &= 0 \end{aligned}$$

6. $$\begin{aligned} 2x - 3y + z &= -10 \\ 3x - 2y + 4z &= -5 \\ x + y + 3z &= 5 \end{aligned}$$

Solve the systems in Exercises 7 through 9 by the Gauss-Jordan method

7. $$\begin{aligned} 2x_1 - 4x_2 - 14x_3 &= 50 \\ x_1 - x_2 - 5x_3 &= 17 \\ 2x_1 - 4x_2 - 17x_3 &= 65 \end{aligned}$$

8. $$\begin{aligned} 3x_1 + 2x_2 &= 3 \\ 6x_1 - 6x_2 &= 1 \end{aligned}$$

9. Find the value of x that makes the matrices equal.

$$\begin{bmatrix} 4 & 3 \\ 3x + 2 & 6 \end{bmatrix} = \begin{bmatrix} 4 & 3 \\ 5 - x & 6 \end{bmatrix}$$

Perform the indicated matrix operations in Exercises 10 through 17, when possible.

10. $$-3\begin{bmatrix} 1 & 4 \\ -2 & 7 \end{bmatrix}$$

11. $$-1\begin{bmatrix} 3 & 2 \\ -6 & -7 \end{bmatrix}$$

12. $$\begin{bmatrix} 1 & 5 \\ -2 & 6 \end{bmatrix} + \begin{bmatrix} 3 & 1 \\ 0 & -4 \end{bmatrix}$$

13. $$\begin{bmatrix} 3 & 2 \\ 6 & -4 \\ 1 & 1 \end{bmatrix} + \begin{bmatrix} 8 & -5 \\ 1 & 3 \\ 2 & -1 \end{bmatrix}$$

14. $$\begin{bmatrix} 2 & 1 & 5 \\ 3 & 0 & 2 \end{bmatrix} + \begin{bmatrix} 1 & 1 \\ -2 & 4 \\ 3 & 1 \end{bmatrix}$$

15. $$\begin{bmatrix} 3 & 1 & -2 \end{bmatrix}\begin{bmatrix} 4 \\ 1 \\ 5 \end{bmatrix}$$

16. $$\begin{bmatrix} 1 & 0 & 2 \\ 3 & 1 & 1 \end{bmatrix}\begin{bmatrix} 6 & 4 & -2 \\ 3 & 5 & -3 \\ -1 & 0 & 1 \end{bmatrix}$$

17. $$\begin{bmatrix} 5 & 9 & 1 \\ 6 & -2 & 4 \end{bmatrix}\begin{bmatrix} 3 & 5 \\ -7 & 2 \end{bmatrix}$$

Find the inverse, when possible, of the matrices in Exercises 18 through 22.

18. $$\begin{bmatrix} 5 & -7 \\ -3 & 4 \end{bmatrix}$$

19. $$\begin{bmatrix} 8 & 6 \\ 7 & 5 \end{bmatrix}$$

20. $$\begin{bmatrix} 5 & -2 \\ -10 & 4 \end{bmatrix}$$

21. $$\begin{bmatrix} 1 & 0 & 3 \\ 2 & -5 & 4 \\ 1 & -2 & 2 \end{bmatrix}$$

22. $$\begin{bmatrix} 1 & 1 & 2 \\ 0 & 1 & -4 \\ 3 & 2 & 10 \end{bmatrix}$$

23. Write the augmented matrix of the system

$$\begin{aligned} 6x_1 + 4x_2 - 5x_3 &= 10 \\ 3x_1 - 2x_2 \quad\quad &= 12 \\ x_1 + x_2 - 4x_3 &= -2 \end{aligned}$$

Find the reduced echelon form of the matrices in Exercises 24 through 26.

24. $\begin{bmatrix} 1 & 3 & 2 & 1 \\ 2 & 4 & -2 & 6 \\ 3 & 1 & 4 & -3 \end{bmatrix}$

25. $\begin{bmatrix} 2 & 4 & 6 & -2 \\ 3 & 1 & 0 & 5 \\ -2 & 1 & 3 & -11 \end{bmatrix}$

26. $\begin{bmatrix} 3 & -1 & 2 \\ 1 & 4 & -1 \\ 4 & 3 & 1 \\ 1 & -9 & 4 \end{bmatrix}$

27. A basketball player scored 59 points in a game with a total of 36 field goals and free throws. How many of each did he make? (A field goal is 2 points, and a free throw is 1 point.)

28. Determine the equilibrium solutions of the following. The demand equation is given first and the supply equation second.

(a) $y = -4x + 241$
$y = 3x - 158$

(b) $y = -7x + 1544$
$y = 5x - 832$

29. An investor wants to earn \$5000 per year by investing \$50,000. She can earn 7% from bonds and 12% from stocks. How much should she invest in each?

3

Topics in this chapter can be applied to:

Optimum Allocation of Resources ● Minimizing Costs ● Maximizing Revenues ● Purchasing Decisions ● Manufacturing Strategies ● Transportation Management

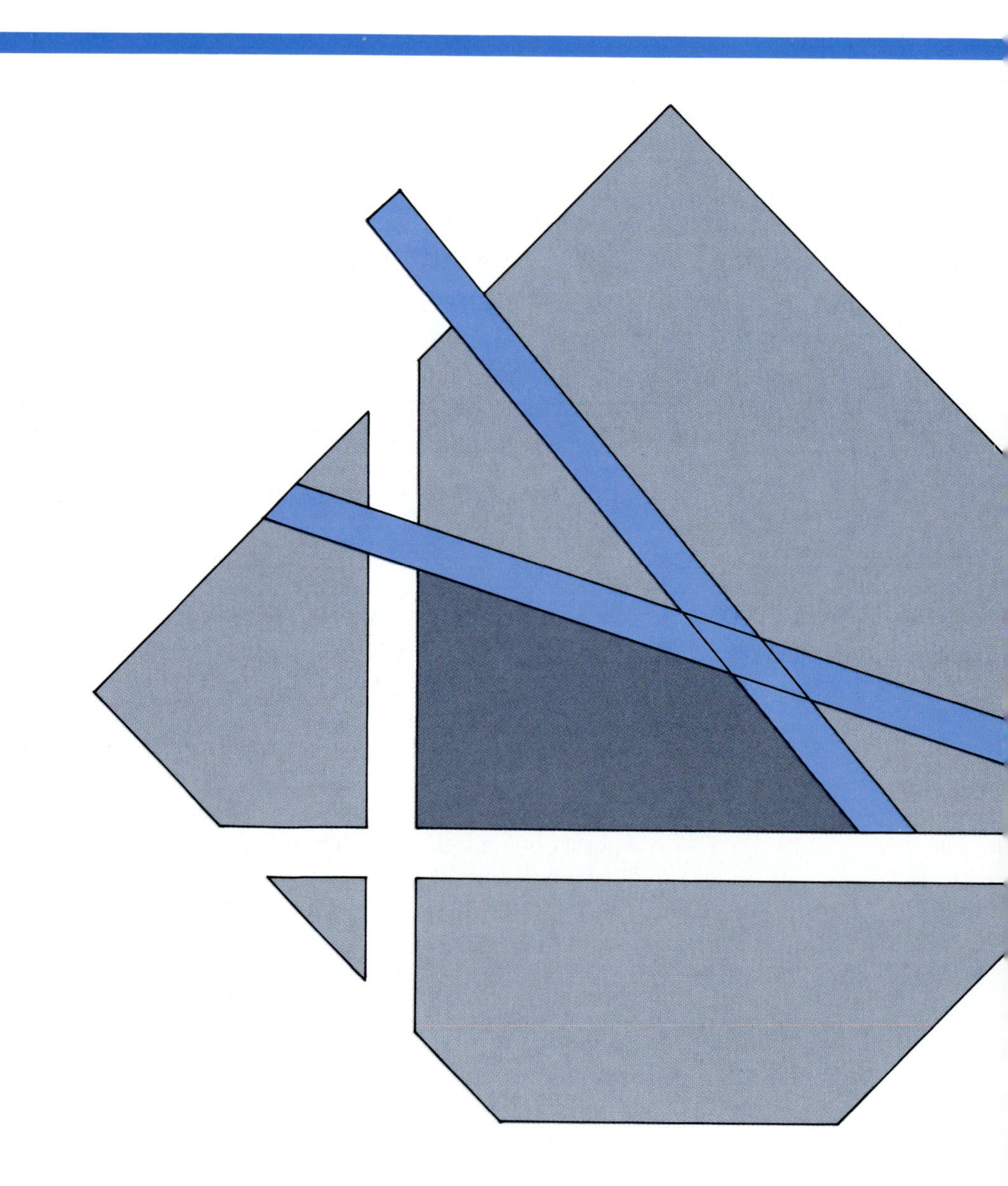

Linear Programming

- **Systems of Linear Inequalities**
- **Linear Programming—A Geometrical Introduction**

Early applications of linear programming were in the military. George B. Dantzig, Marshall Wood, and their associates in the U.S. Department of the Air Force first developed and applied linear programming in 1947 to solve certain military logistic problems. However, the emphasis in applications quickly moved to industry. In 1975, the Nobel Prize committee recognized the importance of linear programming by awarding a Nobel Prize in Economic Science to the scientists Professors Leonid Kantorovich of the Soviet Union and Tjalling C. Koopmans of the United States for their "contributions to the theory of optimum allocation of resources."

Both economists worked independently on the problem of optimum allocation of scarce resources. Kantorovich showed how linear programming can be used to improve economic planning in Russia. He analyzed efficiency conditions for an economy as a whole, demonstrating the connection between the allocation of resources and the price system. His analysis demonstrated how the decentralization of decisions in a planned economy like the Soviet Union depends on the existence of a rational price system.

Koopmans developed his linear programming theory while planning optimal transportation of ships back and forth across the Atlantic Ocean during World War II.

Today, linear programming helps to determine best diets, most efficient production scheduling, and the most economical transportation of goods. The term **programming** predates computers. **Linear programming** refers to a precise procedure that will solve a certain type of problem that involves linear conditions.

3–1 Systems of Linear Inequalities

Feasible Region

Linear programming problems are described mathematically by systems of **linear inequalities** rather than a system of *linear equations*. The solution to a linear programming problem depends on the ability to solve and graph such systems. You will use the material on graphing a linear inequality heavily. See Chapter 1, Section 1–4.

Example 1 (*Compare Exercise 1*)
Graph the following system of inequalities:

$$x + y \leq 2$$
$$x + 4y < 4$$

Solution The solution involves three basic steps. The results are drawn in Figure 3–1.
First, graph $x + y \leq 2$.

(a) Graph the line $x + y = 2$. The $\leq$ symbol implies that the line itself is a part of the solution, so use a solid line. (Recall from Chapter 1 that this line divides the plane into two half planes, one of which is included in the solution of the inequality.)

(b) Select a test point not on the line, say (0, 0). (If the point selected satisfies $x + y = 2$, then it lies on the line, and you need to select another one.)

(c) Substitute $x = 0$ and $y = 0$ into $x + y \leq 2$, that is, $0 + 0 \leq 2$. Since this is true, the point (0, 0) is in the correct half plane. Shade that half plane.

Next, graph $x + 4y < 4$.

(a) Graph the line $x + 4y = 4$. The $<$ symbol implies that the line itself is *not* a part of the solution, so use a dotted line.

(b) Select a test point, say (2, 2).

(c) Substitute $x = 2$ and $y = 2$ into $x + 4y$, that is, $2 + 4(2)$. Since this is not less than 4, the point (2, 2) does *not* make the inequality $x + 4y < 4$ true. Because the inequality is false, the point (2, 2) is not in the correct half plane. The other half plane is correct, so shade it as the solution.

The points that satisfy both inequalities at once make up the **solution** to the system. These points lie in the region where both half planes overlap. This region of intersection (the solution set of the system) is called the **feasible region** (Figure 3–1(b)).

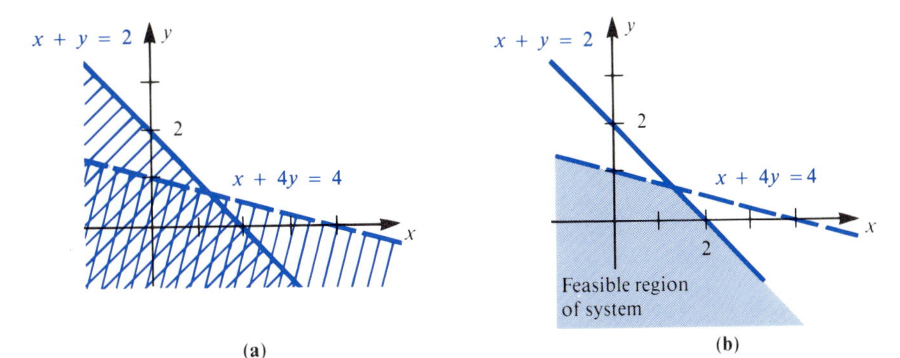

Figure 3–1 (a) (b)

Example 2 (*Compare Exercise 5*)
Graph the solutions to the following system:

$$-x + y \leq 1$$
$$x + y \leq 3$$
$$x \geq 0$$

Solution The lines $x = 0$, $-x + y = 1$, and $x + y = 3$ determine half planes that form solutions to the given inequalities. The graphs of the lines and the appropriate half planes are shown in Figure 3–2. The shaded region common to all three half planes forms the solution to the system of inequalities. Just as a check, substitute x and y from the point (2, 0) into each of the original inequalities. You will find that those values satisfy all three inequalities, so the point belongs to the solution set.

Boundaries and Corners

The lines $x = 0$, $-x + y = 1$, and $x + y = 3$ determine the **boundaries** of the solution set in Example 2. The points contained within the boundaries, and the appropriate points on a boundary, form the solution set of the system. In linear programming terminology, a point that satisfies all inequalities of the system is called a **feasible solution**, and the solution set is called the **feasible region**. Points A and B in Figure 3–2 are **corners** of the region, points in the feasible region where boundary lines intersect.

In linear programming, you will learn that the corners of the solution set (feasible region) contain the optimal solution. We find the corners by solving pairs of simultaneous equations, using equations of lines forming the boundary.

To find the corner A, the point of intersection of the lines $x = 0$ and $-x + y = 1$, we solve the system

$$x = 0$$
$$-x + y = 1$$

Corner A is the point (0, 1).

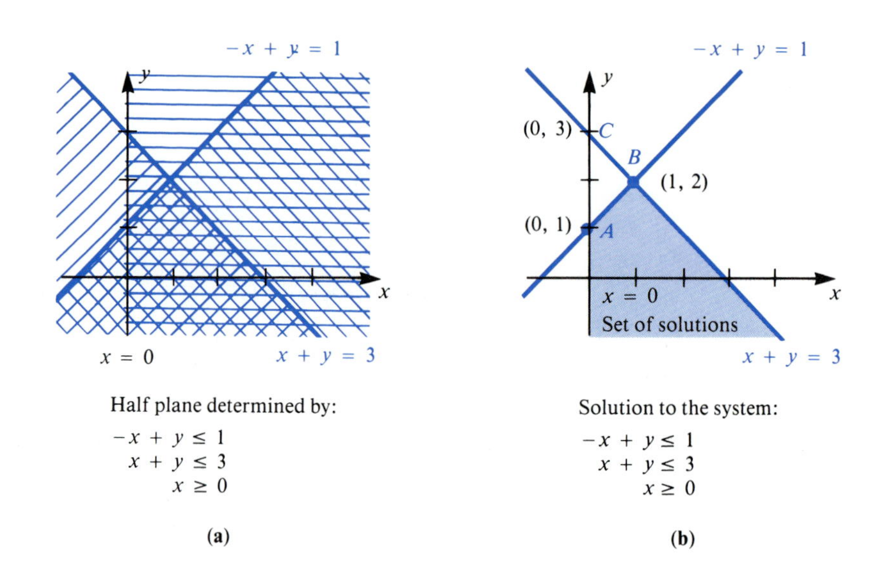

Figure 3–2

Corner B is the point of intersection of the lines $-x + y = 1$ and $x + y = 3$. To determine B, solve the system

$$-x + y = 1$$
$$x + y = 3$$

You should find that corner B is the point (1, 2).

The point C, (0, 3), is the solution of the system

$$x = 0$$
$$x + y = 3$$

and so it is the intersection of two boundary lines. However, C is not a corner point because it lies outside the feasible region. This shows that ***you cannot pick two boundary lines arbitrarily*** and expect their intersection to form a corner point. You need to determine if the point lies in the feasible region.

Example 3 (*Compare Exercise 18*)
Sketch the feasible region (solution set) of the system

$$x + y \leq 3$$
$$-x + 2y \leq 3$$
$$x \geq 0, y \geq 0$$

Solution The nonnegative restrictions $x \geq 0$ and $y \geq 0$ often enter into linear programming problems since the variables usually represent quantities of things such as TV sets or hours a production line is available. Other inequalities represent restrictions associated with these quantities, such as capital limitations or available labor. The graph is the shaded region with corners A, B, C, and O in Figure 3–3.

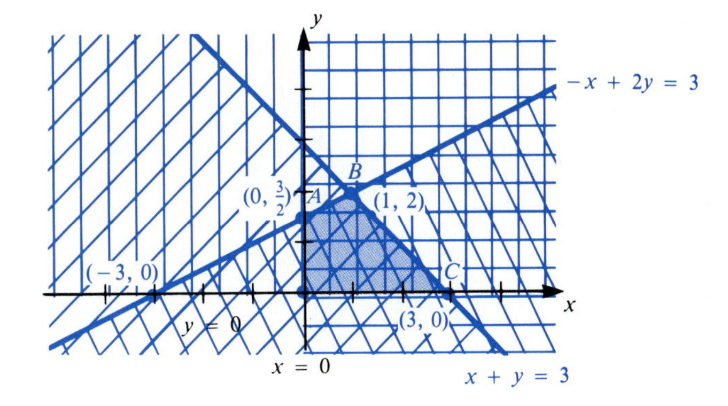

Figure 3–3

To find the Corner A, solve the system of equations

$$\begin{aligned} -x + 2y &= 3 \\ x \qquad &= 0 \end{aligned}$$

Corner A is the point $(0, \frac{3}{2})$.

Find B by solving the system

$$\begin{aligned} x + \ \ y &= 3 \\ -x + 2y &= 3 \end{aligned}$$

On solving this system, we find that B is the point (1, 2).

Corner C is the solution to the system

$$\begin{aligned} x + y &= 3 \\ y &= 0 \end{aligned}$$

C is the point (3, 0).

The corners of the solution set are

$$A\text{: } (0, \tfrac{3}{2}),\ B\text{: } (1, 2),\ C\text{: } (3, 0), \text{ and } O\text{: } (0, 0)$$

No Feasible Solution

Some systems of inequalities have no solution set, as illustrated in the following example.

Example 4 (*Compare Exercise 19*)
Find the solution set (feasible region) of the system

$$\begin{aligned} 5x + 7y &\geq 35 \\ 3x + 4y &\leq 12 \\ x &\geq 0 \\ y &\geq 0 \end{aligned}$$

Solution The inequalities $x \geq 0$ and $y \geq 0$ force the solutions to be in the first quadrant. The test point (0, 0) shows that the points that satisfy $5x + 7y > 35$ lie above the

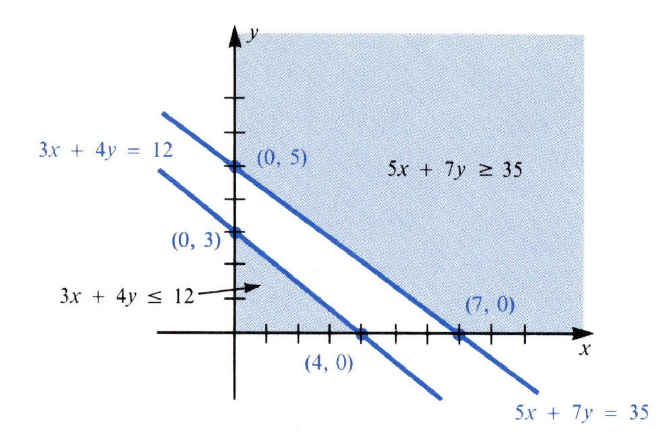

Figure 3–4

line $5x + 7y = 35$, and the points that satisfy $3x + 4y < 12$ lie below the line $3x + 4y = 12$. As shown in Figure 3–4, these two regions do not intersect in the first quadrant. The system then has no solution.

Example 5 (*Compare Exercise 26*)
The organizer of a conference has \$5000 expense money available to participants. Some participants may receive \$25 each for expenses and others may receive \$75 each. Attendance at the conference is limited to 80 participants. State the inequalities that represent this information.

Solution Let x equal the number of participants who may receive \$25, and let y equal the number of participants who may receive \$75. Then, $25x + 75y$ is the total amount of expense money distributed. The inequality $25x + 75y \leq 5000$ states that total expenses are limited to \$5000. The inequality $x + y \leq 80$ states that the conference is limited to 80.

Graphing a System of Inequalities

To graph a system of inequalities,

1. Replace each inequality symbol with "=" to obtain a linear equation.
2. Graph each line. Use a solid line if it is a part of the solution. Use a dotted line if it is not a part of the solution. The line is a part of the solution when $\leq$ or $\geq$ is used. The line is not a part of the solution when $<$ or $>$ is used.
3. Select a test point not on the line.
4. If the test point satisfies the original inequality, it is in the correct half plane. If it does not satisfy the inequality, the other half plane is the correct one.
5. Shade the correct half plane.
6. When the above steps are completed for each inequality, determine where the shaded half planes overlap. This region is the graph of the system of inequalities.

3–1 Exercises

I.

Graph the systems of inequalities in Exercises 1 through 8.

1. (*See Example 1*)
$x + y \leq 3$
$2x - y < -2$

2. $2x + y > 4$
$4x - y \geq 8$

3. $x \geq 3$
$y \geq 2$
$3x + 2y < 18$

4. $-x + 3y < 6$
$2x + y < 7$
$y \geq 0$

5. (*See Example 2*)
$4x + 6y \leq 18$
$x + 3y \leq 6$
$x \geq 0$

6. $2x + y \leq 50$
$4x + 5y \leq 160$
$x \geq 0, y \geq 0$

7. $2x + y \leq 60$
$2x + 3y \leq 120$
$x \geq 0, y \geq 0$

8. $4x + 3y \leq 72$
$4x + 9y \leq 144$
$x \geq 2, y \geq 0$

Find the feasible regions of the systems of inequalities in Exercises 9 through 24. Determine the corners of each feasible region.

9. $x + y > 4$
$2x - 3y \geq 8$

10. $-x + y \leq 3$
$2x + y \leq 6$
$x \geq 0, y \geq 0$

11. $-2x + y < 4$
$x + y \geq 3$

12. $-x + 2y \geq 3$
$2x + y \geq 2$

13. $x \geq 0$
$y < 0$

14. $-x + y \leq 3$
$x + y \leq 4$
$x \geq 0, y > 0$

15. $2x + y \leq 1$
$-x + y > 4$
$y \geq 0$

16. $x - y \leq 2$
$2x + y \leq 4$
$x + 2y \leq 4$

II.

17. $3x + 4y > 24$
$4x + 5y < 20$
$x > 0$

18. (*See Example 3*)
$x + y \leq 2$
$3x + y \leq 6$
$x \geq 0, y \geq 0$

19. (*See Example 4*)
$x + y \leq 2$
$2x + y \geq 6$
$y \geq 0$

20. $2x + y \geq 2$
$x - 3y \leq 6$
$-4x + y > -3$

21. $x + y > 1$
$-x + y \geq 2$
$5x - y < 4$

22. $2x + y \leq 3$
$5x + 2y \leq 7$
$x \geq 0, y \geq 0$

23. $8x + y \geq -10$
$-2x + y \leq 6$
$4x + y \leq 2$
$3x - y \leq -1$

24. $4x - 3y \geq 60$
$x + y \leq 10$
$-x + 2y \geq -50$

III.

25. Determine the feasible region and corners of this system.

$$\begin{aligned} -2x + y &\leq 2 \\ 3x + y &\leq 3 \\ -x + y &\geq -4 \\ x + y &\geq -3 \end{aligned}$$

26. (*See Example 5*) The seating capacity of a theater is 250. Tickets are \$3 for children and \$5 for adults. The theater must take in at least \$1000 per performance. Express this information as a system of inequalities.

27. High Fibre and Corn Bits cereals contain the following amounts of minimum daily requirements of Vitamins A and D for each ounce of cereal.

	Vitamin A	**Vitamin D**
High Fibre	25%	4%
Corn Bits	2%	10%

Use inequalities to express how much of each Mrs. Smith should eat to obtain at least 40% of her minimum daily requirements of Vitamin A and 25% of that for Vitamin D.

28. High Fibre and Corn Bits cereals contain the following amounts of sodium and calories per ounce.

	Calories	**Sodium (mg)**
High Fibre	90	160
Corn Bits	120	200

Mr. Brown's breakfast should provide at least 600 calories, but less than 800 mg of sodium. Find the constraints on the amount of each cereal.

29. The Musical Group has two admission prices for its concerts, \$8 for adults and \$3 for students. They will not book a concert unless they are assured an audience of at least 500 people and total ticket sales of \$2500 or more. Write the constraints on the number of each kind of tickets sold.

30. To be eligible for a University Scholarship, a student must score at least 600 on the SAT verbal test and 600 on the SAT mathematics test, and must have a combined verbal-mathematics score of 1325 or more. Write this as a system of inequalities.

31. A test is scored by giving 4 points for each correct answer and -1 for each incorrect answer. To obtain an acceptable score, a student must answer at least 60 questions and attain a score of 200 points. Express this information with inequalities.

3–2 Linear Programming—A Geometrical Introduction

Constraints and Objective Function
Geometrical Solution
Unusual Linear Programming Situations

Constraints and Objective Function

Managers in business and industry often make decisions in an effort to maximize or minimize some quantity. For example, a plant manager wants to minimize overtime pay for production workers, a store manager makes an effort to maximize revenue, or a stockbroker tries to maximize the return on his investments. Most of these decisions are complicated by restrictions that limit choices. The plant manager may not be able to eliminate all overtime and still meet the contract deadline. A store may be swamped by customers because of its low prices, but it may be losing money.

Linear programming is a mathematical technique that solves certain problems of this kind. Let's start with a simple example.

Example 1 (*Compare Exercise 1*)
An appliance-store manager plans to offer a special on washers and dryers. His storeroom capacity is limited to 50 items. Each washer requires two hours to unpack and set up, and each dryer requires one hour. He has 80 hours of employee time available for unpacking and set-up. Washers sell for \$300 each, and dryers sell for \$200 each. How many of each should he order to obtain the maximum revenue?

Solution Convert the given information to mathematical statements.

Let x = the number of washers
y = the number of dryers

The total number to be placed in the store room is

$$x + y$$

and the total set-up time is

$$2x + 1y$$

Since the manager has space for only 50 items and set-up time of 80 hours, we have the restrictions

$$x + y \leq 50$$
$$2x + y \leq 80$$

Since x and y cannot be negative, we also have

$$x \geq 0 \quad \text{and} \quad y \geq 0$$

Since washers sell for \$300 and dryers sell for \$200, we want to find values of x and y that maximize the total revenue

$$z = 300x + 200y$$

Here is the problem stated in concise form.

Maximize z where

$$z = 300x + 200y$$

subject to

$$x + y \leq 50$$
$$2x + y \leq 80$$
$$x \geq 0$$
$$y \geq 0$$

The inequalities impose restrictions on the problem. We call these inequalities **constraints**. It is customary to call the function $z = 300x + 200y$ the **objective function**. Find the values of x and y that satisfy the system of constraints (inequalities) by the methods from the last section. You should obtain the feasible region as shown in Figure 3–5(a). Corner A is (30, 20).

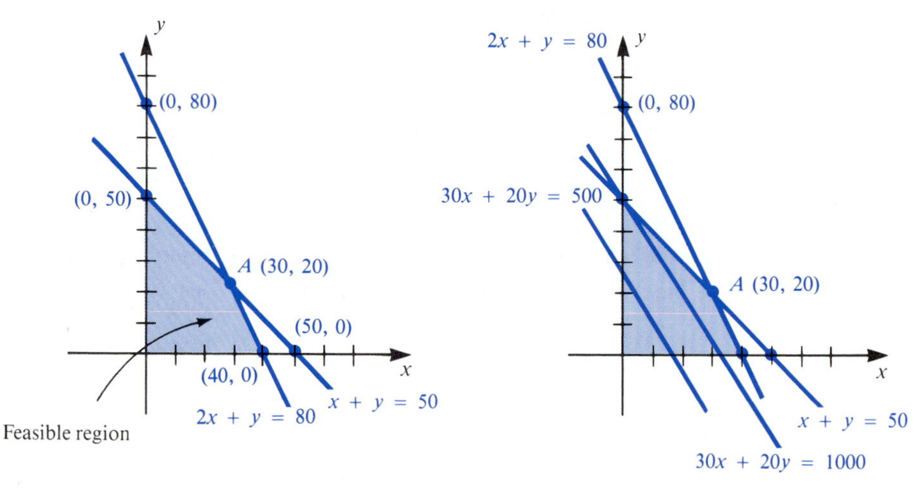

Figure 3–5

Each point in the feasible region determines a value for the objective function. We want to find the point in the feasible region that maximizes the objective function. The point (10, 10) gives $z = 300x + 200y$ the value $z = 5000$, while the point (20, 20) gives the value $z = 10{,}000$. Since there is an infinite number of points in the feasible region, we will not easily find the maximum value of z by a haphazard trial-and-error process.

Geometrical Solution

A basic theorem makes it easier to find a maximum or minimum value of an objective function. The term **optimal** refers to either maximum or minimum.

Theorem 1

Given a linear objective function subject to linear inequality constraints, if the objective function has an optimal value (maximum or minimum), it must occur at a corner point of the feasible region.

The feasible region of the above example has corner points (0, 0), (0, 50), (30, 20), and (40, 0). The value of the objective function in each case is $z = 0$ at (0, 0), $z = 10{,}000$ at (0, 50), $z = 13{,}000$ at (30, 20), and $z = 12{,}000$ at (40, 0). The maximum value of z within the feasible region is therefore 13,000, and the minimum value is 0.

Graphical Solution to Linear Programming Problem

1. Use each constraint (linear inequality) in turn to sketch the boundary of the feasible region.
2. Determine the corner points of the feasible region by solving pairs of linear equations (equations obtained from the constraints).
3. Evaluate the objective function at each corner point.
4. The maximum (minimum) of the objective function at corner points yields the desired maximum (minimum).

Some observations will help us to an intuitive understanding of this theorem. We will continue to use the example above with the objective function $z = 300x + 200y$. When the constraints involve two variables, we can graph the feasible region in the plane. The objective function, like $z = 300x + 200y$, has a third variable z, so we can't graph it in the plane. However, when z is a particular value, like 5000, the equation $300x + 200y = 5000$ graphs as a straight line in the plane. Any point on this line gives $z = 5000$. Let's graph the line $300x + 200y = 5000$. Observe that it divides the feasible region into two parts (Figure 3–5(b)). Using (0, 0) as a test point, we see that points below the line satisfy $300x + 200y < 5000$ and that the points above the line satisfy $300x + 200y > 5000$. Since the point (20, 20) lies above the line $300x + 200y = 5000$, we expect it to give a larger value of $300x + 200y$. It does—namely 10,000. Does any point in the feasible region give a larger value than 10,000? To help answer this, graph the line $300x + 200y = 10{,}000$. It is parallel to $300x + 200y = 5000$, and it too intersects the feasible region. A quick check of the test point (0, 0) shows that the points above the line satisfy $300x + 200y > 10{,}000$. Thus, we expect to obtain a larger value of $z = 300x + 200y$ from any point in the feasible region above the line $300x + 200y = 10{,}000$.

When we select a point that is in the feasible region and is above the line, we can graph the line through that point and parallel to the line $300x + 200y = 10{,}000$. However, as before, any point in the feasible region that lies above the line yields a larger value of $300x + 200y$. [Try (30, 30), for example.] When will we reach the largest value of $z = 300x + 200y$? The answer: when we graph a line parallel to $300x + 200y = 10{,}000$ that has no points in the feasible region above it. That will occur when the line passes through the corner point (30, 20) at which $z = 13{,}000$, the largest possible value. Points beyond the line will give larger values, but the points lie outside the feasible region.

In this example, the feasible region is **bounded** because it can be enclosed in a rectangle. For a bounded feasible region, the objective function will have both a maximum and a minimum.

In some cases, the constraints lead to an inconsistency so there are no points in the feasible region—it is **empty** (see Example 4 of Section 3–1).

Theorem 2

When the feasible region is not empty and is bounded, the objective function has both a maximum and a minimum value, and they must occur at corner points.

Example 2 (*Compare Exercise 3*)

Find the maximum value of the objective function

$$z = 10x + 15y$$

subject to the constraints

$$\begin{aligned} x + 4y &\le 360 \\ 2x + y &\le 300 \\ x \ge 0,\ y &\ge 0. \end{aligned}$$

Solution Graph the feasible region of the system of inequalities (Figure 3–6). The corner points of the feasible region are (0, 90), (0, 0), (150, 0), and (120, 60). The point (120, 60) is found by solving the system

$$\begin{aligned} x + 4y &= 360 \\ 2x + y &= 300 \end{aligned}$$

Find the value of z at each corner point.

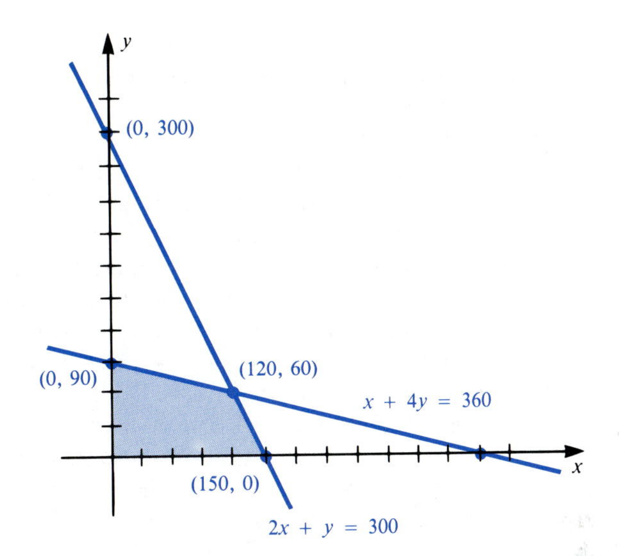

Figure 3–6

Corner	$z = 10x + 15y$
(0, 90)	1350
(0, 0)	0
(150, 0)	1500
(120, 60)	2100

The maximum value of z is 2100 and occurs at the corner (120, 60).

Example 3 (*Compare Exercise 8*)
Find the maximum and minimum values of

$$z = 4x + 6y$$

subject to the constraints

$$5x + 3y \geq 15$$
$$x + 2y \leq 20$$
$$7x + 9y \leq 105$$
$$x \geq 0,\ y \geq 0$$

Solution The feasible region of this system and its corners are shown in Figure 3–7. Compute the value of z at each corner to determine the maximum and minimum values of z.

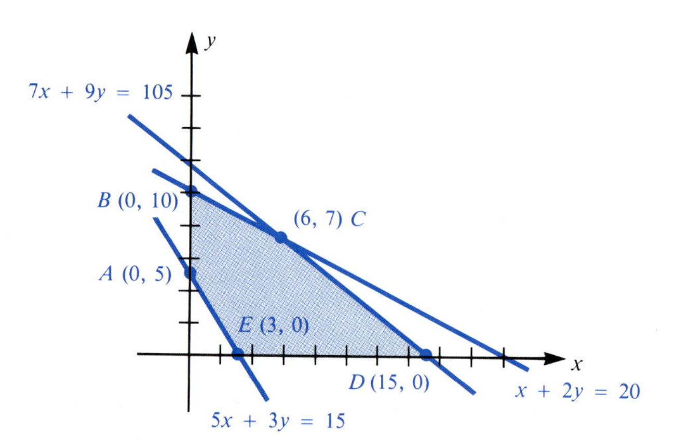

Figure 3–7

Corner	Value of z
(0, 5)	30
(0, 10)	60
(6, 7)	66
(15, 0)	60
(3, 0)	12

The maximum value of z is 66 and occurs at the corner (6, 7). The minimum value of z is 12 and occurs at the corner (3, 0).

Example 4 (*Compare Exercise 18*)
A company makes two products, one on production line A and the other on production line B. Available resources include a labor force equivalent to 900 hours per week and $2800 for weekly operating costs. It takes 5 hours labor to produce each item on line A and 2 hours labor to produce each item on line B. Each item produced on line A costs $8, and those on line B cost $10. The company wishes to maximize its profits. If the profit on each item produced on line A is $3, and the profit for each item produced on line B is $2, how many of each item should be produced in order to achieve maximum profit?

Solution This problem contains two restrictions: time and funds available. The company wants to maximize profit under these restrictions. The problem may be solved in four steps:

1. Specify the variables used.
2. Use mathematical statements to describe the situation.
3. Sketch the graph.
4. Use the graph to determine the solution.

Let

$$x = \text{the number of items produced on line } A$$
$$y = \text{the number of items produced on line } B$$

The time required to produce x items on line A is $5x$, and that required for line B is $2y$, so the total hours labor involved is

$$5x + 2y$$

The inequality

$$5x + 2y \leq 900$$

states that no more than 900 hours of labor is available. The cost of x items on line A is $8x$ and on line B is $10y$, so

$$8x + 10y$$

represents the total weekly operating costs. The inequality

$$8x + 10y \leq 2800$$

describes the restriction on operating costs. The total profit is

$$3x + 2y$$

The objective function is the quantity to be maximized or minimized, so $z = 3x + 2y$ is the objective function. Because x and y represent the number of items produced, they cannot be negative, so we have $x \geq 0$ and $y \geq 0$.

We state this linear programming problem as follows: Maximize the objective function

$$z = 3x + 2y$$

subject to the constraints

$$\begin{aligned} 5x + 2y &\leq 900 \\ 8x + 10y &\leq 2800 \\ x \geq 0, y &\geq 0 \end{aligned}$$

We want to find the solutions of this system of inequalities that gives the maximum value of $3x + 2y$. Graph this system as shown in Figure 3–8. Its corners are the points where pairs of lines intersect, and are found from the systems:

$$\begin{aligned} 8x + 10y &= 2800 \\ x &= 0 \end{aligned} \qquad \begin{aligned} 8x + 10y &= 2800 \\ 5x + 2y &= 900 \end{aligned} \qquad \begin{aligned} 5x + 2y &= 900 \\ y &= 0 \end{aligned} \qquad \begin{aligned} x &= 0 \\ y &= 0 \end{aligned}$$

giving the corners

$$(0, 280) \qquad (100, 200) \qquad (180, 0) \qquad (0, 0)$$

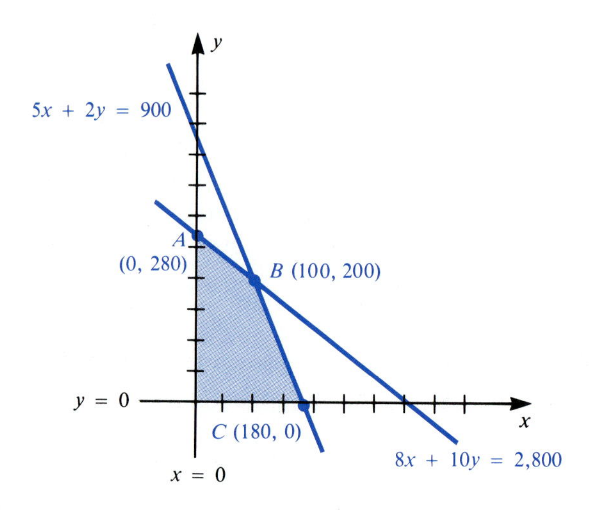

Figure 3–8

The maximum value of z is attained at one of these corner points. Examine each one:

At (0, 280), $z = 3(0) + 2(280) = 560$;
at (100, 200), $z = 3(100) + 2(200) = 700$;
at (180, 0), $z = 3(180) + 2(0) = 540$;
at (0, 0), $z = 0$.

The maximum value of $z = 3x + 2y$, 700, occurs when $x = 100$ and $y = 200$. These results indicate that \$700 is the maximum weekly profit that can be attained under the given constraints, and the company should produce 100 items on line A and 200 items on line B.

Unusual Linear Programming Situations

Multiple Optimal Solutions. In the proceding example, there was just one corner point that gave an optimal solution. It is possible for more than one optimal solution to exist. In other cases, there may be no solution at all. First, look at an example with multiple solutions.

Example 5 (*Compare Exercise 9*)
Determine the maximum value of

$$z = 8x + 2y$$

subject to the constraints

$$4x + y \leq 32$$
$$4x + 3y \leq 48$$
$$x \geq 0, y \geq 0$$

Solution The feasible region of these constraints is shown in Figure 3–9. The corners of the feasible region are (0, 16), (6, 8), (8, 0), and (0, 0).

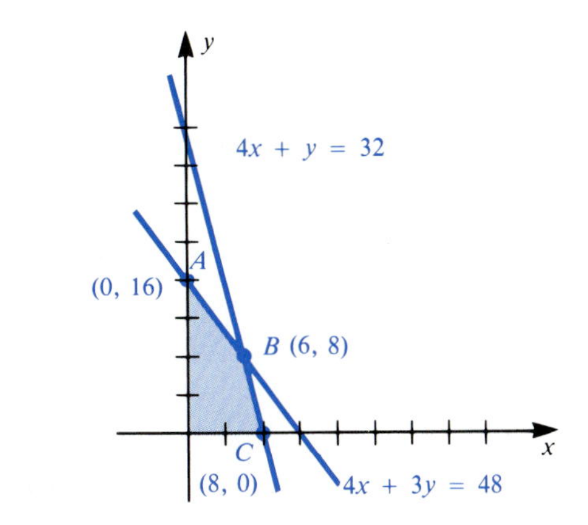

Figure 3–9

The value of $z = 8x + 2y$ at each of these points is as follows:

At $(0, 16)$, $z = 32$;
at $(6, 8)$, $z = 64$;
at $(8, 0)$, $z = 64$;
at $(0, 0)$, $z = 0$.

Thus, the maximum value of $8x + 2y$ is 64, and it occurs at two points, $(6, 8)$ and $(8, 0)$. When this happens in a linear programming problem, the objective function has the same value at every point on the line segment joining the two corners. In this case, the objective function has the maximum value of 64 along the boundary line $4x + y = 32$ from the point $(6, 8)$ to the point $(8, 0)$. For example, the point $(7, 4)$ lies on the line $4x + y = 32$ and gives z the value $8(7) + 2(4) = 64$.

Multiple solutions allow a number of choices of x and y that yield the same optimal value for the objective functions. For example, if x and y represent production quantities, then management can achieve optimal production with a variety of production levels. In case problems occur with one production line, they might adjust production on the other line and still achieve optimal production. When there is just one optimal solution, management has no flexibility; they have one choice of x and y by which they can achieve their objective.

You may wonder how you tell if there are multiple solutions. It turns out that this happens when a boundary line has the same slope as the objective function. (*See Exercise 31*)

Unbounded Feasible Region. The constraints of a linear programming problem might define an unbounded feasible region for which the objective function has no maximum value. In such a case, the problem has no solution. Here is an illustration.

Example 6 (*Compare Exercise 14*)
Determine the maximum value of the objective function $z = x + 4y$, subject to the constraints

$$-4x + y \leq 2$$
$$2x - y \leq 1$$
$$x \geq 0, y \geq 0$$

Solution Figure 3–10 shows the feasible region of these constraints. Observe that the feasible region extends upward indefinitely. Suppose someone claims that the maximum value of the objective function is 20. This is equivalent to stating that

$$x + 4y = 20$$

and no other values of x and y in the feasible region give larger values. You recognize this as the equation of a straight line. Figure 3–10 shows that this line crosses the feasible region. Next, observe that the test point (0, 0) does *not* satisfy the inequality $x + 4y > 20$, so the points in the half plane above the line *must* satisfy it. Thus, every point in the feasible region that lies above the line $x + 4y = 20$ will give a larger value of the objective function. If we substitute larger values, say 100, 5000, and so on, instead of 20 in the equation, we essentially determine lines that are parallel to $x + 4y = 20$ but are further away from the origin. In each case, points in the feasible region that lie above the line give an even larger value of $x + 4y$. Since the feasible region extends upward indefinitely, we can never find a largest value.

We point out that the objective function $z = x + 4y$ does have a *minimum* value at (0, 0). So, an unbounded feasible region does not rule out an optimal solution. It depends on the region and what kind of optimal solution is sought.

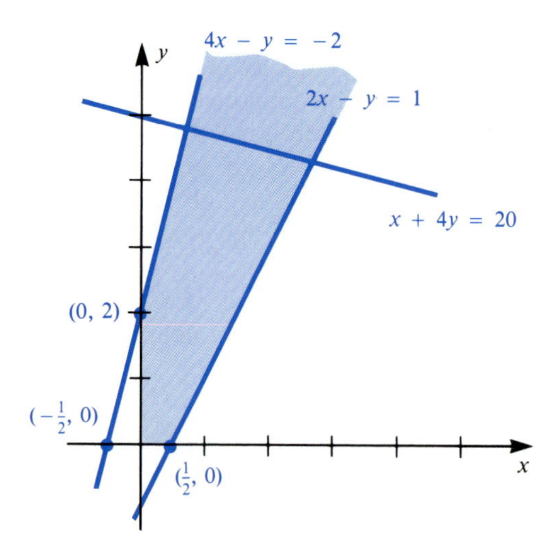

Figure 3–10

No Optimal Solution Because There Is No Feasible Region. The system of inequalities

$$5x + 7y \geq 35$$
$$3x + 4y \leq 12$$
$$x \geq 0,\ y \geq 0$$

is a system of inequalities that has no solution. (See Example 4, Section 3–1.) Whenever the constraints of a linear programming problem do not define a feasible region, there can be no optimal solution.

3–2 Exercises

I.

1. (*See Example 1*) Set up the constraints and objective function for the following linear programming problem.
   A discount store is offered two styles of slightly damaged coffee tables. The store has storage space for 80 tables and 110 hours labor for repairing the defects. Each table of style A requires one hour labor to repair, and style B requires two hours. Style A is priced at \$50 each, and style B at \$40 each. How many of each style should be ordered to maximize gross sales?
2. The organizer of a day conference has \$2000 expense money available for distribution to participants. The participants fall into two categories: those having \$30 expenses, and those having \$10 expenses. Facilities are available to host 100 participants at the conference. Describe the constraints on each type participant.
3. (*See Example 2*) Find the maximum value of the objective function

$$z = 20x + 12y$$

subject to the constraints

$$3x + 2y \leq 18$$
$$3x + y \leq 15$$
$$x \geq 0,\ y \geq 0$$

4. Maximize $z = 2x + y$, subject to
   $4x + y \leq 36$
   $4x + 3y \leq 60$
   $x \geq 0,\ y \geq 0$
5. Maximize $z = x + 4y$, subject to
   $x + 2y \leq 4$
   $x + 6y \leq 8$
   $x \geq 0,\ y \geq 0$
6. Maximize $z = 2x + y$, subject to
   $-3x + y \leq 4$
   $x - y \leq 2$
   $x \geq 0,\ y \geq 0$
7. Maximize $z = 3x + y$, subject to
   $2x - 3y \geq 10$
   $x \leq 8$
   $x \geq 0,\ y \geq 0$

II.

8. (*See Example 3*) Find the maximum and minimum values of $z = 20x + 30y$, subject to
$2x + 10y \le 80$
$6x + 2y \le 72$
$3x + 2y \ge 6$
$x \ge 0, y \ge 0$

9. (*See Example 5*) Maximize $z = 15x + 9y$, subject to
$5x + 3y \le 30$
$5x + y \le 20$
$x \ge 0, y \ge 0$

10. Maximize $z = 4x + 2y$, subject to
$x + 3y \le 15$
$2x + y \le 10$
$x \ge 0, y \ge 0$

11. Maximize $z = x + 5y$, subject to
$x + y \le 10$
$2x + y \ge 10$
$x + 2y \ge 10$

12. Maximize $z = x + 2y$, subject to
$x \ge -2$
$x - y \ge -4$
$x + 2y \le 6$
$2x + y \le 6$

13. Maximize $z = -8x + 10y$, subject to
$x \ge -20$
$x \le 5$
$y \ge 0$
$4x + 3y \le 40$
$-4x + 5y \le 120$

14. (*See Example 6*) Maximize $z = 8x + 3y$, subject to
$2x - 5y \le 10$
$-2x + y \le 2$
$x \ge 0, y \ge 0$

15. Maximize $z = 9x + 13y$, subject to
$3x - y \le 8$
$x - 2y \ge 2$
$x \ge 0, y \ge 0$

16. Maximize $z = 5x + 4y$, subject to
$-5x + y \le 3$
$-2x + y \ge 3$
$x \ge 0, y \ge 0$

17. Minimize $z = x + 2y$, subject to
$-3x + y \le 4$
$-2x - y \ge 1$
$x \ge 0, y \ge 0$

18. (*See Example 4*) A company makes one item on production line *A* and another item on production line *B*. They have 1500 hours of labor and \$4000 operating funds available per week. It takes four hours labor to produce each item on line *A* and six hours labor for each item on line *B*. Each item on line *A* costs \$12 to produce, and each item on line *B* costs \$8. The company makes a profit of \$3 on each item from line *A* and \$4 on each one from line *B*. How many of each item should they produce to maximize total profit?

19. Farmer Jones wants to grow strawberries and tomatoes on a 40-acre plot. There will be 300 hours of labor available for the picking. It takes eight hours to pick an acre of strawberries and six hours for an acre of tomatoes. The per-acre profit on the strawberries is \$700 compared to \$600 for the tomatoes. How many acres of each should be grown in order to maximize total profit?

III.

20. Professor *X* gives a speed exam, limited to 60 minutes, with more problems than any student can work. There are two types of problems. Type *A*

problems are worth five points each, and a student is allowed a maximum of three minutes each to work. Type *B* problems are worth ten points each, and eight minutes maximum is allowed for each one. Each student must work at least three type *B* problems. How many of each type should a student work in order to maximize the total points?

21. Acme Inc. manufacturers two grades of tires, premium and regular. Two machines, I and II, are needed to manufacture each grade of tire. It takes three minutes on each machine to produce a premium tire; and one minute on machine I, and two minutes on machine II to produce a regular tire. The total time available on machine I is 3000 minutes and on machine II, 4500 minutes. The company realizes a profit of $15 on each premium and $7 on each regular tire. How should the manufacturing of the tires be arranged to obtain the largest possible total profit?

22. A company manufactures two types of hand calculators, model C1 and model C2. It takes one hour and 4 hours in labor time to manufacture the C1 and C2, respectively. The cost of manufacturing the C1 is $30, and that of manufacturing the C2 is $20. The company has 1600 hours per week available in labor, and $18,000 available for operating costs. The profit on the C1 model is $10, and on the C2 model it is $8. What weekly production schedule maximizes total profit?

23. A refrigerator company has two plants, one in Cincinnati and one in Akron. It has a retail outlet in Toledo. It takes 20 hours (packing, transportation, and so on) to transport a refrigerator from Cincinnati to the retail outlet and 10 hours from Akron to the retail outlet. It costs $60 to transport each refrigerator from Cincinnati to the retail outlet and $10 per refrigerator from Akron to the retail outlet. There are 1200 hours available (man hours for packing, transportation, and so on) and $2400 budgeted for transportation costs. The profit on each refrigerator manufactured at Cincinnati is $40 and on each manufactured at Akron is $20 (the plant at Cincinnati is newer and more efficient than that at Akron). How should the company allocate the transportation of refrigerators so as to maximize total profits?

24. A company makes a single product on two separate production lines, *A* and *B*. Its labor force is equivalent to 1000 hours per week, and it has $3000 outlay weekly on operating costs. It takes one hour and four hours to produce a single item on *A* and *B*, respectively. The cost of producing a single item on *A* is $5 and on *B* is $4. How many items should be produced on each line in order to maximize total output?

25. The maximum daily production of an oil refinery is 1400 barrels. The refinery can produce two types of fuel: gasoline and heating oil. The production cost per barrel is $6 for gasoline and $8 for heating oil. The daily production budget is $9600. The profit is $3.50 per barrel on gasoline and $4 per barrel on heating oil.

(a) What is the maximum total profit that can be realized daily?

(b) What quantities of each type of fuel are then produced?

26. A manufacturer makes two types of fertilizer, Lawn, and Garden, using chemicals A and B. Lawn fertilizer is made up of 80% of chemical A and 20% of chemical B. Garden fertilizer is made up of 60% of chemical A and 40% of chemical B. The manufacturer requires at least 30 tons of Lawn and at least 50 tons of Garden, and has available 100 tons of A and 50 tons of B. How many units each of Lawn and Garden should be produced in order to maximize the total amount of fertilizer?

27. A city has $600,000 to purchase cars. Two models, the B250 and the X100, are under consideration, costing $8000 and $10,000, respectively. The estimated annual maintenance cost on the B250 is $600 and on the X100, $500. The city will allocate $40,000 for the total maintenance of these cars. The B250 gives 26 miles per gallon and the X100 gives 24 miles per gallon. The city wants to maximize "the gasoline efficiency number" of this group of cars. For x B250's and y X100's, this number would be $26x + 24y$. How many of each model should be purchased?

28. A tailor has 80 square yards of cotton material and 120 square yards of woolen material. A suit requires two yards of cotton and one yard of wool. A dress requires one yard of cotton and three yards of wool. How many of each garment should the tailor make to maximize his income if a suit and a dress each sell for $20? What is the maximum income?

29. A school district is buying new buses. It has a choice of two kinds. The 20-passenger bus costs $18,000, whereas the 30-passenger costs $22,000. $572,000 has been budgeted for the new buses. A maximum of 30 drivers will be available to drive the buses. At least 17 30-passenger buses must be ordered because of the desirability of having a certain number of larger capacity buses. How many of each type should be purchased to carry a maximum number of students?

30. A company is buying lockers. It has narrowed down the choice to two sizes, medium and large. The medium has a volume of 36 cubic feet, whereas the large has a volume of 44 cubic feet. The medium occupies an area of 6 square feet and costs $54, whereas the large occupies an area of 8 square feet and costs $60. Space is limited to 256 square feet of floor space, and funds are limited to $2,100. At least 8 of the large lockers are wanted. In order to maximize volume, how many of each should be purchased?

31. The general form of the objective function is $z = Ax + By$. For a given value of z, the resulting line has a slope of $-A/B$. Show that each of the following linear programming problems has multiple optimal solutions. Verify in each case that the objective function has the same slope as one of the boundary lines. (The constraint $Cx + Dy \leq E$ has the boundary $Cx + Dy = E$, and its slope is $-C/D$)

(a) Maximize $z = 10x + 4y$, subject to
$5x + 2y \leq 50$
$x + 4y \leq 28$
$x \geq 0,\ y \leq 0$

(b) Maximize $z = 5x + 6y$, subject to
$5x + 12y \leq 300$
$10x + 12y \leq 360$
$10x + 6y \leq 300$
$x \geq 0,\ y \geq 0$

(c) Minimize $z = 10x + 15y$,
subject to
$3x + 2y \geq 50$
$2x + 3y \geq 60$
$x + 4y \geq 40$
$x \geq 0, y \geq 0$

(d) Maximize $z = 10x + 24y$,
subject to
$5x + 12y \leq 1200$
$5x + 4y \leq 600$
$x \geq 0, y \geq 0$

IMPORTANT TERMS

3–1 **System of linear inequalities** — **Solutions to a system of linear inequalities** — **Half planes**
Graph of a system of linear inequalities — **Feasible region**

3–2 **Bounded feasible region** — **Unbounded feasible region** — **Boundary of a feasible region**
Corners of a feasible region — **Linear programming** — **Constraints**
Objective function — **Maximize objective function** — **Minimize objective function**
Optimal solution — **Multiple optimal solution**

REVIEW EXERCISES

Graph the following systems of inequalities.

1. $2x + y \leq 4$
$x + 3y < 9$

2. $x + y \leq 5$
$x - y > 3$
$x \geq 1$
$y \leq 3$

Find the feasible region and corner points of the systems in Exercises 3 through 7.

3. $x - 3y \geq 6$
$x - y \leq 4$
$y \geq -5$

4. $5x + 2y \leq 50$
$x + 4y \leq 28$
$x \geq 0$

5. $-3x + 4y \leq 20$
$x + y \geq -2$
$8x + y \leq 40$
$y \geq 0$

6. $3x + 10y \leq 150$
$2x + y \leq 32$
$x \leq 14$
$x \geq 0, y \geq 0$

7. $x - 2y \leq 0$
$-2x + y \leq 2$
$x \leq 2, y \leq 2$

8. Maximize $z = x + 2y$, subject to
$x + y \leq 8$
$x \leq 5$
$x \geq 0, y \geq 0$

9. Maximize $z = 5x + 4y$, subject to
$3x + 2y \leq 12$
$x + y \leq 5$
$x \geq 0, y \geq 0$

10. Find the maximum and minimum values of $z = 2x + 5y$, subject to
$2x + y \geq 9$
$4x + 3y \geq 23$
$x \geq 0, y \geq 0$

11. **(a)** Find the minimum value of $z = 5x + 4y$, subject to
$3x + 2y \geq 18$
$x + 2y \geq 10$
$5x + 6y \geq 46$
$x \geq 0, y \geq 0$

(b) Find the minimum value of $z = 10x + 12y$, subject to the constraints of part (a).

12. A building supplies truck has a load capacity of 25,000 pounds. A delivery requires at least 12 pallets of brick, weighing 950 pounds each and at least 15 pallets of roofing material, weighing 700 pounds each. Express these restrictions with a system of inequalities.

13. A theater has a seating capacity of 275 seats. Adult tickets sell for \$4.50 each and children's tickets sell for \$3.00 each. The theater must have ticket sales of at least \$1100 to break even for the night. Write these constraints as a system of inequalities.

14. A tailor makes suits and dresses. A suit requires 1 yard of polyester and 4 yards of wool. Each dress requires 2 yards of polyester and 2 yards of wool. She has a supply of 80 yards of polyester and 150 yards of wool. What restrictions does this place on the number of suits and dresses she can make?

15. A steel mill produces two grades of stainless steel that is sold in 100 pound bars. The standard grade is 90% steel and 10% chromium by weight, and the premium grade is 80% steel and 20% chromium. The company has 80,000 pounds of steel and 12,000 pounds of chromium on hand. If the price per bar is \$90 for the standard grade and \$100 for the premium grade, how much of each grade should they produce to maximize revenue?

16. The Nut Factory produces a mixture of peanuts and cashews. They guarantee that at least one third of the total weight is cashews.

A retailer wants them to fill an order of 1200 pounds or more of the mixture. The peanuts cost the Nut Factory \$.75 per pound, and the cashews cost \$1.40 per pound. Find the amount of each kind of nut they should use to minimize the cost

(a) if 600 pounds of peanuts are available.

(b) if 900 pounds of peanuts are available.

4

Topics in this chapter can be applied to:

Maximizing Profits ● Structuring Nutrition ● Resource Management and Cost Control ● Transportation Management ● Division of Labor ● Production Strategies ● Resource Allocations ● Plant and Space Management

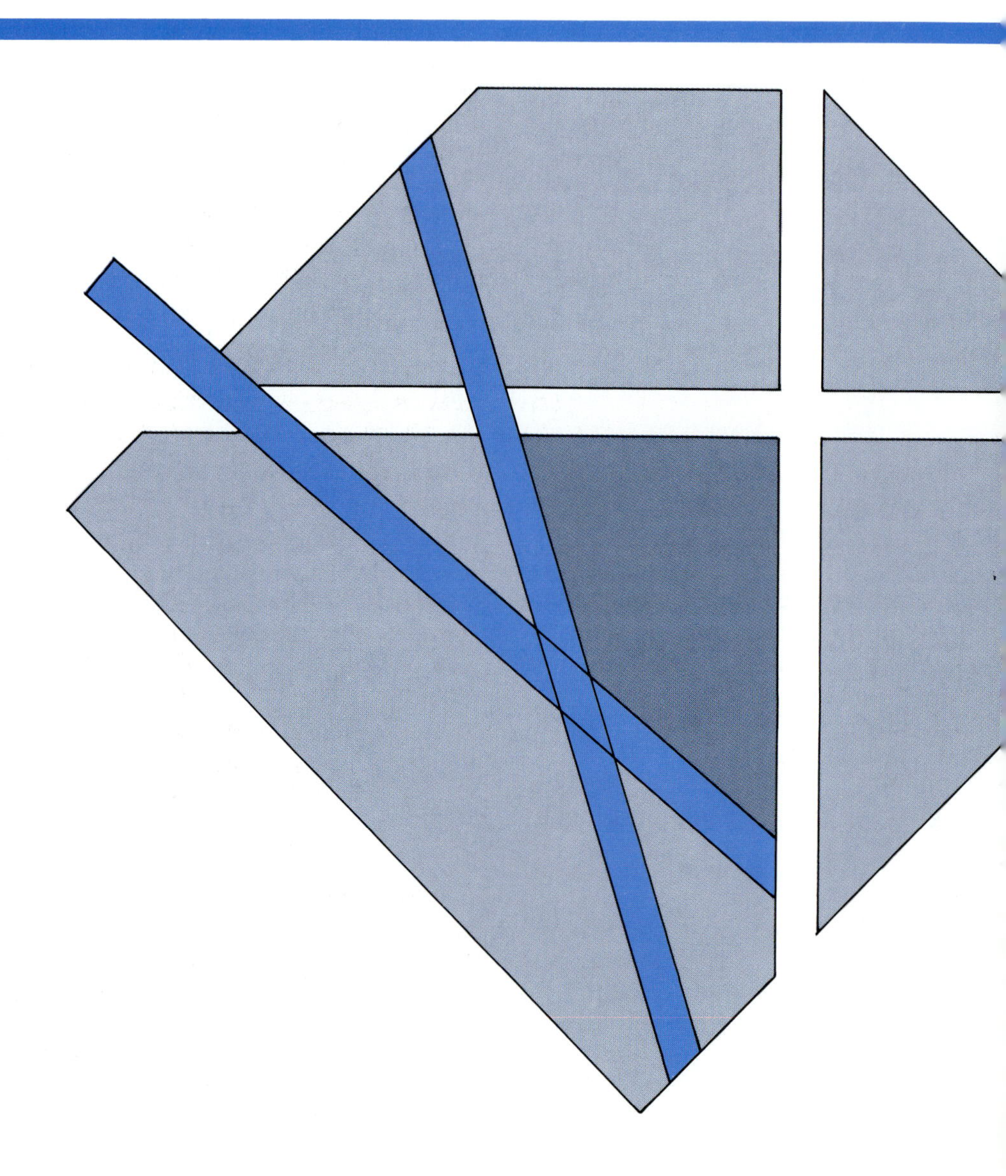

Linear Programming—The Simplex Method

- **Setting up the Simplex Method**
- **The Simplex Method**
- **What's Happening in the Simplex Method?**
- **The Standard Minimum Problem: Duality**
- **Mixed Constraints**

4–1 Setting Up the Simplex Method

Standard Maximum
A Matrix Form of the Standard Maximum Problem
Slack Variables
Simplex Tableau

Chapter 3 introduced you to the basic ideas of linear programming. Perhaps you noticed that all the examples and problems involved two variables. In practice, linear programming problems involve dozens of variables. The graphical method is not practical in problems with more than two variables. There is an algebraic technique that applies to any number of variables and enables us to solve larger linear programming problems. It has the added advantage that it is well suited to a computer, thereby making it possible to avoid tedious pencil and paper solutions. This technique is called the **simplex method**. Basically, it involves modifying the constraints so that one has a system of linear equations and then finding selected solutions of the system. Remember how we solved a system of linear equations using an augmented matrix and reducing it with row operations? The simplex method follows a similar procedure. We introduce the simplex method in several steps, and we shall refer to the graphical method to illustrate the steps involved.

Standard Maximum

The following linear programming problem will be referred to several times.

Illustrative Example

Maximize the objective function

$$z = 4x_1 + 12x_2$$

subject to the constraints

$$\begin{aligned} 3x_1 + x_2 &\le 180 \\ x_1 + 2x_2 &\le 100 \qquad (1) \\ -2x_1 + 2x_2 &\le 40 \\ x_1 \ge 0, x_2 &\ge 0 \end{aligned}$$

Notice that we now use the notation x_1 and x_2 for the variables instead of x and y. This notation allows us to use several variables without running out of letters for variables.

In this section, we deal only with **standard maximum** linear programming problems. They are problems like the above example which have the following properties.

Standard Maximum Problem

1. The objective function is to be maximized.
2. All of the constraint inequalities are $\le$.
3. The constants in the constraints to the right of $\le$ are never negative (180, 100, and 40 in the example).
4. The variables are restricted to nonnegative values.

Figure 4–1 shows the feasible region determined by the constraints. The lines

$$\begin{aligned} 3x_1 + x_2 &= 180 \\ x_1 + 2x_2 &= 100 \\ -2x_1 + 2x_2 &= 40 \\ x_1 = 0, x_2 &= 0 \end{aligned}$$

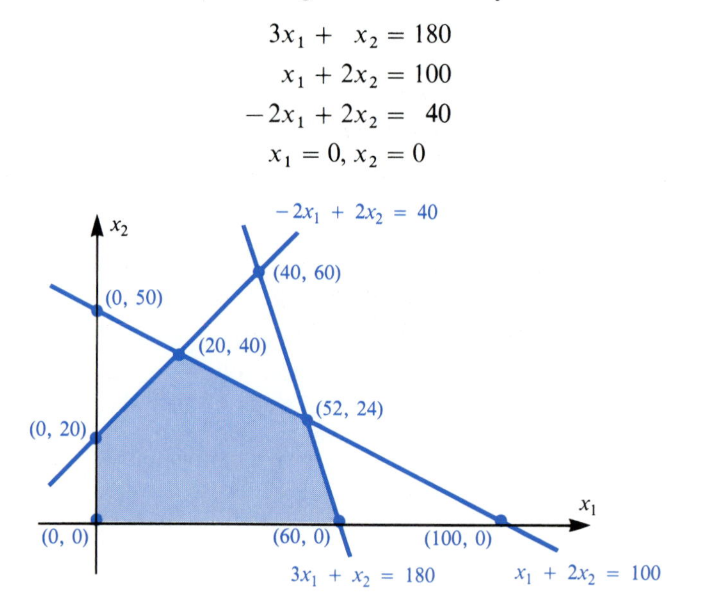

Figure 4–1

form the boundaries of the feasible region. The corners of the feasible region are obtained from the points where pairs of these lines intersect. Not all pairs intersect at a corner point, however. The point (40, 60) is not a corner; it is outside the feasible region.

A Matrix Form of the Standard Maximum Problem

In Chapter 2, we wrote a system of equations as a matrix equation. Similarly, the standard maximum problem can be expressed using matrices.

Example 1 (*Compare Exercise 1*)
Maximize $z = 4x_1 + 12x_2$, subject to the constraints

$$\begin{aligned} 3x_1 + x_2 &\le 180 \\ x_1 + 2x_2 &\le 100 \\ -2x_1 + 2x_2 &\le 40 \\ x_1 \ge 0, x_2 &\ge 0 \end{aligned}$$

Solution This problem can be expressed in matrix form as follows.

The objective function $z = 4x_1 + 12x_2$ can be written

$$z = [4 \quad 12]\begin{bmatrix} x_1 \\ x_2 \end{bmatrix}$$

The constraints

$$\begin{aligned} 3x_1 + x_2 &\le 180 \\ x_1 + 2x_2 &\le 100 \\ -2x_1 + 2x_2 &\le 40 \end{aligned}$$

can be written

$$\begin{bmatrix} 3 & 1 \\ 1 & 2 \\ -2 & 2 \end{bmatrix}\begin{bmatrix} x_1 \\ x_2 \end{bmatrix} \le \begin{bmatrix} 180 \\ 100 \\ 40 \end{bmatrix}$$

and the nonnegative conditions $x_1 \ge 0$, $x_2 \ge 0$ can be written

$$\begin{bmatrix} x_1 \\ x_2 \end{bmatrix} \ge \begin{bmatrix} 0 \\ 0 \end{bmatrix}$$

The notation $A \le B$ means that every element of A is less than or equal to the corresponding element of B.

If we let X be the column matrix of variables

$$\begin{bmatrix} x_1 \\ x_2 \end{bmatrix}$$

let A be the coefficient matrix of the constraints

$$\begin{bmatrix} 3 & 1 \\ 1 & 2 \\ -2 & 2 \end{bmatrix}$$

let B be the column matrix of constant terms of the constraints

$$\begin{bmatrix} 180 \\ 100 \\ 40 \end{bmatrix}$$

and let $C = [4 \quad 12]$ be the row matrix of the coefficients of the objective function, then we can state the standard maximum problem as: Maximize CX subject to $AX \leq B$ where $X \geq 0$.

Slack Variables

The first step in the simplex method converts the constraints to linear equations. We do this by introducing additional variables called **slack variables**. The first constraint, $3x_1 + x_2 \leq 180$, is true for pairs of numbers such as $x_1 = 10$ and $x_2 = 20$, because $3(10) + 20 \leq 180$. Observe that $3x_1 + x_2 + 130 = 180$ when $x_1 = 10$ and $x_2 = 20$. In general, for any pair of values for x_1 and x_2 that make $3x_1 + x_2 \leq 180$ true, there is a number s_1 such that $3x_1 + x_2 + s_1 = 180$. We call s_1 the slack variable because it takes up the slack between $3x_1 + x_2$ and 180. The value of s_1 is never negative. We can add a nonnegative slack variable to each constraint to obtain a set of linear equations.

Example 2 With the addition of slack variables, the constraints (1) become

$$\begin{aligned} 3x_1 + x_2 + s_1 \qquad\qquad &= 180 \\ x_1 + 2x_2 \qquad + s_2 \qquad &= 100 \qquad (2) \\ -2x_1 + 2x_2 \qquad\qquad + s_3 &= 40 \\ x_1 \geq 0, x_2 \geq 0, s_1 \geq 0, s_2 &\geq 0, s_3 \geq 0 \end{aligned}$$

The latter are nonnegative conditions that apply to the variables.

Example 3 (*Compare Exercise 6*)
Write the following constraints as a system of equations using slack variables:

$$\begin{aligned} 5x_1 + 3x_2 + 17x_3 &\leq 140 \\ 7x_1 + 2x_2 + 4x_3 &\leq 256 \\ 3x_1 + 9x_2 + 11x_3 &\leq 540 \\ 2x_1 + 16x_2 + 8x_3 &\leq 99 \end{aligned}$$

Solution Introduce a slack variable for each equation.

$$\begin{aligned} 5x_1 + 3x_2 + 17x_3 + s_1 \qquad\qquad\qquad &= 140 \\ 7x_1 + 2x_2 + 4x_3 \qquad + s_2 \qquad\qquad &= 256 \\ 3x_1 + 9x_2 + 11x_3 \qquad\qquad + s_3 \qquad &= 540 \\ 2x_1 + 16x_2 + 8x_3 \qquad\qquad\qquad + s_4 &= 99 \end{aligned}$$

You may wonder why we are "complicating" the situation by introducing more variables. We do so because it gives a system of equations, not inequalities, making the optimal solution sought a solution to a system of equations.

The objective function needs to be included in the system of equations because we want to find the value of z that comes from the solution of the above system. Its form may need to be modified by writing all terms on the left side. For example,

$$z = 3x_1 + 5x_2$$

is modified to

$$-3x_1 - 5x_2 + z = 0$$

and

$$z = 6x_1 + 7x_2 + 15x_3 + 2x_4$$

is modified to

$$-6x_1 - 7x_2 - 15x_3 - 2x_4 + z = 0$$

Example 4 Include the objective function

$$z = 20x_1 + 35x_2 + 40x_3$$

with the constraints of Example 3 and write them as a system of equations.

Solution

$$\begin{aligned} 5x_1 + 3x_2 + 17x_3 + s_1 &= 140 \\ 7x_1 + 2x_2 + 4x_3 + s_2 &= 256 \\ 3x_1 + 9x_2 + 11x_3 + s_3 &= 540 \\ 2x_1 + 16x_2 + 8x_3 + s_4 &= 99 \\ -20x_1 - 35x_2 + 40x_3 + z &= 0 \end{aligned}$$

Example 5 (*Compare Exercise 9*)
Write the following as a system of equations. Maximize $z = 4x_1 + 12x_2$, subject to

$$\begin{aligned} 3x_1 + x_2 &\leq 180 \\ x_1 + 2x_2 &\leq 100 \\ -2x_1 + 2x_2 &\leq 40 \\ x_1 \geq 0, x_2 &\geq 0 \end{aligned}$$

Solution

$$\begin{aligned} 3x_1 + x_2 + s_1 \quad &= 180 \\ x_1 + 2x_2 \quad + s_2 \quad &= 100 \\ -2x_1 + 2x_2 \quad + s_3 \quad &= 40 \\ -4x_1 - 12x_2 \quad + z &= 0 \end{aligned}$$

Let's pause to make some observations about the systems of equations that we obtain.

1. One slack variable is introduced for each constraint (4 in Example 4 and 3 in Example 5).
2. The total number of variables in the system is the number of original variables (number of x's) plus the number of constraints (number of slack

variables) plus one for z. (The total is $3 + 4 + 1 = 8$ in Example 4 and is $2 + 3 + 1 = 6$ in Example 5.)

3. The introduction of slack variables always results in fewer equations than variables. Recall (the summary in Chapter 2) that this situation generally yields an infinite number of solutions. In fact, one variable for each equation can be written in terms of some other variables.

Look back at the system of equations in Example 5. One solution occurs when we solve for s_1, s_2, s_3, and z in terms of x_1 and x_2.

$$\begin{aligned} s_1 &= 180 - 3x_1 - x_2 \\ s_2 &= 100 - x_1 - 2x_2 \\ s_3 &= 40 + 2x_1 - 2x_2 \\ z &= 4x_1 + 12x_2 \end{aligned}$$

You find a particular solution when you substitute values for x_1 and x_2. The first step of the simplex method always uses zero for that substitution because it gives corner points. So, when $x_1 = 0$ and $x_2 = 0$, we obtain $s_1 = 180$, $s_2 = 100$, $s_3 = 40$, and $z = 0$.

Simplex Tableau

The simplex method uses matrices and row operations on matrices to determine an optimal solution. Let's use the problem from Example 5 to set up the matrix that is called the *simplex tableau.*

Example 6 (*Compare Exercise 13*)
Maximize $z = 4x_1 + 12x_2$, subject to

$$\begin{aligned} 3x_1 + x_2 &\le 180 \\ x_1 + 2x_2 &\le 100 \\ -2x_1 + 2x_2 &\le 40 \\ x_1 \ge 0, x_2 &\ge 0 \end{aligned}$$

Solution First, write the problem as a system of equations using slack variables

$$\begin{aligned} 3x_1 + x_2 + s_1 &= 180 \\ x_1 + 2x_2 + s_2 &= 100 \\ -2x_1 + 2x_2 + s_3 &= 40 \\ -4x_1 - 12x_2 + z &= 0 \end{aligned}$$

Next, form the augmented matrix of this system.

$$\begin{array}{cccccc|c} x_1 & x_2 & s_1 & s_2 & s_3 & z & \\ \hline 3 & 1 & 1 & 0 & 0 & 0 & 180 \\ 1 & 2 & 0 & 1 & 0 & 0 & 100 \\ -2 & 2 & 0 & 0 & 1 & 0 & 40 \\ \hline -4 & -12 & 0 & 0 & 0 & 1 & 0 \end{array}$$

The line is drawn above the bottom row to emphasize that the bottom row is the objective function and because each **pivot** element will be chosen from above the line. (You will learn to find the pivot element in the next section.)

4–1 Exercises

I.

In Exercises 1 through 4, write the standard maximum problems in matrix form.

1. (*See Example 1*) Maximize $z = 50x_1 + 80x_2$ subject to

$$\begin{aligned} 7x_1 + 15x_2 &\le 30 \\ 3x_1 + 14x_2 &\le 56 \\ x_1 \ge 0, x_2 &\ge 0 \end{aligned}$$

2. Maximize $z = 130x_1 + 210x_2 + 110x_3$, subject to

$$\begin{aligned} 5x_1 + x_2 + 17x_3 &\le 48 \\ 6x_1 + 7x_2 + 22x_3 &\le 94 \\ 12x_1 + 8x_2 + x_3 &\le 66 \\ x_1 \ge 0, x_2 \ge 0, x_3 &\ge 0 \end{aligned}$$

3. Maximize $z = 42x_1 + 26x_2 + 5x_3$, subject to

$$\begin{aligned} 13x_1 + 4x_2 + 23x_3 &\le 88 \\ 4x_1 + 15x_2 + 7x_3 &\le 92 \\ 3x_1 - 2x_2 + 32x_3 &\le 155 \\ x_1 \ge 0, x_2 \ge 0, x_3 &\ge 0 \end{aligned}$$

4. Maximize $z = 500x_1 + 340x_2 + 675x_3 + 525x_4$, subject to

$$\begin{aligned} 12x_1 + 5x_2 + 9x_3 + 21x_4 &\le 85 \\ 31x_1 + 14x_2 + 2x_3 &\le 65 \\ x_1 + 43x_2 + 4x_3 + 17x_4 &\le 90 \\ x_1 \ge 0, x_2 \ge 0, x_3 \ge 0, x_4 &\ge 0 \end{aligned}$$

In Exercises 5 through 8, convert the systems of inequalities to systems of equations using slack variables.

5.
$$\begin{aligned} 2x_1 + 3x_2 &\le 9 \\ x_1 + 5x_2 &\le 16 \end{aligned}$$

6. (*See Example 3*)
$$\begin{aligned} 3x_1 - 4x_2 &\le 24 \\ 9x_1 + 5x_2 &\le 16 \\ -x_1 + x_2 &\le 5 \end{aligned}$$

7.
$$\begin{aligned} x_1 + 7x_2 - 4x_3 &\le 150 \\ 5x_1 + 9x_2 + 2x_3 &\le 435 \\ 8x_1 - 3x_2 + 16x_3 &\le 345 \end{aligned}$$

8.
$$\begin{aligned} x_1 + x_2 + x_3 + x_4 &\le 78 \\ 3x_1 + 2x_2 + x_3 - x_4 &\le 109 \end{aligned}$$

In Exercises 9 through 12 express the problems as a system of equations.

9. (*See Example 5*) Maximize $z = 3x_1 + 7x_2$, subject to

$$\begin{aligned} 2x_1 + 6x_2 &\le 9 \\ x_1 - 5x_2 &\le 14 \\ -3x_1 + x_2 &\le 8 \\ x_1 \ge 0, x_2 &\ge 0 \end{aligned}$$

10. Maximize $z = 150x_1 + 280x_2$, subject to

$$\begin{aligned} 12x_1 + 15x_2 &\le 50 \\ 8x_1 + 22x_2 &\le 65 \\ x_1 \ge 0, x_2 &\ge 0 \end{aligned}$$

11. Maximize $z = 420x_1 + 260x_2 + 50x_3$, subject to

$$\begin{aligned} 6x_1 + 7x_2 + 12x_3 &\le 50 \\ 4x_1 + 18x_2 + 9x_3 &\le 85 \\ x_1 - 2x_2 + 14x_3 &\le 66 \\ x_1 \ge 0, x_2 \ge 0, x_3 &\ge 0 \end{aligned}$$

12. Maximize $z = 3x_1 + 4x_2 + 7x_3 + 2x_4$, subject to

$$\begin{aligned} x_1 + 5x_2 + 7x_3 + x_4 &\le 82 \\ 3x_1 + 6x_2 + 12x_3 &\le 50 \\ 2x_1 + 15x_3 + 19x_4 &\le 240 \\ x_1 \ge 0, x_2 \ge 0, x_3 \ge 0, x_4 &\ge 0 \end{aligned}$$

In Exercises 13 through 16, set up the simplex tableau. Do not solve.

13. (*See Example 6*) Maximize $z = 3x_1 + 17x_2$, subject to

$$\begin{aligned} 4x_1 + 5x_2 &\le 10 \\ 3x_1 + x_2 &\le 25 \\ x_1 \ge 0, x_2 &\ge 0 \end{aligned}$$

14. Maximize $z = 140x_1 + 245x_2$, subject to

$$\begin{aligned} 85x_1 + 64x_2 &\le 560 \\ 75x_1 + 37x_2 &\le 135 \\ 24x_1 + 12x_2 &\le 94 \\ x_1 \ge 0, x_2 &\ge 0 \end{aligned}$$

15. Maximize $z = 20x_1 + 45x_2 + 40x_3$, subject to

$$\begin{aligned} 16x_1 - 4x_2 + 9x_3 &\le 128 \\ 8x_1 + 13x_2 + 22x_3 &\le 144 \\ 5x_1 + 6x_2 - 15x_3 &\le 225 \\ x_1 \ge 0, x_2 \ge 0, x_3 &\ge 0 \end{aligned}$$

16. Maximize $z = 18x_1 + 24x_2 + 95x_3 + 50x_4$, subject to

$$\begin{aligned} x_1 + 2x_2 + 5x_3 + 6x_4 &\le 48 \\ 4x_1 + 8x_2 - 15x_3 + 9x_4 &\le 65 \\ 3x_1 - 2x_2 + x_3 - 8x_4 &\le 50 \\ x_1 \ge 0, x_2 \ge 0, x_3 \ge 0, x_4 &\ge 0 \end{aligned}$$

II.

Set up the simplex tableau for each of Exercises 17 through 24. Do not solve.

17. A company manufactures three items, screwdrivers, chisels, and putty knives, on three production lines, *A*, *B*, and *C*, respectively. It takes three hours labor per carton to produce screwdrivers on line *A*, four hours labor per carton of chisels on line *B*, and five hours labor per carton of putty knives on line *C*. Each carton of screwdrivers costs $15 to produce, each carton of chisels costs $12, and each carton of putty knives costs $11. The profit per carton is $5 for screwdrivers, $6 for chisels, and $5 for putty knives. If the company has 2200 hours of labor and $8500 operating funds available per week, how many cartons of each item should they produce to maximize profit?

18. The maximum daily production of an oil refinery is 1900 barrels. The refinery can produce three types of fuel: gasoline, diesel, and heating oil. The production cost per barrel is $6 for gasoline, $5 for diesel, and $8 for heating oil. The daily production budget is $13,400. The profit is $7 per barrel on gasoline, $6 on diesel, and $9 on heating oil. How much of each should be produced to maximize profit?

19. A fellow eats a meal of steak, baked potato with butter, and salad with dressing. One ounce of each contains the indicated calories, protein, and fat.

	Calories	Protein (mg)	Fat (mg)
Salad (oz)	20	0.5	1.5
Potato (oz)	50	1.0	3.0
Steak (oz)	56	9.0	2.0

He is to consume no more than 1000 calories and 35 mg fat. How much of each should he eat in order to maximize the protein consumed?

20. A manufacturing company uses three machines, I, II, and III, to produce three products: note pads, loose-leaf paper, and spiral notebooks. It takes 2, 4, and 0 minutes of time on each of the machines, respectively, to manufacture a carton of notepads. It takes 3, 0, and 6 minutes on the machines, respectively, to produce a carton of loose-leaf paper. Manufacturing a carton of spiral notebooks involves 1, 2, and 3 minutes, respectively, on each machine. The total time available on each machine per day is six hours. The profits are $10, $8, and $12 on each carton of notepads, loose-leaf paper, and spiral notebooks, respectively. How many of each item should be produced to maximize total profit?

21. An industrial furniture company manufactures desks, cabinets, and chairs. These items involve metal, wood, and plastic. The following table represents, in convenient units, the amounts that go into each product and the profit on each item.

	Metal	Wood	Plastic	Profit
Desk	3	4	2	$16
Cabinet	9	1	1	$12
Chair	1	2	2	$ 6

If the company has available 810 units of metal, 400 units of wood, and 100 units of plastic, how many desks, cabinets, and chairs should be produced to maximize total profit?

22. A washing machine manufacturing company produces its machines at three factories, A, B, and C. The washing machines are sold in a certain city, P. It costs $10, $20, and $40 to transport each washing machine, on an average, from A, B, and C, respectively, to P. It has been estimated that it involves 6, 4, and 2 man hours time in packing and transportation to get a washing machine from A, B, and C, respectively, to P. There is $9000 budgeted weekly for transportation of the washing machines to P, and a total of 4000 man hours labor available. How should the company schedule its weekly transportation arrangements to maximize total profit if the profit on each machine from A is \$12, on each machine from B is \$20, and on each machine from C is \$16?

23. A manufacturer makes three lines of tents, the Alpine, the Cub, and the Aspen. They are all made out of the same material; the Alpine requires 30 square yards; the Cub, 15 square yards; and the Aspen, 60 square yards. The manufacturing cost of the Alpine is \$36, the Cub \$24, and the Aspen \$32. The material is available in amounts of 8400 square yards weekly, and the weekly working budget of the company is \$7680. If the profit is \$8 on each Alpine, \$4 on each Cub, and \$12 on each Aspen, what should the weekly production schedule be to maximize total profit?

24. A furniture company finishes two kinds of tables, X and Y. There are three steps in the finishing process: sanding, staining, and varnishing. The various times, in minutes, of each of these processes for the two tables are:

	Sanding	**Staining**	**Varnishing**
X	10	8	4
Y	5	4	8

The three types of equipment needed for sanding, staining, and varnishing are each available 6 hours, 5 hours, and 5 hours 20 minutes, respectively, per day. Each type of equipment can only handle one table at a time. The profit on each X table is \$8 and on each Y table is \$4. How many of each should be finished daily to maximize total profit?

4–2 The Simplex Method

Basic Solution
Pivot Column, Row, Element
Final Tableau

Basic Solution

The simplex method is a process of finding a sequence of selected solutions to the system of equations. The selections are made so that the optimal solution is found in a relatively small number of steps.

Let's look at the example in Section 4–1 again. It is the following. Maximize the objective function $z = 4x_1 + 12x_2$, subject to the constraints

$$\begin{aligned} 3x_1 + x_2 &\leq 180 \\ x_1 + 2x_2 &\leq 100 \\ -2x_1 + 2x_2 &\leq 40 \\ x_1 \geq 0, x_2 &\geq 0 \end{aligned} \qquad (1)$$

We introduce slack variables to obtain the following. Maximize $z = 4x_1 + 12x_2$, subject to

$$\begin{aligned} 3x_1 + x_2 + s_1 &= 180 \\ x_1 + 2x_2 + s_2 &= 100 \\ -2x_1 + 2x_2 + s_3 &= 40 \end{aligned} \qquad (2)$$

where x_1, x_2, s_1, s_2, and s_3 are all nonnegative.

The simplex tableau is

$$\left[\begin{array}{cccccc|c} 3 & 1 & 1 & 0 & 0 & 0 & 180 \\ 1 & 2 & 0 & 1 & 0 & 0 & 100 \\ -2 & 2 & 0 & 0 & 1 & 0 & 40 \\ \hline -4 & -12 & 0 & 0 & 0 & 1 & 0 \end{array}\right] \qquad (3)$$

In this example, we first find solutions which have *two* variables set to zero. The number two is used because the problem involves two x's. Solutions like this are called **basic solutions**.

In the above system of equations, (2), the simplest basic solution is obtained by setting x_1 and x_2 to zero. Then solve for the other variables to obtain $s_1 = 180$, $s_2 = 100$, $s_3 = 40$, and $z = 0$.

Definition

If a linear programming problem has k x's in the constraints, then a **basic solution** is obtained by setting k variables (except z) to zero and solving for the others.

If we set $s_1 = 0$ and $s_3 = 0$, we can obtain another basic solution by solving for the other variables from the simplex tableau. If $s_1 = 0$ and $s_3 = 0$, the system (2) reduces to

$$\begin{aligned} 3x_1 + x_2 &= 180 \\ x_1 + 2x_2 + s_2 &= 100 \\ -2x_1 + 2x_2 &= 40 \\ -4x_1 - 12x_2 + z &= 0 \end{aligned}$$

We have four equations in four unknowns, x_1, x_2, s_2, and z, to solve. This system has the solution (we omit the details)

$$x_1 = 40, x_2 = 60, s_2 = -60, z = 880$$

Notice that s_2 is *negative*. This violates the nonnegative constraint on the x's and slack variables. While the solution

$$x_1 = 40, x_2 = 60, s_1 = 0, s_2 = -60, s_3 = 0, z = 880$$

is a basic solution, it is not feasible. Properly carried out, the simplex method finds only basic feasible solutions.

Definition

Call the number of x variables in a linear programming problem k. A **basic feasible solution** of the system of equations is a solution with k variables (except z) set to zero and with none of the slack variables or x's negative.

With this background, let us proceed to solve this example by finding the appropriate basic feasible solutions.

STEP 1 For the tableau

x_1	x_2	s_1	s_2	s_3	z	
3	1	1	0	0	0	180
1	2	0	1	0	0	100
−2	2	0	0	1	0	40
−4	−12	0	0	0	1	0

find the **initial basic feasible solution**, that is, set all x's to zero. Then read the other variables from the tableau.

$$x_1 = 0, x_2 = 0, s_1 = 180, s_2 = 100, s_3 = 40, z = 0$$

This step is equivalent to using the origin as the corner point of the feasible region.

Pivot Column, Row, Element

STEP 2 Select the **pivot** element and then modify the tableau so that the resulting basic feasible solution will increase the value of z. This step requires the selection of a **pivot** element from the tableau as follows.

(a) To select the column containing the pivot element, do the following. Select the *most negative* entry from the bottom row.

x_1	x_2	s_1	s_2	s_3	z	
3	1	1	0	0	0	180
1	2	0	1	0	0	100
−2	2	0	0	1	0	40
−4	(−12)	0	0	0	1	0

↑ most negative entry gives pivot column

This selects the **pivot column** containing the pivot element. The pivot element itself is an entry in this column *above* the line. We must now determine which row contains the pivot element.

(b) To select the row containing the pivot element, do the following.

To determine the proper row, we must divide each constant above the line in the last column by the corresponding entries in the pivot column. These ratios are written to the right of the tableau.

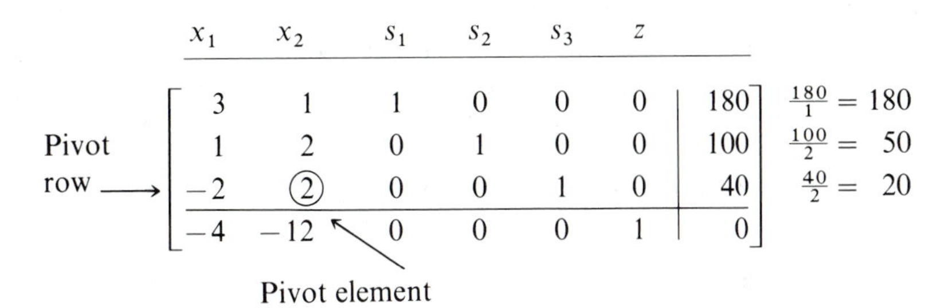

	x_1	x_2	s_1	s_2	s_3	z		
	3	1	1	0	0	0	180	$\frac{180}{1} = 180$
	1	2	0	1	0	0	100	$\frac{100}{2} = 50$
Pivot row →	−2	(2)	0	0	1	0	40	$\frac{40}{2} = 20$
	−4	−12	0	0	0	1	0	

Now select the smallest **positive** ratio, 20. This determines the **pivot row** containing the pivot element. The entry, 2, in the pivot row and pivot column is the **pivot element**.

STEP 3 Move to the next basic feasible solution.

Use row operations to modify the tableau so that the pivot element becomes a 1 and the rest of the pivot column contains 0's. (You recognize this is part of the Gauss-Jordan method for solving systems.)

Divide each entry in the third row (pivot row) by 2, so the pivot entry becomes 1. The third row becomes

$$[-1 \quad 1 \quad 0 \quad 0 \quad \tfrac{1}{2} \quad 0 \quad 20]$$

giving the tableau

x_1	x_2	s_1	s_2	s_3	z	
3	(1)	1	0	0	0	180
1	(2)	0	1	0	0	100
−1	1	0	0	$\frac{1}{2}$	0	20
−4	(−12)	0	0	0	1	0

We now need 0's in the circled locations of the pivot column.

This is accomplished as follows:

Replace Row 1 with (Row 1 − Row 3)

$$= [4 \quad 0 \quad 1 \quad 0 \quad -\tfrac{1}{2} \quad 0 \quad 160].$$

Replace Row 2 with (Row 2 + (−2)Row 3)

$$= [3 \quad 0 \quad 0 \quad 1 \quad -1 \quad 0 \quad 60].$$

Replace Row 4 with (Row 4 + (12)Row 3)

$$= [-16 \quad 0 \quad 0 \quad 0 \quad 6 \quad 1 \quad 240].$$

This gives the tableau

$$\begin{array}{cccccc|c} x_1 & x_2 & s_1 & s_2 & s_3 & z & \\ \hline 4 & 0 & 1 & 0 & -\frac{1}{2} & 0 & 160 \\ 3 & 0 & 0 & 1 & -1 & 0 & 60 \\ -1 & 1 & 0 & 0 & \frac{1}{2} & 0 & 20 \\ \hline -16 & 0 & 0 & 0 & 6 & 1 & 240 \end{array}$$

To determine the basic feasible solution from this tableau, observe that the columns under x_2, s_1, s_2, and z contain all 0's, except for a single 1 in each column. Furthermore, the 1's occur in different rows. These variables are the *basic* variables, the ones we solve for. The other two variables, x_1 and s_3, are the nonbasic variables, the variables we set to 0. Since basic variables correspond to the columns formed by a single 1 and 0's, the nonbasic variables correspond to the columns that differ from this.

The basic feasible solution from this tableau is

$$x_1 = 0, x_2 = 20, s_1 = 160, s_2 = 60, s_3 = 0, z = 240$$

The initial solution gave $z = 0$, and this solution gave $z = 240$, so we do indeed have a larger value of the objective function.

The simplex tableau tells whether the maximum value has been achieved.

STEP 4 Is z maximum?

If the last row contains any negative coefficients, z is not maximum. Since -16 is a coefficient from the last row, 240 is not the maximum value of z, so we proceed to move to another basic feasible solution.

Final Tableau

STEP 5 Find another basic feasible solution.

Proceed as in Steps 2 and 3 with the most recent tableau.

$$\begin{array}{cccccc|c|l} x_1 & x_2 & s_1 & s_2 & s_3 & z & & \\ \hline 4 & 0 & 1 & 0 & -\frac{1}{2} & 0 & 160 & \frac{160}{4} = 40 \\ \textcircled{3} & 0 & 0 & 1 & -1 & 0 & 60 & \frac{60}{3} = 20 \leftarrow \\ -1 & 1 & 0 & 0 & \frac{1}{2} & 0 & 20 & \frac{20}{(-1)} = -20 \\ \hline \textcircled{-16} & 0 & 0 & 0 & 6 & 1 & 240 & \text{Smallest positive ratio} \end{array}$$

Pivot element (→ 3); Pivot Row (row 2). The circled -16: Most negative element. It is in the pivot column. The arrow at 20: Smallest positive ratio.

We now use row operations to obtain a 1 in the pivot element position and 0's in the rest of the pivot column. Replace Row 2 with $\frac{1}{3}$ Row 2 = $[1 \quad 0 \quad 0 \quad \frac{1}{3} \quad -\frac{1}{3} \quad 0 \quad 20]$ to obtain a 1 in the pivot position.

need zeros here (annotations pointing to the circled entries 4, −1, −16 in the x_1 column)

$$\begin{array}{cccccc|c} x_1 & x_2 & s_1 & s_2 & s_3 & z & \\ \hline \textcircled{4} & 0 & 1 & 0 & -\frac{1}{2} & 0 & 160 \\ 1 & 0 & 0 & \frac{1}{3} & -\frac{1}{3} & 0 & 20 \\ \textcircled{-1} & 1 & 0 & 0 & \frac{1}{2} & 0 & 20 \\ \hline \textcircled{-16} & 0 & 0 & 0 & 6 & 1 & 240 \end{array}$$

We obtain the desired 0's by replacing the rows as follows:
Replace Row 1 with (Row 1 + (−4)Row 2)

$$= [0 \quad 0 \quad 1 \quad -\tfrac{4}{3} \quad \tfrac{5}{6} \quad 0 \quad 80]$$

Replace Row 3 with (Row 3 + Row 2)

$$= [0 \quad 1 \quad 0 \quad \tfrac{1}{3} \quad \tfrac{1}{6} \quad 0 \quad 40]$$

Replace Row 4 with (Row 4 + (16)Row 2)

$$= [0 \quad 0 \quad 0 \quad \tfrac{16}{3} \quad \tfrac{2}{3} \quad 1 \quad 560]$$

giving the tableau

$$\begin{array}{cccccc|c} x_1 & x_2 & s_1 & s_2 & s_3 & z & \\ \hline 0 & 0 & 1 & -\frac{4}{3} & \frac{5}{6} & 0 & 80 \\ 1 & 0 & 0 & \frac{1}{3} & -\frac{1}{3} & 0 & 20 \\ 0 & 1 & 0 & \frac{1}{3} & \frac{1}{6} & 0 & 40 \\ \hline 0 & 0 & 0 & \frac{16}{3} & \frac{2}{3} & 1 & 560 \end{array}$$

The basic variables are x_1, x_2, s_1, and z because those columns each have a single 1 and 0's elsewhere, and the 1's are in different rows. The basic feasible solution is obtained by setting s_2 and $s_3 = 0$ and solving for the others. The solution is

$$x_1 = 20,\ x_2 = 40,\ s_1 = 80,\ s_2 = 0,\ s_3 = 0,\ z = 560$$

This solution gives the maximum value of z because there are no negative values in the last row of this tableau, the **final simplex tableau**.

Example 1 *(Compare Exercise 1)*
Use the simplex method to maximize $z = 2x_1 + 3x_2 + 2x_3$, subject to

$$\begin{aligned} 2x_1 + x_2 + 2x_3 &\le 13 \\ x_1 + x_2 - 3x_3 &\le 8 \\ x_1 \ge 0,\ x_2 \ge 0,\ x_3 &\ge 0 \end{aligned}$$

Solution We first write the problem as a system of equations

$$\begin{aligned} 2x_1 + x_2 + 2x_3 + s_1 \qquad &= 13 \\ x_1 + x_2 - 3x_3 \qquad + s_2 \qquad &= 8 \\ -2x_1 - 3x_2 - 2x_3 \qquad\qquad + z &= 0 \end{aligned}$$

The initial simplex tableau is

$$\begin{array}{cccccc|c} x_1 & x_2 & x_3 & s_1 & s_2 & z & \\ \hline 2 & 1 & 2 & 1 & 0 & 0 & 13 \\ 1 & 1 & -3 & 0 & 1 & 0 & 8 \\ \hline -2 & -3 & -2 & 0 & 0 & 1 & 0 \end{array}$$

Since there are three x's, all basic solutions will have three variables set to zero. From the initial tableau, the initial basic feasible solution is

$$x_1 = 0, x_2 = 0, x_3 = 0, s_1 = 13, s_2 = 8, z = 0$$

Since there are negative entries in the last row, the solution is not optimal.

Find the pivot element.

$$\begin{array}{cccccc|c|c} x_1 & x_2 & x_3 & s_1 & s_2 & z & & \\ \hline 2 & 1 & 2 & 1 & 0 & 0 & 13 & \frac{13}{1} \\ 1 & \textcircled{1} & -3 & 0 & 1 & 0 & 8 & \frac{8}{1} \\ \hline -2 & -3 & -2 & 0 & 0 & 1 & 0 & \end{array}$$

Pivot element → circled 1 in row 2, column x_2

Pivot row, smallest positive ratio

Pivot column, most negative entry

Find the next tableau using the following row operations:
Replace Row 1 with (Row 1 − Row 2)

$$= [1 \quad 0 \quad 5 \quad 1 \quad -1 \quad 0 \quad 5]$$

Replace Row 3 with (Row 3 + (3)Row 2)

$$= [1 \quad 0 \quad -11 \quad 0 \quad 3 \quad 1 \quad 24]$$

giving

$$\begin{array}{cccccc|c} x_1 & x_2 & x_3 & s_1 & s_2 & z & \\ \hline 1 & 0 & 5 & 1 & -1 & 0 & 5 \\ 1 & 1 & -3 & 0 & 1 & 0 & 8 \\ \hline 1 & 0 & -11 & 0 & 3 & 1 & 24 \end{array}$$

Again, the solution is not optimal since a negative entry, −11, occurs in the last row. Find the new pivot element.

$$\begin{array}{cccccc|c|l} x_1 & x_2 & x_3 & s_1 & s_2 & z & & \\ \hline 1 & 0 & \textcircled{5} & 1 & -1 & 0 & 5 & \frac{5}{5} = 1 \text{ smallest positive ratio} \\ 1 & 1 & -3 & 0 & 1 & 0 & 8 & \frac{8}{-3} = -2.67 \\ \hline 1 & 0 & -11 & 0 & 3 & 1 & 24 & \end{array}$$

Pivot element → circled 5 in row 1, column x_3

pivot column

Replace the pivot element, 5, with 1 by dividing Row 1 by 5.

$$\begin{array}{cccccc|c} x_1 & x_2 & x_3 & s_1 & s_2 & z & \\ \hline \frac{1}{5} & 0 & 1 & \frac{1}{5} & -\frac{1}{5} & 0 & 1 \\ 1 & 1 & -3 & 0 & 1 & 0 & 8 \\ \hline 1 & 0 & -11 & 0 & 3 & 1 & 24 \end{array}$$

Obtain the simplex tableau for the next basic feasible solution by the row operations:

Replace Row 2 with (Row 2 + (3)Row 1)

$$= [\tfrac{8}{5} \quad 1 \quad 0 \quad \tfrac{3}{5} \quad \tfrac{2}{5} \quad 0 \quad 11]$$

Replace Row 3 with (Row 3 + (11)Row 1)

$$= [\tfrac{16}{5} \quad 0 \quad 0 \quad \tfrac{11}{5} \quad \tfrac{4}{5} \quad 1 \quad 35]$$

giving the tableau

$$\begin{array}{cccccc|c} x_1 & x_2 & x_3 & s_1 & s_2 & z & \\ \hline \frac{1}{5} & 0 & 1 & \frac{1}{5} & -\frac{1}{5} & 0 & 1 \\ \frac{8}{5} & 1 & 0 & \frac{3}{5} & \frac{2}{5} & 0 & 11 \\ \hline \frac{16}{5} & 0 & 0 & \frac{11}{5} & \frac{4}{5} & 1 & 35 \end{array}$$

The basic feasible solution from the tableau is

$$x_1 = 0, x_2 = 11, x_3 = 1, s_1 = 0, s_2 = 0, z = 35$$

Since there are no negative entries in the last row, $z = 35$ is a maximum.

Summary of the Simplex Method

Standard Maximization Problem

1. Convert the problem to a system of equations:
   (a) Convert each inequality to an equation by adding a slack variable.
   (b) Write the objective function

   $$z = ax_1 + bx_2 + \cdots + kx_n$$

   as

   $$-ax_1 - bx_2 - \cdots - kx_n + z = 0$$

2. Form the initial simplex tableau from the equations.
3. Locate the pivot element of the tableau:
   (a) Locate the most negative entry in the bottom row. It is in the pivot column. If there is a tie for most negative, choose either.
   (b) Divide each entry in the last column (above the line) by the corresponding entry in the pivot column. Choose the smallest positive one. It is in the pivot row. In the case of a tie, choose either.

(c) The element where the pivot column and pivot row intersect is the pivot element.

4. Modify the simplex tableau using row operations to obtain a new basic feasible solution.

(a) Divide each entry in the pivot row by the pivot element to obtain a 1 in the pivot positon.

(b) Convert all other entries in the pivot column to 0 by using row operations.

5. Determine whether z has reached its maximum.

(a) If there is a negative entry in the last row of the tableau, z is not maximum. Repeat the process in Steps 3 and 4.

(b) If the bottom row contains no negative entries, z is maximum and the solution is available from the final tableau.

6. Determine the solution from the final tableau.

(a) Set k variables to 0, where k is the number of x's used in the constraints. These are the nonbasic variables. They correspond to the columns that contain more than one nonzero entry.

(b) Determine the values of the basic variables. They correspond to those columns consisting of a single 1 entry and other entries 0.

Example 2 (*Compare Exercise 9*)
Maximize $z = 9x_1 + 5x_2 + 9x_3$, subject to

$$\begin{aligned} 6x_1 + x_2 + 4x_3 &\leq 72 \\ 3x_1 + 4x_2 + 2x_3 &\leq 30 \\ x_1 \geq 0, x_2 \geq 0, x_3 &\geq 0 \end{aligned}$$

Solution Form the system of equations

$$\begin{aligned} 6x_1 + x_2 + 4x_3 + s_1 \qquad\qquad &= 72 \\ 3x_1 + 4x_2 + 2x_3 \qquad + s_2 \qquad &= 30 \\ -9x_1 - 5x_2 - 9x_3 \qquad\qquad + z &= 0 \end{aligned}$$

From this system, we write the initial tableau

$$\begin{array}{cccccc|c} x_1 & x_2 & x_3 & s_1 & s_2 & z & \\ \hline 6 & 1 & 4 & 1 & 0 & 0 & 72 \\ 3 & 4 & 2 & 0 & 1 & 0 & 30 \\ \hline -9 & -5 & -9 & 0 & 0 & 1 & 0 \end{array}$$

Since there is a tie in the last row for the most negative entry, we have two choices for the pivot column. We use the first one.

$$
\begin{array}{l}
\text{Pivot} \\
\text{element}
\end{array}
\quad
\begin{array}{cccccc}
x_1 & x_2 & x_3 & s_1 & s_2 & z
\end{array}
$$

$$
\left[\begin{array}{cccccc|c}
6 & 1 & 4 & 1 & 0 & 0 & 72 \\
\textcircled{3} & 4 & 2 & 0 & 1 & 0 & 30 \\
\hline
-9 & -5 & -9 & 0 & 0 & 1 & 0
\end{array}\right]
\begin{array}{l}
\frac{72}{6} = 12 \\
\frac{30}{3} = 10 \\
\\
\end{array}
$$

We give the sequence of tableaux to find the optimal solution, but leave out some of the details. Be sure you follow each step.

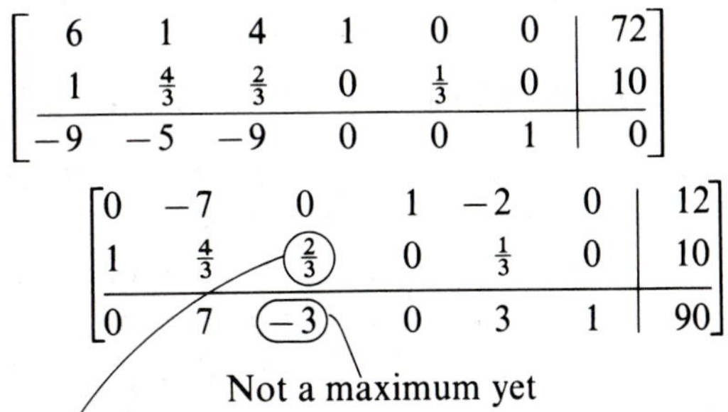

$$
\left[\begin{array}{cccccc|c}
6 & 1 & 4 & 1 & 0 & 0 & 72 \\
1 & \frac{4}{3} & \frac{2}{3} & 0 & \frac{1}{3} & 0 & 10 \\
\hline
-9 & -5 & -9 & 0 & 0 & 1 & 0
\end{array}\right]
$$

$$
\left[\begin{array}{cccccc|c}
0 & -7 & 0 & 1 & -2 & 0 & 12 \\
1 & \frac{4}{3} & \textcircled{\frac{2}{3}} & 0 & \frac{1}{3} & 0 & 10 \\
\hline
0 & 7 & \textcircled{-3} & 0 & 3 & 1 & 90
\end{array}\right]
$$

$$
\left[\begin{array}{cccccc|c}
0 & -7 & 0 & 1 & -2 & 0 & 12 \\
\frac{3}{2} & 2 & 1 & 0 & \frac{1}{2} & 0 & 15 \\
\hline
0 & 7 & -3 & 0 & 3 & 1 & 90
\end{array}\right]
$$

$$
\left[\begin{array}{cccccc|c}
0 & -7 & 0 & 1 & -2 & 0 & 12 \\
\frac{3}{2} & 2 & 1 & 0 & \frac{1}{2} & 0 & 15 \\
\hline
\frac{9}{2} & 13 & 0 & 0 & \frac{9}{2} & 1 & 135
\end{array}\right]
$$

This is the final tableau with the solution

$$x_1 = 0,\ x_2 = 0,\ x_3 = 15,\ s_1 = 12,\ s_2 = 0,\ z = 135$$

Now let's look at an example that has a tie for pivot row.

Example 3 (*Compare Exercise 11*)
Maximize $z = 3x_1 + 8x_2$, subject to

$$
\begin{aligned}
x_1 + 2x_2 &\le 80 \\
4x_1 + x_2 &\le 68 \\
5x_1 + 3x_2 &\le 120 \\
x_1 \ge 0,\ x_2 &\ge 0
\end{aligned}
$$

Solution The initial tableau is

$$
\left[\begin{array}{cccccc|c}
1 & 2 & 1 & 0 & 0 & 0 & 80 \\
4 & 1 & 0 & 1 & 0 & 0 & 68 \\
5 & 3 & 0 & 0 & 1 & 0 & 120 \\
\hline
-3 & -8 & 0 & 0 & 0 & 1 & 0
\end{array}\right]
\begin{array}{l}
\frac{80}{2} = 40 \\
\frac{68}{1} = 68 \\
\frac{120}{3} = 40 \\
\\
\end{array}
$$

$\uparrow$ Pivot column (under -8)

Since there are two ratios of 40 each, either Row 1 or Row 3 may be selected as the pivot row. If we select Row 1, then 2 is the pivot element, and our sequence of tableaux is the following. You should work the row operations so you see that each tableau is correct.

$$\left[\begin{array}{cccccc|c} \frac{1}{2} & 1 & \frac{1}{2} & 0 & 0 & 0 & 40 \\ 4 & 1 & 0 & 1 & 0 & 0 & 68 \\ 5 & 3 & 0 & 0 & 1 & 0 & 120 \\ \hline -3 & -8 & 0 & 0 & 0 & 1 & 0 \end{array}\right]$$

$$\left[\begin{array}{cccccc|c} \frac{1}{2} & 1 & \frac{1}{2} & 0 & 0 & 0 & 40 \\ \frac{7}{2} & 0 & -\frac{1}{2} & 1 & 0 & 0 & 28 \\ \frac{7}{2} & 0 & -\frac{3}{2} & 0 & 1 & 0 & 0 \\ \hline 1 & 0 & 4 & 0 & 0 & 1 & 320 \end{array}\right]$$

This final tableau gives the optimal solution

$$x_1 = 0,\ x_2 = 40,\ s_1 = 0,\ s_2 = 28,\ s_3 = 0,\ z = 320$$

We now demonstrate how the simplex method handles a linear programming problem that has no solution, where the feasible region is unbounded.

Example 4 (*Compare Exercise 16*)
Maximize $x_1 + 4x_2$, subject to

$$\begin{aligned} x_1 - x_2 &\leq 3 \\ -4x_1 + x_2 &\leq 4 \\ x_1 \geq 0,\ x_2 &\geq 0 \end{aligned}$$

Solution The graph of the feasible region is shown in Figure 4–2. This problem converts to the system

$$\begin{array}{rrrrrl} x_1 & - x_2 & + s_1 & & & = 3 \\ -4x_1 & + x_2 & & + s_2 & & = 4 \\ -x_1 & - 4x_2 & & & + z & = 0 \end{array}$$

and to the initial simplex tableau.

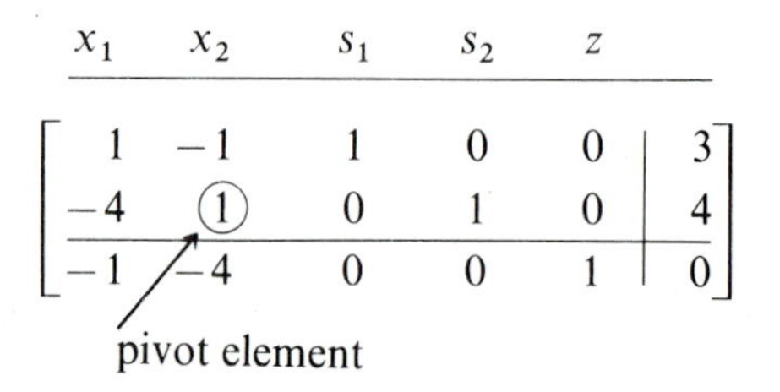

$$\begin{array}{ccccc} x_1 & x_2 & s_1 & s_2 & z \end{array}$$
$$\left[\begin{array}{ccccc|c} 1 & -1 & 1 & 0 & 0 & 3 \\ -4 & \textcircled{1} & 0 & 1 & 0 & 4 \\ \hline -1 & -4 & 0 & 0 & 1 & 0 \end{array}\right]$$

Except for the pivot element, convert all entries in the pivot column to 0 to obtain the next tableau.

$$\left[\begin{array}{rrrrr|r} -3 & 0 & 1 & 1 & 0 & 7 \\ -4 & 1 & 0 & 1 & 0 & 4 \\ \hline -17 & 0 & 0 & 4 & 1 & 16 \end{array}\right]$$

Because of the -17 in the last row, we know that z is not maximal. When we check for the pivot row, we get the ratios $-7/3$ and -1. We cannot proceed with the simplex method because all ratios are negative. When this occurs, there is no maximum. Because the feasible region is unbounded (Figure 4–2) a maximum value of the objective function does not exist.

> When you arrive at a simplex tableau that has no positive entries in the pivot column, the feasible region is unbounded, and the objective function is unbounded. There is no maximum value.

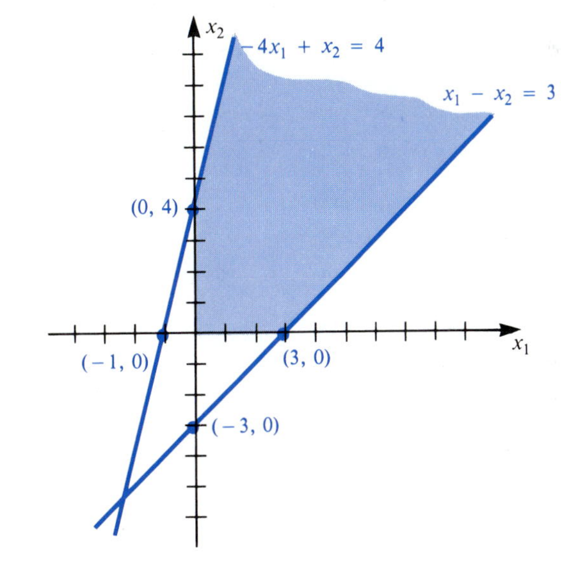

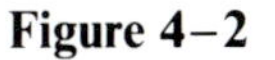

Figure 4–2 Unbound feasible region

4–2 Exercises

I.

Use the simplex method to solve the exercises in this section.

1. (*See Example 1*) Maximize $2x_1 + x_2$, subject to
$4x_1 + x_2 \leq 36$
$4x_1 + 3x_2 \leq 60$
$x_1 \geq 0, x_2 \geq 0$

2. Maximize $x_1 - 4x_2$, subject to
$x_1 + 2x_2 \leq 4$
$x_1 + 6x_2 \leq 8$
$x \geq 0, x_2 \geq 0$

3. Maximize $4x_1 + 6x_2$, subject to

$$\begin{aligned} x_1 + 3x_2 &\le 5 \\ 3x_1 + x_2 &\le 7 \\ x_1 \ge 0, x_2 &\ge 0 \end{aligned}$$

4. Maximize $2x_1 + 3x_2$, subject to

$$\begin{aligned} -4x_1 + 2x_2 &\le 7 \\ 2x_1 - 2x_2 &\le 5 \\ x_1 \ge 0, x_2 &\ge 0 \end{aligned}$$

5. Maximize $10x_1 + 5x_2$, subject to

$$\begin{aligned} x_1 + x_2 &\le 180 \\ 3x_1 + 2x_2 &\le 480 \\ x_1 \ge 0, x_2 &\ge 0 \end{aligned}$$

6. Maximize $x_1 + 2x_2 + x_3$, subject to

$$\begin{aligned} 3x_1 + x_2 + x_3 &\le 21 \\ x_1 + 10x_2 - 4x_3 &\le 140 \\ x_1 \ge 0, x_2 \ge 0, x_3 &\ge 0 \end{aligned}$$

7. Maximize $z = 100x_1 + 200x_2 + 50x_3$, subject to

$$\begin{aligned} 5x_1 + 5x_2 + 10x_3 &\le 1000 \\ 10x_1 + 8x_2 + 5x_3 &\le 2000 \\ 10x_1 + 5x_2 &\le 500 \\ x_1 \ge 0, x_2 \ge 0, x_3 &\ge 0 \end{aligned}$$

8. Maximize $z = 2x_1 + x_2 + x_3$, subject to

$$\begin{aligned} x_1 + 2x_2 + 4x_3 &\le 20 \\ 2x_1 + 4x_2 + 4x_3 &\le 60 \\ 3x_1 + 4x_2 + x_3 &\le 90 \\ x_1 \ge 0, x_2 \ge 0, x_3 &\ge 0 \end{aligned}$$

9. (*See Example 2*) Maximize $z = 3x_1 + 5x_2 + 8x_3$, subject to

$$\begin{aligned} x_1 + x_2 + x_3 &\le 100 \\ 3x_1 + 2x_2 + 4x_3 &\le 200 \\ x_1 + 2x_2 &\le 150 \\ x_1 \ge 0, x_2 \ge 0, x_3 &\ge 0 \end{aligned}$$

10. Maximize $z = 8x_1 + 2x_2$, subject to

$$\begin{aligned} 4x_1 + x_2 &\le 32 \\ 4x_1 + 3x_2 &\le 48 \\ x_1 \ge 0, x_2 &\ge 0 \end{aligned}$$

11. (*See Example 3*) Maximize $z = 33x_1 + 9x_2$, subject to

$$\begin{aligned} x_1 + 8x_2 &\le 66 \\ 3x_1 + 9x_2 &\le 72 \\ 2x_1 + 6x_2 &\le 48 \\ x_1 \ge 0, x_2 &\ge 0 \end{aligned}$$

12. Maximize $z = 4x_1 + 3x_2$, subject to

$$\begin{aligned} 2x_1 + 3x_2 &\le 12 \\ x_1 + 2x_2 &\le 6 \\ 2x_1 + 5x_2 &\le 20 \\ x_1 \ge 0, x_2 &\ge 0 \end{aligned}$$

13. Maximize $z = 22x_1 + 20x_2 + 18x_3$, subject to

$$\begin{aligned} 2x_1 + x_2 + 2x_3 &\le 100 \\ x_1 + 2x_2 + 2x_3 &\le 100 \\ 2x_1 + 2x_2 + x_3 &\le 100 \\ x_1 \ge 0, x_2 \ge 0, x_3 &\ge 0 \end{aligned}$$

14. Maximize $z = x_1 + 2x_2 + 3x_3$, subject to

$$\begin{aligned} 2x_1 + x_2 + 2x_3 &\le 330 \\ x_1 + 2x_2 + 2x_3 &\le 330 \\ -2x_1 - 2x_2 + x_3 &\le 132 \\ x_1 \ge 0, x_2 \ge 0, x_3 &\ge 0 \end{aligned}$$

15. Maximize $z = 8x_1 + 9x_2 + 15x_3$, subject to

$$\begin{aligned} 2x_1 + x_2 + 4x_3 &\le 340 \\ 2x_1 + 5x_2 + 10x_3 &\le 850 \\ 4x_1 + 3x_2 + x_3 &\le 510 \\ x_1 \ge 0, x_2 \ge 0, x_3 &\ge 0 \end{aligned}$$

16. (*See Example 4*) Maximize $z = 2x_1 + 3x_2$, subject to

$$\begin{aligned} -5x_1 + x_2 &\le 5 \\ x_1 - 4x_2 &\le 8 \\ x_1 \ge 0, x_2 &\ge 0 \end{aligned}$$

17. Maximize $z = x_1 + 4x_2$, subject to

$$\begin{aligned} -4x_1 + x_2 &\le 2 \\ 2x_1 - x_2 &\le 1 \\ x_1 \ge 0, x_2 &\ge 0 \end{aligned}$$

18. Maximize $z = 8x_1 + 6x_2 + 2x_3$, subject to

$$\begin{aligned} x_1 - 3x_2 + 2x_3 &\le 50 \\ -2x_1 + 4x_2 + 5x_3 &\le 40 \\ x_1 \ge 0, x_2 \ge 0, x_3 &\ge 0 \end{aligned}$$

19. Maximize $z = 2x_1 + 4x_2 + x_3$, subject to

$$\begin{aligned} -x_1 + 2x_2 + 3x_3 &\le 6 \\ -x_1 + 4x_2 + 5x_3 &\le 5 \\ -x_1 + 5x_2 + 7x_3 &\le 7 \\ x_1 \ge 0, x_2 \ge 0, x_3 &\ge 0 \end{aligned}$$

20. Maximize $z = x_1 + 2x_2 + 4x_3$, subject to

$$\begin{aligned} 8x_1 + 5x_2 - 4x_3 &\le 30 \\ -2x_1 + 6x_2 + x_3 &\le 5 \\ -2x_1 + 2x_2 + x_3 &\le 15 \\ x_1 \ge 0, x_2 \ge 0, x_3 &\ge 0 \end{aligned}$$

II.

21. Maximize $x_1 + 2x_2 + 4x_3 - x_4$, subject to

$$\begin{aligned} 5x_1 \qquad\quad 4x_3 + 6x_4 &\le 20 \\ 4x_1 + 2x_2 + 2x_3 + 8x_4 &\le 40 \\ x_1 \ge 0, x_2 \ge 0, x_3 \ge 0, x_4 &\ge 0 \end{aligned}$$

22. Maximize $4x_1 + 6x_2 + 9x_3 + 7x_4$, subject to

$$\begin{aligned} 2x_1 + 4x_2 + 8x_3 + 6x_4 &\le 168 \\ 4x_1 + 4x_2 + 2x_3 + 2x_4 &\le 112 \\ x_1 \ge 0, x_2 \ge 0, x_3 \ge 0, x_4 &\ge 0 \end{aligned}$$

III.

Solve Exercises 23 through 29 using the simplex method. Note that these exercises are from Section 4–1. You may use the tableaux formed in that section.

23. A company manufactures three items, screwdrivers, chisels, and putty knives, on three production lines, *A*, *B*, and *C*, respectively. It takes three hours labor per carton to produce screwdrivers on line *A*, four hours labor per carton of chisels on line *B*, and five hours labor per carton of putty knives on line *C*. Each carton of screwdrivers costs $15 to produce, each carton of chisels costs $12, and each carton of putty knives costs $11. The profit per carton is $5 for screwdrivers, $6 for chisels, and $5 for putty knives. If the company has 2200 hours of labor and $8500 operating funds available per week, how many cartons of each item should they produce to maximize profit?

24. The maximum daily production of an oil refinery is 1900 barrels. The refinery can produce three types of fuel: gasoline, diesel, and heating oil. The production cost per barrel is $6 for gasoline, $5 for diesel, and $8 for heating oil. The daily production budget is $13,400. The profit is $7 per barrel on gasoline, $6 on diesel, and $9 on heating oil. How much of each should be produced to maximize profit?

25. A manufacturing company uses three machines, I, II, and III, to produce three products: note pads, loose-leaf paper, and spiral notebooks. It takes 2, 4, and 0 minutes of time on each of the machines, respectively, to manufacture a carton of notepads. It takes 3, 0, and 6 minutes on the machines, respectively, to produce a carton of loose-leaf paper. Manufacturing a carton of spiral notebooks requires 1, 2, and 3 minutes, respectively, on each machine. The total time available on each machine per day is six hours. The profits are $10, $8, and $12 on each carton of notepads, loose-leaf paper, and spiral notebooks, respectively. How many of each item should be produced in order to maximize total profit?

26. An industrial furniture company manufactures desks, cabinets, and chairs. These items consist of metal, wood, and plastic. The following table represents, in convenient units, the amounts that go into each product and the profit on each item.

	Metal	Wood	Plastic	Profit
Desk	3	4	2	$16
Cabinet	9	1	1	$12
Chair	1	2	2	$ 6

If the company has available 810 units of metal, 400 units of wood, and 100 units of plastic, how many desks, cabinets, and chairs should be produced to maximize total profit?

27. A washing machine manufacturing company produces its machines at three factories, A, B, and C. The washing machines are sold in a certain city, P. It costs $10, $20, and $40 to transport each washing machine, on an average, from A, B, and C, respectively, to P. It has been estimated that it involves 6, 4, and 2 man hours time in packing and transportation to get a washing machine from A, B, and C, respectively, to P. There is $9000 budgeted weekly for transportation of the washing machines to P and a total of 4000 man hours labor available. How should the company schedule its weekly transportation arrangements to maximize total profit if the profit on each machine from A is $12, on each machine from B is $20, and on each machine from C is $16?

28. A manufacturer makes three lines of tents, the Alpine, the Cub, and the Aspen. They are all made of the same material; the Alpine requires 30 square yards; the Cub, 15 square yards; and the Aspen, 60 square yards. The manufacturing cost of the Alpine is $36, the Cub $24, and the Aspen $32. The material is available in amounts of 8400 square yards weekly, and the weekly working budget of the company is $7680. If the profit is $8 on each Alpine, $4 on each Cub, and $12 on each Aspen, what should the weekly production schedule be to maximize total profit?

29. A furniture company finishes two kinds of tables, X and Y. There are three steps in the finishing process: sanding, staining, and varnishing. The various times, in minutes, of each of these processes for the two tables are:

	Sanding	Staining	Varnishing
X	10	8	4
Y	5	4	8

The three types of equipment needed for sanding, staining, and varnishing are each available 6 hours, 5 hours, and 5 hours 20 minutes, respectively, per day. Each type of equipment can only handle one table at a time. The profit on each X table is \$8 and on each Y table is \$4. How many of each should be finished daily to maximize total profit?

4–3 What's Happening in the Simplex Method

Let's look at what is happening in the simplex method. We will outline the steps of the simplex method and explain why we perform these steps. We will use Example 5, Section 4–1 again.

Maximize $z = 4x_1 + 12x_2$, subject to

$$\begin{aligned} 3x_1 + x_2 &\leq 180 \\ x_1 + 2x_2 &\leq 100 \\ -2x_1 + 2x_2 &\leq 40 \\ x_1 \geq 0, x_2 &\geq 0 \end{aligned}$$

The graph of the feasible region and the lines forming its boundary is shown in Figure 4–3.

1. We convert the problem to a system of equations by adding a nonnegative slack variable to each inequality

$$\begin{aligned} 3x_1 + x_2 + s_1 \qquad\qquad &= 180 \\ x_1 + 2x_2 \quad + s_2 \qquad &= 100 \\ -2x_1 + 2x_2 \qquad + s_3 \quad &= 40 \\ -4x_1 - 12x_2 \qquad\qquad + z &= 0 \end{aligned}$$

   where x_1, x_2, s_1, s_2, and s_3 are all nonnegative.

2. The simplex method searches for solutions to this system of equations. Each simplex tableau gives a basic feasible solution. Recall that we set a variable to zero for each x in the system. This gives points where two of the boundary lines of the feasible region intersect.

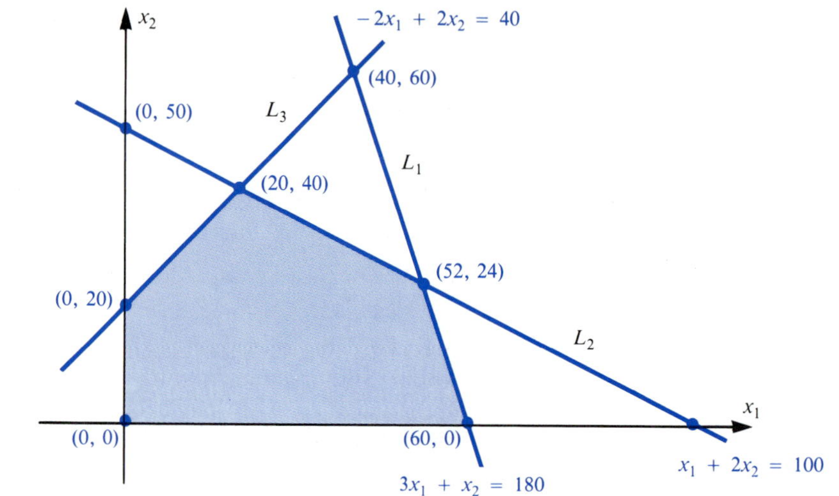

Figure 4–3

Look at Figure 4–3 to illustrate this. The boundary lines are

$$
\begin{aligned}
L_1: &\quad 3x_1 + x_2 = 180 \\
L_2: &\quad x_1 + 2x_2 = 100 \\
L_3: &\quad -2x_1 + 2x_2 = 40
\end{aligned}
$$

$$
\begin{aligned}
x_1\text{-axis:} &\quad x_2 = 0 \\
x_2\text{-axis:} &\quad x_1 = 0
\end{aligned}
$$

Now, look at points where two of these boundary lines intersect. Find the points (0, 0), (0, 20), (0, 50), (20, 40), (40, 60), (52, 24), (60, 0), and (100, 0).

We want to observe how these points relate to the system of equations

$$
\begin{aligned}
3x_1 + x_2 + s_1 \qquad\qquad &= 180 \\
x_1 + 2x_2 \qquad + s_2 \qquad &= 100 \\
-2x_1 + 2x_2 \qquad\qquad + s_3 &= 40
\end{aligned}
$$

(The objective function is not included because it does not enter into determining the feasible region.)

This system has three equations and five variables. When there are more variables than equations, some of the variables can be chosen in any manner whatever. In this case, two variables are arbitrary (two = number of variables minus number of equations). Let us choose the two variables x_1 and x_2. Now we must decide what values we will use for each. We use x_1 and x_2 that come from points of intersection of boundary lines. For each point of intersection, we substitute those values of x_1 and x_2 into each equation of the system. Then we find the corresponding values of the s_1, s_2, and s_3 that form a solution to the system. We omit these computations, but show the results in Table 4–1.

Now let's observe some properties of Table 4–1 and see how they relate to Figure 4–3.

Table 4–1

Point	x_1	x_2	s_1	s_2	s_3
(0, 0)	0	0	180	100	40
(0, 20)	0	20	160	60	0
(0, 50)	0	50	130	0	−60
(20, 40)	20	40	80	0	0
(40, 60)	40	60	0	−60	0
(52, 24)	52	24	0	0	96
(60, 0)	60	0	0	40	160
(100, 0)	100	0	−120	0	240

Observation 1. Notice that $s_1 = 0$ for points (40, 60), (52, 24), and (60, 0). Look at Figure 4–3. All of these points are on L_1. Similarly, $s_2 = 0$ for all points on L_2 ((0, 50), (20, 40), (52, 24), and (100, 0)); and $s_3 = 0$ for all points on L_3.

In general, when a point lies on a boundary line of the feasible region, that slack variable is zero. For a point not on a boundary line, that slack variable is *not* zero.

Since the point (52, 24) is on both L_1 and L_2, $s_1 = 0$ and $s_2 = 0$.

Observation 2. For each of the points in the table, two of the five variables are zero. Remember that we set two variables to zero to get basic solutions in the simplex method. This was done so that we would be using corner points. If we set two variables to zero in an arbitrary fashion, we will get a point where two boundary lines meet, but it may not be a corner.

Observation 3. Some points listed in Table 4–1 are not corner points of the feasible region; namely, (0, 50), (40, 60), and (100, 0). We can observe this by looking at Figure 4–3. However, the table can be used to distinguish a point that is a corner point from one that is not. Look at each point that is not a corner; one of the slack variables is negative. A negative slack variable violates the basic condition that none of the variables are negative. So, (0, 50), (40, 60), and (100, 0) all lie outside the feasible region. The other points listed are in the feasible region.

Now let's put this information together.

(a) When we set two (in general, it is the number of x's) variables to zero, we get a basic solution, a point where two boundaries intersect.

(b) If the basic solution is **feasible**, it is a corner point.

(c) Since the maximum value of the objective function occurs at a corner point, we want the simplex method to give only basic feasible solutions (corner points).

3. The procedure for choosing the pivot element actually accomplishes two things.

(a) It increases z as much as possible; and

(b) it restricts basic solutions to corner points.

Let's look at the initial tableau to illustrate this.

x_1	x_2	s_1	s_2	s_3	z	
3	1	1	0	0	0	180
1	2	0	1	0	0	100
−2	2	0	0	1	0	40
−4	−12	0	0	0	1	0

(a) The choice of the pivot column increases z as much as possible. The objective function in this problem is

$$z = 4x_1 + 12x_2$$

If values of x_1 and x_2 are given, and if you are allowed to increase either one of them by a specified amount, say 1, which one would you change in order to increase z the most? The coefficients of x_1 and x_2 hold the key to your response. If x_1 is increased by 1, then the coefficient 4 causes z to increase by 4. Similarly, an increase of 1 in x_2 causes z to increase by 12. The greatest increase in z is gained by increasing the variable with the largest positive coefficient, 12 in this case. In the tableau, $z = 4x_1 + 12x_2$ is written as $-4x_1 - 12x_2 + z = 0$. In this form, the choice of the *most negative* coefficient is equivalent to choosing the variable that will increase z the most. So, in the simplex method, the pivot column is chosen by the most negative entry in the bottom row because this gives the variable that will increase z the most.

(b) The choice of the pivot row restricts basic solutions to corner points.

In the initial tableau the basic solution assumes that both x_1 and x_2 are zero. Then, the x_2 column becomes the pivot column to obtain the next tableau because that column contains the most negative entry of the last row. Since this means that we want to increase x_2, x_1 remains zero. Using $x_1 = 0$, let's write each row of the initial tableau in equation form.

$$[3 \quad 1 \quad 1 \quad 0 \quad 0 \quad 0 \quad 180] \text{ becomes}$$

$$x_2 + s_1 = 180$$

$$[1 \quad 2 \quad 0 \quad 1 \quad 0 \quad 0 \quad 100] \text{ becomes}$$

$$2x_2 + s_2 = 100, \qquad \text{and}$$

$$[-2 \quad 2 \quad 0 \quad 0 \quad 1 \quad 0 \quad 40] \text{ becomes}$$

$$2x_2 + s_3 = 40$$

(The first number in each row doesn't appear in the equation because it is the coefficient of x_1, which we are using as 0.)

We can write these three equations in the following form:

$$s_1 = 180 - x_2$$
$$s_2 = 100 - 2x_2$$
$$s_3 = 40 - 2x_2$$

Keep in mind that we are going to increase x_2 to achieve the largest increase in z. The larger the increase in x_2, the more z increases; but be careful, we must remain in the feasible region. Since s_1, s_2, and s_3 must not be negative, x_2 must be chosen to avoid making any one of them negative.

The equations

$$s_1 = 180 - x_2$$
$$s_2 = 100 - 2x_2$$
$$s_3 = 40 - 2x_2$$

and the nonnegative condition $s_1 \geq 0$, $s_2 \geq 0$, and $s_3 \geq 0$ indicate that

$$180 - x_2 \geq 0$$
$$100 - 2x_2 \geq 0$$

and

$$40 - 2x_2 \geq 0$$

must all be true. Solving each of these inequalities for x_2 gives

$$\frac{180}{1} \geq x_2$$

$$\frac{100}{2} \geq x_2$$

$$\frac{40}{2} \geq x_2$$

In order for *all three* of s_1, s_2, and s_3 to be nonnegative, the smallest value of x_2, 20, must be used. The ratios

$$\frac{180}{1}, \frac{100}{2}, \frac{40}{2}$$

are exactly the ratios used in the simplex method to determine the pivot row. The selection of the smallest positive ratio is what makes a basic solution a feasible basic solution, that is, a corner point is chosen.

4. Recall that z is maximum when the last row of the simplex tableau contains no negative entries. The final tableau of this problem was

$$\begin{array}{cccccc|c} x_1 & x_2 & s_1 & s_2 & s_3 & z & \\ \hline \end{array}$$

$$\left[\begin{array}{cccccc|c} 0 & 0 & 1 & -\frac{4}{3} & \frac{5}{6} & 0 & 80 \\ 1 & 0 & 0 & \frac{1}{3} & -\frac{1}{3} & 0 & 20 \\ 0 & 1 & 0 & \frac{1}{3} & \frac{1}{6} & 0 & 40 \\ \hline 0 & 0 & 0 & \frac{16}{3} & \frac{2}{3} & 1 & 560 \end{array}\right]$$

This tableau tells us that s_2 and s_3 are set to zero (their columns have more than one nonzero entry) in the optimal solution, and since the last row contains no negative entries, we know that we cannot increase z further. Here's why: Write the last row of this tableau in equation form. It is:

$$\frac{16}{3}s_2 + \frac{2}{3}s_3 + z = 560$$

which can be written as

$$z = 560 - \frac{16}{3}s_2 - \frac{2}{3}s_3$$

This form tells us that if we use any positive number for s_2 or s_3, we will subtract something from 560, thereby making z smaller. So, we stop because another tableau will move us to another corner point and s_2 or s_3 will become positive and therefore reduce z.

Let's compare this situation with the next to the last tableau. It's last row was

$$[-16 \quad 0 \quad 0 \quad 0 \quad 6 \quad 1 \quad 240]$$

This row represents the equation

$$-16x_1 + 6s_3 + z = 240$$

which may be written

$$z = 240 + 16x_1 - 6s_3$$

If x_1 is increased from 0 to a positive number, then a positive quantity will be added to z, thereby increasing it. If you look back at this step in the solution, you will find that x_1 was increased from 0 to 10.

The value of z can be increased as long as there is a negative entry in the last row. It can be increased no further when no entry is negative.

Based on the preceding discussion, the simplex method can be summarized this way.

> The simplex method maximizes the objective function by computing the objective function at selected corner points of the feasible region until the optimal solution is reached. The method begins at the origin and moves at each stage to a corner point determined by the variable that yields the largest increase in z.

4–3 Exercises

I.

1. The constraint $4x_1 + 3x_2 \leq 17$ is written as

$$4x_1 + 3x_2 + s_1 = 17$$

using a slack variable.

(a) Find the value of s_1 for each of the points (0, 0), (2, 2), (5, 10), (2, 1), and (2, 3)

(b) Which points are in the feasible region?

(c) Which points lie on the line $4x_1 + 3x_2 = 17$?

2. The constraint $6x_1 + 5x_2 \leq 25$ is written as

$$6x_1 + 5x_2 + s_1 = 25$$

using a slack variable. Find the value of s_1 for each of the points

(0, 5), (2, 2), (1.5, 3), and (2.5, 2)

3. Find the values of the slack variable, s_1, in the constraint

$$3x_1 + 4x_2 + x_3 \leq 40$$

for the points

(0, 0, 0), (1, 2, 3), (0, 10, 0), and (4, 2, 7)

4. Find the values of the slack variables, s_1 and s_2, in the constraints

$$\begin{aligned} 2x_1 + x_2 + 5x_3 + s_1 &= 30 \\ x_1 + 6x_2 + 4x_3 + s_2 &= 28 \end{aligned}$$

for the points

(1, 1, 1), (2, 1, 5), (0, 4, 1), and (5, 10, 2)

5. The constraints of a linear programming problem are

$$
\begin{aligned}
x_1 + 5x_2 &\le 70 \\
6x_1 + x_2 &\le 72 \\
7x_1 + 6x_2 &\le 113
\end{aligned}
$$

Using slack variables, this reduces to

$$
\begin{aligned}
x_1 + 5x_2 + s_1 \qquad\qquad &= 70 \\
6x_1 + x_2 \qquad + s_2 \qquad &= 72 \\
7x_1 + 6x_2 \qquad\qquad + s_3 &= 113
\end{aligned}
$$

For each point listed in the table below, find the missing entries.

Point	s_1	s_2	s_3	Is point on boundary?	Is point in feasible region?
(5, 10)					
(8, 10)					
(5, 13)					
(11, 13)					
(10, 12)					
(15, 11)					

6. Which single variable contributes the most to increasing z in the following objective functions?

(a) $z = 6x_1 + 5x_2 + 14x_3$
(b) $z = 8x_1 - 12x_2 + 3x_3$
(c) $z = 8x_1 + 4x_2 - 7x_3 + 5x_4$

7. The linear programming problem
Maximize $z = 3x_1 + 2x_2$, subject to

$$
\begin{aligned}
5x_1 + 2x_2 &\le 900 \\
8x_1 + 10x_2 &\le 2800 \\
x_1 \ge 0, \; x_2 &\ge 0
\end{aligned}
$$

has a feasible region bounded by

$$
\begin{aligned}
5x_1 + 2x_2 &= 900 \\
8x_1 + 10x_2 &= 2800 \\
x_1 = 0, \; x_2 &= 0
\end{aligned}
$$

The simplex method solution to this problem includes the following tableaux:

$$
\begin{array}{c}
\\
\textbf{(i)} \\
\\
\\
\end{array}
\begin{array}{cccccc}
x_1 & x_2 & s_1 & s_2 & z & \\
\end{array}
$$

$$
\textbf{(i)}\quad \left[\begin{array}{ccccc|c}
5 & 2 & 1 & 0 & 0 & 900 \\
8 & 10 & 0 & 1 & 0 & 2800 \\
\hline
-3 & -2 & 0 & 0 & 1 & 0
\end{array}\right]
$$

$$
\textbf{(ii)}\quad \left[\begin{array}{ccccc|c}
1 & \frac{2}{5} & \frac{1}{5} & 0 & 0 & 180 \\
0 & \frac{34}{5} & -\frac{8}{5} & 1 & 0 & 1360 \\
\hline
0 & -\frac{4}{5} & \frac{3}{5} & 0 & 1 & 540
\end{array}\right]
$$

$$
\textbf{(iii)}\quad \left[\begin{array}{ccccc|c}
1 & 0 & \frac{25}{85} & -\frac{1}{17} & 0 & 100 \\
0 & 1 & -\frac{4}{17} & \frac{5}{34} & 0 & 200 \\
\hline
0 & 0 & \frac{7}{17} & \frac{2}{17} & 1 & 700
\end{array}\right]
$$

(a) Write the basic solution from each tableau.

(b) Determine the two boundary lines whose intersection gives the basic solution.

II.

8. Given the constraint

$$x_1 + 5x_2 + s_1 = 48$$

(a) When $x_1 = 0$, what is the largest value x_2 can have so that s_1 meets the nonnegative condition?

(b) When $x_1 = 6$, what is the largest value x_2 can have so that s_1 meets the nonnegative condition?

(c) When $x_2 = 0$, what is the largest possible value of x_1 so that the point is in the feasible region?

III.

9. Given the constraints

$$
\begin{aligned}
6x_1 + 7x_2 + s_1 \qquad &= 36 \\
2x_1 + 5x_2 \qquad + s_2 &= 32
\end{aligned}
$$

(a) When $x_1 = 0$, what is the largest possible value of x_2 so that the point is in the feasible region?

(b) When $x_2 = 0$, what is the largest possible value of x_1 so that the point is in the feasible region?

10. Given the constraint

$$4x_1 + 5x_2 + x_3 + s_1 = 45$$

When $x_1 = 0$, what is the largest possible value of x_2 so that s_1 is nonnegative?

4–4 The Standard Minimum Problem: Duality

Standard Minimum Problem

So far the simplex method solved standard maximum problems. Some problems may seek to minimize the objective functions. For example, a dietitian may want to provide a menu that minimizes the number of calories; a plant manager may try to minimize operating costs; or a wholesale company may want to minimize shipping costs. One method for solving the **standard minimum** problem converts it to a standard maximum problem. Before we show you this method, we need a definition of a **standard minimum** problem.

Definition

A linear programming problem is **standard minimum** if

1. the objective function is to be minimized.
2. all the inequalities are $\geq$.
3. the constants to the right of the inequalities are nonnegative.

Dual Problem

We solve the standard minimum problem by converting it to a **dual** maximum problem. Let's look at an example to describe how you set up the **dual problem**.

Example 1 A doctor specifies that a patient's diet contain certain minimum amounts of iron and calcium, but calories are to be held to a minimum.

Two foods, A and B, are used in a meal; and the amounts of iron, calcium, and calories are given in this table.

	Amount Provided by One Unit of		Amount Required
	A	B	
Iron	4	1	12 or more
Calcium	2	3	10 or more
Calories	90	120	

We convert this information into a linear programming form as follows.

Let x_1 = the number of units of A
x_2 = the number of units of B

The iron requirement is

$$4x_1 + x_2 \geq 12$$

The calcium requirement is

$$2x_1 + 3x_2 \geq 10$$

and the calorie count is

$$90x_1 + 120x_2$$

where $x_1 \geq 0$ and $x_2 \geq 0$
The problem is
Minimize $z = 90x_1 + 120x_2$, subject to

$$\begin{aligned} 4x_1 + x_2 &\geq 12 \\ 2x_1 + 3x_2 &\geq 10 \\ x_1 \geq 0, x_2 &\geq 0 \end{aligned}$$

Solution To form the dual problem we first write the minimum problem in a matrix form using an augmented matrix of the constraints and objective function. For this example, the matrix takes the following form. Notice its similarity to the augmented matrix of a system of equations.

$$\left[\begin{array}{cc|c} 4 & 1 & 12 \\ 2 & 3 & 10 \\ \hline 90 & 120 & 1 \end{array}\right]$$

We emphasize: This is *not* a simplex tableau because it does not contain slack variables, and the objective function has not been rewritten. We wrote the matrix in this form because it is used to obtain a maximization problem first.

Next, we obtain a new matrix by taking each *row* of

$$A = \left[\begin{array}{cc|c} 4 & 1 & 12 \\ 2 & 3 & 10 \\ \hline 90 & 120 & 1 \end{array}\right]$$

and making it the *column* of the new matrix. (The new matrix is called the **transpose** of this matrix.) The new matrix is

$$B = \left[\begin{array}{cc|c} 4 & 2 & 90 \\ 1 & 3 & 120 \\ \hline 12 & 10 & 1 \end{array}\right]$$

From the new matrix, B, set up a standard maximum problem. We introduce new variables, y_1 and y_2, because they play a different role from the original ones. We use the rows above the line to form constraints for the maximum problem. The row below the line forms the new objective function. Since we want this matrix to give a standard maximum problem, all inequalities are $\leq$. We write the new constraints and objective functions next to their rows.

$$\begin{array}{c} \begin{matrix} y_1 & y_2 \end{matrix} \\ \left[\begin{array}{cc|c} 4 & 2 & 90 \\ 1 & 3 & 120 \\ \hline 12 & 10 & 1 \end{array}\right] \end{array} \quad \begin{array}{l} \text{new constraint } 4y_1 + 2y_2 \leq 90 \\ \text{new constraint } y_1 + 3y_2 \leq 120 \\ \text{new objective function } w = 12y_1 + 10y_2 \end{array}$$

This gives the **dual** problem

Maximize $w = 12y_1 + 10y_2$, subject to

$$4y_1 + 2y_2 \leq 90$$
$$y_1 + 3y_2 \leq 120$$
$$y_1 \geq 0,\ y_2 \geq 0$$

Example 2 (*Compare Exercise 5*)

Set up the dual problem to the following standard minimum problem.

Minimize $z = 30x_1 + 40x_2 + 50x_3$, subject to

$$10x_1 + 14x_2 + 5x_3 \geq 220$$
$$5x_1 + 3x_2 + 9x_3 \geq 340$$
$$x_1 \geq 0,\ x_2 \geq 0,\ x_3 \geq 0$$

Solution Form the augmented matrix of the problem with the objective function written in the last row.

$$\begin{array}{c} \quad\ \begin{matrix} x_1 & x_2 & x_3 \end{matrix} \\ A = \left[\begin{array}{ccc|c} 10 & 14 & 5 & 220 \\ 5 & 3 & 9 & 340 \\ \hline 30 & 40 & 50 & 1 \end{array}\right] \end{array}$$

Form the transpose of A.

$$\begin{array}{c} \quad\ \begin{matrix} y_1 & y_2 \end{matrix} \\ B = \left[\begin{array}{cc|c} 10 & 5 & 30 \\ 14 & 3 & 40 \\ 5 & 9 & 50 \\ \hline 220 & 340 & 1 \end{array}\right] \end{array}$$

Set up the dual problem from this matrix using $\leq$ on all constraints.

Maximize $w = 220y_1 + 340y_2$, subject to

$$10y_1 + 5y_2 \le 30$$
$$14y_1 + 3y_2 \le 40$$
$$5y_1 + 9y_2 \le 50$$
$$y_1 \ge 0,\ y_2 \ge 0$$

Set Up the Dual Problem of a Standard Minimum Problem

1. Start with a standard minimum problem.
2. Write the augmented matrix, A, of the minimum problem. Write the objective function in the last row.
3. Write the transpose of the matrix A to obtain matrix B. Each row of A becomes the corresponding column of B.
4. Form a constraint for the dual problem from each row of B (except the last) using the new variables and $\le$.
5. Form the objective function of the dual problem from the last row. It is to be maximized.

Solve the Minimization Problem via the Dual Problem

The theory relating a standard minimum problem and its dual problem is beyond the level of this course. The relationship between the solution of a minimum problem and its dual problem is a fundamental theorem, which we will use.

Theorem 1

A standard minimum problem has a solution if and only if its dual problem has a solution. If a solution exists, the standard minimum problem and its dual problem *have the same* optimal solution.

This theorem states that the maximum value of the dual problem objective function is the minimum value of the objective function for the minimum problem. To help see this, let's work through the diet example at the beginning of the section. The problem and its dual are:

Standard Minimum Problem	Dual Problem
Minimize $z = 90x_1 + 120x_2$, subject to	Maximize $w = 12y_1 + 10y_2$, subject to
$4x_1 + x_2 \ge 12$	$4y_1 + 2y_2 \le 90$
$2x_1 + 3x_2 \ge 10$	$y_1 + 3y_2 \le 120$
$x_1 \ge 0,\ x_2 \ge 0$	$y_1 \ge 0,\ y_2 \ge 0$

The procedure is straightforward; solve the dual problem by the simplex method. We first write the dual problem as a system of equations using slack variables and obtain

$$\begin{aligned} 4y_1 + 2y_2 + x_1 \qquad\qquad &= 90 \\ y_1 + 3y_2 \qquad + x_2 \qquad &= 120 \\ -12y_1 - 10y_2 \qquad\qquad + w &= 0 \end{aligned}$$

Notice that x_1 and x_2 are used for slack variables. These are intended to be the same as the variables in the original minimum problem because it turns out that certain values of the slack variables of the dual problem give the desired values of the original variables in the minimum problem. Let's set up the simplex tableau and work through the solution.

Initial Tableau

$$\begin{array}{ccccc|c} y_1 & y_2 & x_1 & x_2 & w & \\ \hline 4 & 2 & 1 & 0 & 0 & 90 \\ 1 & 3 & 0 & 1 & 0 & 120 \\ \hline -12 & -10 & 0 & 0 & 1 & 0 \end{array}$$

We now proceed to find the pivot element and perform row operations in the usual manner. You should fill in details that are omitted.

$$\begin{array}{ccccc|cl} y_1 & y_2 & x_1 & x_2 & w & & \\ \hline 4 & 2 & 1 & 0 & 0 & 90 & \frac{1}{4}R1 \to R1 \\ 1 & 3 & 0 & 1 & 0 & 120 & \\ \hline -12 & -10 & 0 & 0 & 1 & 0 & \end{array}$$

$$\begin{array}{ccccc|cl} y_1 & y_2 & x_1 & x_2 & w & & \\ \hline 1 & \frac{1}{2} & \frac{1}{4} & 0 & 0 & \frac{90}{4} & \\ 1 & 3 & 0 & 1 & 0 & 120 & -R1 + R2 \to R2 \\ \hline -12 & -10 & 0 & 0 & 1 & 0 & 12R1 + R3 \to R3 \end{array}$$

$$\begin{array}{ccccc|cl} y_1 & y_2 & x_1 & x_2 & w & & \\ \hline 1 & \frac{1}{2} & \frac{1}{4} & 0 & 0 & \frac{90}{4} & \\ 0 & \frac{5}{2} & -\frac{1}{4} & 1 & 0 & \frac{390}{4} & \frac{2}{5}R2 \to R2 \\ \hline 0 & -4 & 3 & 0 & 1 & 270 & \end{array}$$

$$\begin{array}{ccccc|cl} y_1 & y_2 & x_1 & x_2 & w & & \\ \hline 1 & \frac{1}{2} & \frac{1}{4} & 0 & 0 & \frac{90}{4} & -\frac{1}{2}R2 + R1 \to R1 \\ 0 & 1 & -\frac{1}{10} & \frac{2}{5} & 0 & 39 & \\ \hline 0 & -4 & 3 & 0 & 1 & 270 & 4R2 + R3 \to R3 \end{array}$$

$$\begin{array}{ccccc|c} y_1 & y_2 & x_1 & x_2 & w & \\ \hline 1 & 0 & \frac{6}{20} & -\frac{1}{5} & 0 & 3 \\ 0 & 1 & -\frac{1}{10} & \frac{2}{5} & 0 & 39 \\ \hline 0 & 0 & \frac{26}{10} & \frac{8}{5} & 1 & 426 \end{array}$$

Since no entries of the last row are negative, the solution is optimal and the maximum value is 426. The maximum value occurs when

$$y_1 = 3,\ y_2 = 39$$

By Theorem 1, the *minimum* value of the original objective function, $z = 90x_1 + 120x_2$, is also 426. The values of x_1 and x_2 that yield this minimum value are found in the bottom row of the final tableau of the dual problem. That bottom row is

$$\begin{array}{cccccc} y_1 & y_2 & x_1 & x_2 & w & \\ \hline [0 & 0 & 2.6 & 1.6 & 1 & 426] \end{array}$$

The numbers under x_1 and x_2 are the values of x_1 and x_2 that give the optimal value of the original minimization problem. So, the objective function $z = 90x_1 + 120x_2$ has the minimum value of 426 at $x_1 = 2.6$, $x_2 = 1.6$.

It is important to remember: To find the solution of a standard minimum problem, *look at the bottom row of the final tableau of the dual problem.*

Example 3 (*Compare Exercise 9*)
Solve the minimization problem by the dual problem method.
Minimize $z = 8x_1 + 15x_2$, subject to

$$\begin{aligned} 4x_1 + 5x_2 &\geq 80 \\ 2x_1 + 5x_2 &\geq 60 \\ x_1 \geq 0,\ x_2 &\geq 0 \end{aligned}$$

Solution The augmented matrix of this problem is

$$A = \left[\begin{array}{cc|c} 4 & 5 & 80 \\ 2 & 5 & 60 \\ \hline 8 & 15 & 1 \end{array}\right]$$

The transpose of A is

$$B = \left[\begin{array}{cc|c} 4 & 2 & 8 \\ 5 & 5 & 15 \\ \hline 80 & 60 & 1 \end{array}\right]$$

B represents the maximization problem.

Maximize $w = 80y_1 + 60y_2$, subject to

$$\begin{aligned} 4y_1 + 2y_2 &\le 8 \\ 5y_1 + 5y_2 &\le 15 \\ y_1 \ge 0,\ y_2 &\ge 0 \end{aligned}$$

The initial simplex tableau of this problem is

$$\begin{array}{ccccc|c} y_1 & y_2 & x_1 & x_2 & w & \\ \hline 4 & 2 & 1 & 0 & 0 & 8 \\ 5 & 5 & 0 & 1 & 0 & 15 \\ \hline -80 & -60 & 0 & 0 & 1 & 0 \end{array}$$

Now, proceed with the pivot and row operations to obtain the sequence of tableaux.

$$\begin{array}{ccccc|c} y_1 & y_2 & x_1 & x_2 & w & \\ \hline \textcircled{4} & 2 & 1 & 0 & 0 & 8 \\ 5 & 5 & 0 & 1 & 0 & 15 \\ \hline -80 & -60 & 0 & 0 & 1 & 0 \end{array} \quad \tfrac{1}{4}\text{R1} \to \text{R1}$$

Pivot element (arrow to the circled 4)

$$\left[\begin{array}{ccccc|c} 1 & \frac{1}{2} & \frac{1}{4} & 0 & 0 & 2 \\ 5 & 5 & 0 & 1 & 0 & 15 \\ \hline -80 & -60 & 0 & 0 & 1 & 0 \end{array}\right] \quad \begin{array}{l} -5\text{R1} + \text{R2} \to \text{R2} \\ 80\text{R1} + \text{R3} \to \text{R3} \end{array}$$

$$\left[\begin{array}{ccccc|c} 1 & \frac{1}{2} & \frac{1}{4} & 0 & 0 & 2 \\ 0 & \textcircled{\frac{5}{2}} & -\frac{5}{4} & 1 & 0 & 5 \\ \hline 0 & -20 & 20 & 0 & 1 & 160 \end{array}\right] \quad \tfrac{2}{5}\text{R2} \to \text{R2}$$

Pivot element (arrow to the circled 5/2)

$$\left[\begin{array}{ccccc|c} 1 & \frac{1}{2} & \frac{1}{4} & 0 & 0 & 2 \\ 0 & 1 & -\frac{1}{2} & \frac{2}{5} & 0 & 2 \\ \hline 0 & -20 & 20 & 0 & 1 & 160 \end{array}\right] \quad \begin{array}{l} -\frac{1}{2}\text{R2} + \text{R1} \to \text{R1} \\ 20\,\text{R2} + \text{R3} \to \text{R3} \end{array}$$

$$\left[\begin{array}{ccccc|c} 1 & 0 & \frac{1}{2} & -\frac{1}{5} & 0 & 1 \\ 0 & 1 & -\frac{1}{2} & \frac{2}{5} & 0 & 2 \\ \hline 0 & 0 & 10 & 8 & 1 & 200 \end{array}\right]$$

This is the final tableau of the dual problem. The last row gives the solution to the minimum problem:

$$x_1 = 10,\ x_2 = 8,\ z = 200$$

Example 4 (*Compare Exercise 15*)

A tire company has plants in Chicago and Detroit. The Chicago plant can make 600 radials and 100 standard tires per day. The Detroit plant can make 300 radial and 100 standard tires per day. It costs $20,000 per day to operate the Chicago plant and \$15,000 per day to operate the Detroit plant. The company has a contract to make at least 24,000 radial and 5000 standard tires. How many days should each plant be scheduled in order to minimize operating costs?

Solution

Let x_1 = number of days the Chicago plant operates
x_2 = number of days the Detroit plant operates

The number of radial tires produced is $600x_1 + 300x_2$, the number of standard tires produced is $100x_1 + 100x_2$, and the operating expenses are $20{,}000x_1 + 15{,}000x_2$. The linear programming problem is

Minimize $z = 20{,}000x_1 + 15{,}000x_2$, subject to

$$600x_1 + 300x_2 \geq 24{,}000$$
$$100x_1 + 100x_2 \geq 5{,}000$$
$$x_1 \geq 0,\ x_2 \geq 0$$

This is a standard minimization problem, so we solve it using its dual problem. The augmented matrix of the problem is

$$A = \left[\begin{array}{cc|c} 600 & 300 & 24{,}000 \\ 100 & 100 & 5{,}000 \\ \hline 20{,}000 & 15{,}000 & 1 \end{array}\right]$$

The transpose of A is

$$B = \left[\begin{array}{cc|c} 600 & 100 & 20{,}000 \\ 300 & 100 & 15{,}000 \\ \hline 24{,}000 & 5{,}000 & 1 \end{array}\right]$$

This matrix represents the dual problem

Maximize $w = 24{,}000y_1 + 5000y_2$, subject to

$$600y_1 + 100y_2 \leq 20{,}000$$
$$300y_1 + 100y_2 \leq 15{,}000$$
$$y_1 \geq 0,\ y_2 \geq 0$$

The tableaux that solve this dual problem follow:

Initial Tableau

$$\begin{array}{c} \begin{array}{ccccc} y_1 & y_2 & x_1 & x_2 & w \end{array} \\ \left[\begin{array}{ccccc|c} 600 & 100 & 1 & 0 & 0 & 20{,}000 \\ 300 & 100 & 0 & 1 & 0 & 15{,}000 \\ \hline -24{,}000 & -5{,}000 & 0 & 0 & 1 & 0 \end{array}\right] \end{array} \quad \tfrac{1}{600}\,\text{R1} \to \text{R1}$$

$$\left[\begin{array}{ccccc|c} 1 & \frac{1}{6} & \frac{1}{600} & 0 & 0 & \frac{200}{6} \\ 300 & 100 & 0 & 1 & 0 & 15{,}000 \\ \hline -24{,}000 & -5{,}000 & 0 & 0 & 1 & 0 \end{array}\right] \quad \begin{array}{l} -300\,\text{R1} + \text{R2} \to \text{R2} \\ 24{,}000\,\text{R1} + \text{R3} \to \text{R3} \end{array}$$

$$\left[\begin{array}{ccccc|c} 1 & \frac{1}{6} & \frac{1}{600} & 0 & 0 & \frac{200}{6} \\ 0 & 50 & -\frac{1}{2} & 1 & 0 & 5{,}000 \\ \hline 0 & -1000 & 40 & 0 & 1 & 800{,}000 \end{array}\right] \quad \tfrac{1}{50}\text{R2} \to \text{R2}$$

$$\left[\begin{array}{ccccc|c} 1 & \frac{1}{6} & \frac{1}{600} & 0 & 0 & \frac{200}{6} \\ 0 & 1 & -\frac{1}{100} & \frac{1}{50} & 0 & 100 \\ \hline 0 & -1000 & 40 & 0 & 1 & 800{,}000 \end{array}\right] \quad \begin{array}{l} -\tfrac{1}{6}\text{R2} + \text{R1} \to \text{R1} \\ \\ 1000\text{ R2} + \text{R3} \to \text{R3} \end{array}$$

$$\left[\begin{array}{ccccc|c} 1 & 0 & \frac{1}{300} & -\frac{1}{300} & 0 & \frac{100}{6} \\ 0 & 1 & -\frac{1}{100} & \frac{1}{50} & 0 & 100 \\ \hline 0 & 0 & 30 & 20 & 1 & 900{,}000 \end{array}\right]$$

The last row gives the solution to the original problem

$$x_1 = 30,\ x_2 = 20,\ z = 900{,}000$$

The minimum operating costs are \$900,000 when the Chicago plant operates 30 days and the Detroit plant operates 20 days.

4–4 Exercises

I.

Write the transpose of the matrices in Exercises 1 through 4.

1. $\begin{bmatrix} 2 & 1 & 3 \\ 4 & 0 & 2 \end{bmatrix}$

2. $\begin{bmatrix} 5 & -1 & 6 \\ 2 & 1 & 2 \\ 3 & -3 & 5 \end{bmatrix}$

3. $\begin{bmatrix} 4 & 3 & 2 \\ 1 & 8 & -2 \\ 6 & -7 & 1 \\ 2 & 4 & 6 \end{bmatrix}$

4. $\begin{bmatrix} 2 & 1 & 5 & 4 & 3 \\ 10 & -2 & 7 & 9 & 14 \\ 8 & 15 & -3 & 6 & 1 \end{bmatrix}$

For each of the minimization problems in Exercises 5 through 8,

(a) set up the augmented matrix of the problem,
(b) find the transpose of the matrix in (a), and
(c) set up the initial tableau for the dual problem.

5. (*See Example 2*) Minimize $z = 25x_1 + 30x_2$, subject to

$$\begin{aligned} 6x_1 + 5x_2 &\geq 30 \\ 8x_1 + 3x_2 &\geq 42 \\ x_1 \geq 0,\ x_2 &\geq 0 \end{aligned}$$

6. Minimize $z = 14x_1 + 27x_2 + 9x_3$, subject to

$$\begin{aligned} 7x_1 + 9x_2 + 4x_3 &\geq 60 \\ 10x_1 + 3x_2 + 6x_3 &\geq 80 \\ 4x_1 + 2x_2 + x_3 &\geq 48 \\ x_1 \geq 0,\ x_2 \geq 0,\ x_3 &\geq 0 \end{aligned}$$

7. Minimize $z = 500x_1 + 700x_2$, subject to

$$\begin{aligned} 22x_1 + 30x_2 &\geq 110 \\ 15x_1 + 40x_2 &\geq 95 \\ 20x_1 + 35x_2 &\geq 68 \\ x_1 \geq 0, x_2 &\geq 0 \end{aligned}$$

8. Minimize $z = 40x_1 + 60x_2 + 50x_3 + 35x_4$, subject to

$$\begin{aligned} 7x_1 + 6x_2 - 5x_3 + x_4 &\geq 45 \\ 12x_1 + 18x_2 + 4x_3 + 6x_4 &\geq 86 \\ x_1 \geq 0, x_2 \geq 0, x_3 \geq 0, x_4 &\geq 0 \end{aligned}$$

Solve the following by solving the dual problem.

9. (*See Example 3*) Minimize $3x_1 + 7x_2$, subject to

$$\begin{aligned} x_1 + 2x_2 &\geq 2 \\ 3x_1 + 4x_2 &\geq 1 \\ x_1 \geq 0, x_2 &\geq 0 \end{aligned}$$

10. Minimize $36x_1 + 66x_2$, subject to

$$\begin{aligned} 4x_1 + 6x_2 &\geq 3 \\ 2x_1 + 2x_2 &\geq 2 \\ x_1 \geq 0, x_2 &\geq 0 \end{aligned}$$

11. Minimize $40x_1 + 16x_2 + 20x_3$, subject to

$$\begin{aligned} 2x_1 + 4x_2 + x_3 &\geq 2 \\ 4x_1 \qquad\quad + x_3 &\geq 1 \\ 8x_2 + 2x_3 &\geq 2 \\ x_1 \geq 0, x_2 \geq 0, x_3 &\geq 0 \end{aligned}$$

12. Minimize $12x_1 + 24x_2$, subject to

$$\begin{aligned} 4x_1 + 6x_2 &\geq 2 \\ 6x_1 + 8x_2 &\geq 4 \\ x_1 + 4x_2 &\geq 6 \\ 8x_1 + 8x_2 &\geq 3 \\ x_1 \geq 0, x_2 &\geq 0 \end{aligned}$$

13. Minimize $6x_1 + 20x_2 + 16x_3 + 4x_4$, subject to

$$\begin{aligned} x_1 + 4x_2 + 2x_3 + 8x_4 &\geq 2 \\ 2x_1 + 8x_2 + 4x_3 + 2x_4 &\geq 4 \\ x_1 \geq 0, x_2 \geq 0, x_3 \geq 0, x_4 &\geq 0 \end{aligned}$$

14. Minimize $7x_1 + 5x_2 + 7x_3 + 8x_4$, subject to

$$\begin{aligned} x_1 + 3x_2 + 2x_3 + 3x_4 &\geq 3 \\ 2x_1 + 7x_2 + 7x_3 + 5x_4 &\geq 2 \\ 5x_1 + 8x_2 + 5x_3 + 2x_4 &\geq 6 \\ x_1 \geq 0, x_2 \geq 0, x_3 \geq 0, x_4 &\geq 0 \end{aligned}$$

II.

15. (*See Example 4*) A tire company has two plants—one in Dallas and one in New Orleans. The Dallas plant can make 800 radials and 280 standard tires per day. The New Orleans plant can make 500 radial and 150 standard tires per day. It costs $22,000 per day to operate the Dallas plant and $12,000 per day to operate the New Orleans plant. They have a contract to make at least 28,000 radial and 9000 standard tires. How many days should each plant be scheduled in order to minimize operating costs?

16. A plant makes two models of an item. Each model *A* requires 3 hours skilled labor and 6 hours unskilled labor. Each model *B* requires 5 hours skilled labor and 4 hours unskilled labor. Their labor contract requires that they employ at least 3000 hours skilled labor and at least 4200 hours unskilled labor. Each model *A* costs $21, and each model *B* costs $25. How many of each should they produce to minimize costs if they must produce a total of 900 items or more?

4–5 Mixed Constraints

The simplex method has been used to solve standard maximum problems (all constraints have $\leq$ inequalities and nonnegative constant terms) and standard minimum problems (all constraints have $\geq$ inequalities and nonnegative coefficients in the objective function).

In this section we study more general problems. The constraints may be a mixture of $\leq$, $\geq$, or $=$, and we may wish to either maximize or minimize the objective function. Because such problems contain a mixture of $\leq$, $\geq$, or $=$, they are referred to as having **mixed constraints**.

Surplus and Artificial Variables

A simple problem illustrates the procedure for mixed constraints.

Example 1 (*Compare Exercise 1*)
Maximize $20x_1 + 70x_2$, subject to

$$\begin{aligned} 2x_1 + 5x_2 &\leq 58 \\ x_1 - x_2 &\geq 15 \\ x_1 \geq 0, x_2 &\geq 0 \end{aligned}$$

Solution Notice that this problem varies from the standard maximum problem in that it has a $\geq$ constraint as well as a $\leq$ constraint. We must first convert this problem to a system of equations. We convert

$$x_1 - x_2 \geq 15$$

to an equation by use of a **surplus variable**, s_2, and write

$$x_1 - x_2 - s_2 = 15$$

Notice that we have subtracted s_2 from $x_1 - x_2$. Like other variables, the surplus variable is not negative, and it is the surplus of $x_1 - x_2$ over 15 (the amount by which $x_1 - x_2$ exceeds 15).

We can now describe the problem with the system of equations

$$\begin{aligned} 2x_1 + 5x_2 + s_1 \qquad &= 58 \\ x_1 - x_2 \qquad - s_2 &= 15 \\ -20x_1 - 70x_2 \qquad + z &= 0 \\ x_1 \geq 0, x_2 \geq 0, s_1 \geq 0, s_2 &\geq 0 \end{aligned}$$

If we follow the usual procedure of getting a basic feasible solution, we obtain $x_1 = 0$, $x_2 = 0$, $s_1 = 58$, $s_2 = -15$, and $z = 0$.

This is not a feasible solution because s_2 is negative. This means that we cannot directly apply the usual procedure of constructing a simplex tableau and performing pivot operations.

However, problems with mixed constraints can be adjusted by use of an **artificial variable** in each constraint that has a surplus variable. Introduce the artificial variable into the equation

$$x_1 - x_2 - s_2 = 15$$

by writing it as

$$x_1 - x_2 - s_2 + a_1 = 15$$

where a_1 is the artificial variable. Like other variables, an artificial variable is not negative. It is called artificial because it eventually enables us to carry out a simplex procedure, but has nothing to do with the amount by which $x_1 - x_2$ exceeds 15.

Yet another adjustment must be made; subtract Ma_1 from the objective function

$$20x_1 + 70x_2$$

giving $z = 20x_1 + 70x_2 - Ma_1$.

If you are still alert at this point, you may wonder, why M? M is a number for which we do not need an exact value, we need only describe it as an arbitrary large constant. In fact, we may consider M to be as large as we please. The name, Big M Method, is given to this technique because of the large value of M. With the two adjustments involving a_1 and M, the original problem becomes the **modified problem**.

Modified Problem.
Maximize $z = 20x_1 + 70x_2 - Ma_1$, subject to

$$\begin{aligned} 2x_1 + 5x_2 + s_1 \qquad &= 58 \\ x_1 - x_2 - s_2 + a_1 &= 15 \\ x_1 \geq 0, x_2 \geq 0, s_1 \geq 0, s_2 &\geq 0, a_1 \geq 0 \end{aligned}$$

Written in standard form, this becomes

$$\begin{aligned} 2x_1 + 5x_2 + s_1 \qquad\qquad\qquad &= 58 \\ x_1 - x_2 \qquad - s_2 + a_1 \qquad &= 15 \\ -20x_1 - 70x_2 \qquad\qquad + Ma_1 + z &= 0 \end{aligned}$$

Just as we have done many times before, we obtain the initial basic solution by setting x_1 and x_2 equal to zero. However, a new wrinkle is added; we also set a_1 equal to zero and solve for the other variables. We set a_1 equal to zero for the same reason that we set x_1 and x_2 equal to zero; a_1 appears in more than one equation. With these three variables set to zero, the solution becomes

$$x_1 = 0, x_2 = 0, a_1 = 0, s_1 = 58, s_2 = -15, z = 0$$

This solution is *not* feasible because s_2 is negative. At this point, it appears we have complicated the problem by introducing an artificial variable a_1 and a large constant M, and we are still faced with an undesirable nonfeasible solution. However, stay with us because the problem can be solved in two phases. The first phase makes an adjustment so that the simplex method can then be applied. We now proceed with phase I of the solution.

PHASE I Set up the initial tableau of the modified problem.

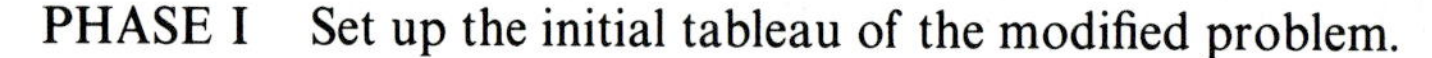

$$\begin{array}{cccccc|c} x_1 & x_2 & s_1 & s_2 & a_1 & z & \\ \hline 2 & 5 & 1 & 0 & 0 & 0 & 58 \\ 1 & -1 & 0 & -1 & 1 & 0 & 15 \\ \hline -20 & -70 & 0 & 0 & M & 1 & 0 \end{array}$$

Since we get a nonfeasible solution from this tableau, we want to modify it to one with a feasible solution. We pivot in the a_1 column to make the desired modification.

Pivot element

$$\begin{array}{cccccc|c} x_1 & x_2 & s_1 & s_2 & a_1 & z & \\ \hline 2 & 5 & 1 & 0 & 0 & 0 & 58 \\ 1 & -1 & 0 & -1 & \textcircled{1} & 0 & 15 \\ \hline -20 & -70 & 0 & 0 & M & 1 & 0 \end{array} \quad (-M)\,R2 + R3 \to R3$$

The pivoting may seem to differ somewhat because M appears in that column instead of a specific number. However, the same procedure applies; multiply Row 2 by $-M$ and add to Row 3 to obtain a 0 in Row 3.

$$\begin{array}{cccccc|c} x_1 & x_2 & s_1 & s_2 & a_1 & z & \\ \hline 2 & 5 & 1 & 0 & 0 & 0 & 58 \\ 1 & -1 & 0 & -1 & 1 & 0 & 15 \\ \hline -M-20 & M-70 & 0 & M & 0 & 1 & -15M \end{array}$$

This tableau has the basic solution

$$x_1 = 0, x_2 = 0, s_2 = 0, s_1 = 58, a_1 = 15, z = -15M$$

This solution is feasible because all variables, except z, are nonnegative. Notice that the pivot operation deleted M from the a_1 column.

We now enter Phase II of the modified problem.

PHASE II The last tableau above has a basic feasible solution, so apply the simplex method to it.

Our first task is to find the most negative entry in the last row:

$$[-M-20 \quad M-70 \quad 0 \quad M \quad 0 \quad 1]$$

Remember, M is a large positive constant. We may assume that it is large enough to make $M - 70$ positive. Then, $-M - 20$ is negative and actually the most negative entry, so the x_1 column is the pivot column. The column ratios indicate Row 2 as the pivot row.

Pivot element → ①

$$\begin{array}{cccccc|c} x_1 & x_2 & s_1 & s_2 & a_1 & z & \\ \hline 2 & 5 & 1 & 0 & 0 & 0 & 58 \\ \textcircled{1} & -1 & 0 & -1 & 1 & 0 & 15 \\ \hline -M-20 & M-70 & 0 & M & 0 & 1 & -15M \end{array} \quad \begin{array}{l} -2R2 + R1 \to R1 \\ \\ (M+20)R2 + R3 \to R3 \end{array}$$

The next tableau is

$$\left[\begin{array}{cccccc|c} 0 & \textcircled{7} & 1 & 2 & -2 & 0 & 28 \\ 1 & -1 & 0 & -1 & 1 & 0 & 15 \\ \hline 0 & -90 & 0 & -20 & M+20 & 1 & 300 \end{array}\right] \quad \tfrac{1}{7}R1 \to R1$$

Next pivot element

$$\left[\begin{array}{cccccc|c} 0 & 1 & \frac{1}{7} & \frac{2}{7} & -\frac{2}{7} & 0 & 4 \\ 1 & -1 & 0 & -1 & 1 & 0 & 15 \\ \hline 0 & -90 & 0 & -20 & M+20 & 1 & 300 \end{array}\right] \quad \begin{array}{l} \\ R1 + R2 \to R2 \\ 90R1 + R3 \to R3 \end{array}$$

$$\left[\begin{array}{cccccc|c} 0 & 1 & \frac{1}{7} & \frac{2}{7} & -\frac{2}{7} & 0 & 4 \\ 1 & 0 & \frac{1}{7} & -\frac{5}{7} & \frac{5}{7} & 0 & 19 \\ \hline 0 & 0 & \frac{90}{7} & \frac{40}{7} & M-\frac{40}{7} & 1 & 660 \end{array}\right]$$

Because M is large, $M - \frac{40}{7}$ is positive and this is the final tableau with the solution

$$x_1 = 19, x_2 = 4, s_1 = 0, s_2 = 0, a_1 = 0, z = 660$$

Don't forget that this is the solution to the modified problem, which is

Maximize $z = 20x_1 + 70x_2 - Ma_1$, subject to

$$\begin{aligned} 2x_1 + 5x_2 + s_1 \qquad\qquad &= 58 \\ x_1 - x_2 \qquad - s_2 + a_1 &= 15 \end{aligned}$$

We simply delete a_1 from the above solution to obtain the optimal solution of the original problem, which is

Maximize $z = 20x_1 + 70x_2$, subject to

$$\begin{aligned} 2x_1 + 5x_2 + s_1 \qquad &= 58 \\ x_1 - x_2 \qquad - s_2 &= 15 \end{aligned}$$

Its optimal solution is $x_1 = 19$, $x_2 = 4$, $s_1 = 0$, $s_2 = 0$, $z = 660$.

If a maximum problem contains constraints with $\geq$, then a surplus variable and an artificial variable are introduced for each $\geq$ constraint. If a constraint includes an $=$, then introduce an artificial variable, but not a slack or surplus variable. The same M is used for all artificial variables, and $-Ma$ is added to the objective function for each artificial variable a.

Summary of the Big M Method

The Big M Method Used to Maximize the Objective Function

1. Set up the problem.
   - **(a)** For each $\leq$ constraint, introduce a slack variable.
   - **(b)** For each $\geq$ constraint, introduce a surplus variable and an artificial variable.
   - **(c)** For each $=$ constaint, introduce an artificial variable. Do not introduce a slack or surplus variable.
   - **(d)** Add $-Ma_i$ to the objective function for each artificial variable a_i.

   These adjustments give the modified problem.
2. Solving the modified problem.

   Phase I

   - **(a)** Form the simplex tableau of the modified problem. The basic solution will not be feasible.
   - **(b)** Convert the tableau to one that has a basic feasible solution. To do this, pivot in each column corresponding to an artificial variable. This will eliminate the M from the bottom row of that column.

   Phase II

   Apply the simplex method to the Phase I tableau to obtain an optimal solution.
3. Optimal solution of the original problem
   - **(a)** If all artificial variables are zero in the optimal solution of the modified solution, then the original problem has an optimal solution. Delete all artificial variables from the optimal solution of the modified problem to obtain the optimal solution of the original problem.
   - **(b)** If one or more of the artificial variables is not zero in the optimal solution of the modified problem, then the original problem has no optimal solution.

We now give several variations of linear programming problems.

Example 2 (*Compare Exercise 3*)
Maximize $4x_1 + 12x_2$, subject to

$$\begin{aligned} 3x_1 + x_2 &\leq 180 \\ x_1 + 2x_2 &= 100 \\ -2x_1 + 2x_2 &\leq 40 \\ x_1 \geq 0, x_2 &\geq 0 \end{aligned}$$

Solution Slack variables are introduced into the first and third constraints, whereas an artificial variable is introduced into the second constraint. This gives the modified problem with constraints

$$\begin{aligned} 3x_1 + x_2 + s_1 \qquad\qquad &= 180 \\ x_1 + 2x_2 \qquad\qquad + a_1 &= 100 \\ -2x_1 + 2x_2 \qquad + s_2 \qquad &= 40 \end{aligned}$$

and the objective function

$$z = 4x_1 + 12x_2 - Ma_1$$

Form the simplex tableau for the modified problem.

$$\begin{array}{cccccc|c} x_1 & x_2 & s_1 & s_2 & a_1 & z & \\ \hline 3 & 1 & 1 & 0 & 0 & 0 & 180 \\ 1 & 2 & 0 & 0 & 1 & 0 & 100 \\ -2 & 2 & 0 & 1 & 0 & 0 & 40 \\ \hline -4 & -12 & 0 & 0 & M & 1 & 0 \end{array}$$

We are now ready to go to the first phase.

Phase I

Pivot in the a_1 column to eliminate M from the last row.

$$\left[\begin{array}{cccccc|c} 3 & 1 & 1 & 0 & 0 & 0 & 180 \\ 1 & 2 & 0 & 0 & 1 & 0 & 100 \\ -2 & 2 & 0 & 1 & 0 & 0 & 40 \\ \hline -M-4 & -2M-12 & 0 & 0 & 0 & 1 & -100M \end{array}\right]$$

This tableau has a feasible basic solution, so we go to Phase II and apply the simplex method.

Phase II

Since M is a large positive number, $-2M - 12$ is more negative than $-M - 4$. The pivot element is the 2 in Row 3, Column 2. Dividing the third row by 2 gives

$$\left[\begin{array}{cccccc|c} 3 & 1 & 1 & 0 & 0 & 0 & 180 \\ 1 & 2 & 0 & 0 & 1 & 0 & 100 \\ -1 & 1 & 0 & \frac{1}{2} & 0 & 0 & 20 \\ \hline -M-4 & -2M-12 & 0 & 0 & 0 & 1 & -100M \end{array}\right] \begin{array}{l} -R3 + R1 \rightarrow R1 \\ -2R3 + R2 \rightarrow R2 \\ \\ (2M + 12)R3 + R4 \rightarrow R4 \end{array}$$

$$\left[\begin{array}{cccccc|c} 4 & 0 & 1 & -\frac{1}{2} & 0 & 0 & 160 \\ 3 & 0 & 0 & -1 & 1 & 0 & 60 \\ -1 & 1 & 0 & \frac{1}{2} & 0 & 0 & 20 \\ \hline -3M-16 & 0 & 0 & M+6 & 0 & 1 & -60M+240 \end{array}\right] \begin{array}{l} \\ \frac{1}{3}R2 \rightarrow R2 \\ \\ \\ \end{array}$$

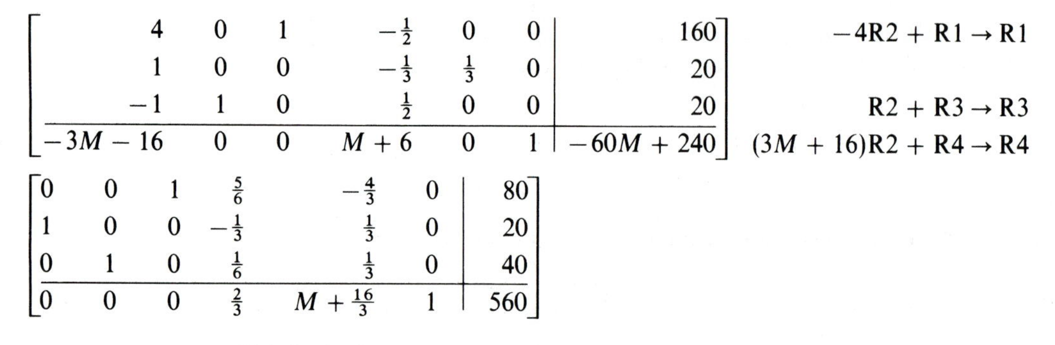

$$\left[\begin{array}{cccccc|c} 4 & 0 & 1 & -\frac{1}{2} & 0 & 0 & 160 \\ 1 & 0 & 0 & -\frac{1}{3} & \frac{1}{3} & 0 & 20 \\ -1 & 1 & 0 & \frac{1}{2} & 0 & 0 & 20 \\ \hline -3M-16 & 0 & 0 & M+6 & 0 & 1 & -60M+240 \end{array}\right] \begin{array}{l} -4R2+R1\to R1 \\ \\ R2+R3\to R3 \\ (3M+16)R2+R4\to R4 \end{array}$$

$$\left[\begin{array}{cccccc|c} 0 & 0 & 1 & \frac{5}{6} & -\frac{4}{3} & 0 & 80 \\ 1 & 0 & 0 & -\frac{1}{3} & \frac{1}{3} & 0 & 20 \\ 0 & 1 & 0 & \frac{1}{6} & \frac{1}{3} & 0 & 40 \\ \hline 0 & 0 & 0 & \frac{2}{3} & M+\frac{16}{3} & 1 & 560 \end{array}\right]$$

This is the final tableau with the optimal solution

$$x_1 = 20,\ x_2 = 40,\ s_1 = 80,\ s_2 = 0,\ a_1 = 0,\ z = 560$$

Since the artificial variable is zero in the optimal solution of the modified problem, the original problem has the optimal solution

$$x_1 = 20,\ x_2 = 40,\ s_1 = 80,\ s_2 = 0,\ z = 560$$

The following example contains the complete mixture of constraints.

Example 3 *(Compare Exercise 5)*
Set up the modified problem for this problem.
Maximize $z = 4x_1 + 12x_2$, subject to

$$\begin{aligned} 3x_1 + x_2 &\le 180 \\ x_1 + 2x_2 &\ge 100 \\ -2x_1 + 2x_2 &= 100 \\ x_1 \ge 0,\ x_2 &\ge 0 \end{aligned}$$

Solution Introduce a slack variable into the first constraint, a surplus and artificial variable into the second constraint, and an artificial variable into the third constraint.

Modified Problem.
Constraints:

$$\begin{aligned} 3x_1 + x_2 + s_1 \qquad\qquad\qquad &= 180 \\ x_1 + 2x_2 \qquad - s_2 + a_1 \qquad &= 100 \\ -2x_1 + 2x_2 \qquad\qquad\qquad + a_2 &= 40 \end{aligned}$$

Objective function:
Maximize

$$z = 4x_1 + 12x_2 - Ma_1 - Ma_2$$

The following example illustrates a situation in which the original problem has no optimal solution.

Example 4 (*Compare Exercise 7*)
Maximize $5x_1 + 4x_2$, subject to

$$-x_1 + x_2 \geq 15$$
$$2x_1 + x_2 \leq 10$$
$$x_1 \geq 0, x_2 \geq 0$$

Solution The simplex tableau for the modified problem is

$$\begin{array}{cccccc|c} x_1 & x_2 & s_1 & s_2 & a_1 & z & \\ \hline -1 & 1 & -1 & 0 & 1 & 0 & 15 \\ 2 & 1 & 0 & 1 & 0 & 0 & 10 \\ \hline -5 & -4 & 0 & 0 & M & 1 & 0 \end{array}$$

Pivot in the a_1 column to eliminate M from the last row.

$$\left[\begin{array}{cccccc|c} -1 & 1 & -1 & 0 & 1 & 0 & 15 \\ 2 & 1 & 0 & 1 & 0 & 0 & 10 \\ \hline M-5 & -M-4 & M & 0 & 0 & 1 & -15M \end{array}\right] \begin{array}{l} -R2 + R1 \rightarrow R1 \\ \\ (M+4)R2 + R3 \rightarrow R3 \end{array}$$

$$\begin{array}{cccccc|c} x_1 & x_2 & s_1 & s_2 & a_1 & z & \\ \hline -3 & 0 & -1 & -1 & 1 & 0 & 5 \\ 2 & 1 & 0 & 1 & 0 & 0 & 10 \\ \hline 3M+3 & 0 & M & M+4 & 0 & 1 & -5M+40 \end{array}$$

This is the final tableau with the optimal solution

$$x_1 = 0, x_2 = 10, s_1 = 0, s_2 = 0, a_1 = 5, z = -5M + 40$$

Since $a_1 \neq 0$, the original problem has no optimal solution. Figure 4–4 shows the

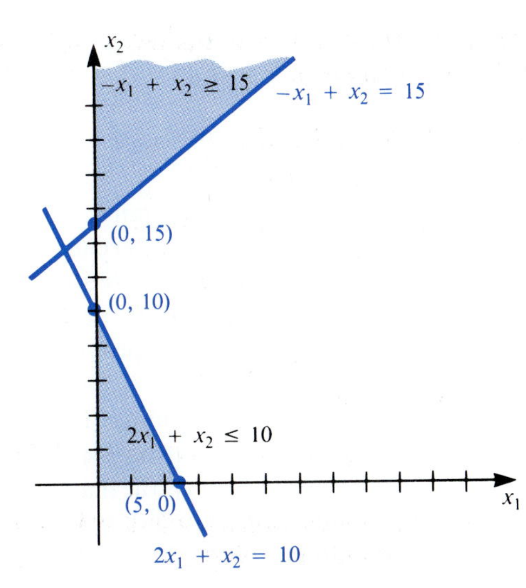

Figure 4–4

reason. The region in quadrant I defined by

$$-x_1 + x_2 \geq 15$$

does not intersect the region defined by

$$2x_1 + x_2 \leq 10$$

in quadrant I. Since there is no feasible region, there is no optimal solution.

Negative Constants in Constraints

The simplex method assumes that all the constants on the right side of the constraints are nonnegative. If a constant is negative, the problem can be handled with an adjustment like the following.

Example 5 (*Compare Exercise 10*)
Set up the modified problem for the problem.
Maximize $40x_1 + 65x_2$, subject to

$$\begin{aligned} 6x_1 - 5x_2 &\leq -16 \\ 4x_1 + 8x_2 &\leq 9 \\ x_1 \geq 0, x_2 &\geq 0 \end{aligned}$$

Solution We can obtain a positive number on the right side of the first constraint by multiplying through by -1. (This reverses the inequality.)

We then replace

$$6x_1 - 5x_2 \leq -16$$

with

$$-6x_1 + 5x_2 \geq 16$$

The modified problem now becomes a problem with the constraints

$$\begin{aligned} -6x_1 + 5x_2 - s_1 \quad + a_1 &= 16 \\ 4x_1 + 8x_2 \quad + s_2 \quad &= 9 \end{aligned}$$

and objective function $z = 40x_1 + 65x_2 - Ma_1$.

Minimum Problem

In Section 4–4, standard minimization problems were solved by converting them to dual problems, which became standard maximization problems. The dual problem could then be solved directly. This technique does not apply to a nonstandard minimization problem, but another technique can be used. It converts a minimization problem to a maximization problem, but in another way.

The adjustment is simple. If z is the objective function to be minimized, then solve the maximization problem using $w = -z$ as the objective function. This works because when k is the maximum value of w, $-k$ is the minimum value of z. For example, when you multiply a set of numbers by -1, you reverse the order. The set of numbers $\{1, 5, 7, 16\}$ has 1 as the smallest and 16 as the largest number.

The smallest number in $\{1, 5, 7, 16\}$ is 1.

The largest number in $\{-1, -5, -7, -16\}$ is -1.

Figure 4–5

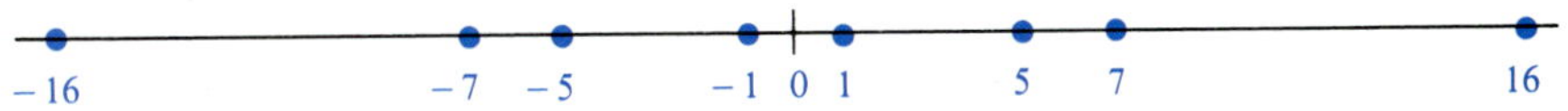

$-16 \quad -7 \quad -5 \quad -1 \quad 0 \quad 1 \quad 5 \quad 7 \quad 16$

The set made of the negatives of these numbers is $\{-1, -5, -7, -16\}$. It has -1 as the *largest* number and -16 as the *smallest*. (See Figure 4–5.)

Example 6 (*Compare Exercise 12*)
Minimize $z = 2x_1 - 3x_2$, subject to

$$x_1 + 2x_2 \leq 10$$
$$2x_1 + x_2 \leq 11$$
$$x_1 \geq 0, x_2 \geq 0$$

Solution Convert the objective function to $w = -2x_1 + 3x_2$.
The tableaux for the solution are as follows:

$$\begin{array}{ccccc|c} x_1 & x_2 & s_1 & s_2 & w & \\ \hline \end{array}$$

$$\left[\begin{array}{ccccc|c} 1 & 2 & 1 & 0 & 0 & 10 \\ 2 & 1 & 0 & 1 & 0 & 11 \\ \hline 2 & -3 & 0 & 0 & 1 & 0 \end{array}\right] \quad \tfrac{1}{2}R1 \rightarrow R1$$

$$\left[\begin{array}{ccccc|c} \frac{1}{2} & 1 & \frac{1}{2} & 0 & 0 & 5 \\ 2 & 1 & 0 & 1 & 0 & 11 \\ \hline 2 & -3 & 0 & 0 & 1 & 0 \end{array}\right] \quad \begin{array}{l} -R1 + R2 \rightarrow R2 \\ 3R1 + R3 \rightarrow R3 \end{array}$$

$$\left[\begin{array}{ccccc|c} \frac{1}{2} & 1 & \frac{1}{2} & 0 & 0 & 5 \\ \frac{3}{2} & 0 & -\frac{1}{2} & 1 & 0 & 6 \\ \hline \frac{7}{2} & 0 & \frac{3}{2} & 0 & 1 & 15 \end{array}\right]$$

The optimal solution (maximum) is $w = 15$ when $x_1 = 0$, $x_2 = 5$. The original problem then has as its optimal (minimum) solution:

$$z = -15 \text{ at } x_1 = 0, x_2 = 5$$

Comments on the Big *M* Method

The technique of introducing an artificial variable and a large number M into a problem accomplishes two things:

(a) Pivoting in columns corresponding to an artificial variable to eliminate M from the last row gives a tableau with a basic feasible solution.

(b) The inclusion of M in the objective function ensures that the artificial variable is zero when the objective function attains its maximum. When the artificial variables are zero, the optimal solution of the original problem follows directly from the optimal solution of the modified problem.

Summary of the Simplex Method

The simplex method maximizes, in a direct fashion, an objective function when the constraints are $\leq$ and all constants to the right of the $\leq$ are nonnegative. Other cases, such as mixed constraints or minimization problems, can be converted so the simplex method applies.

The following summarizes several types of problems that can eventually be solved by the simplex method.

1. Standard Maximum Problem. All constraints are $\leq$, all constant terms to the right of $\leq$ are nonnegative, and the objective function is to be maximized. Method used: Simplex Method.
2. Standard Minimum Problem. All constraints are $\geq$, the objective function is to be minimized, and all coefficients of the objective function are nonnegative. Method used: Find the dual problem and apply the simplex method to it. The optimal value of the dual problem is the optimal value of the original problem.
3. Mixed Constraints, Maximum Problem. There is a mixture of $\geq$, $\leq$, $=$ in the constraints, all constant terms in the constraints are nonnegative, and the objective function is to be maximized. Method used: Introduce slack variables for each $\leq$ constraint, a surplus and artificial variable for each $\geq$ constraint, and an artificial variable for each $=$ constraint. Subtract Ma from the objective function for each artificial variable, a, to obtain the modified problem. Apply Phase I to the modified problem to obtain a tableau with a feasible basic solution.

   Apply Phase II, the simplex method. If all artificial variables in the optimal solution are zero, delete them to obtain the optimal solution of the original problem. If some artificial variable is not zero, the original problem has no optimal solution.
4. Minimum Problem, not standard. Method used: Replace the objective function z with $w = -z$. Then maximize w, subject to the same constraints. The minimum value of z is the negative of the maximum value of w.
5. Some of the constraint constants are negative. Method used: Multiply by -1 each constraint that has a negative constant on the right. Then apply the appropriate method.

4–5 Exercises

I.

1. (*See Example 1*) Maximize $z = 15x_1 + 22x_2$, subject to

$$5x_1 + 11x_2 \leq 350$$
$$15x_1 + 8x_2 \geq 300$$
$$x_1 \geq 0, x_2 \geq 0$$

2. Maximize $z = x_1 - 2x_2$, subject to

$$x_1 + x_2 \geq 10$$
$$2x_1 + 5x_2 \leq 60$$
$$x_1 \geq 0, x_2 \geq 0$$

3. (*See Example 2*) Maximize $z = 5x_1 + 20x_2$, subject to

$$3x_1 + 2x_2 \le 48$$
$$2x_1 + 4x_2 \le 64$$
$$5x_1 + 6x_2 = 104$$
$$x_1 \ge 0, x_2 \ge 0$$

4. Maximize $z = 20x_1 + 25x_2$, subject to

$$2x_1 + 3x_2 \le 160$$
$$-6x_1 + 5x_2 = 80$$

5. (*See Example 3*) Set up the modified problem for this problem:
Maximize $z = 20x_1 + 17x_2$, subject to

$$5x_1 + 3x_2 \le 220$$
$$4x_1 + 10x_2 \ge 80$$
$$-3x_1 + 7x_2 = 42$$
$$x_1 \ge 0, x_2 \ge 0$$

6. Set up the modified problem to this problem:
Maximize $z = 10x_1 + 30x_2 + 50x_3$, subject to

$$6x_1 + 11x_2 + 4x_3 \le 250$$
$$5x_1 + 14x_2 + 8x_3 \ge 460$$
$$x_1 + x_2 + 3x_3 \ge 380$$
$$x_1 \ge 0, x_2 \ge 0, x_3 \ge 0$$

7. (*See Example 4*) Maximize $z = 5x_1 + 8x_2$, subject to

$$x_1 + x_2 \le 20$$
$$5x_1 + 4x_2 \ge 200$$
$$x_1 \ge 0, x_2 \ge 0$$

8. Maximize $z = 4x_1 + 12x_2$, subject to

$$10x_1 + 15x_2 \le 150$$
$$6x_1 + 3x_2 \ge 180$$
$$x_1 \ge 0, x_2 \ge 0$$

9. Maximize $z = 6x_1 + 4x_2$, subject to

$$3x_1 + 2x_2 \le 60$$
$$2x_1 + 3x_2 \ge 24$$
$$x_1 + x_2 = 25$$
$$x_1 \ge 0, x_2 \ge 0$$

10. (*See Example 5*) Set up the modified problem for the following:
Maximize $z = 35x_1 + 60x_2$, subject to

$$8x_1 - 14x_2 \le -20$$
$$5x_1 + 12x_2 \le 40$$
$$x_1 \ge 0, x_2 \ge 0$$

11. Set up the modified problem for this problem:
Maximize $z = 3x_1 + 7x_2 + x_3$, subject to

$$2x_1 + 4x_2 + 7x_3 \le 42$$
$$x_1 - 3x_2 + x_3 \le -10$$
$$x_1 \ge 0, x_2 \ge 0, x_3 \ge 0$$

12. (*See Example 6*) Minimize $z = 2x_1 - 5x_2$, subject to

$$4x_1 + 3x_2 \le 120$$
$$2x_1 + x_2 \le 50$$
$$x_1 \ge 0, x_2 \ge 0$$

13. Minimize $z = 15x_1 + 40x_2$, subject to

$$7x_1 + 3x_2 \ge 168$$
$$5x_1 + 3x_2 \le 210$$
$$x_1 \ge 0, x_2 \ge 0$$

14. Maximize $x_1 + 3x_2$, subject to
$x_1 + 2x_2 \le 4$
$2x_1 - 3x_2 \le -3$
$x_1 \ge 0, x_2 \ge 0$

15. Maximize $x_1 + 4x_2$, subject to
$x_1 + 3x_2 \le 4$
$-x_1 + x_2 \le -2$
$x_1 \ge 0, x_2 \ge 0$

16. Maximize $2x_1 + x_2$, subject to
$-x_1 + 2x_2 \le 4$
$-x_1 + x_2 \ge 1$
$x_1 \ge 0, x_2 \ge 0$

17. Maximize $x_1 + x_2$, subject to
$-x_1 + x_2 \leq 4$
$-2x_1 + x_2 \geq 2$
$x_1 \geq 0, x_2 \geq 0$

18. Maximize $x_1 + 2x_2$, subject to
$2x_1 + x_2 \leq 4$
$x_1 + x_2 = 3$
$x_1 \geq 0, x_2 \geq 0$

19. Maximize $2x_1 - x_2$, subject to
$x_1 + 3x_2 \leq 6$
$x_1 - x_2 = 2$
$x_1 \geq 0, x_2 \geq 0$

20. Maximize $2x_1 + x_2$, subject to
$3x_1 + 2x_2 \leq 6$
$x_1 - 2x_2 \leq 2$
$x_1 \geq 0, x_2 \geq 0$

21. Maximize $3x_1 - 2x_2$, subject to
$x_1 + x_2 \leq 3$
$-x_1 + 2x_2 \leq 3$
$x_1 \geq 0, x_2 \geq 0$

22. Maximize $x_1 + 2x_2$, subject to
$x_1 + x_2 \leq 1$
$2x_1 + 3x_2 \geq 6$
$x_1 \geq 0, x_2 \geq 0$

23. Maximize $2x_1 - x_2$, subject to
$x_1 + 2x_2 \leq 2$
$x_1 + 3x_2 \geq 6$
$x_1 \geq 0, x_2 \geq 0$

II.

24. A college wishes to offer admission to exactly 1000 incoming freshmen. They will give $3000 scholarships based on need and $2000 scholarships based on merit, but the total scholarships cannot exceed $900,000. They will admit at least 100 students based on need. Past experience indicates that the students who receive merit scholarships will have an average SAT score of 1200, those who receive need scholarships will have an average SAT score of 1000, and those receiving no scholarships will have an average SAT score of 900. How many need, merit, and other freshmen should they admit in order to maximize the total SAT scores of all entering freshmen?

25. A distributor offers a store a special on two models of nightstands, the Custom and the Executive, if the store buys at least 100. The Custom costs $70 each, and the store will sell them for $90 each. The Executive costs $80 each and sells for $120 each. The store has 800 square feet of storage space available. The Custom requires 4 square feet each, and the Executive requires 5 square feet each. The store manager wants gross sales of at least $10,800. How many of each should be ordered so that the total cost will be minimized?

IMPORTANT TERMS

4–1 **Maximum standard problem** — **Slack variable** — **Simplex tableau**

4–2 **Basic solution** — **Basic feasible solution** — **Initial basic solution**
Pivot column — **Pivot row** — **Pivot element**
Final tableau

4–4	**Minimum standard problem**	**Dual problem**	**Transpose of a matrix**
4–5	**Mixed constraints**	**Surplus variable**	**Artificial variable**
	Big M Method Phase I	**Phase II**	**Modified problem**

REVIEW EXERCISES

In Exercises 1 through 3, write the constraints as a system of equations using slack variables.

1. $6x_1 + 4x_2 + 3x_3 \le 220$
   $x_1 + 5x_2 + x_3 \le 162$
   $7x_1 + 2x_2 + 5x_3 \le 139$

2. $5x_1 + 3x_2 \le 40$
   $7x_1 + 2x_2 \le 19$
   $6x_1 + 5x_2 \le 23$

3. $6x_1 + 5x_2 + 3x_3 + 3x_4 \le 89$
   $7x_1 + 4x_2 + 6x_3 + 2x_4 \le 72$

In Exercises 4 through 6, write the constraints and objective function as a system of equations.

4. Objective function
   $z = 3x_1 + 7x_2$
   constraints
   $7x_1 + 5x_2 \le 14$
   $3x_1 + 6x_2 \le 25$
   $4x_1 + 3x_2 \le 29$

5. Objective function
   $z = 20x_1 + 36x_2 + 19x_3$
   constraints
   $10x_1 + 12x_2 + 8x_3 \le 24$
   $7x_1 + 13x_2 + 5x_3 \le 35$

6. Objective function
   $z = 5x_1 + 12x_2 + 8x_3 + 2x_4$
   constraints
   $9x_1 + 7x_2 + x_3 + x_4 \le 84$
   $x_1 + 3x_2 + 5x_3 + x_4 \le 76$
   $2x_1 + x_2 + 6x_3 + 3x_4 \le 59$

In Exercises 7 through 16, write the following as a system of equations.

7. Maximize $z = 9x_1 + 2x_2$, subject to
$$3x_1 + 7x_2 \le 14$$
$$9x_1 + 5x_2 \le 18$$
$$x_1 - x_2 \le 21$$
$$x_1 \ge 0, x_2 \ge 0$$

8. Maximize $z = x_1 + 5x_2 + 4x_3$, subject to
$$x_1 + x_2 + x_3 \le 20$$
$$4x_1 + 5x_2 + x_3 \le 48$$
$$2x_1 - 6x_2 + 5x_3 \le 38$$
$$x_1 \ge 0, x_2 \ge 0, x_3 \ge 0$$

9. Maximize $z = 6x_1 + 8x_2 + 4x_3$, subject to
$$x_1 + x_2 + x_3 \le 15$$
$$2x_1 + 4x_2 + x_3 \le 44$$
$$x_1 \ge 0, x_2 \ge 0, x_3 \ge 0$$

10. Maximize $z = 5x_1 + 5x_2$, subject to
$$5x_1 + 3x_2 \le 15$$
$$2x_1 + 3x_2 \le 12$$
$$x_1 \ge 0, x_2 \ge 0$$

11. Maximize $z = 4x_1 + 5x_2$, subject to
$$x_1 + 3x_2 \le 12$$
$$2x_1 + 4x_2 \le 16$$
$$x_1 \ge 0, x_2 > 0$$

12. Maximize $z = 3x_1 + 5x_2 + 2x_3$, subject to
$$2x_1 + 4x_2 + 2x_3 \le 34$$
$$3x_1 + 6x_2 + 4x_3 \le 57$$
$$2x_1 + 5x_2 + x_3 \le 31$$
$$x_1 \ge 0, x_2 \ge 0, x_3 \ge 0$$

13. Maximize $z = 3x_1 + 4x_2$, subject to
$$x_1 - 3x_2 \le 6$$
$$-x_1 + x_2 \le 8$$
$$x_1 \ge 0, x_2 \ge 0$$

14. Maximize $z = 10x_1 + 15x_2$, subject to
$$-4x_1 + x_2 \le 3$$
$$x_1 - 2x_2 \le 12$$
$$x_1 \ge 0, x_2 \ge 0$$

15. Maximize $z = 3x_1 + 6x_2 + x_3$, subject to

$$\begin{aligned} 4x_1 + 4x_2 + 8x_3 &\le 800 \\ 8x_1 + 6x_2 + 4x_3 &\le 1800 \\ 8x_1 + 4x_2 \qquad &\le 400 \\ x_1 \ge 0, x_2 \ge 0, x_3 &\ge 0 \end{aligned}$$

16. Maximize $z = x_1 + 3x_2 + x_3$, subject to

$$\begin{aligned} 4x_1 + x_2 + x_3 &\le 372 \\ x_1 + 8x_2 + 6x_3 &\le 1116 \end{aligned}$$

17. Find the values of the slack variables in the constraints.

$$\begin{aligned} 4x_1 + 3x_2 + 6x_3 + s_1 &= 68 \\ x_1 + 2x_2 + 5x_3 + s_2 &= 90 \end{aligned}$$

for the points (3, 2, 1) and (5, 10, 3).

18. Which variable contributes the most to increasing z in the following objective functions?

(a) $z = 7x_1 + 2x_2 + 9x_3$

(b) $z = x_1 + 8x_2 + x_3 + 7x_4$

19. Given the constraint

$$7x_1 + 4x_2 + 17x_3 + s_1 = 56$$

When $x_2 = 0$, what is the largest possible value of x_3 so that s_1 is nonnegative?

20. Write the transpose of the following matrices

$$\begin{bmatrix} 3 & 1 & -2 \\ 4 & 0 & 6 \\ 5 & 7 & 8 \end{bmatrix} \qquad \begin{bmatrix} 4 & 3 & 2 & 1 \\ -5 & 6 & 12 & 9 \end{bmatrix}$$

21. For each of the following minimization problems, set up the augmented matrix and the initial tableau for the dual problem. Do no solve.

(a) Minimize $x = 30x_1 + 17x_2$, subject to

$$\begin{aligned} 4x_1 + 5x_2 &\ge 52 \\ 7x_1 + 14x_2 &\ge 39 \\ x_1 \ge 0, x_2 &\ge 0 \end{aligned}$$

(b) Minimize $z = 100x_1 + 225x_2 + 145x_3$, subject to

$$\begin{aligned} 20x_1 + 35x_2 + 15x_3 &\ge 130 \\ 40x_1 + 10x_2 + 6x_3 &\ge 220 \\ 35x_1 + 22x_2 + 18x_3 &\ge 176 \\ x_1 \ge 0, x_2 \ge 0, x_3 &\ge 0 \end{aligned}$$

Solve the following

22. Minimize $z = 18x_1 + 36x_2$, subject to
$3x_1 + 2x_2 \ge 24$
$5x_1 + 4x_2 \ge 46$
$4x_1 + 9x_2 \ge 60$
$x_1 \ge 0, x_2 \ge 0$

23. Maximize $z = 5x_1 + 15x_2$, subject to
$4x_1 + x_2 \le 200$
$x_1 + 3x_2 \ge 120$
$x_1 \ge 0, x_2 \ge 0$

24. Maximize $z = 5x_1 + 15x_2$, subject to

$$\begin{aligned} 4x_1 + x_2 &\leq 200 \\ x_1 + 3x_2 &\geq 120 \\ -x_1 + 3x_2 &= 150 \\ x_1 \geq 0, x_2 &\geq 0 \end{aligned}$$

25. Minimize $z = 3x_1 - 2x_2$, subject to

$$\begin{aligned} x_1 + 3x_2 &\leq 30 \\ 3x_1 + x_2 &\leq 21 \\ x_1 \geq 0, x_2 &\geq 0 \end{aligned}$$

26. Maximize $z = 2x_1 + x_2$, subject to

$$\begin{aligned} x_1 + 3x_2 &\leq 9 \\ x_1 - x_2 &\leq -2 \\ x_1 \geq 0, x_2 &\geq 0 \end{aligned}$$

27. Set up the initial simplex tableau for the following. Do not solve.

A company manufactures three items, hunting jackets, all-weather jackets, and ski jackets. It takes 3 hours labor per dozen to produce hunting jackets, 2.5 hours per dozen for all-weather jackets, and 3.5 hours per dozen for ski jackets. The cost per dozen is \$26 for hunting jackets, \$20 for all-weather jackets, and \$22 for ski jackets. Their profit, per dozen, is \$7.50 for hunting jackets, \$9 for all-weather jackets, and \$11 for ski jackets. The company has 3200 hours labor and \$18,000 operating funds available. How many of each jacket should they produce to maximize profits?

28. A fertilizer company produces two kinds of fertilizers, Lawn and Tree. They have orders on hand that call for the production of at least 20,000 bags of Lawn and 5000 bags of Tree fertilizer. Plant *A* can produce 500 bags of Lawn and 100 bags of Tree fertilizer per day. Plant *B* can produce 250 bags of Lawn and 90 bags of Tree fertilizer per day. It costs \$18,000 per day to operate plant *A* and \$12,000 per day to operate plant *B*. How many days should they operate each plant to minimize operating costs?

5

Topics in this chapter can be applied to problems such as:

Sampling Populations ● Sampling Preferences and Opinions ● Constructing Questionnaires ● Routing Alternatives ● Product Combinations ● Product Selections ● Display Planning ● Conducting Surveys ● Developing Polls

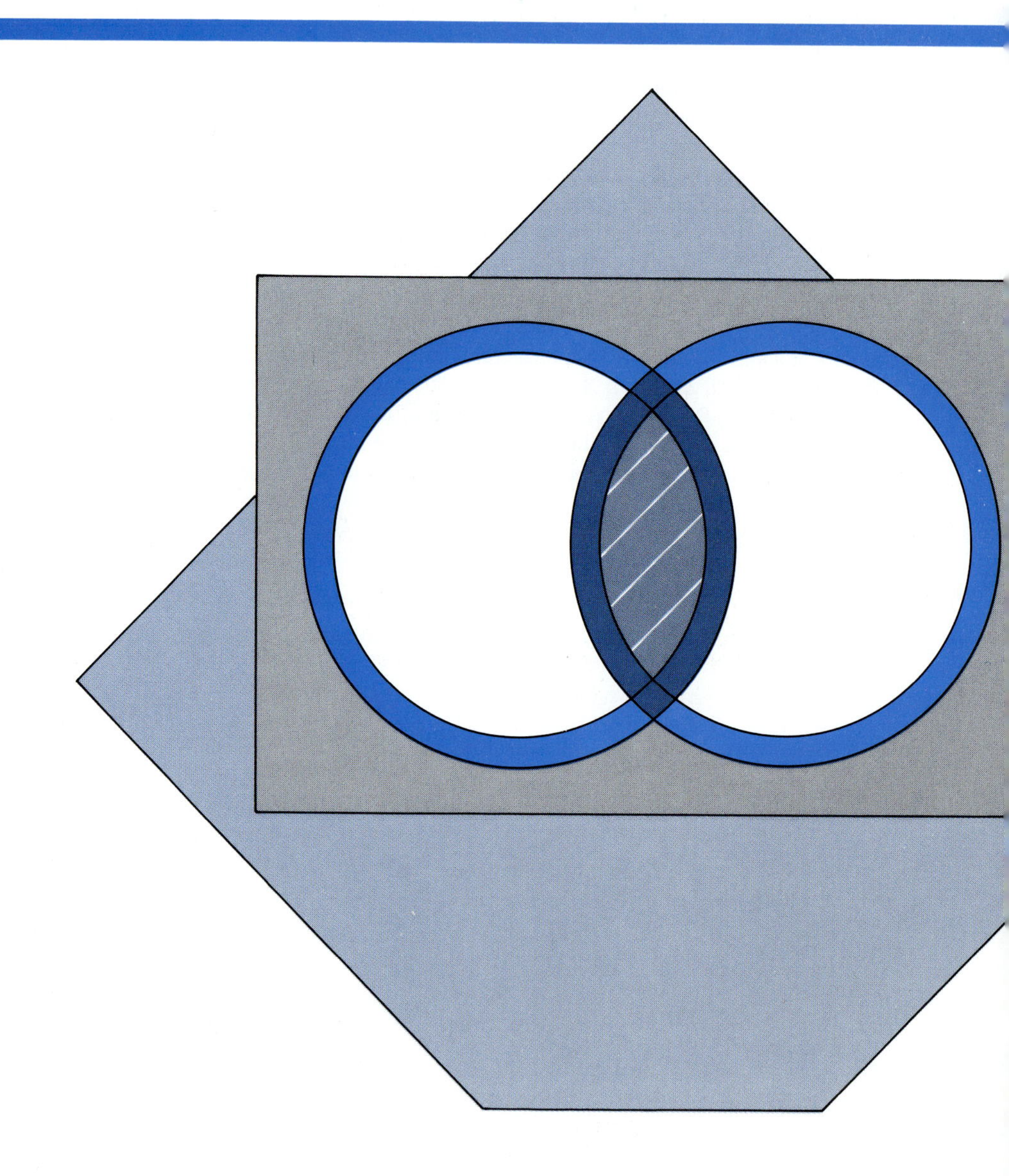

Sets and Counting

- Sets
- Venn Diagrams
- Fundamental Counting Principle
- Permutations and Combinations
- Partitions

5–1 Sets

One of the basic concepts in mathematics is that of a set. Even in everyday life we talk about sets. We talk about a set of dishes, a set of spark plugs, a reading list (a set of books), the people invited to a party (a set of people), the vegetables served at the cafeteria (a set of vegetables), and the teams in the NFL playoffs (a set of football teams).

The concept of a set is so basic that proposed definitions of it tend to use words that mean essentially the same thing. We may speak of a set as a **collection** of objects. If we attempt to define a collection, we tend to think of the same words as we use in defining a set. Rather than go in circles trying to define a set, we will not make a definition. However, you should have an intuitive feeling of the meaning of a set.

While we do not define the concept of a set, any particular set should be **well defined** in the sense that there should be no ambiguity as to the objects that make up the set. For example, the collection of all states in the United States is a set containing 50 elements. There is no ambiguity as to the objects that make up this collection. On the other hand, the collection of great U.S. presidents is not a well-defined set; it depends on who makes the selection and what they mean by "great."

Let's look at some of the traditional terminology used when speaking about sets. The objects that form a set may be varied. We can have a set of people, a set of numbers, a set of ideas, or a set of raindrops. In mathematics the general term for an object in a set is **element**. A set may have people, numbers, ideas, and raindrops as elements. So that the contents of a set will be clearly understood, we need to be rather precise in describing a set of elements. One way to describe a set is to list explicitly all its elements. We usually enclose the list with braces. When we write

$$A = \{\text{Tom, Dick, Harry}\}$$

we are stating that we have named the set "A," and its elements are Tom, Dick, and Harry. Customarily, capital letters designate sets and small letters represent elements in a set. The notation

$$x \in A$$

is read "x is an element of A." We may also say that "x is a member of the set A." The statement "x is not an element of the set A" is written

$$x \notin A$$

For the set

$$B = \{2, 4, 6, 8\}$$

we may write $4 \in B$ to specify that 4 is one of the elements of the set B and $5 \notin B$ to specify that 5 is not one of the elements of B. Here are some examples of other sets:

Example 1 The set of positive integers less than 7 is $\{1, 2, 3, 4, 5, 6\}$.

Example 2 The vowels of the English alphabet form the set $\{a, e, i, o, u\}$.

Example 3 The countries of North America form the set {Canada, USA, Mexico}.

Example 4 The set of all natural numbers is $\{1, 2, 3, 4, \ldots\}$. Notice that we used three dots to indicate that the natural numbers continue, without ending, beyond 4. This is an infinite set. The three dots are commonly used to indicate that the preceding pattern continues. This notation is the mathematical equivalent of *et cetera*. The context in which the three dots are used determines which elements are missing. The notation$\{20, 22, 24, \ldots, 32\}$ indicates that 26, 28, and 30 are missing. This set is finite because it contains a finite number of elements. When "..." is used in representing a finite set, the last element of the set is usually given. An infinite set gives no indication of a last element, like the set of positive even numbers $\{2, 4, 6, \ldots\}$.

Set-Builder Notation

Another way of describing a set uses what is called **implicit** or **set-builder** notation. The set of the vowels of the English alphabet may be described by

$$\{x \mid x \text{ is a vowel of the English alphabet}\}$$

This is read, "The set of all x such that x is a vowel of the English alphabet." Note that the vertical line | is read "such that." The symbol preceding the vertical line, x in this instance, designates a typical element of the set. The statement to the right of the vertical line *describes* a typical element of the set; it tells how to find a specific instance of x.

Example 5 (*Compare Exercise 2*)

(a) The elements of the set $A = \{x \mid x \text{ is an odd integer between 10 and 20}\}$ are 11, 13, 15, 17, and 19.

(b) The set $\{x \mid x \text{ is a male student at Community College and } x \text{ is over 6 ft tall}\}$ is the set of all male students at Community College who are over 6 ft tall.

(c) The set $\{x \mid x \text{ is a member of the United States Senate}\}$ is the set of all senators of the United States.

(d) The set $\{x \mid x = 3n - 2 \text{ where } n \text{ is a positive integer}\}$ contains $1, 4, 7, 10, \ldots$.

Equal Sets

Two sets A and B are said to be **equal** if they consist of exactly the same elements; this is denoted by $A = B$. The sets $A = \{2, 4, 6, 8\}$ and $B = \{6, 6, 2, 8, 4\}$ are equal because they consist of the same elements. Listing the elements in a different order or with repetition does not create a different set.

Empty Set

Sometimes a set is described and it has no elements. For example, the set of golfers who made nine consecutive holes-in-one at the Richland Golf Course has no elements. Such a set, consisting of no elements, plays an important role in set theory. It is called the **empty set** and is denoted by $\varnothing$. A set is called **nonempty** if it contains one or more elements. The set of people 10 ft tall is empty. The set of blue-eyed people is not empty. The set of numbers that belong to both $\{1, 3, 5\}$ and $\{2, 4, 6\}$ is empty.

Subset

The set B is said to be a **subset** of A if every element of B is also an element of A. This is written $B \subset A$.

Example 6 (*Compare Exercise 12*)

$\{2, 5, 9\} \subset \{1, 2, 4, 5, 7, 9\}$ because every element of $\{2, 5, 9\}$ is in the set $\{1, 2, 4, 5, 7, 9\}$. The set $\{3, 5, 9\}$ is not a subset of $\{1, 2, 4, 5, 7, 9\}$ because 3 is in the first set but not in the second.

Example 7 (*Compare Exercise 19*)
The subsets of $\{1, 2, 3\}$ are

$$\varnothing, \{1\}, \{2\}, \{3\}, \{1, 2\}, \{1, 3\}, \{2, 3\}, \{1, 2, 3\}$$

Notice that a set is a subset of itself; $\{1, 2, 3\}$ is a subset of $\{1, 2, 3\}$. Because $\varnothing$ contains no elements, it is true that every element of $\varnothing$ is in $\{1, 2, 3\}$, so $\varnothing \subset \{1, 2, 3\}$. In the same sense, $\varnothing$ is a subset of every set. The set $\{1, 2\}$ is a **proper** subset of $\{1, 2, 3\}$ because it is a subset of and not equal to $\{1, 2, 3\}$. We point out that if $A \subset B$ and $B \subset A$, then $A = B$.

The Union of Two Sets

Sometimes, we wish to construct a set using elements from two given sets. One way to obtain a new set is to combine the two given sets into one.

Definition 1

The *union* of two sets, A and B, is the set whose elements are from A or from B, or from both. Denote this set by $A \cup B$.

In set-builder notation, this is

$$A \cup B = \{x \mid x \in A \text{ or } x \in B \text{ or } x \text{ in both}\}$$

If you ask a number of people how they interpret a statement like "It is Joe or it is Jane," some will interpret it as either Joe or Jane, but not both. Others may include the possibility of both. In standard mathematical terminology the phrase "$x \in A$ or $x \in B$" includes the possibility of x existing in both. Thus, the phrase "or x in both" is mathematically redundant.

Example 8 (*Compare Exercise 30*)
Given $A = \{2, 4, 6, 8, 10\}$ and $B = \{6, 7, 8, 9, 10\}$. To determine $A \cup B$, list all elements of A and add those from B that are not already listed to obtain

$$A \cup B = \{2, 4, 6, 8, 10, 7, 9\}$$

Remember: The order in which the elements in a set are listed is not important. It may seem more natural to list the elements as $\{2, 4, 6, 7, 8, 9, 10\}$. Notice that an element that appears in both A and B is listed only once in $A \cup B$.

Example 9 **(a)** Given the sets

$$A = \{\text{Tom, Dick, Harry}\} \quad \text{and} \quad B = \{\text{Sue, Ann, Jo, Cathy}\}$$

$$A \cup B = \{\text{Tom, Dick, Harry, Sue, Ann, Jo, Cathy}\}$$

(b) Given the sets

$$A = \{x \mid x \text{ is a letter of the word } \textit{radio}\}$$

and

$$B = \{x \mid x \text{ is one of the first six letters of the English alphabet}\}$$

$$A \cup B = \{a, b, c, d, e, f, r, i, o\}$$

The Intersection of Two Sets

Another way to construct a set from two sets is by performing the operation called the **intersection** of two sets:

Definition 2

The **intersection** of two sets, A and B, is the set of all elements contained in both sets A and B, that is, those elements which A and B have in common. The intersection of A and B is denoted by

$$A \cap B = \{x \mid x \in A \text{ and } x \in B\}$$

Example 10 (*Compare Exercise 34*)

(a) If $A = \{2, 4, 5, 8, 10\}$ and $B = \{3, 4, 5, 6, 7\}$, then

$$A \cap B = \{4, 5\}$$

(b) If $A = \{a, b, c, d, e, f, g, h\}$ and $B = \{b, d, e\}$, then

$$A \cap B = \{b, d, e\}$$

Example 11 (*Compare Exercise 52*)
Given the sets

$$A = \text{set of all positive even integers}$$

and

$$B = \text{set of all positive multiples of } 3$$

Describe and list the elements of $A \cap B$ and $A \cup B$.

Solution Since an element of $A \cap B$ must be even (a multiple of 2) and a multiple of 3, then each such element must be a multiple of 6. Thus, $A \cap B$ is the set of all positive multiples of 6, that is, $\{6, 12, 18, 24, \ldots\}$. Since this is an infinite set, only the pattern of numbers can be listed. $A \cup B$ is the set of positive integers that are multiples of 2 or multiples of 3.

In a finite set you can count the elements and finish at a definite number. You can never count all the elements in an infinite set. No matter how high you count, there will be some left uncounted. The set of whole numbers between 10 and 20 is finite; there are 9 of them. The set of fractions between 10 and 20 is infinite; there is no end to them.

Disjoint Sets

If A and B do not have any elements in common, they are said to be **disjoint** and $A \cap B = \varnothing$. The sets $A = \{2, 4, 6, 8\}$ and $B = \{1, 3, 5, 7\}$ are disjoint.

5–1 Exercises

I.

1. Establish whether each of the following statements is true or false for $A = \{1, 2, 3, 4\}$, $B = \{3, 4, 5, 6\}$, $C = \{5, 6, 7, 8\}$, and $D = \{1, 3, 5, 7, \ldots\}$.

   (a) $2 \in A$ **(b)** $5 \in A$ **(c)** $8 \in B$
   (d) $5 \notin C$ **(e)** $6 \notin A$ **(f)** $5 \in D$
   (g) $20 \in D$ **(h)** $49 \in D$ **(i)** $200 \in D$
   (j) $201 \in D$

List the elements of each of the sets in Exercises 2 through 6.

2. (*See Example 5*) $A = \{x \mid x \text{ is a positive odd integer less than } 10\}$
3. $B = \{x \mid x \text{ is a letter in the word Mississippi}\}$
4. $A = \{x \mid x \text{ is an integer larger than 13 and less than } 20\}$
5. $C = \{x \mid x \text{ is an even integer larger than } 15\}$
6. $A = \{x \mid x \text{ is a prime integer less than } 20\}$

Which of the following pairs of sets in Exercises 7 through 11 are equal?

7. $A = \{1, 2, 3, 4\}$, $B = \{3, 1, 2, 5\}$
8. $A = \{1, -1, 0, 4\}$, $B = \{-1, 0, 1, 4\}$
9. $A = \{a, e, i, o, u\}$, $B = \{x \mid x \text{ is a vowel of the English alphabet}\}$
10. $A = \{x \mid x \text{ is a prime integer less than } 20\}$, $B = \{2, 3, 5, 7, 11, 13, 17, 19\}$
11. $A = \{5, 10, 15, 20\}$, $B = \{x \mid x \text{ is a multiple of } 5\}$

In which of the Exercises 12 through 18 is A a subset of B?

12. (*See Example 6*) $A = \{2, 4, 6, 8\}$, $B = \{1, 2, 3, 4, 5, 6, 7, 8, 9, 10\}$
13. $A = \{a, b, c\}$, $B = \{a, e, i, o, u\}$
14. $A = \{1, 2, 3, 5\}$, $B = \{1, 2, 3, 4, 5\}$
15. $A = \{3, 8, 2, 5\}$, $B = \{2, 3, 5, 8\}$
16. $A = \{1, 3, 7, 9, 11, 13\}$, $B = \{1, 9, 13\}$
17. A = set of even integers, B = set of integers
18. A = set of male college students, B = set of college students
19. (*See Example 7*) Determine all the subsets of A, B, and C.

    (a) $A = \{-1, 2, 4\}$ **(b)** $B = \{4\}$ **(c)** $C = \{-3, 5, 6, 8\}$

Which of the following sets in Exercises 20 through 29 are empty?

20. female presidents of the United States
21. a flock of extinct birds
22. the integers larger than 10 and less than 5

23. English words that begin with the letter k and end with the letter e
24. families with five children
25. odd integers divisible by 2
26. integers larger than 0 and less than 37
27. integers larger than 5 and less than 6
28. fractions larger than 5 and less than 6
29. $\{2, 4, 6\} \cap \{1, 8, 13\}$

List the elements of the sets in Exercises 30 through 34.

30. (*See Example 8*) $\{2, 1, 7\} \cup \{4, 6, 7\}$
31. $\{a, b, c, x, y, z\} \cup \{b, a, d\}$
32. $\{h, i, s, t, o, r, y\} \cup \{m, u, s, i, c\}$
33. $\{5, 2, 9, 4\} \cup \{3, 0, 8\}$
34. (*See Example 10*) $\{h, i, s, t, o, r, y\} \cap \{e, n, g, l, i, s, h\}$

To determine the sets indicated in Exercises 35 through 45 use

$$A = \{1, 2, 3\}$$
$$B = \{1, 2, 3, 6, 9\}$$
$$C = \{-3, -1, 0, 2, 3, 6, 7\}$$

35. $A \cap B$
36. $A \cap C$
37. $A \cup B$
38. $A \cup C$
39. $A \cap B \cap C$
40. $A \cup B \cup C$
41. $A \cap \varnothing$
42. $B \cap \varnothing$
43. $(A \cup C) \cap B$
44. $A \cup (B \cap C)$
45. $(A \cap C) \cup B$

II.

List three elements of each of the sets in Exercises 46 through 48.

46. $\{x \mid x = 2n + 1, n \text{ a positive integer}\}$
47. $\{x \mid x = n^2, n \text{ a positive integer}\}$
48. $\{x \mid x = n^3, n \text{ a nonnegative integer}\}$

List all the elements of the sets in Exercises 49 and 50.

49. $\{x \mid (x + 2)(x - 3)(x - 4) = 0\}$
50. $\{x \mid (x - 1)(x - 2)(x + 3)(x + 4) = 0\}$
51. Replace the asterisk in each of the following statements with $=$, $\subset$, or "neither" as appropriate.

    (a) $\{2, 4, 6\} * \{1, 2, 3, 4, 5, 6\}$
    (b) $\{2, 4, 8\} * \{1, 3, 5, 9\}$
    (c) $\{2, 5, 17\} * \{17, 5, 2\}$
    (d) $\{8, 10, 12\} * \{8, 10\}$
    (e) $\{2, 4, 6, 8\} * \{x \mid x \text{ is an even integer}\}$

52. (*See Example 11*) Find $A \cap B$ where

$$A = \{x \mid x \text{ is an integer larger than } 12\}$$
$$B = \{x \mid x \text{ is an integer less than } 21\}$$

53. Find $A \cap B$ where

$$A = \{x \mid x \text{ is an integer that is a multiple of } 5\}$$
$$B = \{x \mid x \text{ is an integer that is a multiple of } 7\}$$

Determine which of the pairs of sets in Exercises 54 through 57 are disjoint.

54. $A = \{1, 2, 3, 4, 5, 6\}$, $B = \{2, 4, 6, 8, 10\}$

55. A = the set of prime numbers greater than 100, B = the set of even integers

56. $A = \{3, 6, 9, 12\}$, $B = \{5, 10, 15, 20\}$

57. A = all multiples of 4, B = all multiples of 3

58. Sets A, B, C, and D are the following:

$A = \{1, 2, 3, 4\}$ $\quad$ $B = \{-5, 7, 8, 10\}$

$C = \{1, 2, 3, 4, 5, 9, 10, 11\}$ $\quad$ $D = \varnothing$

Find the following:

(a) $A \cup B$ **(b)** $B \cup D$ **(c)** $A \cup B \cup C$

(d) $A \cap C$ **(e)** $A \cap D$ **(f)** $A \cap B \cap D$

59. Let A be the set of students at Miami Bay University who are taking finite mathematics. Let B be the set of students at Miami Bay University who are taking American history. Describe $A \cap B$.

5–2 Venn Diagrams

Venn Diagrams

It usually helps to use a diagram to represent an idea. For sets, it is customary to use a rectangular area to represent the **universe** to which a set belongs. We have not used the term universe before. The universe is not some all-inclusive set that contains everything. When I talk about a set of students, I generally am thinking of college students, so the elements of my set of students are restricted to college students. Thus, in that context, the universe is the set of college students. When a student at Midway High speaks of a set of students, she restricts the elements of her set to students at her school, so the set of all students at Midway High School

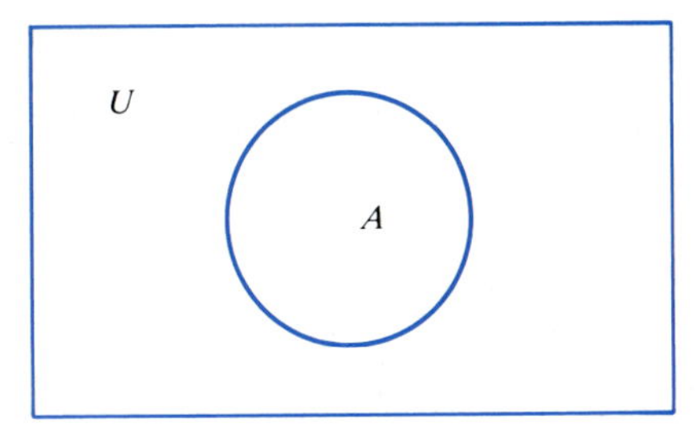

Figure 5–1

forms her universe. In a given context, the sets under discussion have elements that come from some restricted set. We call the set from which the elements come the universe or the **universal set**. We usually represent the universe by a rectangle. A circular area within the rectangle represents a set of elements in the universe. The diagram in Figure 5–1 represents a set A in a universe U. It is called a **Venn diagram**.

Example 1 Figure 5–2(a) shows the Venn diagram of the intersection of two sets. The shaded area is $A \cap B$.

Figure 5–2(b) shows a Venn diagram that represents the union of two sets. The shaded area represents $A \cup B$.

Figure 5–2(c) shows two disjoint sets. While their intersection is empty, their union is not, and it is shown by the shaded area.

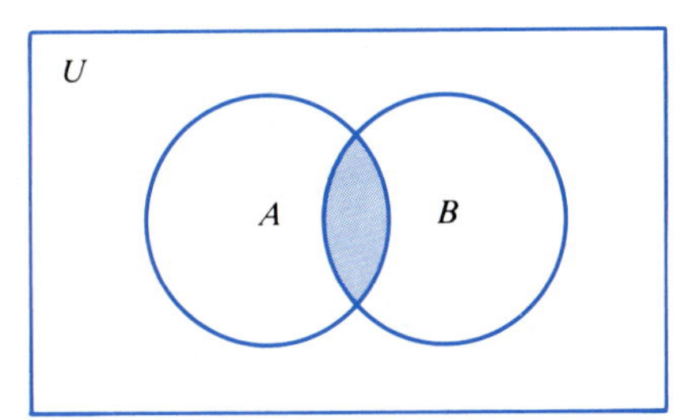

(a) $A \cap B$

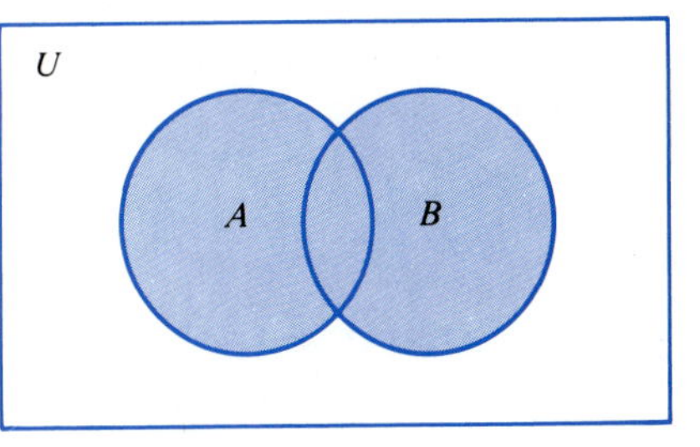

(b) $A \cup B$

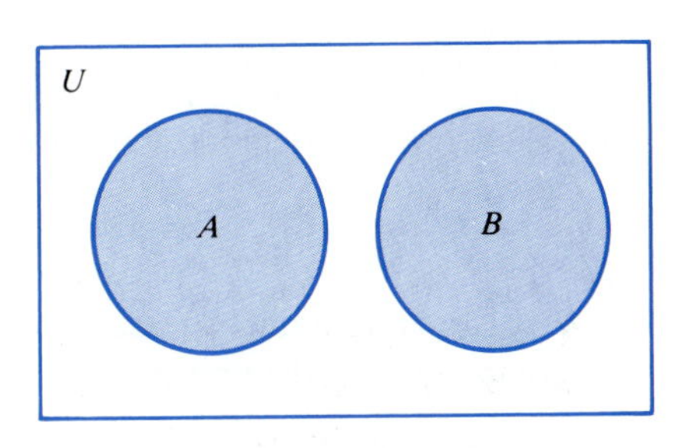

(c) A and B are disjoint. $A \cap B$ has no elements; it is the empy set.

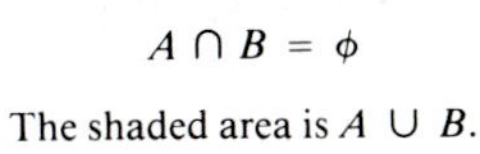
$A \cap B = \phi$

The shaded area is $A \cup B$.

Figure 5–2

Example 2 Let U = set letters of the English alphabet

A = set of vowels

B = set of the first nine letters of the alphabet

Place each of the letters a, c, d, e, i, u, x, y in the appropriate region in a Venn diagram.

Solution The letters a, e, i, and u are elements of A; a, c, d, e, and i are elements of B; and x and y are in neither set. Notice that a, e, and i are in both A and B, so they are in their intersection. We may indicate this as shown in Figure 5–3.

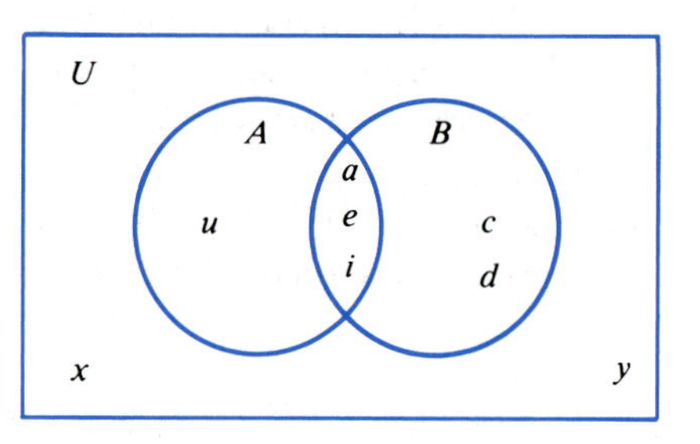

Figure 5–3

Complement

The elements in the universe U that lie outside A form a set called the **complement** of A, denoted by A'. (See Figure 5–4.)

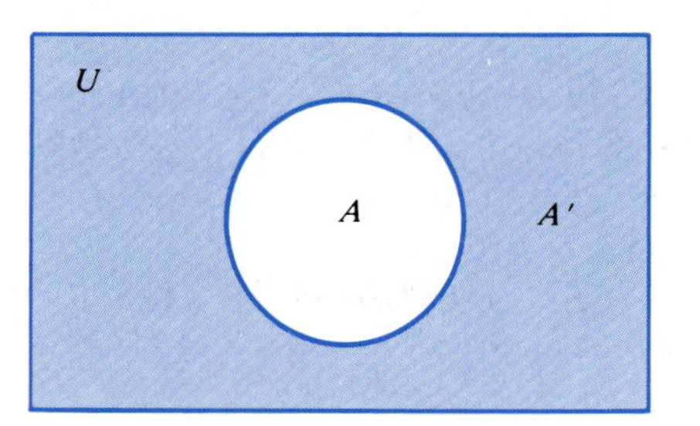

Figure 5–4

Example 3 (*Compare Exercise 1*)

Let $U = \{1, 2, 3, 4, 5, 6, 7, 8\}$ and $A = \{1, 2, 8\}$. Then $A' = \{3, 4, 5, 6, 7\}$.

From the definition of complement, it follows that for any set A in a universe U,

$$A \cap A' = \varnothing \qquad \text{and} \qquad A \cup A' = U$$

A set and its complement have no elements in common. The union of a set and its complement is the universal set.

Number of Elements in a Subset

We now turn our attention to counting the elements in a set. We want to be able to use information about a set and determine the number of elements in it, but avoid actually counting the elements one by one.

We use the notation $n(A)$, read "n of A," to indicate the number of elements in set A. If A contains 23 elements, we write $n(A) = 23$.

Suppose that you count 10 people in a group that like brand X cola and 15 that like brand Y. We denote this by $n(\text{brand } X) = 10$ and $n(\text{brand } Y) = 15$. How many people are involved in the count? The answer depends on the number who like both brands. In set terminology, the people who like brand X form one set, and those who like brand Y form another set. The totality of all people involved in the count form the union of the two sets, and those who like both brands form the intersection of the two sets. If we attempt to determine the total count by adding the number who like brand X and the number who like brand Y, then we count those who like both X and Y twice. We need to subtract the number who like both from the sum of those who like X and those who like Y in order to obtain the total number involved. If there are 4 people who like both brands, then the total count is $10 + 15 - 4 = 21$.

In general, the relationship between the number of elements in each of two sets, their union, and their intersection is given by the following theorem.

Theorem 1

$$n(A \cup B) = n(A) + n(B) - n(A \cap B)$$

where $n(A)$ represents the number of elements in set A, $n(B)$ represents the number of elements in set B, and $n(A \cap B)$ represents the number of elements of $A \cap B$.

Example 4 (*Compare Exercise 6*)
$A = \{a, b, c, d, e, f\}$, $B = \{a, e, i, o, u, w, y\}$. Count $n(A)$, $n(B)$, $n(A \cap B)$, and $n(A \cup B)$.

Solution $n(A) = 6$ and $n(B) = 7$. In this case, $A \cap B = \{a, e\}$, so $n(A \cap B) = 2$. $A \cup B = \{a, b, c, d, e, f, i, o, u, w, y\}$, so $n(A \cup B) = 11$. This checks with the formula, $n(A \cup B) = 6 + 7 - 2 = 11$.

Example 5 (*Compare Exercise 8*)
Set A is the 9 o'clock English class of 15 students, so A contains 15 elements. Set B is the 11 o'clock history class of 20 students, so B contains 20 elements. $A \cap B$ is the set of students in both classes (there are 7), so $A \cap B$ contains 7 elements. The number of elements in $A \cup B$ (a joint meeting of the classes) is

$$n(A \cup B) = n(A) + n(B) - n(A \cup B)$$

so

$$n(A \cup B) = 15 + 20 - 7 = 28$$

Notice where the elements of A and B lie in the Venn diagram in Figure 5–5.

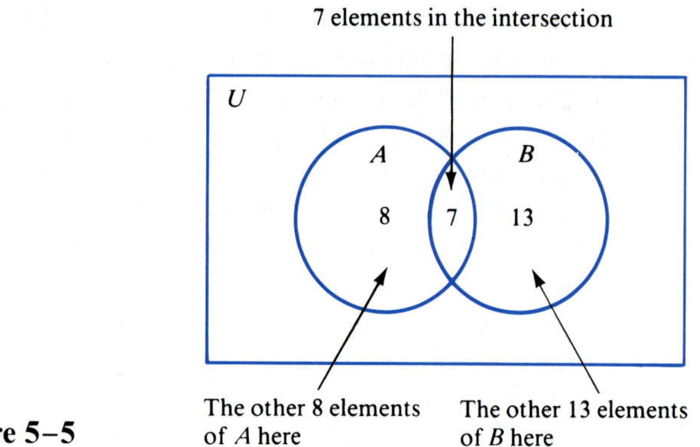

Figure 5–5

Example 6 (*Compare Exercise 9*)
The union of two sets, $A \cup B$, has 48 elements. Set A contains 27 elements and set B contains 30 elements. How many elements are in $A \cap B$?

Solution Using the relationship of Theorem 1, we have

$$\begin{aligned} 48 &= 27 + 30 - n(A \cap B) \\ 48 - 27 - 30 &= -n(A \cap B) \\ -9 &= -n(A \cap B) \end{aligned}$$

So

$$n(A \cap B) = 9$$

Example 7 (*Compare Exercise 24*)
One hundred students were asked if they were taking psychology (P) or biology (B). The responses showed that

61 were taking psychology, that is, $n(P) = 61$;
18 were taking both, that is, $n(P \cap B) = 18$;
12 were taking neither

(a) How many were taking biology?
(b) How many were taking psychology but not biology?
(c) How many were not taking biology?

Solution **(a)** Since 12 were taking neither, the rest, 88, were taking at least one of the courses; so $n(P \cup B) = 88$. We can find $n(B)$ from

$$\begin{aligned} n(P \cup B) &= n(P) + n(B) - n(P \cap B) \\ 88 &= 61 + n(B) - 18 \\ n(B) &= 45 \end{aligned}$$

(b) Since 18 students were taking both psychology and biology, the remainder of the 61 psychology students were taking only psychology. So, $61 - 18 = 43$ students were taking psychology but not biology.

(c) The students not taking biology were those 12 taking neither and the 43 taking only psychology, a total of 55 not taking biology. (See Figure 5–6.)

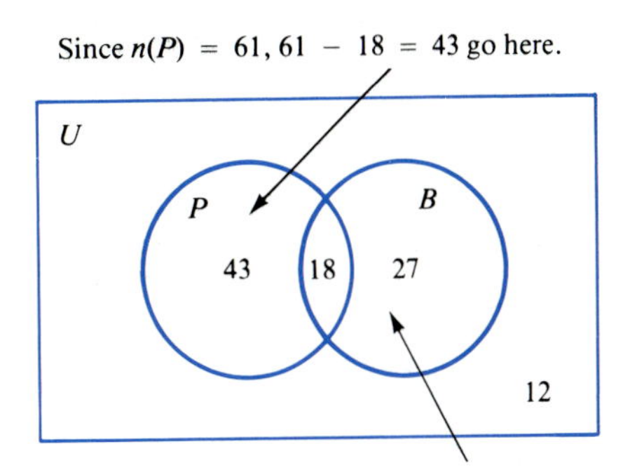

Figure 5–6

Venn Diagrams Using Three Sets

A Venn diagram of three sets divides a universe into as many as eight regions. We can use information about the number of elements in some of the regions (subsets) to obtain the number of elements in other subsets. (See Figure 5–7.)

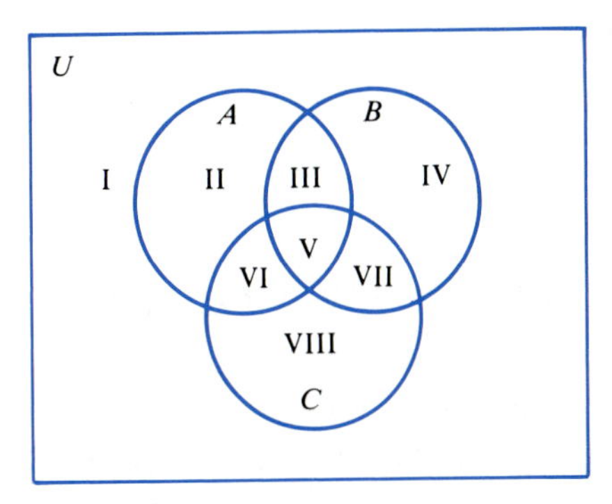

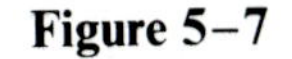
Figure 5–7 Three sets may divide the Universe into eight regions.

Example 8 The sets A, B, and C intersect as shown in Figure 5–8. The numbers in each region indicate the number of elements in that subset.

The number of elements in other subsets may be obtained from this diagram. For example,

$$n(A) = 9 + 2 + 3 + 7 = 21$$
$$n(B) = 2 + 3 + 1 + 4 = 10$$
$$n(A \cap B) = 2 + 3 = 5$$
$$n(A \cap B \cap C) = 3$$
$$n(A \cup B) = 9 + 2 + 3 + 7 + 4 + 1 = 26$$

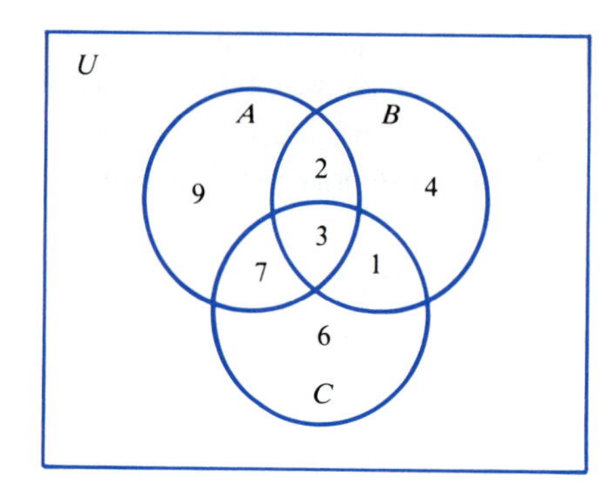

Figure 5–8

Example 9 (*Compare Exercise 27*)
A survey yields the following information about the musical preferences of students:

30 like classical,
24 like country,
31 like jazz,
9 like country and classical,
12 like country and jazz,
10 like classical and jazz,
4 like all three,
6 like none of the three.

Draw a diagram that shows this breakdown of musical tastes. Determine the total number of students interviewed.

Solution Begin by drawing a Venn diagram as shown in Figure 5–9(a).

The universe is the set of college students interviewed. We want to determine the number of students in each region of the diagram. Since some students may like more than one kind of music, these sets may overlap. We begin where the largest number of sets overlap (where the three sets intersect) and work out to fewer intersecting sets. Using the information that 4 students like all three types of music, place a 4 in the region where all three sets intersect. Of the 9 students who like both country and classical, we have already recorded 4 of them (those who like all three). The other 5 are in the intersection of classical and country that lies outside jazz (Figure 5–9(b)). In a similar fashion, the number who like both jazz and country is broken down into the 4 who like all three and the 8 who like jazz and country but are outside the region of all three. Since these three regions account for 17 of those who like country, the other 7 who like country are in the region where country does not intersect the jazz and classical. Fill in the rest of the regions; the results are shown in the diagram in Figure 5–9(c).

The total number of students interviewed is obtained by adding the number in each region of the Venn diagram. The total is 64.

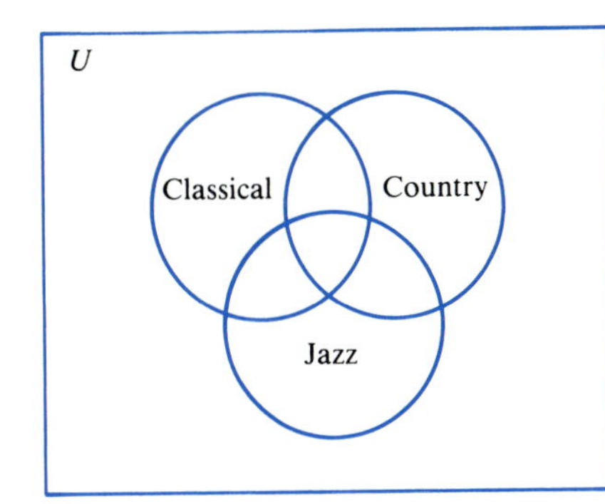

(a) A Venn diagram representing students grouped by musical preferenced.

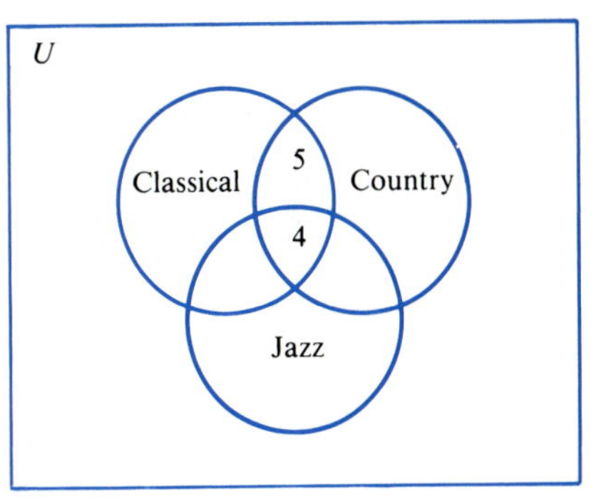

(b) Four students like all 3 types of music. A total of 9 like both classical and country.

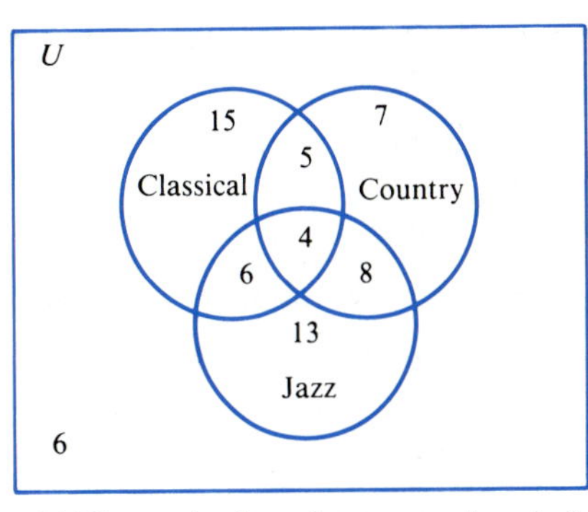

(c) The number in each category of musical preferences.

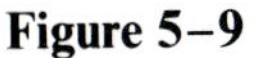

5–2 Exercises

I.

1. (*See Example 3*) If $U = \{1, 2, 3, 4, 5, 6\}$ and $A = \{2, 3, 5\}$, determine A'.
2. If $U = \{-3, 5, 7, 8\}$ and $A = \{5, 7\}$, determine A'.
3. If $U = \{2, 3, 7, 21, 24, 25\}$ and $A = \{2, 7, 21\}$, determine A'.
4. If $U = \{1, 2, 5, 8, 9, 10\}$, $A = \{1, 2\}$, and $B = \{2, 8, 9\}$, determine

   (a) A'. **(b)** B'. **(c)** $(A \cup B)'$. **(d)** $(A \cap B)'$.

5. If $U = \{-3, 4, 7, 12, 14, 21\}$, $A = \{7, 14, 21\}$, and $B = \{-3, 7, 12\}$, determine

   (a) A'. **(b)** B'. **(c)** $(A \cup B)'$. **(d)** $(A \cap B)'$.

6. (*See Example 4*) $A = \{1, 3, 5, 7, 9, 11\}$ and $B = \{3, 6, 9, 12\}$. Find

   (a) $n(A)$. **(b)** $n(B)$. **(c)** $n(A \cap B)$. **(d)** $n(A \cup B)$.

7. $A = \{a, b, c, d, e, f, g, x, y, z\}$ and $B = \{c, r, a, z, y\}$. Find

   (a) $n(A)$. **(b)** $n(B)$. **(c)** $n(A \cap B)$. **(d)** $n(A \cup B)$.

8. (*See Example 5*) Given $n(A) = 26$, $n(B) = 14$, and $n(A \cap B) = 6$, determine $n(A \cup B)$.

9. (*See Example 6*) Given $n(A) = 15$, $n(B) = 22$, and $n(A \cup B) = 30$, determine $n(A \cap B)$.

10. Given $n(A) = 21$, $n(A \cup B) = 33$, and $n(\mathrm{A} \cap B) = 5$, determine $n(B)$.

11. For two sets A and B, $n(A) = 14$, $n(A \cup B) = 28$, and $n(A \cap B) = 5$. Find $n(B)$.

12. If $n(A \cup B) = 249$, $n(A \cap B) = 36$, and $n(B) = 98$, find $n(A)$.

13. Let $A = \{-1, 2, 4\}$, $B = \{3, 4, 7, 8\}$ and $C = \{7, 8, 9\}$. Compute each of the following.

    (a) $n(A)$ **(b)** $n(A \cap B)$ **(c)** $n(A \cap C)$
    (d) $n(A \cup B)$ **(e)** $n(A \cup B \cup C)$

14. Let $A = \{1, 2, 4, 7\}$, $B = \{6, 8, 9\}$, and $C = \{1, 2, 8, 9\}$. Compute each of the following.

    (a) $n(B)$ **(b)** $n(B \cap C)$ **(c)** $n(A \cap B \cap C)$
    (d) $n(A \cup C)$ **(e)** $n(A \cup B \cup C)$

15. Let $A = \{a, b, c, d\}$, $B = \{b, r, s\}$, and $C = \{p, q, r, s\}$. Compute each of the following.

    (a) $n(C)$ **(b)** $n(A \cup B)$ **(c)** $n(A \cap C)$
    (d) $n(A \cup B \cup C)$ **(e)** $n(A \cap B \cap C)$

16. If $U = \{-7, -6, -5, 0, 1, 2, 8, 10\}$, $A = \{-7, -5, 2, 8\}$, $B = \{0, 1, 2\}$, and $D = \varnothing$, determine

    (a) A'. **(b)** B'. **(c)** D'. **(d)** $(A \cup B)'$.
    (e) $(A \cap B)'$. **(f)** $(A \cup D)'$. **(g)** $(A \cap D)'$.

II.

17. Two sets are formed using 100 elements. There are 60 elements in one set and 75 in the other. How many are in the intersection of the two sets?

18. Consider the Venn diagram in Figure 5–10. The number of elements in each subset is given. Compute

    (a) $n(A \cap B)$. **(b)** $n(A \cap B \cap C)$. **(c)** $n(A \cup B)$.
    (d) $n(A \cup B \cup C)$. **(e)** $n(A')$.

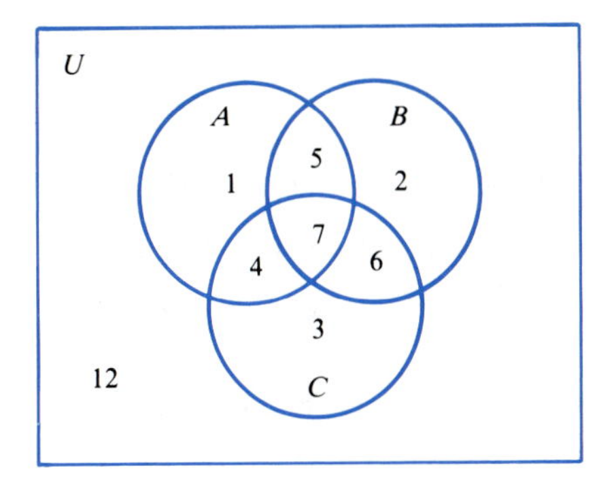

Figure 5–10

19. Consider the Venn diagram in Figure 5–11. The number of elements in each subset is given. Compute

(a) $n(A \cup B)$. **(b)** $n(A \cap B)$. **(c)** $n(B \cup C)$.

(d) $n(A \cup B \cup C)$. **(e)** $n(A \cap B \cap C)$. **(f)** $n(A' \cap B)$.

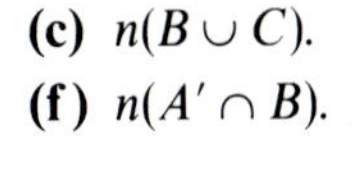

(g) $n(A' \cup B')$.

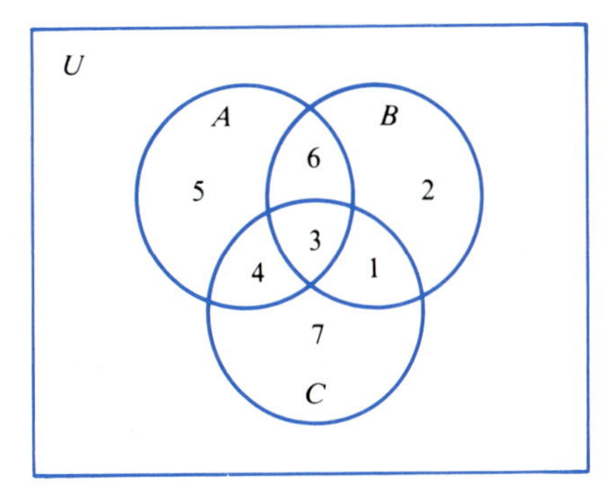

Figure 5–11

20. Construct a Venn diagram that represents the following information.

$$n(A \cap B \cap C) = 4, \quad n(A \cap B) = 12, \quad n(A \cap C) = 14,$$
$$n(B \cap C) = 10, \quad n(A) = 30, \quad n(B) = 25, \quad n(C) = 30$$

21. Construct a Venn diagram that represents the following information.

$$n(A \cup B \cup C) = 60, \quad n(A) = 20, \quad n(B) = 18, \quad n(A \cap B) = 5,$$
$$n(A \cap B \cap C) = 1, \quad n(A \cap C) = 7, \quad n(B \cap C) = 8$$

22. A marketing class polled 150 people at a shopping center to determine how many read the *Daily News* or the *Weekly Gazette*. They found the following:

126 read the *Daily News*,

31 read both,

10 read neither.

How many read the *Weekly Gazette*?

23. If a universal set contains 500 elements, $n(A) = 240$, $n(A \cup B) = 460$, and $n(A \cap B) = 55$, find $n(B')$.

24. (*See Example 7*) If 20 people belong to the Alpha Club, 30 people belong to the Beta Club, and 6 belong to both clubs,

(a) how many belong only to the Alpha Club?
(b) how many belong only to the Beta Club?
(c) how many belong to just one club?
(d) how many belong to one or both clubs?

25. Of 50 students surveyed, 19 owned a microcomputer, 42 owned a calculator, and 15 owned both.

(a) How many students did not own a calculator?
(b) How many students owned a calculator but did not own a microcomputer?
(c) How many owned neither?
(d) How many owned one or the other but not both?

26. The Venn diagram in Figure 5–12 gives information about the number of students in the freshman class at a college who take mathematics (M), English (E), and business courses (B).

(a) How many students take math and English?
(b) How many students take business and math but not English?
(c) How many students take math, English, and business?
(d) How many students take business or math?
(e) How many students take business, and English or math?

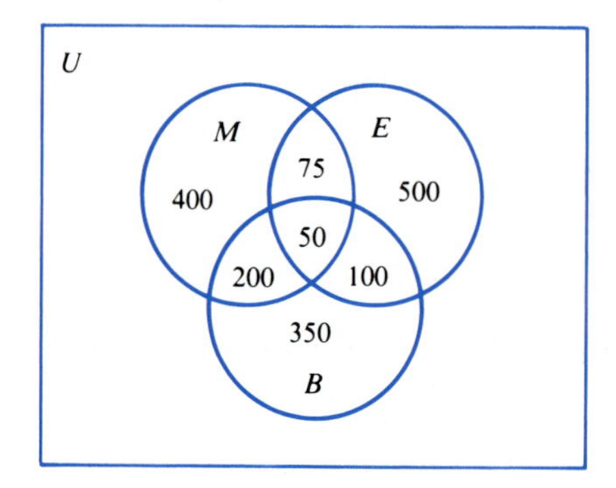

Figure 5–12

27. (*See Example 9*) A survey of 100 students at New England College showed the following:

48 take English,
49 take history,
38 take language,

17 take English and history,
15 take English and language,
18 take history and language,
7 take all three.

How many students

(a) take history but neither of the other two?
(b) take English and history, but not language?
(c) take none of the three?
(d) take just one of the three?
(e) take exactly two of the three?
(f) do not take language?

28. A company has carried out a survey of households having washers (W), dryers (D), or dishwashers (DW). The results are represented in the Venn diagram in Figure 5–13. Determine

(a) the total number of households contacted.
(b) the number with a washer but no dryer.
(c) the number having a washer and dryer.
(d) the number having all three appliances.
(e) the number having a washer and dryer, but no dishwasher.
(f) the number having a dishwasher or a washer.

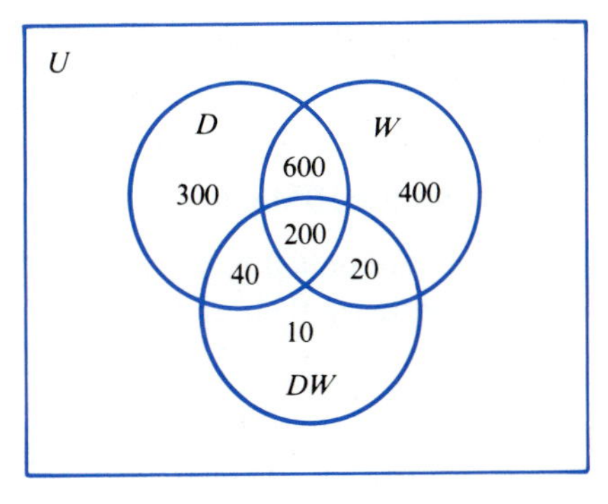

Figure 5–13

29. In a survey of 75 college students, it was found that of the three student publications, *The Lariat*, *The Rope*, and *The Roundup*,

23 read *The Lariat*,
18 read *The Rope*,
14 read *The Roundup*,
10 read *The Lariat* and *The Rope*,
9 read *The Lariat* and *The Roundup*,
8 read *The Rope* and *The Roundup*,
5 read all three.

(a) How many read none of the publications?
(b) How many read only *The Lariat*?
(c) How many read neither *The Lariat* nor *The Rope*?
(d) How many read *The Rope* or *The Roundup* or both?

30. Businesses in a certain community were surveyed to find the types of computers being used. The survey focused on three types, namely IBM (I), Apple (A), and Digital Equipment Corporation (E). The results were as follows:

Computer	Number of Users
I	50
A	30
E	25
I and A	4
I and E	5
E and A	3
I, A, and E	2

Construct a Venn diagram giving the complete numerical picture of computer usage in the businesses. From the Venn diagram determine

(a) the number of businesses that use only IBM computers.
(b) the number of businesses that use either IBM or Digital Equipment Corporation computers.
(c) the number of businesses that do not use IBM computers.

31. Graduating seniors at a certain college were polled to see whether they had taken physics (P), history (H), or sociology (S) in their undergraduate classes. The results were as follows:

Subject	Number Studied
P	90
H	120
S	150
P and H	40
P and S	50
S and H	60
P, H, and S	25

Construct a Venn diagram showing this information. From the Venn diagram determine

(a) the number of students that took history or sociology.
(b) the number of students that took history and sociology.
(c) the number of students that took history and sociology, but not physics.
(d) the number of students that did not take history.

III.

32. Given sets A and B, where $n(A) = 5$, $n(B) = 10$, and $n(A \cup B) = 15$,

(a) find $n(A \cap B)$.
(b) find $A \cap B$.

33. Mr. X taught freshman English and Mrs. X taught freshman history. They hosted a party at their home for their students and their dates. Twenty-eight of the students were in Mr. X's class, 33 were in Mrs. X's class, 7 were in both Mr. and Mrs. X's classes, and 14 were in neither class. How many students were at the party?

34. A group of 63 music students were comparing notes on their high school activities. They found that

40 played in the band,
31 sang in a choral group,
9 neither played in the band nor sang in a choral group.

How many played in the band and sang in a choral group?

35. What is inconsistent about the following?

$$n(A) = 25, \qquad n(B) = 10, \qquad n(A \cup B) = 23$$

36. A survey was carried out among 85 customers at a fast food restaurant. They were asked whether they liked or disliked the hamburgers (H), french fries (F), and coffee (C). Their reactions were as follows:

Item	**Liked**
H	57
F	55
C	55
H and F	45
H and C	47
F and C	46
H, F, and C	40

Determine the number of customers that

(a) disliked the coffee.
(b) disliked the french fries.
(c) disliked the hamburgers.
(d) liked the hamburgers or fries but disliked the coffee.
(e) liked the hamburgers only.

37. A class polled 320 people as to their memberships in the local Jaycees and Rotary. Of these, 210 people were Jaycees, 60 of these Jaycees were also Rotarians, and 90 people said they were Rotarians.

(a) How many Rotarians were not Jaycees?
(b) How many of the 320 polled were not Jaycees or Rotarians?

38. A questionnaire was distributed to 175 businessmen at an international conference to determine the languages that each spoke fluently. It was found that 10 spoke French and Spanish, 5 spoke English, French, and Spanish, and 15 spoke English and Spanish. Forty spoke more than one of the three languages. There were 115 people who spoke no Spanish and 100 who spoke no French. All 175 businessmen spoke at least one of the three languages.
Determine the number of people who

(a) spoke English.
(b) spoke only French.
(c) spoke only Spanish.
(d) spoke only English.
(e) spoke only English and French.

39. A group of 35 politicians in Washington were polled as to how many read *The New York Times*, *The Washington Post*, and *The Miami Herald*. All of them read at least one of the papers. It was found that 3 read only *The New York Times*, 4 read only *The Washington Post*, and 1 reads only *The Miami Herald*. Three read *The New York Times* and *The Miami Herald* only, 20 read *The New York Times* and *The Washington Post* only, and 1 person reads *The Miami Herald* and *The Washington Post* only.

(a) How many people read all three newspapers?
(b) How many do not read *The Washington Post*?
(c) How many read either *The New York Times* or *The Washington Post*?

40. There are 29 different elements in the sets A, B, and C. Is the following information consistent?

$$n(A) = 15, \quad n(B) = 12, \quad n(C) = 15,$$
$$n(A \cap B) = 5, \quad n(A \cap C) = 6,$$
$$n(B \cap C) = 4, \quad n(A \cap B \cap C) = 2$$

41. Charles Tally gave the following summary of his interviews with 135 students for a sociology project:

    65 said they like to attend movies,
    77 said they like to attend football games,
    61 said they like to attend the theater,
    28 said they like to attend movies and football games,
    25 said they like to attend movies and the theater,
    29 said they like to attend football games and the theater,
    8 said they like to attend all three,
    4 said they do not like to attend any of these.

    The professor refused to accept Charlie's paper because the information was inconsistent. Was the professor justified in claiming that the information was inconsistent?

42. A staff member of the news service of Great Lakes University gave the following information in a news release:
    Out of a class of 500 students at Great Lakes

    281 are taking English,
    196 are taking English and math,
    87 are taking math and a foreign language,
    143 are taking English and a foreign language,
    and 36 are taking all three.

    The staff member was told that the information must not be released. Why?

43. Use a Venn diagram to establish that if $A \subset B$, then $A \cap B = A$.

5–3 Fundamental Counting Principle

Tree Diagrams
Fundamental Counting Principle

A teenager posed the following problem: How many different outfits can she wear if she has two skirts and three blouses? This is a simple counting problem that we can solve by listing the different outfits in a systematic manner:

1. first skirt with first blouse,
2. first skirt with second blouse,
3. first skirt with third blouse,
4. second skirt with first blouse,
5. second skirt with second blouse,
6. second skirt with third blouse.

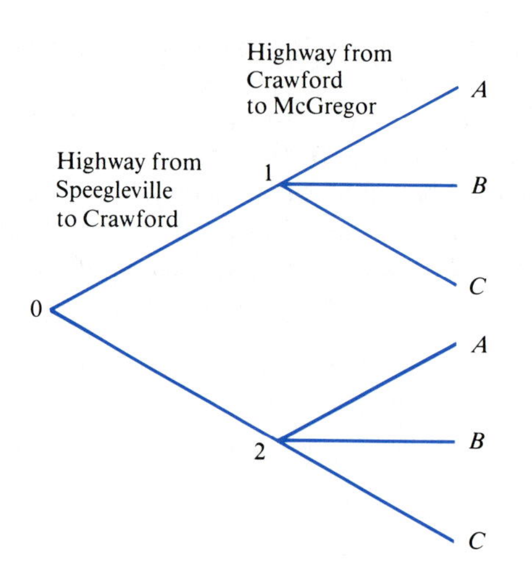

Figure 5–14

Notice the pattern of the list. For each skirt she can obtain three outfits by selecting each of the different blouses, so the total number of outfits is simply three times the number of skirts.

If you are to select 1 book from a list of 5 books and a second book from a list of 7 books, you can determine the number of selections of 2 books by writing the first book from list one with each of the 7 books from list two. You will list the 7 books five times, one with each of the books from list one. So, you will have a list that is $5 \times 7 = 35$ long.

We can list all possible selections in another way using tree diagrams.

Tree Diagrams

Let's look at another problem. Suppose that there are two highways from Speegleville to Crawford and three highways from Crawford to McGregor. How many different routes can we choose to go from Speegleville to McGregor through Crawford? A **tree diagram** provides a visual means to list all possible routes (Figure 5–14).

Reading from left to right, starting at 0, draw two branches representing the two highways from Speegleville to Crawford (use 1 and 2 to designate the highways). At the end of each of these two branches, draw three branches representing the three highways from Crawford to McGregor. (Use A, B, and C to designate the highways). A choice of a first level branch and a second level branch determines a route from Speegleville to McGregor. Notice that the total number of possible routes is six because each first level branch is followed by three second level branches.

Example 1 (*Compare Exercise 1*)
A company has two positions to fill, a department manager and an assistant manager. There are three people eligible for the manager position and four people eligible for the assistant-manager position. Use a tree diagram to show the different ways the two positions can be filled.

Solution Label the candidates for department manager as A, B, and C. Label the candidates for assistant manager as D, E, F, and G.

The tree diagram in Figure 5–15 illustrates the 12 possible ways the positions can be filled. Reading from left to right, starting at 0, there are three possible "branches" (managers). A branch for each possible assistant manager is attached to the end of each branch representing a manager. In all, 12 paths begin at 0 and go to the end of a branch. For example, the path $0BD$ represents the selection of B as the manager and D as the assistant manager.

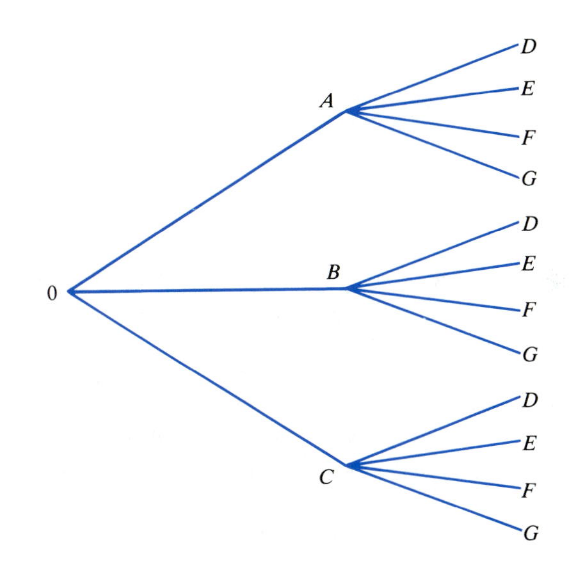

Figure 5–15

A tree diagram shows all possible ways to make a sequence of selections, and it shows the number of different ways the selections can be made.

The total number of selections is the product of the number of first level branches and the number of second level branches, that is, the number of ways the first selection can be made times the number of ways the second selection can be made.

These examples are fundamentally the same kind of problem. Let's make a general statement that includes each one.

Fundamental Counting Principle

Suppose that there are two activities, A_1 and A_2. Each activity can be carried out in several ways. Determine how many different ways the first activity can be performed followed by the second activity.

In the three preceding examples, A_1 and A_2 are the following.

Selections of an outfit: A_1 is the activity of selecting a skirt, and A_2 is the activity of selecting a blouse.

Routes from Speegleville to McGregor: A_1 is the selection of a highway from Speegleville to Crawford, and A_2 is the selection of a highway from Crawford to McGregor.

Filling two positions: A_1 is the selection of a manager, and A_2 is the selection of an assistant manager.

In many cases we do not need a list of all possible selections, but we need their number. In such a case, the problem reduces to a question of how many ways we can carry out the activity A_1 followed by the activity A_2. The solution is quite simple: multiply the number of ways the activity A_1 can be performed by the number of ways the activity A_2 can be performed. This is often called the **Fundamental Counting Principle**.

Theorem 2 ***Fundamental Counting Principle***

Given two activities, A_1 and A_2, that can be performed in N_1 and N_2 different ways, respectively, the total number of ways A_1 followed by A_2 can be performed is

$$N_1 \times N_2$$

Example 2 (*Compare Exercise 5*)

A taxpayers' association is to elect a chairman and a secretary. There are four candidates for chairman and five candidates for secretary. How many different ways can a slate of officers be elected?

Solution Here, A_1 is the activity of selecting a candidate for chairman, and $N_1 = 4$. A_2 is the activity of selecting a candidate for secretary, and $N_2 = 5$. Thus, $4 \times 5 = 20$ slates can be elected.

Example 3 (*Compare Exercise 8*)

The library wishes to display two rare books, one from the history collection and one from the literature collection. If the library has 50 history and 125 literature books to select from, how many different displays are possible?

Solution Number of displays $= 50 \times 125 = 6250$

Example 4 (*Compare Exercise 13*)

Jane selects 1 card from a deck of 52 different cards. The first card is *not* replaced before Joe selects the second one. How many different ways can they select the two cards?

Solution Jane selects from a set of 52 cards, so $N_1 = 52$. Joe selects from the remaining cards, so $N_2 = 51$. Two cards can be drawn in $52 \times 51 = 2652$ ways.

The fundamental counting principle also can be used to apply to several activities.

Corollary

Given activities $A_1, A_2, \ldots, A_k$, which can be performed in $N_1, N_2, \ldots, N_k$ different ways, respectively, the number of ways one can perform A_1 followed by $A_2 \ldots$ followed by A_k is

$$N_1 \times N_2 \times \cdots \times N_k$$

Example 5 (*Compare Exercise 14*)
How many different ways can class officers be selected if there are three candidates for president, four candidates for vice president, and seven candidates for secretary?

Solution The number of different ways is given by

$$3 \times 4 \times 7 = 84$$

Example 6 (*Compare Exercise 19*)
A quiz consists of four multiple-choice questions with five possible responses to each question. How many different ways can the quiz be answered?

Solution In this case, there are four activities, that is, answering each of four questions. Each activity (answering a question) can be performed (choosing a response) in five different ways. The answers can be given in

$$5 \times 5 \times 5 \times 5 = 625$$

different ways.

Example 7 (*Compare Exercise 21*)
Social Security numbers have the following format.

$$ABC - DE - FGHI$$

where A, B, C, D, E, F, G, H, and I each are integers 0 through 9. How many different social security numbers can be formed?

Solution There are ten possibilities for each of the nine digits, $A, B, \ldots, I$ that are selected. There are thus

$$10 \times 10 \times 10 \times 10 \times 10 \times 10 \times 10 \times 10 \times 10 = 1{,}000{,}000{,}000$$

different social security numbers.

In this example, all selections are made from the same set, the digits zero through nine. Since a digit, say 5, can occur more than once in a Social Security number, we say that "repetitions are allowed."

In some situations repetitions are not allowed.

Example 8 (*Compare Exercise 23*)
The digits zero through nine are written one per card. Three different cards are drawn and a three-digit number is formed. How many different three-digit numbers can be formed in this way?

Solution Since the number is formed from the three cards drawn, the digits must all be different, that is, no repetitions of digits occur in a number. There are ten possibilities for the first digit, then any of the nine remaining digits may be used for the second digit, and finally, the third digit may be any of the remaining eight digits, so

$$10 \times 9 \times 8 = 720$$

different numbers can be formed.

5–3 Exercises

I.

1. (*See Example 1*) A student has a red tie and a green tie. He has a white shirt, a blue shirt, and a yellow shirt. Draw a tree diagram that shows all possible ways a tie and shirt can be selected.
2. Draw a tree diagram to show the different ways a boy and then a girl can be selected from the following set of children.

$$\{\text{Amy, Bobby, Carl, Debbie, Eric, Flo}\}$$

3. Draw a tree diagram showing all sequences of heads and tails in two tosses of a coin.
4. For her breakfast, Susan always selects one item from each of the following:

   orange juice or tomato juice,
   cereal or eggs,
   toast or muffins.

   Draw a tree diagram to show all possible menus.
5. (*See Example 2*) Each day a teacher has a boy and a girl summarize the assignment for the day. There are 12 boys and 15 girls in the class. In how many different ways can the teacher select the 2 students?
6. A girl has six necklaces and eight pairs of earrings. In how many different ways can she select a necklace and a pair of earrings to wear?
7. A grocery store has five brands of crackers and nine different varieties of cheese. How many different combinations of one brand of cheese and one brand of crackers can a shopper buy?
8. (*See Example 3*) A reading list for American history contains two groups of books. There are 16 books in the first group and 21 in the second. A student is to read one book from each group. How many ways can the choice of books be made?
9. A man has six suits and seven shirts. How many different outfits can he form?
10. A house has 4 doors and 18 windows. In how many ways can a burglar pass through the house if he enters by a window and leaves by a door?
11. The cafeteria has a selection of four meats and seven vegetables. How many different selections of one meat and one vegetable are possible?
12. A travelling salesperson may take one of five different routes from Brent to Centreville and three different routes from Centreville to Moundville. How many different routes are possible from Brent to Moundville through Centreville?

13. (*See Example 4*)

(a) In how many different ways can a player select a diamond and a club from a deck of 52 bridge cards?

(b) In how many different ways can a player select one card of each suit from a bridge deck?

14. (*See Example 5*) A cafeteria offers a selection of two meats, four vegetables, and three desserts. In how many different ways can a diner select a meal of one meat, one vegetable, and one desert?

15. A car manufacturer provides six exterior colors, five interior colors, and three different trims. How many different color-trim schemes are available?

16. An interior decorator has a choice of three different carpets, six different wallcoverings, and three different upholstery fabrics. How many interiors can she design?

17. One can go from Washington to Philadelphia along three different routes, Philadelphia to Baltimore along two routes, and Baltimore to New York along four routes. How many possible routes are there from Washington to New York that pass through Philadelphia and then Baltimore?

18. A builder developing a subdivision may choose from four different roofing subcontractors, five different electrical subcontractors, three different plumbing subcontractors, two different carpenters, and six different painters. In how many ways can she select one of each?

II.

19. (*See Example 6*) A quiz consists of six multiple-choice questions with four possible responses to each one. How many different ways can the quiz be answered?

20. A coin is flipped three times. How many different sequences of heads and tails are possible?

21. (*See Example 7*) A telephone number consists of seven digits. How many phone numbers are possible if the first digit is not 1 or 0?

22. How many different radio call letters beginning with K and consisting of four letters can be assigned to radio stations?

23. (*See Example 8*) How many four-digit numbers can be formed from the digits $\{1, 2, 3, 4, 5\}$

(a) if a digit can be repeated?

(b) if a digit cannot be repeated?

24. How many different three-digit numbers can be formed from the digits $\{0, 1, 2, 3, 4, 5\}$

(a) if a digit cannot be repeated?

(b) if a digit can be repeated?

25. A license plate number consists of three letters followed by three digits.

(a) How many different license numbers can be formed if repetition of letters and digits is allowed?

(b) How many different license numbers can be formed if repetition of letters and digits is not allowed?

26. Draw a tree diagram showing all sequences of heads and tails that are possible in three tosses of a coin.

27. Use a tree diagram to show the different ways first, second, and third prizes can be awarded to three different contestants—Jones, Allen, and Cooper.

28. Each parent has two genes for a given trait, AA, Aa, or aa. A child will inherit one gene from each parent. Draw a tree diagram to show the possibilities for a child if one parent has AA and the other Aa.

29. The Panthers and Lions baseball teams play a championship series. The first team to win two games (best two out of three) is the winner. Use a tree diagram to show the possible outcomes of the series.

30. A Boy Scout troop hikes from the mall to their camping area at Hilltop Farm. They have three possible routes from the mall to the farm: the highway access road, a gravel road, or through a subdivision. At the farm the scout troop can continue to the camping area by way of a trail through the woods, a lane across the pasture, or along a stream. Draw a tree diagram showing all possible routes from the mall to their camping area.

5–4 Permutations and Combinations

Permutations
Notation for Number of Permutations
Combinations
Special Cases

Permutations

The Fundamental Counting Principle is but one of a number of counting techniques, so we will look at some others. The first of these deals with **permutations**. We illustrate this type of counting technique with the following example.

Example 1 (*Compare Exercise 25*)
A jewelry store has a collection of five silver service sets. It plans to display them three at a time, arranged in a row of display windows. How many different arrangements are possible?

Solution First, we point out that there are two ways to obtain different arrangements:

(a) one or more silver service sets may be replaced by another not in the display;

(b) the location of two or more of the silver service sets on display may be exchanged.

This problem may be viewed as a fundamental counting problem with three activities.

Activity 1: Select a silver service for the first location. This can be done in five ways because any of the five sets can be selected.

Activity 2: Select a silver service for the second location. This can be done in four ways because any one of the four remaining sets can be selected.

Activity 3: Select a silver service for the third location. This can be done in three ways because two sets are already displayed and any one of the three remaining can be selected.

Thus, there are $5 \times 4 \times 3$ possible arrangements.

This example has three characteristics that determine a permutation.

Permutations

1. A permutation is an arrangement of elements from a single set.
2. Repetitions are not allowed.
3. The order in which the elements are arranged is significant.

The statement "order is significant" is meant to be subject to a number of interpretations. Order is significant in situations such as the following:

(a) different positions are to be filled and the exchange of objects in two positions is considered a different way to fill the positions;

(b) objects are arranged in a row or in specified locations;

(c) people are selected to fill different offices in an organization (each office is a position);

(d) prizes are given for first place, second place, and so on (each prize is a position);

(e) items are distributed among several people (each person may be considered a position).

Example 2 (*Compare Exercise 29*)
Ten students each submit one essay for competition. How many ways can first, second, and third prizes be awarded?

Solution This is a permutation problem because

1. each essay selected is from the same set;
2. no essay can be submitted more than once; that is, an essay cannot be awarded two prizes;
3. the order (prize given) of the essays is important.

Any of the ten essays may be chosen for first prize. Then any of the remaining nine may be chosen for the second prize, and any of the other eight may be chosen for third prize. According to the fundamental counting principle, the three prizes may be awarded in

$$10 \times 9 \times 8 = 720$$

different ways.

Example 3 (*Compare Exercise 33*)
In how many different ways can a penny, a nickel, a dime, and a quarter be given to four children if one coin is given to each child?

Solution Each child may be considered a "position" that receives a coin. The number of ways a coin may be given to each child is

first child: four possibilities of a coin,
second child: three possibilities of a coin,
third child: two possibilities of a coin,
fourth child: one possibility of a coin.

Thus, the coins may be distributed in

$$4 \times 3 \times 2 \times 1 = 24$$

different ways.

Example 4 (*Compare Exercise 35*)
At the Cumberland River Festival, four Cumberland "belles" are stationed at historic Fort House; one stands at the entrance, one in the living room, one in the dining room, and one on the back veranda. If there are ten Cumberland belles, how many different ways can four be selected for stations?

Solution This is a permutation since a woman can be selected for, at most, one station, and the order (place stationed) is significant. The permutation can be made in

$$10 \times 9 \times 8 \times 7 = 5040$$

different ways.

If the problem in Example 4 had been to select four belles to be present in the living room with no particular station for each one, then Example 4 would not have been a permutation problem since the belles would not be arranged in any particular order. (The number of selections in this case uses a technique discussed later.)

Notation for Number of Permutations

The notation commonly used to represent the number of permutations for a set is written as $P(8, 3)$ and is read "permutation of eight things taken three at a time." This notation represents the number of permutations of three elements from a set of eight elements. $P(10, 4)$ represents the number of permutations of four elements selected from a set of ten elements. ($P(5, 3)$ is the answer to Example 1, and $P(10, 4)$ is the answer to Example 4.)

We want you to understand the pattern for calculating numbers like $P(10, 4)$ so you can do it routinely. Let's look at some examples.

$P(10, 4) = 10 \times 9 \times 8 \times 7$ permutations of four elements taken from a set of ten elements.

$P(5, 3) = 5 \times 4 \times 3$

$P(7, 2) = 7 \times 6$ permutations of two elements selected from a set of seven elements.

$P(21, 3) = 21 \times 20 \times 19$

In each case, the calculation begins with the first number in the parentheses, 21 in $P(21, 3)$ and 7 in $P(7, 2)$. The second number in $P(21, 3)$, $P(7, 2)$, and so on, determines the number of terms in the product. Since the terms decrease by one to the next term, you only need to know the first term and how many are needed to calculate the answer.

This reasoning lets us know that $P(30, 5)$ is a product of five terms beginning with 30, and each term thereafter decreasing by one, so

$$P(30, 5) = 30 \times 29 \times 28 \times 27 \times 26$$

$$P(105, 4) = 105 \times 104 \times 103 \times 102$$

and

$$P(4, 4) = 4 \times 3 \times 2 \times 1$$

Can you give the last term in $P(52, 14)$ without writing out all the terms? The preceding examples give a pattern that helps. The last term of

$P(5, 3)$ is $5 - 2 = 3$ (Also written $5 - 3 + 1$)

$P(10, 4)$ is $10 - 3 = 7$ (Also written $10 - 4 + 1$)

$P(105, 4)$ is $105 - 3 = 102$ (Also written $105 - 4 + 1$)

so we expect the last term of

$P(52, 14)$ to be $52 - 13 = 39$ (Also written $52 - 14 + 1$)

The numbers in parentheses above are another way of calculating the last terms. That form is more useful in calculating the general case.

In general, $P(n, k)$ indicates the number of arrangements that can be formed by selecting k elements from a set of n elements. Following the observed pattern, it may be written

$$P(n, k) = n(n - 1)(n - 2)\cdots(n - k + 1)$$

There is a special notation for the case when all elements of a set are used in a permutation. Notice that $P(4, 4)$ is just the product of the integers four through one—$4 \times 3 \times 2 \times 1$. In general, $P(n, n)$ is the product of the integers n through one. The following notation is used.

Definition 3 The product of the integers n through one is denoted by $n!$ (called n factorial).

$$n! = n(n-1)(n-2) \times \cdots \times 2 \times 1$$

Example 5 (*Compare Exercise 1*)

$$7! = 7 \times 6 \times 5 \times 4 \times 3 \times 2 \times 1$$
$$2! = 2 \times 1$$
$$6! = 6 \times 5 \times 4 \times 3 \times 2 \times 1$$

Notice that $6! = 6 \times 5!$ and $4! = 4 \times 3!$, and so on.

$$1! = 1$$

$0!$ is defined to be 1

Arithmetic involving factorials can be carried out easily if you are careful to use the factorial as defined.

$$\frac{5!}{3!} = \frac{5 \times 4 \times 3 \times 2 \times 1}{3 \times 2 \times 1}$$
$$= 5 \times 4$$
$$= 20$$
$$3!\,4! = 3 \times 2 \times 1 \times 4 \times 3 \times 2 \times 1 = 144$$

Example 6 (*Compare Exercise 36*)
How many different ways can six people be seated in a row of six chairs?

Solution This is a permutation with six positions to be filled from a set of six, so the number of arrangements is

$$P(6, 6) = 6! = 6 \times 5 \times 4 \times 3 \times 2 \times 1 = 720$$

Factorials allow us to write the expression for the number of permutations in another form that is sometimes useful. For example,

$$P(8, 3) = 8 \times 7 \times 6$$
$$= \frac{3 \times 7 \times 6 \times 5!}{5!}$$

Since $8 \times 7 \times 6 \times 5! = 8 \times 7 \times 6 \times 5 \times 4 \times 3 \times 2 \times 1 = 8!$, we can write

$$P(8, 3) = \frac{8!}{5!}$$

Be sure you understand that

$$P(6, 4) = \frac{6!}{2!} \qquad \text{(2! came from (6 − 4)!)}$$

In general, we can write

$$P(n, k) = \frac{n!}{(n-k)!}$$

Example 7 (*Compare Exercise 39*)
Many auto license plates have three letters followed by three digits. How many ways can the three letters be arranged if

(a) a letter is not repeated on a license plate?
(b) repetitions of the letters are allowed?

Solution **(a)** This is a permutation, since a letter may not be used more than once and a different arrangement of letters gives a different license number. The number of arrangements is

$$P(26, 3) = 26 \times 25 \times 24 = 15{,}600$$

(b) This is not a permutation, since a letter may appear more than once on a license plate. The fundamental counting principle applies, so the number of arrangements is

$$26 \times 26 \times 26 = 17{,}576$$

Combinations

When you pay your bill at the Pizza Place, the cashier is interested in the collection of coins and bills you give her, not in the order in which you present them. When you are asked to answer six out of eight test questions, the collection of questions is important, not the arrangement. If the professor wishes to compute the number of different ways students can choose six questions from eight, she is not dealing with permutations. She is interested in knowing the number of ways a subset of six elements can be obtained.

Definition 4

A subset of elements chosen from a given set without regard to their arrangement is called a **combination**.

The notation $C(n, k)$, read "combinations of n things taken k at a time," represents the number of subsets consisting of k elements taken from a set of n elements.

$C(8, 3)$ is the number of ways 3 elements can be selected from a set of 8. $C(52, 6)$ is the number of ways 6 elements can be selected from a set of 52.

The keys to recognizing a combination are given in the following box.

Combinations

1. A combination selects elements from a single set.
2. Repetitions are not allowed.
3. The order in which the elements are arranged is *not* significant.

Notice that a combination differs from a permutation only in that order is not significant in a combination, whereas it is important in a permutation.

Example 8 Given the set $A = \{a, b, c, d, e, f\}$, the subset $\{b, d, f\}$ is a combination of three elements taken from a set of six elements.

Since the elements of the subset $\{b, d, f\}$ can be arranged in several ways, we expect there to be several permutations for each subset. This indicates that there are more permutations than combinations in a given set.

Example 9 (*Compare Exercise 41*)
List all combinations of two elements taken from the set $\{a, b, c\}$.

Solution Because of the small number of elements involved, it is rather easy to list all subsets consisting of two elements. They are $\{a, b\}$, $\{a, c\}$, and $\{b, c\}$. Thus, $C(3, 2) = 3$.

It is much more difficult to list all five element subsets from a set of 26 elements. If we are interested only in the number of such subsets, not their listing, the problem becomes easier. Let's look at an example that illustrates how we can determine the number of combinations.

Example 10 Let's determine $C(6, 3)$, the number of different ways we can select a subset of three elements from a set of six elements. Let's look at the subset $\{a, b, c\}$. From the material on permutations, you know that we can arrange these three elements from one combination in $3! = 6$ different ways. They are

$$abc,\ acb,\ bac,\ bca,\ cab, \text{ and } cba$$

Each of these six permutations is made of the same set of three elements; that is, they are made from the same **combination** of elements. In fact, if you take any combination of three elements, they can be arranged in 3! different ways (permutations). This gives us a relationship between combinations and permutations. Since each combination of three elements can be arranged in 3! ways, we can obtain all permutations by taking the elements from a combination and finding all six arrangements of those three elements. Thus,

$$P(6, 3) = 3!C(6, 3)$$

Be sure you understand this relationship. To obtain all permutations of three elements, you first select one combination of those three elements and form all permutations of those elements. Then select a combination of three other elements and form all permutations. Each time you select a combination of three elements, you can form 3! permutations.

By dividing both sides of

$$P(6, 3) = 3!C(6, 3)$$

by 3! we get

$$C(6, 3) = \frac{P(6, 3)}{3!}$$

Similarly,

$$C(12, 4) = \frac{P(12, 4)}{4!}$$

$$C(5, 2) = \frac{P(5, 2)}{2!}$$

$$C(101, 14) = \frac{P(101, 14)}{14!}$$

In general we have the following:

Theorem 3

$$P(n, k) = k!\, C(n, k)$$

or

$$C(n, k) = \frac{P(n, k)}{k!}$$

Because $P(n, k)$ can be written as

$$\frac{n!}{(n-k)!}$$

$C(n, k)$ can also be written as

$$\frac{n!}{k!(n-k)!}$$

We now have a convenient way of calculating the number of combinations.

Example 11 (*Compare Exercise 17*)

$$C(5, 2) = \frac{P(5, 2)}{2!} = \frac{5 \times 4}{2 \times 1} = 10$$

$$C(5, 3) = \frac{P(5, 3)}{3!} = \frac{5 \times 4 \times 3}{3 \times 2 \times 1} = 10$$

$$C(10, 3) = \frac{P(10, 3)}{3!} = \frac{10 \times 9 \times 8}{3 \times 2 \times 1} = 120$$

$$C(30, 4) = \frac{P(30, 4)}{4!} = \frac{30 \times 29 \times 28 \times 27}{4 \times 3 \times 2 \times 1}$$

$$= 27{,}405$$

$$C(8, 6) = \frac{8!}{6!2!} = 28$$

$$C(15, 3) = \frac{15!}{3!12!}$$

Example 12 (*Compare Exercise 43*)
A student has seven books on his desk. How many different ways can he select a set of three?

Solution Since the order is not important, this is a combination problem.

$$C(7, 3) = \frac{P(7, 3)}{3!} = \frac{7 \times 6 \times 5}{3 \times 2 \times 1} = 35$$

Example 13 (*Compare Exercise 45*)

(a) In how many ways can a committee of four be selected from a group of ten people?

(b) In how many ways can a slate of officers consisting of a president, vice president, and secretary be selected from a group of ten people?

Solution **(a)** The order of selection is not important in the selection of a committee, so this is a combination problem of taking four elements from a set of ten.

$$C(10, 4) = \frac{P(10, 4)}{4!} = 210$$

(b) In selecting a slate of officers, President Jones, Vice President Smith, and Secretary Allen is a different slate than President Allen, Vice President Smith, and Secretary Jones. Each office is a position to be filled, so order is significant. The number of slates is $P(10, 3)$.

Notice the pattern used in computing combinations. To compute $C(10, 4)$, begin with 10 and write four integers in decreasing order. Then divide by 4!. This is true in general. To compute $C(15, 5)$, form the numerator using the five integers beginning with 15 and decreasing by 1. The denominator is 5!. In general, we can write $C(n, k)$ by forming the numerator from the product of k integers that begin with n and decrease by 1. The denominator is $k!$.

Special Cases

The other form,

$$C(n, k) = \frac{n!}{(n-k)!k!}$$

is also a useful form. Let's use it to look at some special cases.

(a) How many ways can one element be selected from a set? $C(6, 1)$ is the number of ways one element can be selected from a set of six. It is

$$C(6, 1) = \frac{6!}{1!5!} = \frac{6 \cdot 5!}{1!5!} = 6$$

In general,

$$C(n, 1) = \frac{n!}{1!(n-1)!}$$

$$= \frac{n(n-1)!}{1!(n-1)!} = n$$

So, one item can be selected from a set of n items in n ways.

(b) How many ways can zero items be selected from a set? We write $C(6, 0)$ to represent the number of ways no elements can be selected from a set of six. The formula gives

$$C(6, 0) = \frac{6!}{0!6!}$$

Since $0! = 1$, this reduces to $C(6, 0) = 1$. In general,

$$C(n, 0) = \frac{n!}{0!n!} = 1$$

Does your intuition tell you that there is just one way to select zero elements from a set? The one way is to take none.

(c) How many ways can all the elements be selected from a set? Our intuition tells us there is just one way; namely, take all of them. The formula agrees.

$$C(6, 6) = \frac{6!}{6!0!} = 1$$

and

$$C(n, n) = \frac{n!}{n!0!} = 1$$

Example 14 (*Compare Exercise 49*)

A cafeteria offers a selection of four meats, six vegetables, and five desserts. In how many ways can you select a meal consisting of two different meats, three different vegetables, and two different desserts?

Solution Basically, this is a problem that can be solved using the fundamental counting principle. We obtain the possible number of meals by multiplying the number of ways you can select two meats, the number of ways you can select three vegetables, and the number of ways you can select two desserts.

The number of ways you can select meats, vegetables, and desserts each form a combination problem. Thus, we obtain the number of meals as

$$\text{(number of meat selections)} \times \text{(number of vegetable selections)}$$

$$\times \text{(number of dessert selections)} = C(4, 2) \times C(6, 3) \times C(5, 2)$$

$$= \frac{4 \times 3}{2 \times 1} \times \frac{6 \times 5 \times 4}{3 \times 2 \times 1} \times \frac{5 \times 4}{2 \times 1}$$

$$= 6 \times 20 \times 10 = 1200$$

Example 15 (*Compare Exercise 51*)
A club has 14 male and 16 female members. A committee composed of 3 men and 3 women is formed. In how many ways can this be done?

Solution The male members can be chosen in

$$C(14, 3) = \frac{14 \times 13 \times 12}{3 \times 2 \times 1} = 364$$

different ways. The female members can be chosen in

$$C(16, 3) = \frac{16 \times 15 \times 14}{3 \times 2 \times 1} = 560$$

different ways. Thus, by the Fundamental Counting Principle, the committee can be chosen in $364 \times 560 = 203{,}840$ ways.

Instead of counting the number of outcomes for a sequence of activities, some counting problems seek the number of possible outcomes when the outcome selected is from one activity *or* another.

Example 16 (*Compare Exercise 52*)
How many different committees can be selected from eight men and ten women if a committee is composed of three men *or* three women?

Solution For a moment, think of listing all possible selections of a committee. The list has two parts, a list of committees composed of three women and a list of committees composed of three men. The total number of possible committees can be obtained by adding the number of all-female to the number of all-male committees. We get each of these by

Number of all-female committees $= C(10, 3) = 120$
Number of all-male committees $= C(8, 3) = 56$
Total number of committees $= 176$

Do not confuse this problem with the number of ways a committee of three men *and* a committee of three women can be chosen. That calls for the selection of a *pair* of committees. This example calls for the selection of *one* committee.

Example 17 (*Compare Exercise 72*)
One freshman, three sophomores, four juniors, and six seniors apply for five positions on an Honor Council. If the council must have at least two seniors, in how many different ways can the council be selected?

Solution The council has at least two seniors when it has two, three, four, or five seniors. We must compute the number of councils possible with two, with three, and so on, and add.

2 seniors and 3 others: $C(6, 2) \times C(8, 3) = 15 \times 56 = 840$
3 seniors and 2 others: $C(6, 3) \times C(8, 2) = 20 \times 28 = 560$

4 seniors and 1 other: $C(6, 4) \times C(8, 1) = 15 \times 8 = 120$
5 seniors: $C(6, 5) = 6$

The total is $840 + 560 + 120 + 6 = 1526$.

5–4 Exercises

I.

Perform the computations in Exercises 1 through 15.

1. (*See Example 5*) $3!$
2. $7!$
3. $5!$
4. $3!2!$
5. $5!3!$
6. $\frac{5!}{6!}$
7. $\frac{7!}{3!}$
8. $\frac{10!}{4!6!}$
9. $\frac{12!}{7!}$
10. $P(12, 3)$
11. $P(6, 4)$
12. $P(6, 2)$
13. $P(100, 3)$
14. $P(5, 5)$
15. $P(7, 4)$
16. What is the number of permutations of eight objects taken three at a time?

Perform the computations in Exercises 17 through 23.

17. (*See Example 11*) $C(6, 2)$
18. $C(4, 3)$
19. $C(13, 3)$
20. $C(9, 4)$
21. $C(9, 5)$
22. $C(20, 3)$
23. $C(4, 4)$
24. Verify the following:

(a) $C(7, 3) = C(7, 4)$
(b) $C(7, 2) = C(7, 5)$
(c) $C(6, 4) = C(6, 2)$
(d) $C(9, 6) = C(9, 3)$
(e) $C(8, 3) = C(8, 5)$

These are examples of a general fact that

$$C(n, k) = C(n, n - k)$$

II.

25. (*See Example 1*) An artist selects three paintings from a collection of six to display in a row. How many different arrangements of the display are possible?

26. How many arrangements of three people seated along one side of a table are possible if there are eight people to select from?

27. The program committee of a music festival must arrange five numbers for an evening performance. They have seven performers available. How many different arrangements of the evening performance are possible?

28. Five people are to be seated in a row. How many different arrangements of all five people are possible?

29. (*See Example 2*) Eight fellows are candidates for Mr. Ugly. In how many different ways can first, second, and third places be awarded?
30. In how many ways can five essays be ranked in a contest?
31. Seven paintings are exhibited by art students.
    (a) An art appreciation class is asked to rank the paintings one through seven. How many different rankings are possible?
    (b) If the students are asked to rank only the top three, how many rankings are possible?
32. A race has ten horses entered. How many distinct possibilities are there for the first three places?
33. (*See Example 3*) A father bought three different gifts. In how many different ways can he give one to each of his three children?
34. The bookstore distributes four different books, one each to four students. In how many different ways can this be done?
35. (*See Example 4*) There are four Pizza Places in Lorena, and the management must assign a manager to each store.
    (a) If there are four people available, in how many ways can they be assigned?
    (b) If there are seven people available, in how many ways can a manager be assigned?
36. (*See Example 6*) There are 6 horses in a race. In how many different orders can the horses finish?
37. In how many different ways can a manager arrange the batting order in a nine-man baseball team?
38. In how many ways can five children line up in a row to have their picture taken?
39. (*See Example 7*) How many three-digit numbers can be formed using the digits 1, 2, 3, and 4, if
    (a) no repetition is allowed?
    (b) repetition is allowed?
40. How many four-digit numbers can be formed using the digits 1 through 9 if no digit can appear twice in a number?
41. (*See Example 9*) List all combinations of two different elements taken from $\{a, b, c, d\}$.
42. List all combinations of three different elements taken from {Tom, Dick, Harriet, Jane}.
43. (*See Example 12*) The Pizza Place must hire two employees from six applicants. In how many ways can this be done?
44. Students are to answer four out of five exam questions. In how many different ways can the questions be selected?

45. (*See Example 13*) A Boy Scout troop has 15 members. In how many different ways can the scoutmaster appoint 3 members to clean up camp?

46. Blackhawk Tech gives four presidential scholarships. The scholarships are of equal value. If there are 50 nominees, in how many ways can the scholarships be awarded?

47. Seven firms are competing for three different contracts. In how many ways can the contracts be awarded if no firm gets more than one contract?

48. A teacher has a collection of 20 questions. How many different tests of 5 questions each can be made from the set of questions?

49. (*See Example 14*) A cafeteria offers a selection of five meats, six vegetables, and eight desserts. In how many ways can you select a meal of two different meats, three different vegetables, and two different desserts?

50. From a set of seven math books, nine science books, and five literature books, in how many ways can a student select two from each set?

51. (*See Example 15*) A company must hire three truck drivers and four clerks. There are six applicants for the truck-driver positions and ten applicants for the clerk positions. In how many ways can the seven employees be chosen?

52. (*See Example 16*) How many different committees can be formed from a group of 9 women and 11 men if a committee is composed of 3 women or 3 men?

53. A woman is considering the purchases of three paintings for her home. Her decorator advises that the selection be made from a collection of seven landscapes or from a collection of six historical paintings. How many different ways can she select three paintings from one collection or the other?

54. A child has three pennies, five nickels, and four dimes. How many ways can two coins of the same denomination be selected?

55. How many different "words" can be made using all the letters of MATH? A "word" is any string of letters, not just strings that give words in a dictionary.

56. Three children go to the ice cream store that serves 31 flavors. In how many different ways can the children be served one dip of ice cream each

(a) if each one receives a different flavor?

(b) if the same flavor can be served to more than one child?

57. Ten students apply for the position of grader in mathematics. Each of four teachers receives one grader. In how many different ways can the teachers be assigned a grader if a student grades for one teacher?

58. A president, a vice president, and a secretary are to be selected from a group of 30 people. How many selections are possible?

59. How many different 13-card hands can be obtained from a deck of 52 cards?

60. Charlie gets a dish of three dips of ice cream at a store with 31 flavors. In how many different ways can he get his three dips of ice cream?

61. In how many different ways can you select three different letters from the word HISTORY?

62. In how many ways can three aces be selected from a bridge deck?

III.

63. There is room for 30 students in a class; 35 students want to enroll. How many possible combinations are there for the class?

64. In how many ways can one select three different letters from the word HISTORY and two different letters from the word ENGLISH?

65. Write out the following as expressions in n:

(a) $P(n, 2)$ **(b)** $P(n, 3)$
(c) $P(n, 1)$ **(d)** $P(n, 5)$

66. If $P(7, k) = 210$, what is k?

67. How many license plates can be made using first a letter, then three digits (from 0 through 9), and then two letters?

68. A builder has six different house models to choose from. He plans to build four houses on adjoining lots down one side of a street. How many different arrangements are possible if

(a) there are no restrictions on the houses to be built?
(b) all the houses are to be different?
(c) next-door houses must be different?

69. $A = \{a, b, c, d, e, f\}$.

(a) How many subsets does A have?
(b) How many three-element subsets does A have?
(c) How many three-element subsets does A have that contain a and c?

70. $B = \{1, 2, 3, 4, 5\}$.

(a) How many subsets does B have?
(b) How many three-element subsets does B have?
(c) How many three-element subsets does B have that contain the number 4?
(d) How many subsets does B have that contain the numbers 2, 3, and 4?

71. In how many ways can a 5-card hand containing exactly two queens and three kings be dealt from a 52-card bridge deck?

72. (*See Example 17*) A committee of five is selected from five men and six women. How many committees are possible if there must be at least three men on the committee?

73. An English reading list has eight American novels and six English novels. A student must read four novels from the list and at least two must be American. In how many different ways can the four books be selected?

74. Five freshmen, four sophomores, and two juniors are present at a meeting of students.

(a) In how many ways can a six-member committee of three freshmen, two sophomores, and one junior be formed?

(b) In how many ways can a committee be selected with no more than two freshmen members?

75. A quiz team of 5 children is to be selected from a class of 25 children. There are 15 girls and 10 boys in the class.

(a) How many teams, made up of 3 girls and 2 boys, can be selected?

(b) How many teams can be selected with at least 3 girls?

76. From a penny, a nickel, a dime, a quarter, and a half dollar, how many sums of money can be formed

(a) using three coins?

(b) using four coins?

77. A test has ten true-false and eight multiple-choice questions.

(a) In how many ways can a student select six true-false and five multiple-choice questions to answer?

(b) In how many ways can a student select ten questions, at least six of which are multiple-choice?

78. A student club has 75 members, of whom 15 are seniors, 20 juniors, 25 sophomores, and 15 freshmen. A president, vice president, secretary, and treasurer are to be chosen. The president and vice president must be seniors, the secretary must be a junior, and the treasurer must be a sophomore. How many different slates of officers can be formed?

79. Twenty companies are competing for four government contracts. One is worth \$5 million, one is worth \$3 million, a third is worth \$1 milllion, and a fourth is worth $\$\frac{1}{2}$ million. Each company is eligible for only one contract. In how many ways can the contracts be awarded?

80. On a bookshelf there are three history books, two mathematics books, and four biology books, all books being different.

(a) In how many ways can the nine books be arranged?

(b) In how many ways can the books be arranged if all the history books are to be together, all the mathematics book are to be together, and all the biology books are to be together?

81. A test contains 15 questions. A student is instructed to answer 10 of the 15 questions.

(a) In how many ways can the student carry out the instructions?

(b) In how many ways can she complete the test if she has to answer five of the first eight questions and five of the last seven questions?

5–5 Partitions

The following example illustrates a particular type of counting problem, the **partition**.

Example 1 (*Compare Exercise 9*)
A group of 15 students is to be divided into three groups to be transported to a game. The three vehicles will carry four, five, and six students, respectively. In how many different ways can the three groups be formed?

Solution Select the four students that ride in the first vehicle. This can be done in

$$C(15, 4) = \frac{15!}{4!11!}$$

different ways. (***Notice*** the form we use for $C(15, 6)$. It is more useful in this case.) After this selection, five students may be selected for the second vehicle in

$$C(11, 5) = \frac{11!}{5!6!}$$

different ways. (There are 11 students left after the first vehicle is filled.) There are six students left for the last vehicle and they can be chosen in

$$C(6, 6) = \frac{6!}{6!0!}$$

different ways.

By the Fundamental Counting Principle, the total number of different ways is

$$C(15, 4) \times C(11, 5) \times C(6, 6) = \frac{15!}{4!11!} \times \frac{11!}{5!6!} \times \frac{6!}{6!0!}$$

$$= \frac{15!}{4!5!6!} = 630{,}630$$

This partition problem has the following properties that make it a partition.

1. The set is divided into disjoint subsets (no two subsets intersect).
2. Each member of the set is in one of the subsets.

The following is a more formal definition of a partition.

Definition 1

A set S is partitioned into k nonempty subsets $A_1, A_2, \ldots, A_k$ if

1. Every pair of subsets is disjoint: that is, $A_i \cap A_j = \varnothing$ when $i \neq j$.
2. $A_1 \cup A_2 \cup \cdots \cup A_k = S$

Note: We assume the subsets are selected in a certain order or according to k different characteristics; for example, by the number in a subset.

We now determine the number of ways a set can be partitioned.

From Example 1, we see that the number of ways a set of 15 elements can be partitioned into subsets of 4, 5, and 6 elements may be expressed as

$$\frac{15!}{4!5!6!}$$

A commonly used notation for this quantity is

$$\binom{15}{4, 5, 6}$$

This is generalized in the following theorem.

Theorem 1

A set with n elements can be partitioned into k subsets of $r_1, r_2, \ldots, r_k$ elements ($r_1 + r_2 + r_k = n$) in the following number of ways:

$$\binom{n}{r_1, r_2, \ldots, r_k} = \frac{n!}{r_1!r_2!\ldots r_k!}$$

Example 2 (*Compare Exercise 11*)

A set of 12 people ($n = 12$) can be divided into three groups of 3, 4, and 5 (r_1, r_2, and r_3) in

$$\binom{12}{3, 4, 5} = \frac{12!}{3!4!5!} = 27{,}720$$

different ways.

Example 3 (*Compare Exercise 13*)

The United Way Allocations Committee has 14 members. In how many ways can they be divided into the following subcommittees so that no member serves on two subcommittees?

Scouting subcommittee, two members
Salvation Army subcommittee, four members
Health Services subcommittee, five members
Summer Recreational Program subcommittee, three members

Solution The subcommittees form a partition, since no one is on two subcommittees and all 14 members are used. The number of partitions is

$$\binom{14}{2, 4, 5, 3} = \frac{14!}{2!4!5!3!} = 2{,}522{,}520$$

Let's look at a special case of partitions. Suppose a set of eight objects is partitioned into two subsets of three and five objects. The formula for partitions gives

$$\binom{8}{3, 5} = \frac{8!}{3!5!}$$

Notice that the formula for $C(8, 3)$ and $C(8, 5)$ both give

$$C(8, 3) = \frac{8!}{5!3!} = C(8, 5)$$

so the number of partitions into two subsets is just the number of ways a subset of one size can be selected. This result occurs because when one subset of three objects is selected, the remaining five objects automatically form the other subset in the partition. In general, the following is true.

> The number of partitions of a set into two subsets is the number of ways one of the subsets can be formed.

5–5 Exercises

I.

Compute the following.

1. $\binom{12}{3, 3, 3, 3}$ **2.** $\binom{8}{2, 2, 4}$ **3.** $\binom{7}{3, 4}$ **4.** $\binom{10}{4, 6}$

5. $\binom{9}{2, 3, 4}$ **6.** $\binom{8}{4, 4}$ **7.** $\binom{6}{2, 4}$ **8.** $\binom{7}{2, 2, 2, 1}$

II.

9. (*See Example 1*) In how many ways can a lab instructor assign nine students so that three perform experiment A, three perform experiment B, and three perform experiment C?

10. In how many ways can a set of nine objects be divided into subsets of two, three, and four objects?

11. (*See Example 2*) An accounting instructor separates her 18 students into three groups of six each. Each group is assigned a different problem. In how many ways can the class be divided into these groups?

12. A store has 12 items to be displayed in three display windows. In how many ways can they be displayed if six are placed in the first window, four in the second window, and two in the third window?

13. (*See Example 3*) In how many different ways can a 15-person committee be divided into subcommittees having 6, 4, and 5, members?

14. The State University football team plays 11 games. In how many ways can they complete the season with four wins, six losses, and one tie?

15. A scholarship committee will award four \$5,000 scholarships, four \$8,000 scholarships, and two \$10,000 scholarships. Ten students are selected to receive scholarships. In how many ways can the scholarships be awarded?

III.

16. In the game of Bridge the deck contains 52 cards. Each of the four players receives 13 cards. In how many different ways can this be done? Leave your answer in factorial form.

IMPORTANT TERMS

5–1	**Set**	**Element of a set**	**Equal sets**
	Empty set	**Subset**	**Union**
	Intersection	**Disjoint sets**	
5–2	**Universe**	**Venn diagram**	**Complement**
5–3	**Fundamental counting principle**	**Tree diagram**	
5–4	**Permutation**	**Combination**	**Factorial**
5–5	**Partition**		

REVIEW EXERCISES

1. $A = \{6, 10, 15, 21, 30\}$, $B = \{6, 12, 24, 48\}$, and $C = \{x \mid x \text{ is an integer divisible by } 3\}$. Identify the following as true or false.

   (a) $21 \in A$ **(b)** $21 \in B$ **(c)** $25 \in C$
   (d) $30 \notin A$ **(e)** $16 \notin B$ **(f)** $24 \notin C$
   (g) $6 \in A \cap B \cap C$ **(h)** $12 \in A \cap B$ **(i)** $10 \in A \cup B$
   (j) $A \subset B$ **(k)** $B \subset C$ **(l)** $C \subset A$
   (m) $\varnothing \subset B$ **(n)** $A \subset C$ **(o)** A and B are disjoint

2. Let the universe set $U = \{-2, -1, 0, 1, 2, 3, 4\}$, $A = \{-2, 0, 2, 4\}$, and $B = \{-2, -1, 1, 2\}$.
   Find the following:

   A' B' $(A \cap B)'$ $A' \cap B'$
   $A' \cup B'$ $A \cup A'$

3. $n(A) = 27$, $n(B) = 30$, and $n(A \cap B) = 8$. Find $n(A \cup B)$.
4. $n(A \cup B) = 58$, $n(A) = 32$, and $n(B) = 40$. Find $n(A \cap B)$.
5. Draw a tree diagram showing the ways you can select a meat and then a vegetable from roast, fish, chicken, peas, beans, and squash.
6. The freshman class traditionally guards the school mascot the night before homecoming. There are five key locations where a freshman is posted. Nine freshmen volunteer for the 2 AM assignment. In how many different ways can they be assigned?
7. How many different license plates can be made using four digits followed by two letters

   (a) if repetitions of digits and letters are allowed?
   (b) if repetitions are not allowed?

8. A museum has a display case with four display compartments. Eight antique vases are available for display. How many displays are possible putting one vase in each compartment?

9. A medical research team randomly selects 5 patients from a group of 15 for special treatment. In how many different ways can the patients be selected?

10. In how many different ways can a medical research team divide 15 patients into three groups of five each?

11. One student representative is selected from each of four clubs. In how many different ways can the four students be selected given the following number of members in each club: Rodeo Club, 40 members; Kite Club, 27 members; Frisbee Club, 85 members; Canoeing Club, 34 members.

12. In the finale of the University Sing, there are ten people in the first row. Club A has three members on the left end, club B has four members in the center, and club C has three members on the right end. In how many different ways can the line be arranged?

13. In how many different ways can a chairman, a secretary, and four other committee members be formed from a group of ten people?

14. A club has 12 pledges. On a club work day, four pledges are assigned to the Red Cross, six to the Salvation Army, and two are not assigned. In how many ways can the groups be selected?

15. Compute

(a) $P(15, 3)$ (b) $C(15, 3)$ (c) $P(101, 2)$ (d) $P(22, 4)$

(e) $3!$ (f) $\frac{10!}{3!7!}$ (g) $\binom{8}{3, 2, 3}$ (h) $\binom{11}{3, 2, 2, 4}$

16. The Venn diagram in Figure 5–16 represents the number of children participating in the three sports, soccer (S), baseball (B), and football (F), in a city recreation program. Determine

(a) the number of children participating in the sports.
(b) the number of children playing only soccer.
(c) the number of children that participate in soccer or baseball.
(d) the number of children that participate in soccer and football.
(e) the number of children that participate in soccer, but not in baseball.

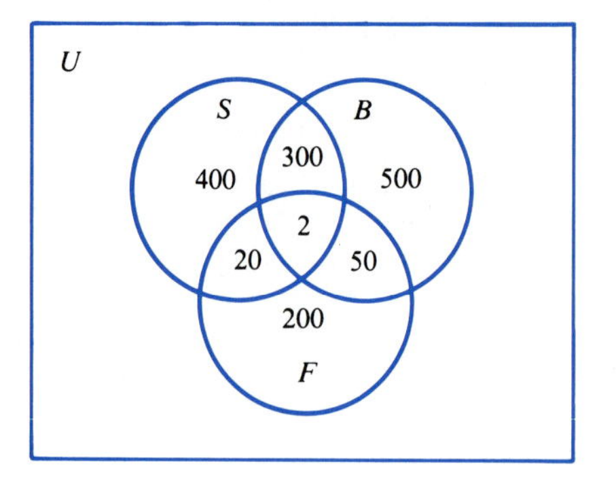

Figure 5–16

17. A survey of 60 people gave the following information:

25 jog regularly,
26 ride a bicycle regularly,
26 swim regularly,
10 both jog and swim,
6 both swim and ride a bicycle,
7 both jog and ride a bicycle,
1 does all three,
3 do none of the three.

Show that there is an error in this information.

18. The bookstore had a sale on records, books, and T-shirts. A cashier observed the purchases of 38 people and found that

16 bought records,
15 bought books,
19 bought T-shirts,
5 bought books and records,
7 bought books and T-shirts,
6 bought records and T-shirts,
3 bought all three.

(a) How many bought records and T-shirts, but no books?
(b) How many bought records, but no books?
(c) How many bought T-shirts, but no books and no records?
(d) How many bought none of the three?

19. A poll was conducted among a group of teenagers to see how many have televisions, radios, and microcomputers. The results were as follows: (T denotes television, R radio, M microcomputer.)

Item	Number of Teenagers Having this Item
T	39
R	73
M	10
T and R	22
M and R	3
T and M	4
T and R and M	2

Determine

(a) how many had a radio and TV, but no microcomputer?
(b) how many had a microcomputer, but no TV?
(c) how many had exactly two of the three items?

20. A man has eight shirts, four pairs of trousers, and three jackets. How many different ensembles can he wear?

21. In how many different ways can a group of 15 people elect a president, vice president, and secretary?

22. From a group of five people, two are to be selected to be delegates to a certain conference. How many selections are possible?

23. Students taking an examination are required to answer four questions out of six. How many selections are possible?

24. Twenty people attend a meeting at which three different door prizes are awarded by drawing names.

(a) If a name is drawn and replaced for the next drawing, in how many ways can the door prizes be awarded?
(b) If a name is drawn and not replaced, in how many ways can the door prizes be awarded?

25. A bag contains six white balls, four red balls, and three green balls. In how many ways can a person draw out two white, three red, and two green balls?

26. A faculty senate at a university is made up of 11 members. The university has three schools: liberal arts, business, and music. Six senators are to be elected from the college of liberal arts, three from the business school, and two from the school of music. The faculties of liberal arts, business, and music have 150, 60, and 15 professors, respectively.
How many different senates can be elected?

27. An artist plans a recital of six classics and four moderns. He plans to play four classics and two moderns before the intermission and the rest following the intermission. How many different arrangements of his program are possible?

28. A club agrees to provide five students to work at the school carnival. One sells balloons, one sells popcorn, one sells cotton candy, one sells candied apples, and one sells soft drinks. Nine students agree to help. In how many ways can the assignments be made?

29. A professor gives a reading list of six books. A student is to read three. In how many ways can the selection be made?

30. For a psychology study, five people are to be chosen at random from a group of 20 people. In how many ways can this be done?

31. How many different five-card hands can be obtained from a deck of 52 cards?

32. Compute

(a) $P(8, 4)$ (b) $C(9, 5)$ (c) $P(7, 7)$ (d) $C(5, 5)$

(e) $4!$ (f) $\frac{7!}{3!4!}$ (g) $\frac{8!}{4!}$

33. A social organization and a service club held a joint meeting. Of the 83 people present, 46 belonged to the social organization and 51 belonged to the service club. How many belonged to both?

34. The digits {2, 3, 4, 5, 6, 7} are used to form three-digit numbers.

    (a) How many can be formed if repetitions are allowed?

    (b) How many can be formed if repetitions are not allowed?

    (c) How many numbers larger than 500 can be formed with repetitions allowed?

35. A department store has ten sportswear outfits for display. How many different displays are possible if a group of four outfits is displayed?

36. A business has 16 female employees and 14 male employees. How many different advisory committees consisting of two males and two females are possible?

37. List all the subsets of {red, white, blue}.

38. A panel of four is selected from eight businessmen and seated in a row behind a table. In how many different orders can they be seated.?

39. A club has ten freshmen and eight sophomore members. At a club picnic the cook and entertainment leader are freshmen and the clean-up crew consists of three sophomores. How many different ways can these five be selected?

40. Draw a tree diagram showing the ways a girl then a boy can be selected from the set: {Tommy, Betty, Cheryl, Gary, Susie}.

41. A student is allowed to check out four books from the reserve room. They must all come from one collection of six books or from another collection of eight books. How many different ways can the selection be made?

6

Topics in this chapter can be applied to:

Predicting Outcomes ● Projecting Results ● Evaluating Preferences ● Interpreting Surveys ● Quality Control ● Sales Techniques ● Traffic Control ● Medical Evaluations ● Advertising ● Voting Behavior ● Managing Charity Funds ● Demographics ● Investment Patterns ● Genetics ● Stock and Bond Analysis

Probability

6–1 Introduction to Probability

When the weatherman predicts the weather, when a coach evaluates his team's chances of winning, or when a businessman projects the success of the big clearance sale, an element of uncertainty exists. The weatherman knows he is often wrong, the coach knows there is no such thing as a sure win, and the businessman knows the best advertised sale sometimes flops. Often in our daily lives we would like to measure the likelihood of an outcome of an event or activity. An area of mathematics, **probability theory**, provides a measure of the likelihood of the outcome of phenomena and events. Insurance companies use it to decide on financial policies, the government uses it to determine its fiscal and economic policies, theoretical physicists use it to understand the nature of atomic-sized systems in quantum mechanics, and public-opinion polls, such as the Harris Poll, have their theoretical acceptability based on probability theory.

Terminology

In order to discuss and use probability theory, we need to understand some terminology used: **experiment**, **trials**, **outcomes**, **sample space**, and **event**.

Definition 1 An activity or phenomenon under consideration is called an **experiment**. The experiment can produce a variety of results, called **outcomes**. We study activities that can be repeated or phenomena that can be observed a number of times. We call each observation or repetition of the experiment a **trial**.

Example 1

(a) Predicting the weather is an experiment with "fair," "cooler," "warmer," "snow," "rain," and so on: as possible outcomes. Each prediction is a trial.

(b) Drawing a number out of a hat is an experiment with the number drawn as an outcome. Each draw of a number is a trial.

(c) Tossing a coin is an experiment with "heads" and "tails" as possible outcomes. A trial occurs each time the coin is tossed.

(d) A test to determine the germination of flower seeds is an experiment with "germinated" and "not germinated" as possible outcomes. Each test conducted is a trial.

In general, experiments involve chance or random results. This means the outcomes do not occur in a set pattern, they vary depending on impartial chance, and the outcome cannot be determined in advance. The selection of a card from a well shuffled deck, the order in which leaves fall off a tree, and the number of cars that pass a checkpoint on the freeway are examples of experiments that have random outcomes.

An experiment need not classify outcomes in a unique way. It depends on how the results are interpreted. When a multiple-choice test of 100 questions is given, the instructor wants to know the number of correct answers given by each student. For this purpose, an outcome can be any of the numbers 0 through 100. When the tests are returned to the students they tend to ask, "What is an A?" From their viewpoint they are interested in the outcomes A, B, C, D, and F. Then there may be the student who only asks "What is passing?" To that student there are just two outcomes of interest, pass and fail.

When asked, "What is today?" a person may respond in several ways, such as, "It is April 1," "It is Friday," "It is payday," or any one of numerous responses. Depending on the focus of the individual, the set of possible responses may be all the days in a year; all the days of the week; or the two outcomes: payday, not payday.

Since the outcomes of an experiment can be classified in a variety of ways, it is important that everyone understand which set of outcomes is used. We call the set of outcomes used a **sample space**.

Definition 2

A sample space is the set of all possible outcomes of an experiment. Each element of the sample space is called a **sample point** or **simple outcome**.

Example 2 (*Compare Exercise 1*)

(a) If the experiment is tossing a coin, the sample space is {heads, tails}.

(b) If the experiment is drawing a card from a bridge deck, one sample space is the set of 52 cards.

(c) If the experiment is drawing a number from the numbers 1 through 10, the sample space can be {1, 2, 3, 4, 5, 6, 7, 8, 9, 10}. If the numbers are drawn to randomly divide ten people into two groups, the set {even, odd} is an appropriate sample space.

(d) If the experiment is tossing a coin twice, a sample space is {HH, HT, TH, TT}.

We do not insist that there is a single, correct sample space for an experiment because the situation dictates how to interpret the results. However, we do insist that a sample space conform to two properties.

Properties of a Sample Space

Let S be the sample space of an experiment.

1. Each element in the set S is an outcome of the experiment.
2. Each outcome of the experiment corresponds to exactly one element in S.

If a student is selected from a group of university students, and the class standing of the student is the outcome of interest, then {freshman, sophomore, junior, seniors} is a valid sample space. If the sex of the student is the outcome of interest, then {male, female} is a valid sample space. You can form other sample spaces using age, GPA, and so on, as the outcomes of interest.

In some instances our interest lies in a collection of outcomes in the sample space, not just one outcome. If I toss a coin twice, I may be interested in the likelihood that the coin will land with the same face up both times. I am interested in the subset of outcomes {HH, TT}, not just one of the possible outcomes. We call such a collection of simple outcomes an **event**.

Definition 3

An event is a subset of a sample space.
An event can be a subset with a single outcome, so a simple outcome is a special case of an event.

Example 3 **(a)** In the experiment of drawing a number from the numbers 1 through 10, the sample space is

$$S = \{1, 2, 3, 4, 5, 6, 7, 8, 9, 10\}$$

The event of drawing an odd number is the subset $\{1, 3, 5, 7, 9\}$. The event of drawing an even number is the subset $\{2, 4, 6, 8, 10\}$. The event of drawing a prime number is the subset $\{2, 3, 5, 7\}$.

(b) A teacher selects one student from a group of six students. The sample space is {Scott, Jane, Mary, Kaye, Ray, Randy}. The event of selecting a student with first initial R is {Ray, Randy}. The event of selecting a student with first initial J is {Jane}. The event of selecting a student with first initial A is the empty set.

We say an *event occurs* when any of the simple outcomes of the event occurs.

Probability Assignment

How do we measure the likelihood of an event? What is the likelihood of getting a prime number when selecting a number from the set 1 through 10? How does that compare to the likelihood of getting an even number? We call the measure of the likelihood of an event the **probability** of the event. The probability of an event is a number assigned to that event. We use 0 for the probability that an event cannot occur (like the probability of a coin standing on edge), and use 1 for the probability of an event that is bound to occur (such as the probability of the sun rising in the east). Other probabilities range from 0 to 1. A larger number indicates a higher likelihood of the event occurring. If we call an event E, we designate the probability of the event by $P(E)$ (called "P of E").

There is no single, predetermined way of assigning probability. It depends on the situation, what features we want to emphasize, and which features we consider unimportant. However, we use some standard properties of probability to assign probabilities to the outcomes in any sample space.

Properties of Probability

Let a sample space be $S = \{e_1, e_2, \ldots, e_n\}$.

1. Each outcome in the sample space, e_i, is assigned a probability denoted by $P(e_i)$.
2. Each probability is a number that is not negative and is no larger than 1 ($0 \leq P(e_i) \leq 1$).
3. If an event E is a set of simple outcomes, then $P(E)$ is the sum of the probabilities of all simple outcomes that make up the event, for example, $P(\{e_1, e_2, e_3\}) = P(e_1) + P(e_2) + P(e_3)$.
4. The sum of probabilities of all the simple outcomes in a sample space is 1, that is, $P(e_1) + P(e_2) + \cdots + P(e_n) = 1$.

Example 4 (*Compare Exercise 3*)
If an experiment has five possible outcomes, and we assign the following probabilities to the outcomes:

$$P_1 = .20,\ P_2 = .10,\ P_3 = .15,\ P_4 = .30,\ P_5 = .25$$

we have a valid probability assignment because:

1. Each outcome is assigned a probability.
2. Each probability is nonnegative and not larger than 1.
3. The sum of all the probabilities is 1.

Example 5 (*Compare Exercise 3*)
Suppose we assign the following probabilities to a five-point sample space:

$$P_1 = .2,\ P_2 = .2,\ P_3 = .3,\ P_4 = .25,\ P_5 = .4$$

This is not a valid probability assignment because the sum of all probabilities is 1.35.

Example 6 The assignment of probabilities

$$P_1 = .3,\ P_2 = -.1,\ P_3 = .2,\ P_4 = .3,\ P_5 = .3$$

is not valid because a negative probability is assigned.

Example 7 (*Compare Exercise 7*)
A bag contains six balls that are colored red, white, or blue. There are heavy balls and light balls. The experiment is to draw one ball from the bag. Assign individual probabilities according to the following table:

	Red	**White**	**Blue**
Heavy	.05	.15	.23
Light	.10	.17	.30

(a) The probability of drawing a heavy, blue ball is .23. Using the $P(E)$ notation this is written $P(\text{heavy blue}) = .23$.

(b) What is the probability of drawing a white ball? Here the event is the subset {heavy white, light white}. We are interested in the probability that the ball drawn is in that subset. According to the third probability property, the probability of drawing a white ball is $.15 + .17 = .32$, the sum of the probabilities of the individual outcomes making up the event.

In a similar fashion we obtain the probability of the following events:

(c) The probability of drawing a light ball is:

$$.10 + .17 + .30 = .57 \ (P(\text{light}) = .57).$$

(d) The probability of drawing a heavy red or a light blue ball is $.05 + .30 = .35$

(e) The probability of drawing a ball that is not blue is

$$.05 + .15 + .10 + .17 = .47$$

($P(\text{not blue}) = .47$).

Example 8 (*Compare Exercise 8*)

A football coach responds to a question of the team's chances of winning the big game with the following statement: "The probability of winning is twice that of losing and the probability of a tie is half that of losing." Find the probability of each outcome: win, lose, or tie.

Solution Let x = probability of losing. Then, $P(\text{win}) = 2x$ and $P(\text{tie}) = 0.5x$. Since these probabilities must add to 1, we have

$$x + 2x + .5x = 1$$

$$3.5x = 1$$

$$\text{so } x = \frac{1}{3.5} = .286 \text{ (rounded)}$$

From this we have $P(\text{win}) = .572$, $P(\text{lose}) = .286$, and $P(\text{tie}) = .143$. (Because we rounded to three decimal places, these add to 1.001.)

Two Important Special Cases

1. If an event is the empty set, the probability that an outcome is in the event is zero, that is, $P(\varnothing) = 0$.
2. If an event is the entire sample space, then $E = S$ and $P(E) = P(S) = 1$.

Empirical Probability

Probabilities may be assigned by observing a number of trials and using the frequency of outcomes to estimate probability. For example, the operator of a concession stand at a park keeps a record of the kinds of drinks children buy. Her records show the following.

Drink	Frequency
Cola	150
Lemonade	275
Fruit Juice	75
	500

In order to estimate the probability that a child will buy a certain kind of drink, we compute the **relative frequency** of each drink. Do this by dividing the frequency of each drink by the total number of drinks.

Drink	Frequency	Relative Frequency
Cola	150	$\frac{150}{500} = .30$
Lemonade	275	$\frac{275}{500} = .55$
Fruit Juice	$\underline{75}$	$\frac{75}{500} = \underline{.15}$
	500	1.00

Notice that the relative frequency has all the properties of a probability assignment, so we use it to estimate probability. The probability that a child will buy lemonade is .55, $P(\text{cola}) = .30$, and $P(\text{fruit juice}) = .15$. Probability based on relative frequency is called **empirical probability**.

Example 9 (*Compare Exercise 15*)
A college has an enrollment of 1210 students. The number in each class is as follows.

Class	Number of Students
Freshman	420
Sophomore	315
Junior	260
Senior	215

A student is selected at random. Estimate the probability that the student is

(a) a freshman
(b) a sophomore
(c) a junior
(d) a senior.

Solution Estimate the probability of each as the relative frequency.

Class	Number of Students	Relative Frequency
Freshman	420	$\frac{420}{1210} = .35$
Sophomore	315	$\frac{315}{1210} = .26$
Junior	260	$\frac{260}{1210} = .21$
Senior	215	$\frac{215}{1210} = .18$

This gives $P(\text{freshman}) = .35$, $P(\text{sophomore}) = .26$, $P(\text{junior}) = .21$, and $P(\text{senior}) = 18$

6–1 Exercises

I.

1. Give a sample space for each of the following experiments: (*See Example 2*)
   (a) answering a true-false question,
   (b) selecting a letter at random from the English alphabet,
   (c) tossing a single die (six-sided),
   (d) tossing a coin three times,
   (e) selecting a day of the week,
   (f) predicting the outcome of a football game between Grand Canyon College and Bosque College, (Give two different ways you could describe the outcome of a game; that is, give two sample spaces.)
   (g) the grade received in a course, (Give two ways the space might be formed.)
   (h) selecting two students from {Susan, Leann, Dana, Julie},
   (i) selecting a day from the month of January, (Give two ways the sample space can be formed.)
   (j) selecting a person from a speech class. (Give two sample spaces.)
2. Consider the experiment of tossing a coin. Give the sample space and list all the events associated with this experiment. Indicate the simple events.

3. (*See Examples 4, 5, and 6*) Which of the following probability assignments are valid? Give reasons for your answer.

   (a) Sample space = $\{A, B, C, D\}$
   $P(A) = .3$, $P(B) = .2$, $P(C) = .1$, $P(D) = .4$

   (b) Sample space = $\{A, B, C\}$
   $P(A) = .4$, $P(B) = 0$, $P(C) = .6$

   (c) Sample space = {Tom, Dick, Harry}
   $P(\text{Tom}) = .35$, $P(\text{Dick}) = .40$, $P(\text{Harry}) = .20$

   (d) Sample space = $\{2, 4, 6, 8, 10\}$
   $P(2) = .2$, $P(4) = .3$, $P(6) = .25$, $P(8) = .25$, $P(10) = .15$

   (e) Sample space $\{A, B, C, D, E\}$
   $P(A) = 1/5$, $P(B) = 1/5$, $P(C) = 1/5$, $P(D) = 1/5$, $P(E) = 1/5$

   (f) Sample space = {Oklahoma, Utah, Maine, Alabama}
   $P(\text{Oklahoma}) = .3$, $P(\text{Utah}) = -.4$, $P(\text{Maine}) = .5$, $P(\text{Alabama}) = .6$

   (g) Sample space = {True, False}
   $P(\text{True}) = 1$, $P(\text{False}) = 0$

   (h) Sample space = {Yes, No, Maybe}
   $P(\text{Yes}) = .8$, $P(\text{No}) = .3$, $P(\text{Maybe}) = 1.1$

   (i) Sample space = {True, False}
   $P(\text{True}) = 0$, $P(\text{False}) = 0$

4. The sample space of an experiment is $\{A, B, C, D\}$, where $P(A) = .1$, $P(B) = .2$, $P(C) = .3$, and $P(D) = .4$

   (a) Find $P(\{A, B\})$.
   (b) Find $P(\{B, D\})$.
   (c) Find $P(\{A, C, D\})$.
   (d) Find $P(\{A, B, C, D\})$.

5. The sample space of an experiment is {Ford, Oldsmobile, Volkswagen, Plymouth, Honda}, where $P(\text{Ford}) = .05$, $P(\text{Oldsmobile}) = .18$, $P(\text{Plymouth}) = .30$, $P(\text{Volkswagen}) = .10$, and $P(\text{Honda}) = .37$.

   (a) Find $P(\{\text{Volkswagen, Honda}\})$.
   (b) Find $P(\{\text{Ford, Oldsmobile, Plymouth}\})$.

6. For the sample space $\{A, B, C\}$, the points A and B are assigned probabilities $P(A) = .2$ and $P(B) = .4$. What is $P(C)$?

7. (*See Example 7*) For the sample space $\{1, 2, 3, 4, 5\}$, the following probabilities are assigned: $P(1) = .15$, $P(2) = .16$, $P(3) = .19$, $P(4) = .30$, and $P(5) = .20$.

   (a) Find the probability of $\{2, 4\}$.
   (b) Find the probability of $\{1, 3, 5\}$.
   (c) Find the probability of selecting a prime.

II.

8. (*See Example 8*) A sample space of an experiment is $\{A, B, C\}$. Find the probability of each simple outcome if $P(A) = P(B)$ and $P(C) = 2\ P(A)$.

9. A sample space has four simple outcomes, A, B, C, and D. Find the probability of each if $P(A) = 2\ P(D)$, $P(B) = 3\ P(D)$, and $P(C) = 4\ P(D)$.

10. The sample space of an experiment is $\{a, b, c\}$, with $P(b) = .25$. If $P(\{a, b\}) = .375$ and $P(\{b, c\}) = .875$, find $P(a)$ and $P(b)$.

11. The sample space of an experiment is $\{A, B, C\}$, where $P(A) + P(B) = .75$, $P(B) + P(C) = .45$. Find $P(A)$, $P(B)$, and $P(C)$.

12. A student is to be selected from a group of six students. For each classification of freshman and sophomore, there is a math major, an art major, and a biology major. The probability of each individual being selected is given in the following table:

	Math	**Art**	**Biology**
Freshman	.10	.08	.17
Sophomore	.22	.30	.13

(a) Find the probability that a freshman is selected.

(b) Find the probability that an art major is chosen.

(c) Find the probability that a freshman math major or a sophomore biology major is chosen.

13. An experiment consists of drawing one card from a bridge deck. The sample space contains 52 points, each of the 52 cards. Suppose we assign a probability of $\frac{1}{52}$ to each of the cards.

(a) Find the probability that an ace is drawn.

(b) Find the probability that the card drawn is a king or a queen.

(c) Find the probability that the card drawn is a diamond.

(d) Find the probability that the card drawn is a red card.

14. A fellow has some change in his pocket: two pennies, four nickels, three dimes, four quarters, and two half dollars. He draws out a coin at random. The probability of drawing any one of the coins is

.05 for each penny,

.04 for each nickel,

.02 for each dime,

.07 for each quarter,

.20 for each half dollar.

(a) Find the probability that the coin drawn is a dime.

(b) Find the probability that the coin drawn is a quarter or a half dollar.

(c) Find the probability that the coin drawn is a nickel.

(d) Find the probability that the value of the coin drawn is less then ten cents.

15. (*See Example 9*) The owner of a hamburger stand found that 800 people bought hamburgers as follows.

Kind of Burger	Frequency
Miniburger	140
Burger	345
Big Burger	315

Find the probability of a customer purchasing each kind of hamburger.

16. An auto dealer sold 120 minivans. His records show the repairs required during the first year:

Repairs	Frequency
Minor	70
Major	28
No repairs	22

A customer purchases a minivan from the dealer. Find the probability that she will return during the first year for

(a) minor repairs. **(b)** major repairs. **(c)** no repairs.

17. A policewoman's radar unit clocked 1800 vehicles on the interstate highway. Her report showed the following information.

Speed	Frequency
Below 40	160
40–49	270
50–55	1025
Over 55	345

Based on this report, estimate the probability of a vehicle driving within each of the speed categories.

18. Customers at a shopping center are given a taste test to determine their preference of coffee. Based on the following information, estimate $P(A)$, $P(B)$, and $P(C)$.

Brand	Number Preferring this Brand
A	88
B	62
C	174

6–2 Equally Likely Events

In general, no unique method exists for assigning probabilities to the outcomes of an experiment. However, there exists a class of experiments in which the assignments are straightforward. Such experiments have the characteristic that the individual outcomes are **equally likely**. That is, each outcome has the same chance, or probability, of occurring as any other outcome. Generally, we say a tossed coin is just as likely to turn up a head as a tail, so heads and tails are equally likely to happen. If a coin is altered so that it comes up heads two thirds of the time, then heads and tails as not equally likely. Unless stated otherwise, we assume a tossed coin is fair; that is, heads and tails are equally likely. All cards drawn from a well shuffled deck have an equal chance of being drawn. We say that the outcomes are equally likely.

If a coin is drawn from a purse with coins of different denominations, it doesn't seem reasonable to say that the outcomes are equally likely because of the variations of size. The small size of a dime makes it less likely to be drawn than a half dollar, for example.

For experiments with a finite number of equally likely outcomes, the probability of each simple outcome is $1/n$, where n is the number of outcomes in the sample space.

If we toss a coin, the sample space is $\{H, T\}$. We intuitively agree that heads and tails are equally likely, so

$$P(H) = \frac{1}{2} \quad \text{and} \quad P(T) = \frac{1}{2}$$

because the sample space has two elements. If we select a name at random from 25 different names, then each name has $1/25$ probability of being drawn.

Since the probability of an event is the sum of probabilities of all simple outcomes in the event, we can compute probabilities of events like the following.

Select a number at random from the set

$$S = \{1, 2, 3, 4, 5, 6, 7, 8, 9\}$$

What is the probability that it is even, that is, the number is in the event

$$E = \{2, 4, 6, 8\}$$

We assume equally likely outcomes, so the probability of each number is $\frac{1}{9}$. Then

$$\begin{aligned} P(E) &= P(2) + P(4) + P(6) + P(8) \\ &= \frac{1}{9} + \frac{1}{9} + \frac{1}{9} + \frac{1}{9} \\ &= \frac{4}{9} \end{aligned}$$

It is no accident that

$$\frac{4}{9} = \frac{\text{number of elements in } E}{\text{number of elements in } S}$$

For an experiment with equally likely outcomes, compute the probability of an event by counting the number of elements in the event, and then dividing by the number of elements in the sample space. We sometimes call the outcomes in the event **successes** and outcomes not in the event **failures**.

Theorem 1

If an event E contains s simple outcomes (successes), and the sample space S contains n simple outcomes, and if the simple outcomes are equally likely (have the same probabilities), then the probability of E is

$$P(E) = \frac{s}{n} = \frac{\text{number of outcomes of interest (success)}}{\text{total number of outcomes possible}} = \frac{n(E)}{n(S)}$$

Reminder: The event E occurs when any of the simple outcomes in E occurs.

If the event can fail in f ways, the probability of failure is

$$q = \frac{f}{n}$$

Since we admit only success or failure, $n = s + f$. We can conclude from this that $P(\text{success}) + P(\text{failure}) =$

$$\frac{s}{n} + \frac{f}{n} = \frac{s + f}{n} = \frac{n}{n} = 1$$

so the probability of success and the probability of failure always add to 1.

Example 1 (*Compare Exercise 2*)

Draw a number at random from the integers 1 through 10. What is the probability that a prime is drawn?

Solution In this case $n = 10$ and $E = \{2, 3, 5, 7\}$, so $s = 4$. This gives $P(\text{prime}) = \frac{4}{10}$.

Example 2 (*Compare Exercise 3*)
An urn contains four red, three green, and five white balls. If a single ball is drawn, what is the probability that it is green?

Solution Since there are a total of 12 balls, $n = 12$. An outcome is successful in three ways, the drawing of any one of the three green balls, so $s = 3$. Thus,

$$P(\text{green}) = \frac{3}{12} = \frac{1}{4}$$

Example 3 (*Compare Exercise 5*)
A pair of dice is rolled. What is the probability of rolling an 8 (the two numbers that turn up add to 8)? Of rolling a 3 (the two numbers add to 3)?

Solution According to the Fundamental Counting Principle, there are $6 \times 6 = 36$ ways the two dice can turn up, so $n = 36$.

Each outcome in the sample space may be thought of as a pair of numbers like (3, 2); the number showing on the first die is 3, and the number showing on the second die is 2. The sample space is composed of the following pairs.

(1, 1), (1, 2), (2, 1), (1, 3), (3, 1), (1, 4), (4, 1),
(1, 5), (5, 1), (1, 6), (6, 1), (2, 2), (2, 3), (3, 2),
(2, 4), (4, 2), (2, 5), (5, 2), (2, 6), (6, 2), (3, 3),
(3, 4), (4, 3), (3, 5), (5, 3), (3, 6), (6, 3), (4, 4),
(4, 5), (5, 4), (4, 6), (6, 4), (5, 5), (5, 6), (6, 5),
(6, 6)

We say that an 8 is rolled when the two numbers showing add to 8. An 8 may be obtained in five different ways:

first die 2, second die 6, that is, (2, 6),
first die 6, second die 2, that is, (6, 2),
first die 3, second die 5, that is, (3, 5),
first die 5, second die 3, that is, (5, 3),
first die 4, second die 4, that is (4, 4).

Thus,

$$P(8) = \frac{5}{36}$$

Since there are just two ways to roll a 3,

$$P(3) = \frac{2}{36} = \frac{1}{18}$$

Example 4 (*Compare Exercise 7*)
Two students are selected at random from a class of eight boys and nine girls. What is the probability that both students selected are girls?

Solution Determine the number of outcomes in the event of interest by the number of different ways two girls can be selected from a group of nine. This is

$$C(9, 2) = \frac{9 \times 8}{2 \times 1} = 36$$

The number of outcomes, n, in the sample space is the number of different ways two students can be selected from the whole group of 17. This is

$$C(17, 2) = \frac{17 \times 16}{2 \times 1} = 136$$

Thus,

$$P(\text{two girls}) = \frac{36}{136} = \frac{9}{34}$$

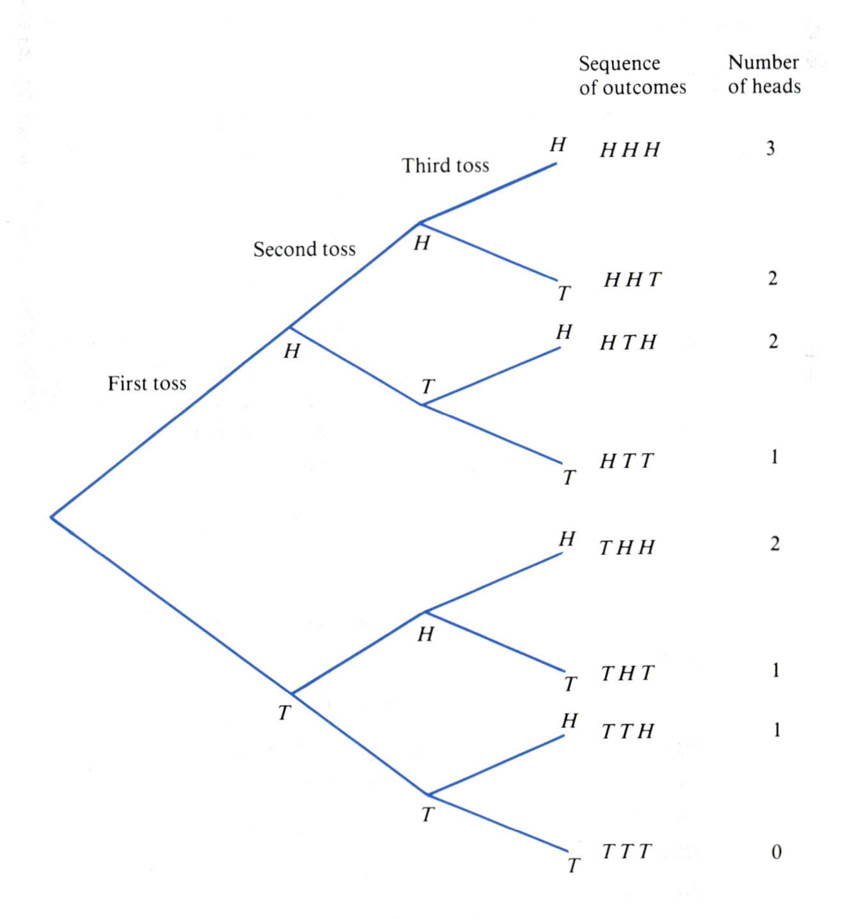

Figure 6–1 Sequence of heads and tails on three tosses of a coin

Example 5 (*Compare Exercise 13*)
Find the probability of at least two heads appearing on three tosses of a coin.

Solution We use a tree diagram to obtain the necessary information.
We observe that there are a total of eight possible outcomes with four of them showing two or more heads. Thus,

$$P(\text{at least 2 heads}) = \frac{4}{8} = \frac{1}{2}$$

Example 6 (*Compare Exercise 15*)
A grocery store stocks eight different sugar-coated cereals and twelve that are not sugar-coated. A child is allowed to select some cereal. He grabs three different cereals. What is the probability two are sugar-coated and one is not?

Solution The total number of outcomes, n, is $C(20, 3)$, since there are 20 different cereals. A success is the selection of two cereals that are sugar-coated and one that is not. These can be selected in $C(8, 2) \times C(12, 1)$ different ways. The probability is

$$P = \frac{C(8, 2) \times C(12, 1)}{C(20, 3)} = \frac{28 \times 12}{1140} = \frac{28}{95}$$

6–2 Exercises

I.

1. The experiment is rolling a pair of dice. Find s, the number of successes, for rolling the following numbers (the two numbers that turn up give that total):

   (a) 2 **(b)** 4 **(c)** 5
   (d) 6 **(e)** 7 **(f)** 9
   (g) 11 **(h)** 12

2. (*See Example 1*) Fifteen names are written one per card. Two of the names begin with a C, four begin with a T, five begin with a B, and four begin with a P. If a name is drawn at random, what is the probability that it begins with a T?
3. (*See Example 2*) An urn contains five red and three green balls. If one ball is drawn, what is the probability that it is red?
4. An urn contains three red, four white, and seven green balls.

   (a) If one ball is drawn, find the probability that it is white.
   (b) Find the probability that it is not white.

5. (*See Example 3*) A pair of dice is rolled. Find the probability of rolling a 7.
6. Find the probability of rolling a 12 with a pair of dice.

7. (*See Example 4*) Two students are selected at random from a class of ten men and seven women. Find the probability two men are chosen.
8. An urn contains four red and five white balls. Two balls are drawn. Find the probability that both are red.
9. Two cards are drawn from a deck of cards. Find the probability that both are aces.
10. A card is drawn at random from a deck of bridge cards. Find the probability that it is

    (a) a three.
    (b) a ten or a jack.
    (c) not a diamond.

11. A bag contains four black marbles and five white marbles that are all exactly alike except in color. If one marble is drawn, find the probability that it will be

    (a) white.
    (b) red.
    (c) black or white.

12. Five red and three green books are placed at random on a shelf. Without looking, a person takes a book. Find the probability that it is a red book.
13. (*See Example 5*) Find the probability of at least one head appearing on two tosses of a coin.
14. Find the probability of at most one head appearing on three tosses of a coin.
15. (*See Example 6*) A class has twelve freshmen and nine sophomores. If three students are selected at random, find the probability that two are freshmen and one is a sophomore.
16. A box contains seven red balls and ten green balls. Three balls are drawn from the box. Find the probability that two are red and one is green.

II.

17. A tray contains fifteen electronic components, four of which are defective. If four components are selected, what is the probability that

    (a) all four are defective?
    (b) three are defective and one is good?
    (c) exactly two are defective?
    (d) none are defective?

18. Two people are selected at random from a group of twelve Republicans and ten Democrats. Find the probability that

    (a) both are Democrats.
    (b) one is a Republican and one is a Democrat.

19. The probability that the university football team will win is .6, and the probability that it will lose is .3. What is the probability of a tie?

III.

20. A teacher writes 15 problems on cards, one per card. The problems include 6 easy problems, 5 medium problems, and 4 hard problems. A student draws three problems for a quiz. Find the probability that all three are hard problems.

21. A child has nine cards numbered 1 through 9. The child places three cards in a row to form a three-digit number. Find the probability that the number is larger than 500.

22. Six children are selected from a group of 10 boys and 12 girls. Find the probability that half are boys and half are girls.

23. A shipment of 14 televisions contains 6 regular and 8 deluxe models. The manufacturer failed to mark the model designation on the cartons. If 4 cartons are selected at random, what is the probability that exactly 3 of them are the deluxe model?

24. Two bad light bulbs get mixed up with seven good ones. If you take two of the bulbs, what is the probability that both are bad?

25. A mathematics class is composed of twelve freshmen, 10 sophomores, and six juniors. Three freshmen, two sophomores, and one junior receive A grades in the course. If a student is selected at random from the class, find the probability that

(a) the student is an A student.

(b) the student is an A freshman student.

(c) the student is a sophomore.

26. Five thousand lottery tickets are sold. Jones buys five tickets. If one ticket is drawn for a $100 prize, what is the probability that Jones wins?

27. An urn contains six blue, five yellow, and three white balls. If a ball is drawn at random, find the probability that

(a) it is yellow.

(b) it is not yellow.

28. In a roll of 50 pennies there are 15 dated 1977, there are 18 with a Denver mint mark, and there are 6 with the Denver mint mark dated 1977. If a penny is drawn at random, what is the probability that it is dated 1977 or has a Denver mint mark?

29. A coin is bent, so heads and tails are not equally likely. The coin is tossed 500 times and heads occurs 400 times.

(a) Estimate the probability of a toss of the coin coming up heads.

(b) Estimate the probability of tails.

30. A die is loaded (weighted on one side so that it is not fair). It is rolled 300 times, and the following outcomes are observed:

1 occurs 21 times	2 occurs 35 times	3 occurs 67 times
4 occurs 51 times	5 occurs 54 times	6 occurs 72 times

Estimate the probability of rolling a

(a) 2. **(b)** 5. **(c)** 6.

31. For several years a mathematics teacher studied the relationship between ACT scores in mathematics and performance in calculus. She observed that of the students who scored above 28 on the ACT mathematics test, 324 made an A grade in calculus, and 246 made below an A in calculus. If an entering freshman with a score above 28 on the ACT mathematics test is chosen at random, what is the probability that the student will make an A grade in calculus?

32. A coin is flipped twice in succession. Use a tree diagram to determine the probability of its landing on heads the first time and tails the second time.

33. A coin is flipped four times in succession. Find the probability of

(a) three heads and one tail, order being unimportant.

(b) heads on the first three tosses and tails on the fourth toss.

34. A die is tossed twice. Find the probability of

(a) a 4 and a 5, order being unimportant.

(b) a 4 the first time and a 5 the second time.

(c) the sum of the numbers obtained being 6.

(d) the same number on both dice.

35. A box contains two black balls and two white balls. Three balls are drawn in succession without replacement. Construct a tree diagram to determine the probability of drawing

(a) two black balls and then a white ball.

(b) two black balls and a white ball, order being unimportant.

36. The probability of randomly selecting a female from a group of 50 people is .4. How many females are in the group?

37. Tommy has a jar containing 250 dimes. If a dime is selected at random, the probability of drawing a dime dated before 1960 is .2. How many dimes are dated before 1960?

38. An urn contains six green balls and balls of other colors. If a ball is selected at random, the probability of drawing a green ball is .3. How many balls are in the urn?

39. A car has six spark plugs, two of which are malfunctioning. If two of the plugs are replaced at random, what is the probability that both malfunctioning plugs are replaced?

40. Two children are selected in succession from a group of five children. Three of them are girls: Alice, Beth, and Cindy. Two of them are boys: Dan and Ed.
Draw a tree diagram and use it to compute the probability of

(a) selecting a girl first, then a boy.
(b) selecting a girl and a boy, order being unimportant.

6–3 Compound Events: Union, Intersection, and Complement

If we are performing an experiment which includes events E and F, we often want to form new events from E and F to gain information about another part of the experiment. These new events are called **compound events**. Three major compound events are formed using familiar set operations.

Compound Events

Let E and F be events in a sample space S.

1. The event $E \cup F$ is the event consisting of those outcomes that are in E or F or both.
2. The event $E \cap F$ is the event consisting of those outcomes that are in both E and F.
3. The event E' (complement of E) is the event consisting of those elements in the sample space that are not in E.

Example 1 (*Compare Exercise 1*)
Let the sample space $S = \{1, 2, 3, 4, 5, 6, 7, 8, 9, 10\}$. Let the event E be "the number is even." Then, $E = \{2, 4, 6, 8, 10\}$. Let the event F be "the number is prime." Then, $F = \{2, 3, 5, 7\}$. From these

$$E \cup F = \{2, 3, 4, 5, 6, 7, 8, 10\}$$
$$E \cap F = \{2\}$$
$$F' = \{1, 4, 6, 8, 9, 10\}$$
$$E' = \{1, 3, 5, 7, 9\}$$

See Figure 6–2.

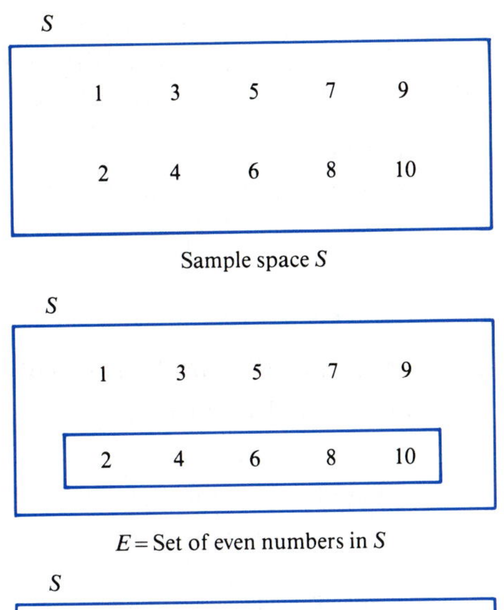

Sample space S

E = Set of even numbers in S

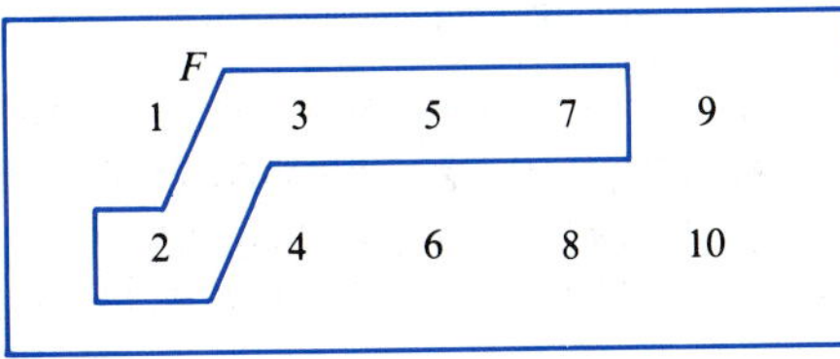

F = Set of prime numbers in S

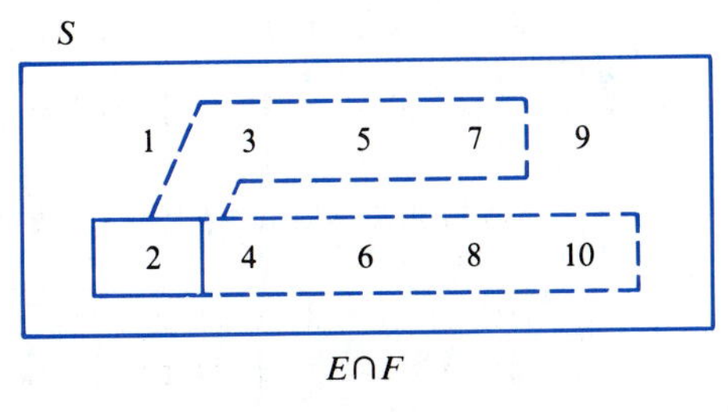

$E \cap F$

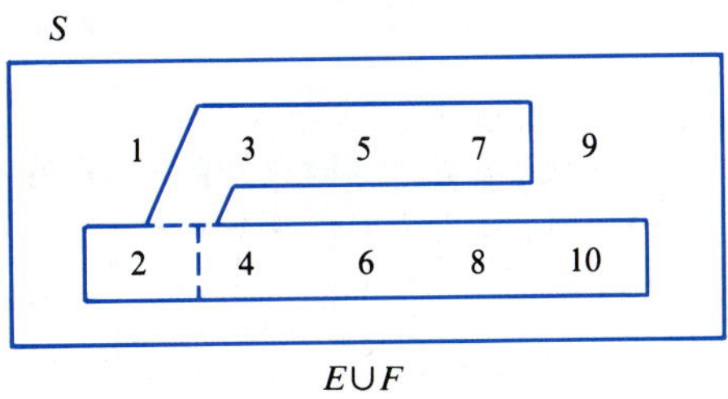

$E \cup F$

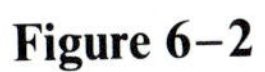

Figure 6–2

Example 2 (*Compare Exercise 2*)
For the experiment of selecting a student at random, let E be the event "the student is taking art," and let F be the event "the student is taking history." Then,

$E \cup F$ is the event "the student is taking art or history or both."

$E \cap F$ is the event "the student is taking both art and history."

E' is the event "the student is not taking art."

F' is the event "the student is not taking history."

Key Words to Compound Events

Let's emphasize the key words that are used in describing compound events. They will help in recognizing the kind of compound event.

The key word for describing and recognizing $E \cup F$ is the word "or."
The key word for describing and recognizing $E \cap F$ is the word "and."
The key word for describing and recognizing E' is the word "not."

Probability of Compound Events

It is often helpful to determine the probability of a compound event by using the probabilities of the individual events making up the compound event.

Probability of E'

You recognize that if 10% of a class receives an A grade, then 100% − 10% = 90% do not receive an A grade. If $\frac{2}{3}$ of a store's customers prefer brand A, then $1 - \frac{2}{3} = \frac{1}{3}$ of the customers do not prefer brand A. These statements are similar to the probability statement in Theorem 1.

Theorem 1 For an event E,

$$P(E') = 1 - P(E)$$
$$P(E) = 1 - P(E')$$

where E' is the complement of E in the sample space S.

Example 3 (*Compare Exercise 5*)
If the probability that Smith wins the door prize at a club meeting is 0.1, then the probability that Smith does not win is $1 - 0.1 = 0.9$.

If the probability that a part is defective is 0.08, then the probability that it is not defective is $1 - .08 = 0.92$.

If the probability that Jones does not get a promotion is 0.35, then the probability that Jones does get a promotion is 0.65.

You will encounter problems in which it may be difficult to find the probability of an event, but rather easy to find the probability of the event failing. Theorem 1 enables us to easily find the probability of the event. The following example illustrates this idea.

Example 4 (*Compare Exercise 31*)
A branch office of a corporation employs six women and five men. If four employees are selected at random to help open a new branch office, find the probability that at least one is a woman.

Solution To select at least one woman succeeds when one woman or two women or three women or four women are selected.

We count the number of successes by counting the number of ways we can select

one woman and three men,
two women and two men,
three women and one man,
four women,

and adding these four results. (*See Examples 16 and 17 in Section 5–4*). The number of successes is

$$C(6, 1)C(5, 3) + C(6, 2)C(5, 2) + C(6, 3)C(5, 1) + C(6, 4)$$

This quantity is divided by $C(11, 4)$ to obtain the probability of at least one woman being selected. You can carry out the above computation if you like, but let's look at an easier way.

The only way the company can *fail* to select at least one woman is to select *all* men. That probability is

$$\frac{C(5, 4)}{C(11, 4)} = \frac{1}{66}$$

By Theorem 1, the probability of success is one minus the probability of failure. So, the probability of at least one woman being selected is $1 - \frac{1}{66} = \frac{65}{66}$. (See Figure 6–3.)

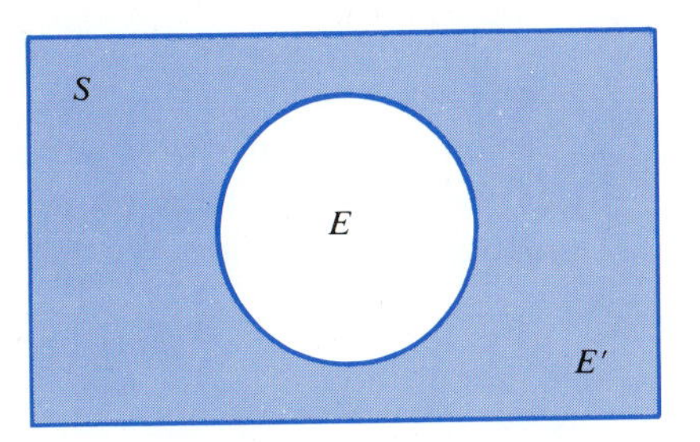

Figure 6–3 E' is the complement of E in S

Probability of $E \cup F$

In a group of 200 students, 40 take English, 50 take math, and 10 take both. If a student is selected at random, the following probabilities hold.

$$P(\text{English}) = \frac{\text{number taking English}}{\text{number in the group}} = \frac{40}{200}$$

$$P(\text{math}) = \frac{\text{number taking math}}{\text{number in the group}} = \frac{50}{200}$$

$$P(\text{English and math}) = \frac{\text{number taking English and math}}{\text{number in the group}} = \frac{10}{200}$$

$$P(\text{English or math}) = \frac{\text{number taking English or math}}{\text{number in the group}} = \frac{80}{200}$$

The number 80 in the last equation was obtained by using the formula from the chapter on sets:

$$n(E \cup F) = n(E) + n(F) - n(E \cap F)$$

If we let S be the group of 200 students, let E be those taking English, and let F be those taking math, then the above equations may be written

$$P(\text{English}) = \frac{n(E)}{n(S)} = \frac{40}{200}$$

$$P(\text{math}) = \frac{n(F)}{n(S)} = \frac{50}{200}$$

$$P(\text{English and math}) = \frac{n(E \cap F)}{n(S)} = \frac{10}{200}$$

$$\begin{aligned} P(\text{English or math}) &= \frac{n(E \cup F)}{n(S)} \\ &= \frac{n(E) + n(F) - n(E \cap F)}{n(S)} \\ &= \frac{40 + 50 - 10}{200} = \frac{80}{200} \end{aligned}$$

(See Figure 6–4.)

You may wonder why we seem to have complicated the last equation. We want to illustrate a basic property of probability, so we need to carry the last equation a little further. We may write

$$P(E \cup F) = \frac{40 + 50 - 10}{200}$$

as

$$P(E \cup F) = \frac{40}{200} + \frac{50}{200} - \frac{10}{200}$$

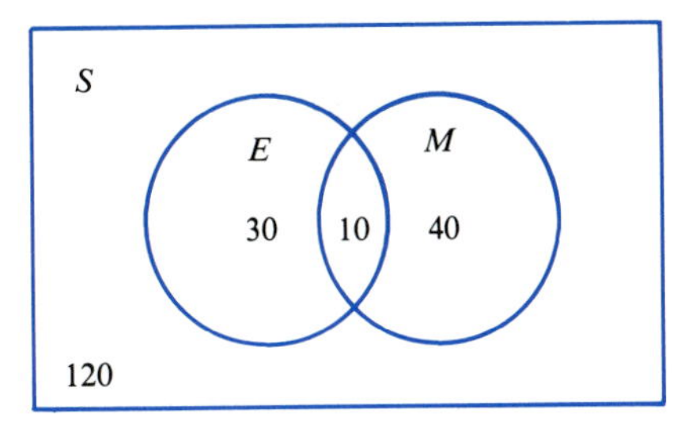

Figure 6–4 $n(S) = 200$, $n(E) = 40$, $n(M) = 50$, $n(E \cap M) = 10$, $n(E \cup M) = 80$

The right side of the last equation is

$$P(E) + P(F) - P(E \cap F)$$

This holds in general.

Theorem 2

> **(a)** $P(E \cup F) = P(E) + P(F) - P(E \cap F)$
> **(b)** If the outcomes are equally likely,
> $$P(E \cup F) = \frac{n(E) + n(F) - n(E \cap F)}{n(S)}$$

We point out that the above example assumes equally likely outcomes. However, Theorem 2(a) is true in other cases as well.

Example 5 (*Compare Exercise 35*)
In a remote jungle village the probability of a child contracting malaria is .45, the probability of contracting measles is .65, and the probability of contracting both is .20. What is the probability of a child contracting malaria or measles?

Solution By Theorem 2,

$$\begin{aligned} P(\text{malaria or measles}) &= P(\text{malaria}) + P(\text{measles}) - P(\text{malaria and measles}) \\ &= .45 + .65 - .20 = .90 \end{aligned}$$

Mutually Exclusive Events

Theorem 2 may not be helpful in some problems because we may not know how to compute $P(E \cap F)$. That will come in the next section. However, the special situation when E and F have no outcomes in common can be solved.

Definition 1

> Two events E and F are said to be **mutually exclusive** if they have no outcomes in common.
> When two events are mutually exclusive, an outcome in one is excluded from the other. As sets, E and F are **disjoint**.

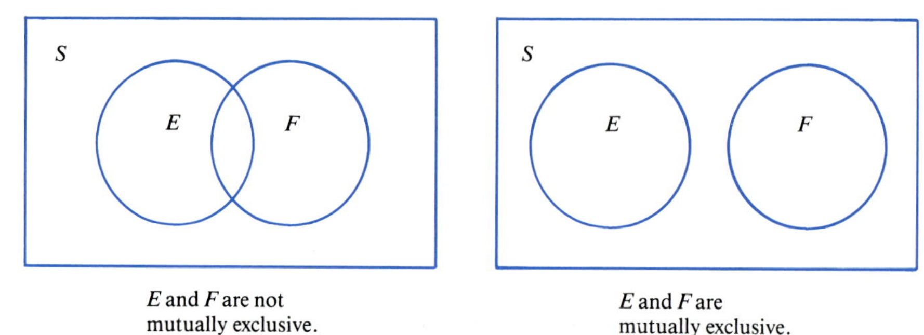

Figure 6–5 E and F are not mutually exclusive. E and F are mutually exclusive.

Example 6 (*Compare Exercises 11–15*)
When a coin is tossed, "heads" and "tails" are mutually exclusive because each one excludes the other.

Rolling a 7 with a pair of dice is mutually exclusive with rolling a 9 because they cannot occur at the same time.

Taking English and taking art are not mutually exclusive because they can both occur at the same time.

When two events are mutually exclusive, the computation of the probability of E or F simplifies because $E \cap F = \varnothing$. Recall that $P(\varnothing) = 0$, so when E and F are mutually exclusive,

$$P(E \cup F) = P(E) + P(F) - P(E \cap F)$$

becomes

$$P(E \cup F) = P(E) + P(F)$$

Theorem 3

> If E and F are mutually exclusive events, then
>
> $$P(E \cup F) = P(E) + P(F)$$

Example 7 (*Compare Exercise 16*)
Two people are selected at random from a group of seven men and five women. What is the probability that both are men or both are women?

Solution The events "both men" and "both women" are mutually exclusive because the existance of one excludes the other. Thus,

$$P(\text{both men or both women}) = P(\text{both men}) + P(\text{both women})$$
$$= \frac{C(7, 2)}{C(12, 2)} + \frac{C(5, 2)}{C(12, 2)}$$
$$= \frac{21}{66} + \frac{10}{66} = \frac{31}{66}$$

6–3 Exercises

I.

In Exercises 1 through 4, describe $E \cup F$, $E \cap F$, and E'.

1. (*See Example 1*) $S = \{1, 2, 3, 4, 5, 6, 7, 8, 9, 10\}$, $E = \{1, 3, 5, 7, 9\}$, and $F = \{1, 2, 3, 4\}$.
2. (*See Example 2*) S is the set of customers who shop at Harvey's Department Store. E is the set of customers who shop in the bargain basement. F is the set of customers who shop in the jewelry department.
3. S is the set of students at State U. E is the set of students passing English. F is the set of students not passing chemistry.
4. S is the set of employees of a major corporation. E is the set of executives. F is the set of female employees.
5. (*See Example 3*) The probability that Jack can work a problem is $\frac{3}{5}$. Find the probability that Jack cannot work the problem.
6. $P(E) = 0.6$. Find $P(E')$.
7. $P(E') = 0.7$. Find $P(E)$.
8. $P(E) = 0$. Find $P(E')$.
9. The probability that a person selected at random likes hamburgers at a local hamburger stand is 0.3. Find the probability that a person selected at random does not like the hamburgers.
10. The Doomsday Life Insurance Company made a study that indicated that a person who reaches 50 years of age has a probability of 0.23 of eventually dying of cancer. Find the probability of that person dying of some other cause.

Are events described in Exercises 11 through 15 mutually exclusive? (*See Example 6*)

11. E: selecting a fellow from a group of students
    F: selecting a girl from a group of students
12. E: drawing a king from a deck of cards
    F: drawing a face card from a deck of cards
13. E: selecting a student who is taking English
    F: selecting a student who is taking math
14. E: selecting a number that is a multiple of five
    F: selecting a number that is a multiple of seven
15. E: a person having a birthday in February
    F: a person having a birthday on the 30th of the month
16. (*See Example 7*) Two people are selected at random from a group of eight women and ten men. Find the probability that both are men or both are women.

17. Four cards are drawn at random from a deck of 52 playing cards.

    (a) Find the probability that they are all aces.
    (b) Find the probability that they are all aces or all jacks.

18. A bag contains six black marbles, three white marbles, and two red marbles. One marble is drawn. Find the probability of drawing

    (a) a black marble.
    (b) a white marble.
    (c) a red marble.
    (d) a white or a red marble.

19. A jeweler's bag contains six rings; one is plain gold, one is plain silver, one is gold with emeralds and rubies, one is silver with diamonds, one is gold with diamonds and emeralds, and one is silver with rubies. If a ring is selected at random, find the probability that

    (a) it is a gold ring.
    (b) it is a ring with diamonds.
    (c) it is a ring with diamonds or emeralds.
    (d) it does not have rubies.

20. A single card is drawn from a deck of 52 bridge cards. Find the probability that it is

    (a) a five.
    (b) a club.
    (c) a red spade.
    (d) an even number.
    (e) a nine or a ten.
    (f) a heart, diamond, or black card.

21. On a single toss of one die, find the probability of tossing

    (a) a number less than 6.
    (b) the number 4.
    (c) 2, 4, or 5.
    (d) an odd number less than 5.
    (e) 1 or 3.

22. A bag contains ten balls: four white, five black, and one red. Find the probability of getting, in one drawing,

    (a) a white ball.
    (b) a black ball.
    (c) a red ball.
    (d) a black or white ball.
    (e) a white or red ball.

23. A box contains ten balls numbered 1 through 10. One ball is selected at random. Find the probability of getting

    (a) number 10.
    (b) number 1.
    (c) an even number.

(d) an odd number.

(e) a number greater than 5.

(f) an even number or a number greater than 4.

(g) an even number or a number less than 8.

24. A card is drawn at random from a deck of 52 playing cards. Find the probability of drawing

(a) an ace or a 7.

(b) a 6, a 7, or an 8.

25. An urn contains two red, one white, three black, and four green balls. One ball is drawn. Find the probability that it is white or black.

II.

Determine whether the events given in Exercises 26 through 29 are mutually exclusive.

26. A person is selected from the customers at Fast Foods. F is the event that a female is selected, and G is the event that a child is selected.

27. Consider the experiment of selecting a student from a college. F is the event of the student attending on a student visa, and G is the event of the student being an American citizen.

28. A child is selected from a group of children who were born prematurely. F is the event that the child suffers some loss of hearing, and G is the event that the child suffers some loss of vision.

29. A coin is tossed three times. F is the event that it lands heads up on the second toss, G is the event that it lands tails up on the second toss.

30. Two dice are thrown. What is the probability that the same number appears on each die?

31. (*See Example 4*) Three people are selected at random from five women and seven men.

(a) Find the probability that at least one man is selected.

(b) Find the probability that at most two men are selected.

32. A coin is tossed four times. Find the probability that heads turn up at least once.

33. A new apartment complex advertises that they will give away three mopeds by a drawing from the first 50 students that sign a lease. Four friends are among the first 50 to sign. What is the probability at least one of them will win a moped?

34. A card is drawn from a deck of 52. Find the probability that it is an ace or a spade.

35. (*See Example 5*) Of 400 college students, 120 are enrolled in math, 220 are enrolled in English, and 55 are enrolled in both. If a student is selected at random, find the probability that

(a) the student is enrolled in mathematics.
(b) the student is enrolled in English.
(c) the student is enrolled in both.
(d) the student is enrolled in mathematics or English.

36. A survey of 100 college faculty shows that 70 own a car, 30 own a bicycle, and 10 own both. Find the probability that a faculty member chosen at random owns a car or a bicycle.

37. In a group of 35 girls, 10 have blonde hair, 14 have brown eyes, and 4 have both blonde hair and brown eyes. If a girl is selected at random,

(a) find the probability that she has blonde hair and brown eyes.
(b) find the probability that she has blonde hair or brown eyes.

38. A person is known to have his birthday in April.

(a) What is the probability that it is April 15th?
(b) What is the probability that it is in the first seven days of April?
(c) What is the probability that it is in the first half of April?

39. The longest place name that exists in the entire world is

Llanfairpwllgwyngyllgogerychwyrndrobwllllantysiliogogogoch.

It is a small village in North Wales. The name means "St. Mary's Church, in a hollow, by the white hazel, near the fierce whirlpool, close to the Church of St. Tysilio, near to the red cave." (The place with the shortest name is a French village called Y.) A letter is selected at random from the word. Determine the probability that it is

(a) g.
(b) l.
(c) o or l.
(d) one of the first five letters of the alphabet.
(e) an English vowel.
(f) a Welsh vowel (they are a, e, i, o, u, w, y).

40. A person is selected at random from a pool of ten people consisting of six men and four women for a psychology study. Find the probability that the person selected is

(a) a man. **(b)** a woman.

41. A letter is selected at random from the word MISSISSIPPI. Find the probability that the letter is

(a) s **(b)** p **(c)** i
(d) a **(e)** m **(f)** s, p, or i

III.

42. From a group of four people, two are to be selected for two awards, a first and second price. Label these people A, B, C, D. Construct a tree diagram for this experiment.

(a) Find the probability of A receiving first prize and B second.
(b) Find the probability that D will get an award.
(c) Find the probability that A will get first prize.
(d) Find the probability that A and B will get awards.

43. Consider the set consisting of three letters $\{o, u, y\}$. What is the probability that a child who cannot read will arrange these letters to form the word "you?"

44. A slate of officers consisting of a president, a vice president, a secretary, and a treasurer is to be selected at random from ten people. Find the probability that Mr. Jones will be president and that Mr. Morris will be vice president, both of whom are members of the group of ten.

45. Eight horses are in a race. Find the probability that Regent will come in first and Conway second, given that both horses are in the race.

46. A hand of 13 cards is dealt from a deck of 52. Find the probability of getting

(a) all four jacks.
(b) all red cards.
(c) all spades.

47. There are 12 signs, or houses, in the Zodiac (astrological chart of the heavens): Capricorn, Aquarius, Pisces, Aries, Taurus, Gemini, Cancer, Leo, Virgo, Libra, Scorpio, and Sagittarius. Assuming all signs to be equally likely, find the probability that

(a) a person selected at random will be a Sagittarian.
(b) two persons selected at random will both be Sagittarians.
(c) two persons selected at random were both born under the same sign.

48. A die is rolled. What is the probability that the number is odd or is a 2?

49. A die is rolled. Use a tree diagram to find the probability of rolling

(a) an even number less than 6.
(b) an odd number less than 5.

50. A coin is tossed and a die is rolled. Find the probability of obtaining

(a) heads and a 4.
(b) heads or a 4.
(c) tails and an even number.
(d) tails or a number less than or equal to 5.

51. A family has four children. Find the probability that there are

(a) either four boys or four girls.
(b) at least one boy and one girl.

52. Two types of stocks, A and B, are owned by an individual. During the previous 30-days of trading, both A and B increased in value on 12 of these days. On ten days A increased while B decreased. On the remaining eight days A decreased while B increased in value. Let E be the event that A increases in value, F that A decreases, G that B increases, and H that B decreases. On the basis of these statistics, compute and interpret

(a) $P(F)$, $P(H)$.
(b) $P(E \cap H)$.
(c) $P(F' \cup G)$.

53. What is the probability that the next person you meet was born on a Saturday?

54. A four-digit number is to be constructed at random from the numbers 1, 2, 3, 4, 5, and 6, with repetition of digits allowed. Find the probability of getting

(a) the number 1234. **(b)** a number less than 3333.

6–4 Conditional Probability

Conditional Probability
Multiplication Rule
Independent Events

Conditional Probability

We compute the probability of an event E based on the outcomes in the sample space and in the event E. If a related event F occurs, it might provide additional information that allows us to adjust the probability of E.

For example, suppose you are taking a test with multiple-choice questions. A question has four possible answers listed, and you have no idea of the correct answer. If you make a wild guess at the answer, the probability of guessing the correct answer is $\frac{1}{4}$. However, suppose you know one of the given answers cannot be correct. This improves your chances of guessing the right answer, and the probability becomes $\frac{1}{3}$.

In general terms, this example is described as follows. The probability of an event E is sought. In case a related event F has occurred, giving reason to change the sample space and consequently the probability of E, we determine what is called **conditional probability** of E given that the event F has occurred. We denote this by $P(E \mid F)$, which we read "the probability of E given F."

We can state the above example as: "The probability of guessing the correct answer given one answer is known incorrect is $\frac{1}{3}$."

If a student guesses wildly at the correct answer from the four given ones, the sample space consists of the four possible answers. When one answer is ruled out, the sample space reduces to three possible answers. Sometimes it helps to look at a conditional probability problem as one where the sample space changes when certain conditions exist or related information is given. Let's look at the following from that viewpoint.

A student has a job testing microcomputer chips. The chips are produced by two machines, I and II. It is known that 5% of the chips produced by machine I are defective and 15% of the chips produced by machine II are defective. The student has a batch of chips that she assumes is a mixture from both machines. If she selects one at random, what is the probability that it is defective? You cannot give a precise answer to this question unless the proportion of chips from each machine is known. It does seem reasonable to say that the probability lies in the interval from 0.05 through 0.15.

Now suppose the student is given more information; the chips are all from machine II. This certainly changes her estimate of the probability of a defective chip, she knows the probability is 0.15. The sample space changes from a set of chips from both machines to a set of chips from machine II. This illustrates the point that when you gain information about the state of the experiment, you may be able to change the probabilities assigned to the outcomes.

Here's how we write some of the information from the above examples.

Example 1 (*Compare Exercise 1*)

(a) Machines I and II produce microchips, with 5% of those from machine I being defective and 15% of those from Machine II being defective. This can be stated

$$P(\text{defective chip} \mid \text{machine I}) = .05$$
$$P(\text{defective chip} \mid \text{machine II}) = .15$$

(b) There are four possible answers to a multiple-choice question, one of which is correct. The probability of guessing the right answer is $\frac{1}{4}$. However, if one incorrect answer can be eliminated, the sample space is reduced from four to three answers, and the probability of guessing correctly becomes $\frac{1}{3}$. This is stated

$$P(\text{guessing correct answer}) = \frac{1}{4}$$

$$P(\text{guessing correct answer}) \text{ one incorrect answer eliminated}) = \frac{1}{3}$$

Let's look at an example of how to compute $P(E \mid F)$.

Example 2 (*Compare Exercise 2*)
Professor X has two sections of philosophy. The regular section has 35 students, and the honors section has 25 students. He gives both sections the same test and 14 students make an A, five in the regular section and nine in the honors section.

(a) If a test paper is selected at random, what is the probability that it is an A paper?

(b) A test paper is selected at random. If it is known that the paper is from the honors section, what is the probability that it is an A paper?

Solution The sample space S is the collection of all 60 papers, the event E is the set of all A papers, and F is the set of papers from the honors class.

(a) $P(E) = \dfrac{14}{60}$

(b) The knowledge that the paper is from the honors section reduces the sample space to those 25 papers, the event F. The problem is to find the probability of an A paper, given that it is from the honors section, that is, find $P(E|F)$. Since the honors section contains 9 A papers and 25 papers total,

$$P(E|F) = \frac{9}{25}$$

(See Figure 6–6.)

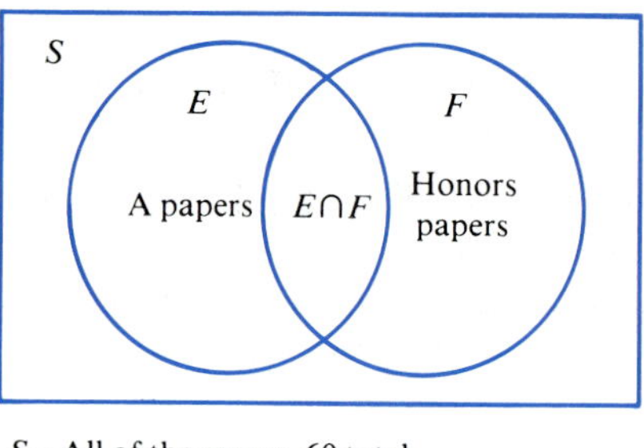

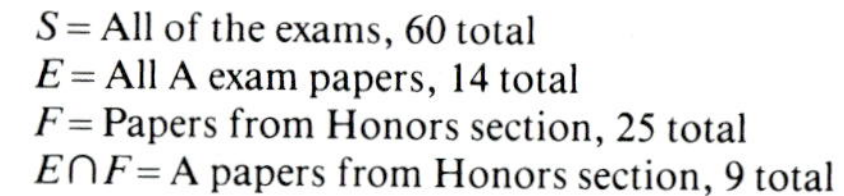

S = All of the exams, 60 total
E = All A exam papers, 14 total
F = Papers from Honors section, 25 total
$E \cap F$ = A papers from Honors section, 9 total

Figure 6–6

We now express the above results in a different form to make a general statement on conditional probability. The conditional probability from this example can be written

$$P(E|F) = \frac{9}{25} = \frac{n(E \cap F)}{n(F)}$$

If we divide the numerator and the denominator of the last fraction by $n(S)$, we have

$$P(E|F) = \frac{\dfrac{n(E \cap F)}{n(S)}}{\dfrac{n(F)}{n(S)}} = \frac{P(E \cap F)}{P(F)}$$

Be sure you understand that, in essence, F becomes the sample space when we compute $P(E \mid F)$, and we focus our attention on the contents of F. F is sometimes called the **reduced sample space**. The definition of conditional probability is similar to the results of the last example.

Rule for Computing $P(E \mid F)$

E and F are events in a sample space S, with $P(F) \neq 0$. The conditional probability of E given F, denoted by $P(E \mid F)$, is

(a)
$$P(E \mid F) = \frac{P(E \cap F)}{P(F)}$$

This holds whether or not the outcomes of S are equally likely.

(b) If the outcomes of S are equally likely, $P(E \mid F)$ may be written

$$P(E \mid F) = \frac{n(E \cap F)}{n(F)}$$

Example 3 (*Compare Exercise 8*)
In a group of 200 students, 40 are taking English, 50 are taking mathematics, and 12 are taking both.

(a) If a student is selected at random, what is the probability that the student is taking English?

(b) A student is selected at random from those taking mathematics. What is the probability that the student is taking English?

(c) A student is selected at random from those taking English. What is the probability that the student is taking mathematics?

Solution **(a)** $P(\text{English}) = \frac{40}{200} = \frac{1}{5}$

(b) This problem is that of finding $P(\text{English} \mid \text{math})$, so

$$P(\text{English} \mid \text{math}) = \frac{n(\text{English and math})}{n(\text{math})} = \frac{12}{50} = \frac{6}{25} = .24$$

This may also be expressed in terms of probability:

$$P(\text{English} \mid \text{math}) = \frac{P(\text{English and math})}{P(\text{math})} = \frac{\frac{12}{200}}{\frac{50}{200}} = \frac{.06}{.25} = .24$$

(c) This seeks

$$P(\text{Math} \mid \text{English}) = \frac{n(\text{Math and English})}{n(\text{English})} = \frac{12}{40} = \frac{3}{10}$$

Parts (b) and (c) illustrate that $P(E \mid F)$ and $P(F \mid E)$ may not be equal.

Multiplication Rule

We obtain a useful formula for the probability of E and F by multiplying the rule for conditional probability throughout by $P(F)$.

Theorem 3 *Multiplication Rule*

E and F are events in a sample space S.

$$P(E \cap F) = P(E)P(F \mid E)$$

This may also be written as

$$P(E \cap F) = P(F)P(E \mid F)$$

This theorem states that we can find the probability of E and F by multiplying the probability of E by the conditional probability of F given E.

Example 4 (*Compare Exercise 9*)
Two cards are drawn from a bridge deck, without replacement. What is the probability that the first is an ace and the second is a king?

Solution According to Theorem 3, we want to find

$$P(\text{Ace first and king second}) = P(\text{Ace first}) \times P(\text{King second} \mid \text{ace first})$$

Since the first card is drawn from the full deck of 52 cards,

$$P(\text{Ace first}) = \frac{4}{52}$$

The first card is not replaced, so the sample space for the second card is reduced to 51 cards. Since we are assuming that the first card was an ace, there are still four kings in the deck. Then,

$$P(\text{King second} \mid \text{ace first}) = \frac{4}{51}$$

It then follows that

$$P(\text{Ace first and king second}) = \left(\frac{4}{52}\right) \times \left(\frac{4}{51}\right) = \frac{4}{663}$$

Example 5 (*Compare Exercise 12*)
A box contains 12 light bulbs, three of which are defective. If three bulbs are selected at random, what is the probability that all three are defective?

Solution According to the Multiplication Rule, applied twice,

$$P(\text{First defective and second defective and third defective})$$
$$= \left(\frac{3}{12}\right)\left(\frac{2}{11}\right)\left(\frac{1}{10}\right) = \frac{1}{220}$$

Independent Events

Sometimes two events, E and F, are related in a way so that the occurrence of one in no way affects the probability of the other. For example, the first toss of a coin has no influence on the outcome of the second toss. If one person selects a card from one deck, and another person selects a card from a second deck, then the card drawn from the first deck has no effect on the outcome of the card drawn from the second deck and vice versa.

If neither of two events, E and F, affects the probability of the occurrence of the other, we intuitively expect that $P(E|F) = P(E)$, since the probability of E is unchanged whether or not we are given F. Similarly, $P(F|E) = P(F)$. This leads to the following definition of independence.

Definition 5

The events E and F are independent if

$$P(E|F) = P(E) \qquad \text{or} \qquad P(F|E) = P(F)$$

Otherwise, E and F are dependent.

Note: If $P(E|F) = P(E)$, then $P(F|E) = P(F)$ also holds and vice versa.

Example 6 (*Compare Exercise 28*)

From the Venn diagram in Figure 6–7, the sample space contains 100 elements, E contains 25, F contains 35, and $E \cap F$ contains 5. Then the following probabilities hold.

$$P(E) = \frac{25}{100} = .25$$

$$P(F) = \frac{35}{100} = .35$$

$$P(E|F) = \frac{5}{35} = .1429$$

$$P(F|E) = \frac{5}{25} = .20$$

Since $P(E) \neq P(E|F)$, E and F are dependent. We could reach the same conclusion by the observation $P(F) \neq P(F|E)$.

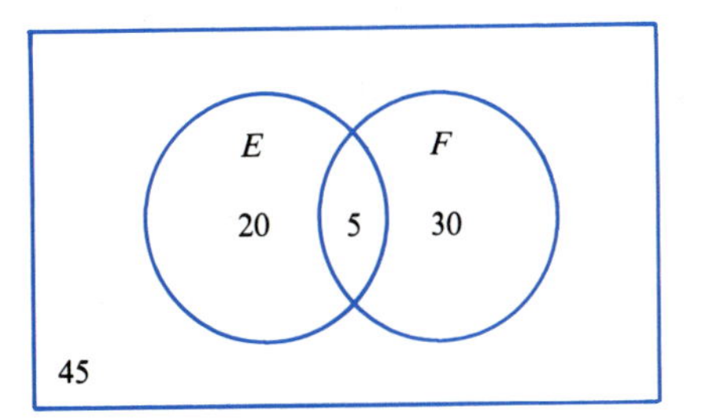

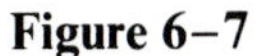

Figure 6–7

When E and F are independent events, the Multiplication Rule simplifies to the following.

Theorem 4

If E and F are **independent** events, then

$$P(E \cap F) = P(E)P(F)$$

We can also use Theorem 4 to determine if two events are independent. For the events in Example 6,

$$P(E \cap F) = \frac{5}{100} = .05$$

$$P(E)P(F) = (.25)(.35) = .0875$$

Since $P(E \cap F) \neq P(E)P(F)$, E and F are not independent—the same conclusion we reached by using Definition 5.

While it may be intuitively clear that two tosses of a coin are independent, in other situations independence may be determined only by using Definition 5 or Theorem 4 (as in Example 6). Actually, any one of the three conditions provides a test for independence.

Test for Independence of Events

If any one of the following holds, then events E and F are independent.

1. $P(E \mid F) = P(E)$
2. $P(F \mid E) = P(F)$
3. $P(E \cap F) = P(E)P(F)$

Example 7 (*Compare Exercise 29*)

A card is selected at random from the following four cards: 10 of diamonds, 10 of spades, 8 of hearts, and 6 of clubs. Let E be the event "select a red card," and let F be the event "select a 10."

(a) Are E and F mutually exclusive?

(b) Are E and F independent?

Solution **(a)** E and F are not mutually exclusive because selecting a red card and selecting a 10 can occur at the same time; for example, selecting the 10 of diamonds.

(b) From the information given,

$$P(E) = P(\text{red card}) = \frac{1}{2}$$

$$P(F) = P(10) = \frac{1}{2}$$

$$P(E \cap F) = P(\text{red } 10) = \frac{1}{4}$$

Since $P(E \cap F) = P(E)P(F) = \frac{1}{4}$, E and F are independent.

This example illustrates two points.

1. Two independent events need not be mutually exclusive. In fact, mutually exclusive events are generally dependent. (*See Exercise 60*)
2. Sometimes the only way to determine if events are independent is to perform the computation in the tests for independence.

We can now compute the compound probability, $P(E \cap F)$, by using Theorems 3 and 4; use Theorem 3 for dependent events and Theorem 4 for independent events.

Example 8 (*Compare Exercise 18*)
One bag contains four red and five white balls. A second bag contains three green and seven yellow balls. If a ball is drawn from each bag, what is the probability that the first is white and the second is green?

Solution Since the balls are drawn from different bags, the two events are independent, so

$$P(\text{first white and second green}) = \left(\frac{5}{9}\right) \times \left(\frac{3}{10}\right) = \frac{1}{6}$$

Example 9 (*Compare Exercise 22*)
Jack and Jill work on a problem independently. The probability that Jack solves it is $\frac{2}{3}$ and the probability that Jill solves it is $\frac{4}{5}$.

(a) What is the probability that both solve it?
(b) What is the probability that neither solves it?
(c) What is the probability that exactly one of them solves it?

Solution **(a)** The probability that both solve the problem is

$$P(\text{Jack solves it and Jill solves it}) = \left(\frac{2}{3}\right) \times \left(\frac{4}{5}\right)$$
$$= \frac{8}{15}$$

(b) The probability that Jack does not solve the problem is $1 - \frac{2}{3} = \frac{1}{3}$, and the probability that Jill does not is $1 - \frac{4}{5} = \frac{1}{5}$. Then

$$P(\text{Jack doesn't and Jill doesn't}) = \left(\frac{1}{3}\right) \times \left(\frac{1}{5}\right) = \frac{1}{15}$$

(c) There are two ways one of them will solve the problem, namely, Jack does and Jill doesn't, or Jack doesn't and Jill does. These two ways are mutually exclusive events, so we need to compute the probability of each of these outcomes and then add.

$$P(\text{Jack does and Jill doesn't}) = \left(\frac{2}{3}\right) \times \left(\frac{1}{5}\right) = \frac{2}{15}$$

$$P(\text{Jack doesn't and Jill does}) = \left(\frac{1}{3}\right) \times \left(\frac{4}{5}\right) = \frac{4}{15}$$

Then,

$$P(\text{one of them solves the problem}) = \frac{2}{15} + \frac{4}{15} = \frac{6}{15} = \frac{2}{5}$$

Recall that $P(E \cup F) = P(E) + P(F) - P(E \cap F)$. When E and F are independent, $P(E \cap F) = P(E)P(F)$, so the above gives the following theorem.

Theorem 5

When E and F are independent,

$$P(E \cup F) = P(E) + P(F) - P(E)P(F)$$

Example 10 (*Compare Exercise 25*)
Jack and Jill work on a problem independently. The probability that Jack solves it is $\frac{2}{3}$, and the probability that Jill solves it is $\frac{4}{5}$. What is the probability that at least one of them solves it?

Solution We will show you two ways to work this problem.

Case I. Using mutually exclusive events.
Case II. Using Theorem 5.

Case I At least one of them solving the problem is equivalent to exactly one solving the problem (E), *or* both solving the problem (F). These events, E and F, are mutually exclusive, so

$$P(E \cup F) = P(E) + P(F).$$

$$P(E) = \frac{2}{5} \text{ from part (c) of Example 9}$$

$$P(F) = \frac{8}{15} \text{ from part (a) of Example 9}$$

so

$$P(\text{at least one}) = P(E \cup F) = \frac{2}{5} + \frac{8}{15} = \frac{14}{15}$$

Case II If we let A = Jack solves the problem and B = Jill solves the problem, then $A \cup B$ is one or both (at least one) solves the problem. From Theorem 5,

$$\begin{aligned} P(A \cup B) &= P(A) + P(B) - P(A)P(B) \\ &= \frac{2}{3} + \frac{4}{5} - \frac{2}{3} \times \frac{4}{5} \\ &= \frac{10 + 12 - 8}{15} \\ &= \frac{14}{15} \end{aligned}$$

Let's illustrate some of these concepts using a tree diagram.

Example 11 (*Compare Exercise 41*)
The city council has money for one public service project, a recreation-sports complex, a performing arts center, or a branch library. They polled 200 citizens for their preference, 120 men and 80 women. The men responded as follows: 45% preferred the recreation-sports complex, 20% the performing arts center, and 35% the branch library. The women responded as follows: 15% favored the recreation-sports complex, 40% the performing arts center, and 45% the branch library.

(a) Represent this information on a tree diagram.

(b) What is the probability that a person selected at random prefers the performing arts center?

(c) What is the probability that a person selected at random is a woman who prefers the performing arts center or the branch library?

(d) Are the events "male" and "prefers the recreation-sports complex" independent?

Solution Use the following abbreviations: M for male and F for female; RS, PA, and BL for recreation-sports complex, performing arts center, and branch library, respectively.

The information provided gives the following probabilities.

$$P(M) = .6,\ P(F) = .4$$

$$P(RS \mid M) = .45,\ P(PA \mid M) = .20,\ P(BL \mid M) = .35$$
$$P(RS \mid F) = .15,\ P(PA \mid F) = .40,\ P(BL \mid F) = .45$$

(a) Figure 6–8 shows the tree diagram with this information.

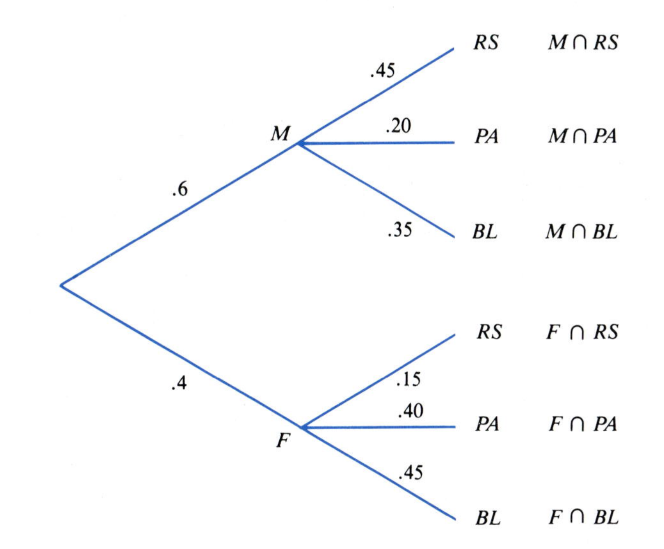

Figure 6–8

(b) The two branches that terminate at PA are $M \cap PA$ and $F \cap PA$, so

$$P(PA) = .12 + .16 = .28$$

(c) The branches $F \cap PA$ and $F \cap BL$ are the two outcomes in this event, so

$$P(F \text{ who prefers } PA \text{ or } BL) = .16 + .18 = .34$$

(d) If the events "male" and "prefers the recreation-sports complex" are independent, then $P(RS \mid M)$ must equal $P(RS)$. Since $P(RS \mid M) = .45$ and $P(RS) = .33(.27 + .06)$, the events are not independent.

6–4 Exercises

1. (*See Example 1*) The probability that Jane will solve a problem is $\frac{3}{4}$, and the probability that Jill will solve the problem is $\frac{1}{2}$.

   (a) What is $P(\text{Problem will be solved} \mid \text{Jane})$?

   (b) What is $P(\text{Problem will be solved} \mid \text{Jill})$?

2. (*See Example 2*) A card is drawn from a deck of 52 playing cards. Determine the probability of its being

   (a) a 10, given that it is not a face card (jack, queen, or king).

   (b) a king, given that it is a red card.

   (c) a 4, given that it is an even number.

3. A ball is selected from an urn containing six red balls, four white balls, and three black balls. Determine the probability of its being

   (a) a red ball, given that it is either red or white.

   (b) a black ball, given that it is not red.

4. Two dice are tossed. Determine the probability that

   (a) their sum is 4, given that one face shows a 1.

   (b) their sum is 7, given that one face is less than 5.

5. A class has 15 boys and 10 girls. One student is selected; F is the event of selecting a girl, and K is the event of selecting Kate, one of the girls in the class. Determine $P(K \mid F)$ and $P(F \mid K)$.

6. A card is drawn from a deck of 52 playing cards. Find the probability that the card will be a king, given that it is a face card. (The face cards are the jacks, queens, and kings.)

7. Two dice are tossed. What is the probability that their sum will be 6, given that one face shows a 2?

8. (*See Example 3*) The table below gives a breakdown of the student population during a certain semester at a college.

	Freshmen	Sophomores	Juniors	Seniors	Total
Male	506	453	412	380	1751
Female	550	434	408	376	1768
Total	1056	887	820	756	3519

A student is selected at random from the student body. Find the probability of that student being

(a) male.
(b) female.
(c) a freshman.
(d) a senior.
(e) a female freshman.
(f) a male senior.
(g) female, given that she is a sophomore.
(h) male, given that he is a sophomore.
(i) a senior, given that he is a male.
(j) a junior, given that she is a female.
(k) female, given that she is a junior or a senior.
(l) male, given that he is a freshman or sophomore.

9. (*See Example 4*) Two cards are drawn from a bridge deck without replacement. Find the probability that the first is a 4 and the second is a 5.
10. Two cards are drawn at random from a deck of playing cards without replacement. Find the probability that the first is a king and the second is a queen.
11. A card is drawn at random from a deck of playing cards. It is replaced and another card is drawn at random.

    (a) Find the probability that the first card is an 8 and the second is a 10.
    (b) Find the probability that both are clubs.
    (c) Find the probability that both are red cards.

12. (*See Example 5*) A person draws three balls in succession from a box containing four red, two white, and six blue balls. Find the probability that the balls drawn are red, white, and blue, in that order.
13. A die is tossed twice. Find the probability of getting an even number on the first throw and a number greater than four on the second.

14. A box contains three black balls and two white balls. Two balls are drawn in succession; the ball is replaced after each drawing. Determine the probability of first drawing a black ball and then a white ball.

15. A box contains two white balls and one red ball. Two balls are drawn in succession without replacement. Find the probability of drawing

(a) two white balls.
(b) a white ball followed by a red ball.
(c) a red ball and a white ball, order being unimportant.

16. A die is tossed five times. Find the probability of its landing on four the first three times and on an even number the last two times.

17. Consider the experiment of tossing a coin followed by the experiment of tossing a die. Find the probability of throwing heads on the coin followed by throwing a number greater than four on the die.

18. (*See Example 8*) Box A contains eight \$1 bills and two \$10 bills. Box B contains five \$1 bills, four \$5 bills, and one \$10 bill. If a bill is drawn at random from each box, find the probability that

(a) no \$10 bills are drawn.
(b) exactly one \$10 bill is drawn.
(c) at least one \$10 bill is drawn.

19. A box contains five balls numbered 1 through 5. Three balls are selected in succession without replacement. Find the probability of getting

(a) numbers 1, 2, and 3 in that order.
(b) numbers 1, 2, and 3 in any order.
(c) three balls, the sum of whose numbers is 4.
(d) three balls, the sum of whose numbers is 6.
(e) three odd-numbered balls.
(f) two odd-numbered balls and one even-numbered ball.

20. A freshman entering college has $\frac{1}{4}$ probability of getting married while at college and $\frac{7}{10}$ probability of graduating. Assuming that the events of marrying and graduating are independent, find the probability of

(a) graduating single.
(b) graduating married.
(c) getting married and not graduating.

21. (*See Example 10*) In each of the following, E and F are independent. Find $P(E \cup F)$.

(a) $P(E) = .3$, $P(F) = .5$
(b) $P(E) = .2$, $P(F) = .6$
(c) $P(E) = .4$, $P(F) = .6$

22. (*See Example 9*) The probabilities that Jack and Jill can solve a homework problem are $\frac{3}{5}$ and $\frac{2}{3}$, respectively. Find the probability that

(a) both will solve the problem.

(b) neither will solve the problem.

(c) Jack will and Jill won't solve the problem.

(d) Jack won't and Jill will solve the problem.

(e) exactly one of them will solve the problem.

(f) at least one of them will solve the problem.

23. The probabilities that two students will not show up for class on a beautiful spring day are .2 and .3, respectively. It is a beautiful spring day.

(a) Find the probability that neither will show up for class.

(b) Find the probability that both will show up for class.

(c) Find the probability that exactly one will show up for class.

(d) Find the probability that at least one will show up for class.

24. The probabilities that three students will not show up for class on a cold winter day are .1, .2, and .4. On a cold winter day, what is the probability that

(a) none of the three will show up for class?

(b) all three will show up for class?

(c) exactly one of the three will show up for class?

25. Three people are shooting at a target. The probabilities that they hit the target are .5, .6, and .8.

(a) Find the probability that all three hit the target.

(b) Find the probability that all three miss the target.

(c) Find the probability that at least one hits the target.

(d) Find the probability that at least one misses the target.

26. A coin is tossed six times. Find the probability that all six tosses will land on heads.

27. A child goes to an ice cream shop with just enough money to buy one dip of ice cream. The probability that he buys vanilla is $\frac{1}{4}$, the probability that he buys chocolate is $\frac{1}{3}$, and the probability that he buys peach is $\frac{1}{6}$.

(a) Find the probability that he buys none of the three.

(b) Find the probability that he buys chocolate or peach.

28. (*See Example 6*) Each of the Venn diagrams in Figure 6–9 shows the number of elements in each region determined by the two events E and F and the sample space S.

In each case, determine

(a) if the events are mutually exclusive.

(b) if the events are independent.

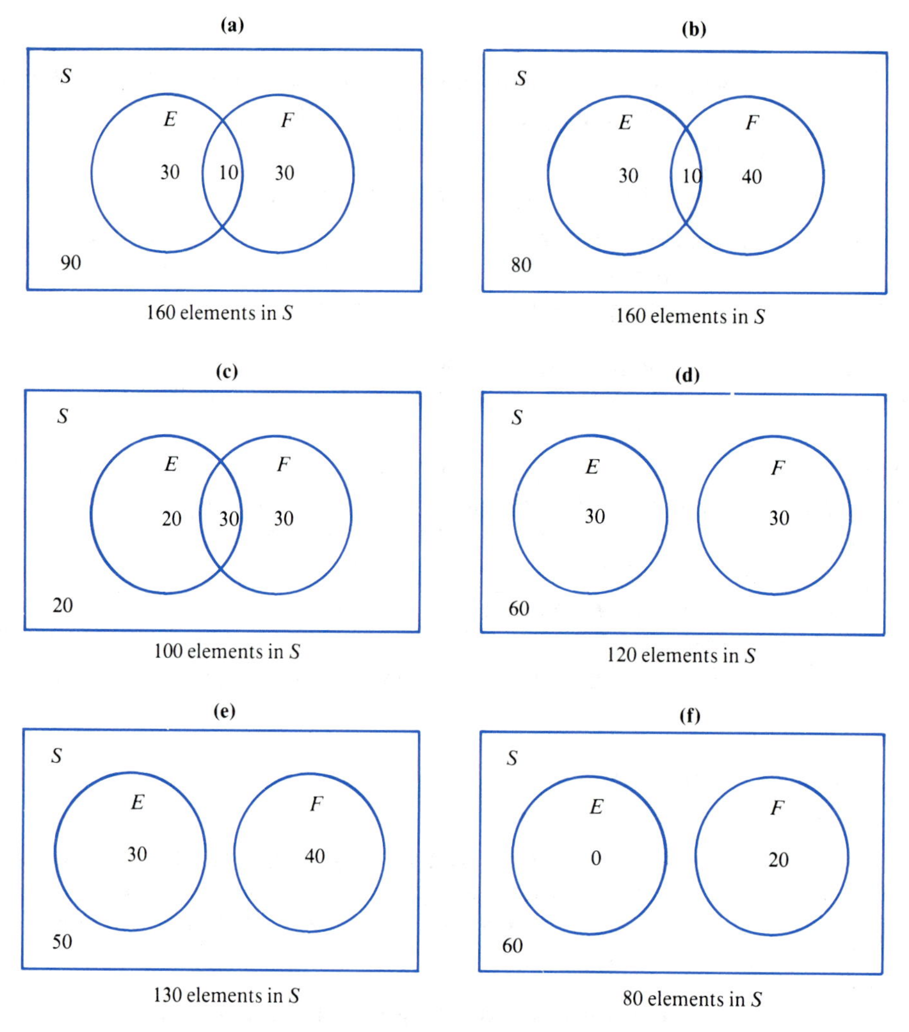

Figure 6–9

29. (*See Example 7*) A jewel box contains four rings;

one has a diamond and an emerald,
one has a diamond and a ruby,
one has a ruby and an emerald,
one has pearls.

A ring is selected at random from the box. Let the event E be "a ring has a diamond," and let the event F be "a ring has a ruby."

(a) Are the events E and F mutually exclusive?
(b) Are the events E and F independent?

30. Streams near industrial plants may suffer chemical or thermal pollution from waste water released in the stream. An environmental task force estimates that 6% of the streams suffer both types of pollution, 40% suffer chemical pollution, and 30% suffer thermal pollution. Determine if chemical pollution and thermal pollution are independent.

31. Let E and F be events such that $P(E) = .4$, $P(F) = .6$, and $P(E \cap F) = .3$.

(a) Find $P(E \cup F)$ and $P(E|F)$.

(b) Are E and F independent?

32. Let E and F be events such that $P(E) = .6$, $P(F) = .5$, and $P(E \cup F) = .8$.

(a) Find $P(E \cap F)$.

(b) Find $P(E|F)$.

(c) Are E and F independent?

33. A car dealer believes that if a customer comes into the showroom, the probability of selling the customer a car is $\frac{1}{10}$. If the dealer can get the customer to sit behind the wheel of a car, the probability of making a sale is $\frac{1}{3}$.

(a) Find the probability of not making a sale to a customer who comes into the showroom.

(b) Find the probability of not making a sale to someone who sits behind the wheel.

34. A garage has 12 tires in stock and one is defective. Four tires are taken from stock and placed on a car. What is the probability that one will be defective?

35. Assume that the U.S. Senate has 40 Republicans and 60 Democrats. Find the probability that a 12-member committee whose members are selected at random contains four Republicans and eight Democrats.

36. In a given supply of 40 CB sets, eight were known to be defective. Of the 40 sets, five have already been sold without being checked out. Find the probability that two of the five sold were defective and three were good.

37. In a group of 35 students, ten are mathematics majors, four are physics majors, 15 are English majors, and six are history majors (none are double majors). From the group of 35 students, eight are to form a committee. Find the probability that the committee will consist of two mathematics majors, one physics major, two English majors, and three history majors.

38. Let F be the event of selecting a face card from a pack of 52 playing cards, and let G be that of selecting a king. Determine $P(F)$, $P(G)$, $P(G|F)$, and $P(F|G)$. Are F and G independent events?

39. A card is picked from a pack of 52 playing cards. Let F be the event of selecting a red card, and let G be that of selecting a face card. Are F and G independent events?

40. A die is rolled twice. Event F corresponds to tossing a three on the first throw, and G corresponds to tossing a six on the second throw. Show that F and G are independent.

41. (*See Example 11*) The Mall polled 300 customers, 180 women and 120 men, about their preference of the use of an open area. Their choices were a driftwood display, a fountain, and an abstract sculpture. The men responded as follows: 45% preferred the driftwood display, 35% preferred the fountain, and 20% preferred the abstract sculpture. The women responded as follows: 30% preferred the driftwood display, 55% preferred the fountain, and 15% preferred the abstract sculpture.

(a) Represent this information with a tree diagram.

(b) What is the probability that a person selected at random prefers the fountain?

(c) What is the probability that a person selected at random is a male who prefers the driftwood display?

(d) Are the events "female" and "prefers the fountain" independent?

42. A record shop surveyed 250 customers (90 high school students, 80 college students, and 80 college graduates) about their preference to the musical groups The Swingers and The Top Brass. They responded as follows:

High school students: 65% preferred The Swingers, and 35% preferred The Top Brass.

College students: 54% preferred The Swingers, and 46% preferred The Top Brass.

College graduates: 48% preferred The Swingers, and 52% preferred The Top Brass.

(a) Represent this information with a tree diagram.

(b) If a person is selected at random, what is the probability that it is a college student who prefers The Top Brass?

(c) What is the probability that a person selected at random prefers The Swingers?

(d) Are the events "high school student" and "prefers The Top Brass" independent?

43. In Exercise 8, let F be the event of selecting a sophomore and let G be the event of selecting a female. Are F and G independent events?

44. In Exercise 8, let F be the event of selecting a junior, and let G be the event of selecting a female. Are F and G independent events?

45. Professor X stimulates interest in homework by holding a drawing. Two students are each required to draw a problem at random and to solve that problem. On a certain day, ten problems were assigned. Ann is to draw first and Al second. Al had solved nine problems but had not solved the other one.

(a) Find the probability that Ann draws the problem that Al did not solve.

(b) Find the probability that neither Ann nor Al draws the problem that Al did not solve.

(c) Find the probability that Al does not draw the problem that he did not solve.

46. Find the probability of a person being dealt all spades in a bridge hand (13 cards).

47. In a club with 15 seniors, 20 juniors, ten sophomores, and five freshmen, a committee of eight members is chosen at random. Find the probability that the committee selected consists of two seniors, two juniors, two sophomores, and two freshmen.

48. Tim has two irregular coins. The nickel comes up heads three fifths of the time, and the quarter comes up heads two thirds of the time. The nickel is tossed. If the nickel comes up heads, then the quarter is tossed twice. If the nickel comes up tails, the quarter is tossed once. Find the probability that tails comes up exactly twice.

49. There are four different flights during the day from Daytona to New York, and six flights during the day from New York to London. Two people have made separate arrangements to fly on the same day from Daytona to London, with a connection in New York. Find the probability that they will be on different flights to New York, but on the same flight from New York to London.

50. A society is to select a president, a vice president, and a secretary from a group of eight people. Find the probability that Mr. Jones and Mr. Thomas, two members of the group, will be selected president and vice president, respectively.

51. A hand of 13 cards is dealt from a pack of 52 bridge cards. Determine the probability of

(a) getting only one ace.

(b) getting all four aces.

(c) getting no aces.

52. A subcommittee consisting of eight students is to be selected from a committee of 22 students: seven seniors, six juniors, five sophomores, and four freshmen. Find the probability that the committee will consist of three seniors, two juniors, two sophomores, and one freshman.

53. A log lies across a small stream. A fox and ten mice use the log to cross the stream at night. When the fox arrives at the log, he waits to prey on any mice that arrive after him. His capacity is five mice; if he catches five, he leaves. Suppose all orders of arrival (one at a time) are equally likely. Find the probability that

(a) four mice will be eaten.

(b) five mice will be eaten.

(c) no mice will be eaten.

54. Consider the experiment of tossing a die twice. Find the probability of getting

(a) a 6 on the first throw and an even number on the second throw.

(b) an even number on the first throw and an odd number on the second throw.

(c) an even number and an odd number, order being unimportant.

(d) two numbers whose difference is 1.

(e) two 6's.

55. A two-digit number is to be constructed at random from the numbers 1, 2, 3, 4, and 5, repetition being allowed. Find the probability of getting

(a) the number 33.

(b) a number whose first digit is smaller than its second.

(c) a number the sum of whose digits is 5.

(d) a number whose digits are the same.

(e) a number greater than 34.

56. Two letters are selected without replacement from the word MISSISSIPPI. Find the probability of getting

(a) m followed by i.

(b) two p's.

(c) two s's.

(d) a p and an i, order being unimportant.

(e) s followed by i.

57. A given box has 20 light bulbs in it, four of which are defective. If three light bulbs are taken out of the box, what is the probability that exactly one of the three is defective?

58. Out of 15 people who apply for jobs, ten are men and five are women. If six people are hired at random, find the probability that three men and three women are hired.

59. A building with 30 floors is serviced by an elevator. If 4 people use the elevator at once, determine the probability that they all get off at

(a) the 20th floor.

(b) the same floor.

(c) different floors.

60. Show that if E and F are mutually exclusive events, neither with probability zero, then they are dependent. (*Hint*: consider $P(E \cap F)$ and $P(E)P(F)$.)

61. Show that $P(E \mid E) = 1$ when $P(E) \neq 0$.

62. Show that if $E \subset F$ and $P(F) \neq 0$, then $P(E \mid F) = P(E)/P(F)$

63. Show that if $E \subset F$ and $P(F) \neq 0$, then $P(F \mid E) = 1$.

6–5 Bayes' Rule

Conditional probability deals with the probability of an event when you have information about something that happened earlier. Let's look at a situation that reverses the information. Imagine the following.

A club separates a stack of bills and places some in box A and the rest in box B for a drawing. Each box contains some \$50 bills and other bills. At the drawing, a person selects one box and draws a bill from that box. Conditional probability answers questions such as, "If box A is selected, what is the probability that a \$50 bill is drawn?" Symbolically, this is, "What is $P(\$50 \text{ bill} \mid \text{given box } A \text{ is selected})$?" This question assumes that the first event (selecting a box) is known and asks for the probability of the second event.

Bayes' rule deals with a reverse situation. It answers a question such as, "If the person ends up with a \$50 bill, what is the probability that it came from box A?" This assumes the second event is known and asks for the probability of the first event. Bayes' rule determines the probability of an earlier event based on information about an event that happened later.

Let's look at an example with some probabilities given to see when and how to use Bayes' rule. We will then make a formal statement of the rule.

The student body at a college is 60% male and 40% female. The Registrar's records show that 30% of the men attended private high schools and 70% attended public high schools. Further, 80% of the women attended private high schools and 20% attended public schools.

Before we go further, let's summarize this information in probability notation. We use M and F for male and female, and we abbreviate private and public with PRI and PUB. Then, for a student selected at random,

$$P(M) = .6,\ P(F) = .4$$
$$P(PRI \mid M) = .3,\ P(PUB \mid M) = .7$$
$$P(PRI \mid F) = .8,\ P(PUB \mid F) = .2$$

Figure 6–10 is a tree diagram of this information, with its four branches terminating at $M \cap PRI$, $M \cap PUB$, $F \cap PRI$, and $F \cap PUB$. Their respective

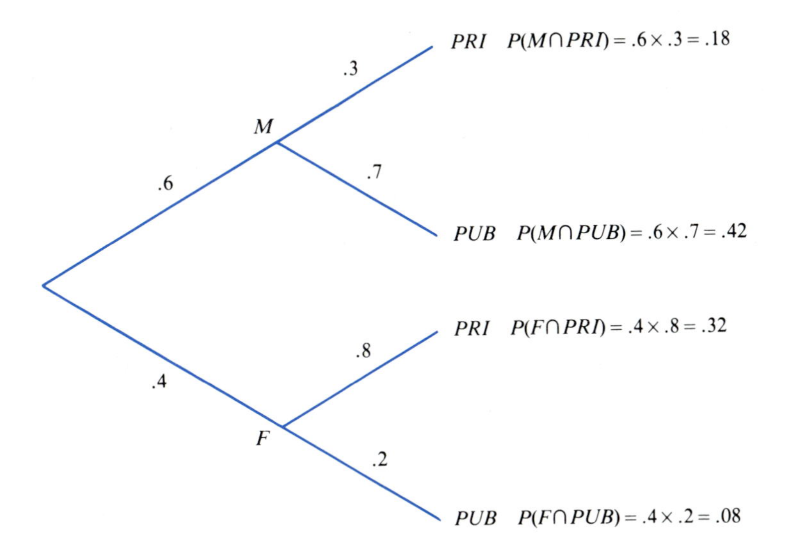

Figure 6–10

probabilities are

$$P(M \cap PRI) = P(M)P(PRI \mid M) = .6 \times .3 = .18$$
$$P(M \cap PUB) = P(M)P(PUB \mid M) = .6 \times .7 = .42$$
$$P(F \cap PRI) = P(F)P(PRI \mid F) = .4 \times .8 = .32$$
$$P(F \cap PUB) = P(F)P(PUB \mid F) = .4 \times .2 = .08$$

We can find the probability that a randomly selected student attended a private school by locating all branches that terminate in *PRI* and adding the probabilities. Thus,

$$P(PRI) = .18 + .32 = .50$$
$$P(PUB) = .42 + .08 = .50$$

Notice that in symbolic notation,

$$P(PRI) = P(M \cap PRI) + P(F \cap PRI)$$

and

$$P(PUB) = P(M \cap PUB) + P(F \cap PUB)$$

All the above information was developed for reference throughout the example. Now let's look at a problem where Bayes' rule is helpful.

Suppose a student selected at random is known to have attended a private school. What is the probability that the student selected is a girl, that is, what is $P(F \mid PRI)$? Notice that this information is missing from above. The definition of conditional probability gives

$$P(F \mid PRI) = \frac{P(F \cap PRI)}{P(PRI)}$$

You will find both $P(F \cap PRI)$ and $P(PRI)$ listed above. They were computed, not given originally. We obtained $P(PRI)$ from

$$P(PRI) = P(M \cap PRI) + P(F \cap PRI)$$

so we can write

$$P(F \mid PRI) = \frac{P(F \cap PRI)}{P(M \cap PRI) + P(F \cap PRI)}$$

This is one form of Bayes' rule. We get a more complicated looking form, but one that uses the given information more directly, when we substitute

$$P(M \cap PRI) = P(M)P(PRI \mid M)$$

and

$$P(F \cap PRI) = P(F)P(PRI \mid F)$$

to get

$$P(F \mid PRI) = \frac{P(F)P(PRI \mid F)}{P(M)P(PRI \mid M) + P(F)P(PRI \mid F)}$$
$$= \frac{.4 \times .8}{.6 \times .3 + .4 \times .8} = \frac{.32}{.18 + .32} = \frac{.32}{.50} = .64$$

This last form has the advantage of using information that was given directly in the problem.

We need to observe one more fact before making a general statement. We used $P(PRI) = P(M \cap PRI) + P(F \cap PRI)$. The events M and F make up *all* the branches in the first stage of the tree diagram, so $M \cup F$ gives all of the sample space ($M \cup F = S$). Furthermore, M and F are mutually exclusive. These two conditions on M and F are needed for Bayes' rule.

Recall This means that M and F form a partition of S. Now we are ready for the general statement.

Bayes' Rule

Let E_1 and E_2 be mutually exclusive events whose union is the sample space ($E_1 \cup E_2 = S$). Let F be an event in S, $P(F) \neq 0$. Then,

(a) $$P(E_1 \mid F) = \frac{P(E_1 \cap F)}{P(F)}$$

(b) $$P(E_1 \mid F) = \frac{P(E_1 \cap F)}{P(E_1 \cap F) + P(E_2 \cap F)}$$ also

(c) $$P(E_1 \mid F) = \frac{P(E_1)P(F \mid E_1)}{P(E_1)P(F \mid E_1) + P(E_2)P(F \mid E_2)}$$

Example 1 (*Compare Exercise 5*)
A microchip company has two machines that produce the chips. Machine I produces 65% of their chips, but 5% of its chips are defective. Machine II produces 35% of the chips and 15% of its chips are defective.

A chip is selected at random and found defective. What is the probability that it came from machine I?

Solution Let I be the set of chips produced by machine I and II be those produced by machine II. We want to find $P(\text{I} \mid \text{defective})$. We are given $P(\text{I}) = 0.65$, $P(\text{II}) = 0.35$, $P(\text{Defective} \mid \text{I}) = 0.05$, and $P(\text{Defective} \mid \text{II}) = 0.15$. We may use the second form of Bayes' rule directly.

$$P(\text{I} \mid \text{defective}) = \frac{P(\text{I})P(\text{Defective} \mid \text{I})}{P(\text{I})P(\text{Defective} \mid \text{I}) + P(\text{II})P(\text{defective} \mid \text{II})}$$

$$= \frac{.65 \times .05}{.65 \times .05 + .35 \times .15}$$

$$= \frac{.0325}{.085} = .38$$

Locate all of these computations on the tree diagram in Figure 6–11.

Bayes' rule is not restricted to the situation where just two mutually exclusive events form all of S. There can be any finite number of mutually exclusive events

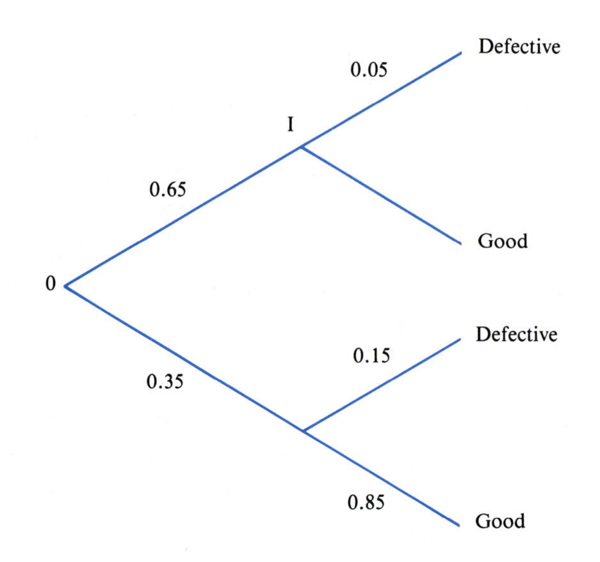

Tree diagram showing the possible situations of defective and good chips from machines I and II.

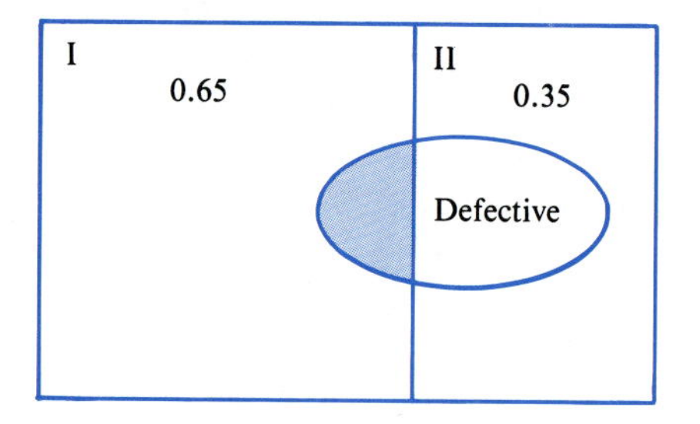

Figure 6–11 The shaded area represents those chips from machine I that are defective.

as long as their union is the sample space. A more general form of Bayes' rule is the following.

Bayes' Rule (General)

Let $E_1, E_2, \ldots, E_n$ be mutually exclusive events whose union is the sample space S (they partition S), and let F be any event where $P(F) \neq 0$. Then,

(a) $P(E_i \mid F) = \dfrac{P(E_i \cap F)}{P(F)}$

(b) $P(E_i \mid F) = \dfrac{P(E_i \cap F)}{P(E_1 \cap F) + P(E_2 \cap F) + \cdots + P(E_n \cap F)}$

(c) $P(E_i \mid F) = \dfrac{P(E_i)P(F \mid E_i)}{P(E_1)P(F \mid E_1) + P(E_2)P(F \mid E_2) + \cdots + P(E_n)P(F \mid E_n)}$

Example 2 (*Compare Exercise 8*)
A manufacturer buys an item from three subcontractors, A, B, and C. A has the better quality control; only 2% of his items are defective. He furnishes the manufacturer with 50% of the items. B furnishes 30% of the items, and 5% of his items are defective. C furnishes 20% of the items, and 6% of his items are defective. The manufacturer finds an item defective (D).

(a) What is the probability that it came from A? (Find $P(A \mid D)$.)
(b) What is the probability that it came from B? (Find $P(B \mid D)$.)
(c) What is the probability that it came from C? (Find $P(C \mid D)$.)

Solution Let

A represent the set of items produced by A
B represent the set of items produced by B
C represent the set of items produced by C
D represent the set of defective items.

The following probabilities are given

$$P(A) = .50, \quad P(B) = .30, \quad P(C) = .20$$
$$P(D \mid A) = .02, \quad P(D \mid B) = .05, \quad P(D \mid C) = .06$$

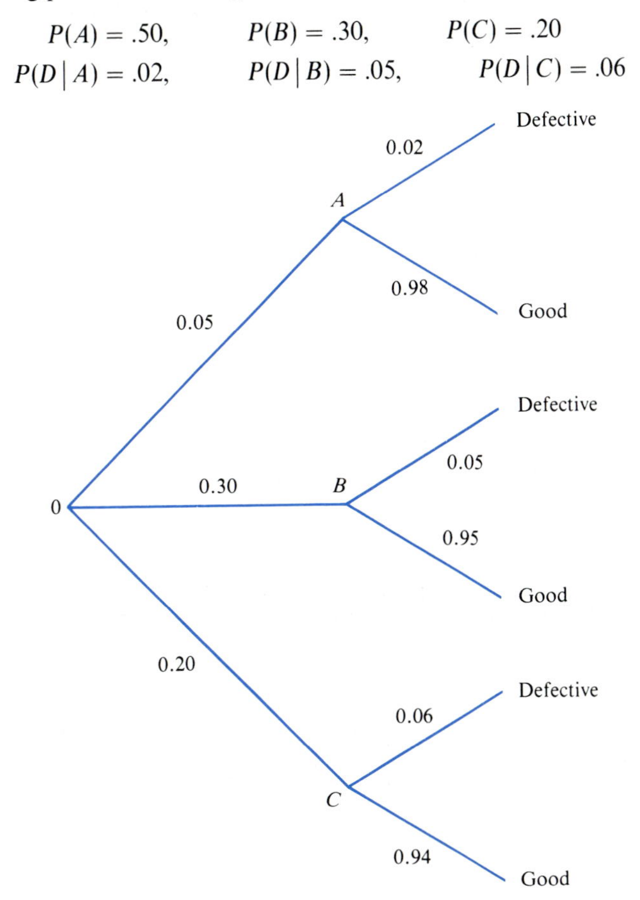

Figure 6–12

Let's use the first form of Bayes' rule to solve the problem. We need the following probabilities, which are computed using the Multiplication Rule:

$$P(D \cap A) = P(A)P(D \mid A) = .50 \times .02 = .010$$
$$P(D \cap B) = P(B)P(D \mid B) = .30 \times .05 = .015$$
$$P(D \cap C) = P(C)P(D \mid C) = .20 \times .06 = .012$$

Then,

(a) $P(A \mid D) = \dfrac{.010}{.010 + .015 + .012} = \dfrac{.010}{.037} = .27$

(b) $P(B \mid D) = \dfrac{.015}{.010 + .015 + .012} = \dfrac{.015}{.037} = .41$

(c) $P(C \mid D) = \dfrac{.012}{.010 + .015 + .012} = \dfrac{.012}{.037} = .32$

Again, trace the computations on the tree diagram in Figure 6–12.

Example 3 (*Compare Exercise 12*)
The probability that a woman exposed to German measles will contract the disease is 0.2. If she is pregnant when exposed, the probability that her child will have a certain birth defect is 0.1; otherwise, the probability of the defect is 0.01.
A child is born with this defect. What is the probability that its mother contracted German measles while pregnant?

Solution Let

M = the set of pregnant women who were exposed to and contracted measles.
M' = the set of pregnant women who were exposed to and did not contract measles.
S = the sample space of pregnant women exposed to measles ($M \cup M'$).
D = the set of mothers who bear a child with this defect.

The Venn diagram of these events is shown in Figure 6–13.

We are given $P(M) = .2$, $P(M') = .8$, $P(D \mid M) = 0.1$, and $P(D \mid M') = 0.01$; and we are to find $P(M \mid D)$.

$$P(M \mid D) = \frac{P(M \cap D)}{P(M \cap D) + P(M' \cap D)}$$

$$= \frac{P(M)P(D \mid M)}{P(M)P(D \mid M) + P(M')P(D \mid M')}$$

$$= \frac{.2 \times .1}{.2 \times .1 + 0.8 \times .01} = \frac{.02}{.028} = .71$$

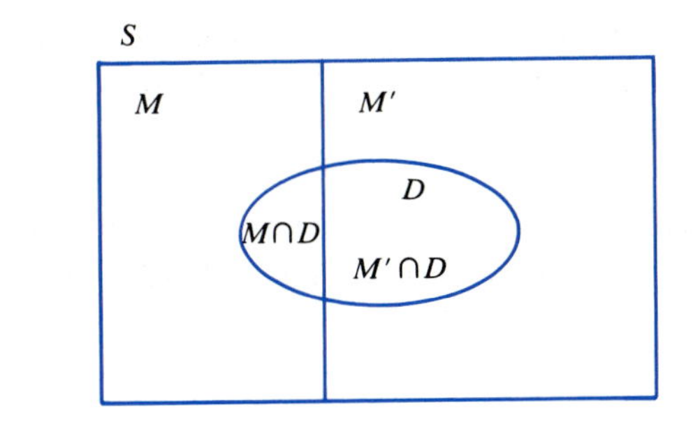

Figure 6–13

6–5 Exercises

1. Draw the tree diagram that shows the following information. A class is composed of 30 girls and 20 boys. On the final exam, 12% of the girls and 9% of the boys made an A.

2. Compute $P(E_1 \mid F)$ for the following:

   (a) $E_1 \cup E_2 = S$, $P(E_1 \cap F) = .7$, $P(F) = .9$
   (b) $E_1 \cup E_2 = S$, $P(E_1) = .75$, $P(E_2) = .25$, $P(F \mid E_1) = .40$, $P(F \mid E_2) = .10$

3. E_1 and E_2 partition the sample space S, and F is an event.

   $$P(E_1) = .3, \qquad P(E_2) = .7$$
   $$P(F \mid E_1) = .25, \qquad P(F \mid E_2) = .15$$

   Find $P(E_1 \mid F)$.

4. F is an event, and E_1, E_2, and E_3 partition S.

   $$P(E_1) = .20, \qquad P(E_2) = .35, \qquad P(E_3) = .45$$
   $$P(F \mid E_1) = .40, \qquad P(F \mid E_2) = .10, \qquad P(F \mid E_3) = .25$$

   (a) Find $P(E_1 \cap F)$, $P(E_2 \cap F)$, and $P(E_3 \cap F)$.
   (b) Find $P(F)$.
   (c) Find $P(E_2 \mid F)$.

5. (*See Example 1*) A manufacturer has two machines that produce television picture tubes. Machine I produces 55% of the tubes, and 3% of its tubes are defective. Machine II produces 45% of the tubes, and 4% of its tubes are defective.

   A picture tube is found defective. Find the probability that it was produced by machine II.

6. A refrigerator is found defective. It could have been manufactured at either of two plants, I or II. Plant I manufactures 40% of the refrigerators, and plant II

manufactures 60% of the refrigerators. Of those refrigerators manufactured at plant I, 3% are defective; and of those manufactured at plant II, 2% are defective. Find the probability that the refrigerator came from plant I.

7. A football team plays 60% of its games at home and 40% away. It typically wins 80% of its home games and 55% of its away games. If the team wins on a certain Saturday, what is the probability that it played at home?

8. (*See Example 2*) A distributor receives 100 boxes of transistor radios. Factory A shipped 35 boxes, factory B shipped 40 boxes, and factory C shipped 25 boxes. The probability of a radio being defective is .04 from factory A, .02 from factory B, and .06 from factory C.

   A box and a radio from that box are both selected at random. The radio is found defective. Find the probability that it came from factory B.

9. A certain pocket calculator is found defective. It could have been manufactured in any one of three factories, I, II, or III. Of all such calculators produced, 15% came from factory I, 45% came from factory II, and 40% came from factory III. Further, it is known that 1% of all calculators produced at factory I are defective, 2% of those produced at factory II are defective, and 4% of those produced at factory III are defective.

   (a) Find the probability that the defective calculator came from factory I.
   (b) Find the probability that it came from factory II.
   (c) Find the probability that it came from factory III.

10. The durations of the traffic light colors at a major intersection are as follows: green, 90 seconds; yellow, 4 seconds; red, 40 seconds.

    (a) A car has just passed legally through the intersection (either a green or yellow light was showing). Find the probability that a green light was showing.
    (b) A car has just passed through the intersection. Find the probability that it came through illegally.

11. Box I contains four red balls and six green balls. Box II contains five red balls and seven green balls. A box is selected, and then a ball is drawn from the selected box. The probability of selecting box I is $\frac{3}{5}$.

    (a) Show all possible outcomes and their probabilities by a tree diagram.
    (b) Find $P(\text{red} \mid \text{box I})$, $P(\text{red and box I})$, and $P(\text{red})$.
    (c) Find $P(\text{box I} \mid \text{red})$.

12. (*See Example 3*) If a patient is allergic to penicillin, the probability of a reaction to a new drug X is 0.4. If the patient is not allergic to penicillin, the probability of a reaction to drug X is 0.1. Eight percent of the population is allergic to penicillin. A patient is given drug X and has a reaction. Find the probability that the patient is allergic to penicillin.

13. It is estimated that 12% of all adults are college graduates, while 88% are not. If 60% of all graduates earn more than $20,000 per annum, and only 20% of nongraduates earn more than $20,000, what proportion of those people who earn more than $20,000 are college graduates?

14. During the hour 9 PM to 10 PM, the following times were devoted to programs and commercials on the three major networks:

	Commercials	Programs
Channel 2	10 min	50 min
Channel 6	12 min	48 min
Channel 9	$11\frac{1}{2}$ min	$48\frac{1}{2}$ min

The three channels are equally likely to be selected. If someone comes into a room between 9 and 10, and a television is on and showing a commercial, find the probability that the television is tuned to channel 2.

15. Urn I has five white balls and four blue balls. Urn II has three white balls and six blue balls. A ball is taken from urn I and put into urn II. A ball is then drawn from urn II and is found to be white.

(a) Draw a tree diagram.
(b) What is the probability that the white ball drawn originally came from urn I?

16. At a certain clinic, a test for strep has been found to be 90% accurate in that 90% of those with strep will have a positive reaction. However, 1% of those without strep also have a positive reaction. Suppose that 80% of those examined have strep. Find the probability that a child who has a positive reaction has strep.

17. At a clinic, a preliminary test for hepatitis has been found to be 95% accurate in that 95% of those with hepatitis have a positive reaction. However, 2% of those without hepatitis also have a positive reaction. Suppose 70% of those examined have hepatitis. Find the probability that a person who has a positive reaction has hepatitis.

18. Authorities testing school children for hearing deficiency have determined that the test is successful 85% of the time; that is, 85% of those with a hearing deficiency will have a positive reaction, while 5% without a hearing deficiency have a positive reaction. Suppose that 10% of those examined have a hearing deficiency. Find the probability that a child who has a positive reaction does not have a hearing deficiency.

19. The following driving-accident statistics have been compiled by an insurance company:

Age Group	Percent of Total Accidents	Percent of Total Population
Under 25	18	12
25–35	22	24
36–50	30	32
51–70	25	25
Over 70	5	7

Of the total population, 15% are involved in accidents while driving.

(a) Find the probability that a driver who is involved in an accident is under 25 years of age.

(b) Find the probability that a driver who is involved in an accident is over 50 years of age.

(c) Given a person is under 25, find the probability that person is involved in an accident.

20. In a major city, 70% of the drivers are over 25 years old, and 12% of them will have a traffic violation during a 12-month period. The drivers 25 years and under comprise 30% of the drivers, and 28% of them will have a traffic violation in a 12-month period. A driver is charged with a traffic violation. Find the probability he is over 25.

21. The IRS checks the deduction for contributions to identify fraudulent tax returns. They believe that if a taxpayer claims more than a certain standard amount, there is a .20 probability that the return is fraudulent. If the deduction does not exceed the IRS standard, the probability of a fraudulent return reduces to .03. About 11% of the returns exceed the IRS standard.

(a) Estimate the percentage of returns that are fraudulent.

(b) A concerned citizen informs the IRS that a certain return is fraudulent. Find the probability that its deductions exceed the IRS standard.

22. A chain has three stores in Knoxville, A, B, and C. The stores have 60, 80, and 110 employees of whom 40%, 55%, and 50% are women, respectively. Economic conditions force the chain to lay-off some employees. They decide to do this by random selection. The first person selected is a woman. Find the probability that she works at store C.

23. A certain television set is found defective. It could have been manufactured at any one of four factories, A, B, C, or D. Of all such television sets, 10% are produced at A, 15% at B, 55% at C, and 20% at D. It is determined that 3% of the sets produced at A, 1% of those produced at B, 2% of those produced at C, and 1% of those produced at D are defective. For each factory, find the probability that the defective set came from that factory.

24. At a certain university, it has been estimated that the probability that a male freshman will major in mathematics is $\frac{1}{10}$, and the probability that a female freshman will major in mathematics is $\frac{1}{15}$. The freshman class is 55% male. If a student chosen at random from the freshman class is a mathematics major, find the probability that the student is male.

25. A certain disease can be detected by a blood test in 95% of those who have it. Unfortunately, the test also has a probability of .02 of showing that a person has the disease when in fact he does not. It has been estimated that 1% of those people routinely tested actually have the disease. If the test shows that a certain person has the disease, find the probability that he actually has it.

26. The voters in a certain state are registered 30% Republican, 60% Democrat, and 10% independent. In a certain election involving three candidates, a Republican, a Democrat, and an independent, the Republican candidate was elected. It was estimated that she gained 85% of the Republican vote, 40% of the Democrat vote, and 8% of the independent vote. What percentage of those who voted Republican were registered Democrats?

27. A company manufactures integrated circuits (IC's) on silicon chips at three different plants, X, Y, and Z. Out of every 1000 chips produced, 400 come from X, 350 come from Y, and 250 come from Z. It has been estimated that of the 400 from X, 10 are defective, whereas 5 of those from Y are defective, and only 2 of those from Z are defective. Determine the probability that a defective chip came from plant Y.

6–6 Bernoulli Experiments

Bernoulli Trials

Suppose that you toss a coin ten times. What is the probability that heads appears seven out of the ten times?

If you guess at the answers of ten multiple-choice questions, what are your chances for a passing grade?

These problems are examples of a certain type of probability problem, **Bernoulli trials**. Such problems involve repeated trials of an experiment with only two possible outcomes; heads or tails, right or wrong, yes or no, win or lose, and so on. We classify the two outcomes as **success** or **failure**.

In order to classify an experiment as a Bernoulli trial experiment several properties must hold.

Bernoulli Experiment

1. The experiment is repeated a fixed number of times (n times).
2. Each trial has only two possible outcomes: success and failure. The outcomes are exactly the same for each trial.
3. The probability of success remains the same for each trial. (We use p for the probability of success and $q = 1 - p$ for the probability of failure.)
4. The trials are independent. (The outcome of one trial has no influence on later trials.)
5. We are interested in the total number of successes, not the order in which they occur.

   There may be 0, 1, 2, 3, ..., or n successes in n trials.

Example 1

(a) We are interested in the number of times heads occurs when a coin is tossed eight times. Each toss of the coin is a trial, so there are eight repeated trials ($n = 8$). We consider the outcome heads a success and tails a failure. The probability of success (heads) on each trial is $p = \frac{1}{2}$, and the probability of failure (tails) is $1 - \frac{1}{2} = \frac{1}{2}$. This is an example of a Bernoulli trial.

(b) A student guesses at all the answers on a ten-question multiple-choice quiz (four choices of an answer on each question). This fulfills the properties of a Bernoulli trial because

1. each guess is a trial ($n = 10$);
2. there are two possible outcomes, correct and incorrect;
3. the probability of a correct answer is $\frac{1}{4}$ ($p = \frac{1}{4}$), and the probability of an incorrect answer is $\frac{3}{4}$ ($q = \frac{3}{4}$) on each trial;
4. and, the guesses are independent because guessing an answer on one question gives no information on other questions.

(c) Suppose eight cards are drawn from a deck and none are replaced. We are interested in the number of spades drawn. This is not a Bernoulli trial because the trials (selecting a card) are not independent (the first card drawn affects the possible choices of the second card), and the probability of drawing a spade changes each time a card is removed.

(d) Suppose a card is drawn from a deck, the card is noted and placed back in the deck, and the deck is shuffled. This is repeated eight times. We are interested in the number of times a spade is drawn.

This experiment is a Bernoulli trial because each trial is the same, the trials are independent, and the probability of obtaining a spade remains the same for each trial.

Technically, each trial has 52 outcomes, each card in the deck. However, we reduce the outcomes to the two possible outcomes "success" or "failure" when we collect all spades into the event defined as success and all other cards into the event defined as failure. Then, $p = \frac{13}{52}$ and $q = \frac{39}{52}$.

The following example illustrates how a tree diagram may be used to count the number of successes when the number of trials is small.

Example 2 (*Compare Exercise 15*)
A coin is tossed three times. What is the probability of exactly two heads in the three tosses?

Solution Look at the paths of Figure 6–14 that begin at 0 and terminate at the end of a branch. There are a total of eight paths, and three of them contain exactly two heads. The probability of terminating at the end of those branches is $\frac{1}{8}$ for each one. (Since the probability of each branch is $\frac{1}{2}$, the probability of taking a sequence of three paths is $(\frac{1}{2}\frac{1}{2}\frac{1}{2} = \frac{1}{8}$.) So, the probability of exactly two heads in three tosses of a coin is $\frac{3}{8}$.

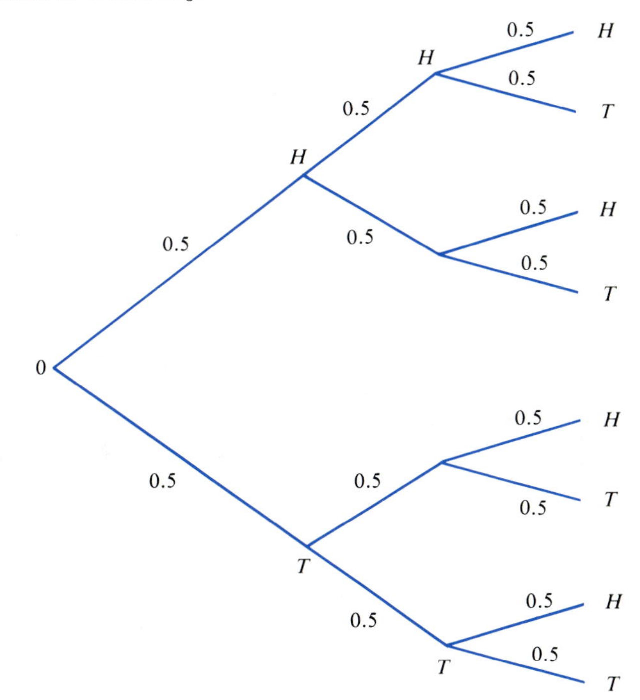

Figure 6–14

Probability of a Bernoulli Experiment

The tree diagram of a Bernoulli trial problem can become unwieldy with a large number of trials; so, an algebraic formula is useful for computing the probability of a specified number of successes. Let's start with a simple example.

Example 3 (*Compare Exercise 18*)
A quiz has five multiple-choice questions with four possible answers to each. A student wildly guesses the answers. What is the probability that he guesses exactly three correctly?

Solution As a Bernoulli trial problem, $n = 5$, $p = \frac{1}{4}$, and $q = \frac{3}{4}$. We are about to give a formula that calculates the desired probability, but you need to be aware that you have no reason, at this point, to know why it is true. That will be explained later. We want you to understand the quantities used in the formula so that you can more easily follow the justification. Now for the formula. The probability of exactly three correct answers is

$$\begin{aligned} P(3 \text{ correct}) &= C(5, 3)\left(\frac{1}{4}\right)^3\left(\frac{3}{4}\right)^2 \\ &= 10\left(\frac{1}{64}\right)\left(\frac{9}{16}\right) \\ &= .088 \qquad \text{(Rounded to 3 decimals)} \end{aligned}$$

Let's make some observations about the computation, because they will hold for the general formula.

1. In $C(5, 3)$, 5 is the number of trials, and 3 is the number of successes.
2. In $(\frac{1}{4})^3$, $\frac{1}{4}$ is the probability of success in a single trial, and 3 is the number of successes in the 5 trials.
3. In $(\frac{3}{4})^2$, $\frac{3}{4}$ is the probability of failure in a single trial, and 2 is the number of failures in 5 trials. It may seem trivial, but note that $2 = 5 - 3$; the number of failures equals the number of trials minus the number of successes.

We now give you the general formula.

Probability of a Bernoulli Experiment

Given a Bernoulli experiment and

n independent repeated trials,
p is the probability of success in a single trial,
$q = 1 - p$ is the probability of failure in a single trial,
x is the number of successes ($x \leq n$).

Then, the probability of x successes in n trials is

$$P(x \text{ successes in } n \text{ trials}) = C(n, x)p^x q^{n-x}$$

Example 4 (*Compare Experiment 21*)
A single die is rolled three times. What is the probability that a five turns up exactly twice?

Solution

$$n = 3, x = 2, p = \frac{1}{6}, q = \frac{5}{6}$$

so

$$\begin{aligned} P(x = 2) &= C(3, 2)\left(\frac{1}{6}\right)^2\left(\frac{5}{6}\right)^1 \\ &= 3\left(\frac{1}{36}\right)\left(\frac{5}{6}\right) \\ &= .0694 \end{aligned}$$

Notice that in the tree diagram (Figure 6–15) there are three branches with exactly two 5's. This number is $C(3, 2)$. The probability of termination at the end of one such branch is

$$\left(\frac{1}{6}\right)^2\left(\frac{5}{6}\right)$$

The notation $P(x = 2)$ is used to indicate the probability of two successes.

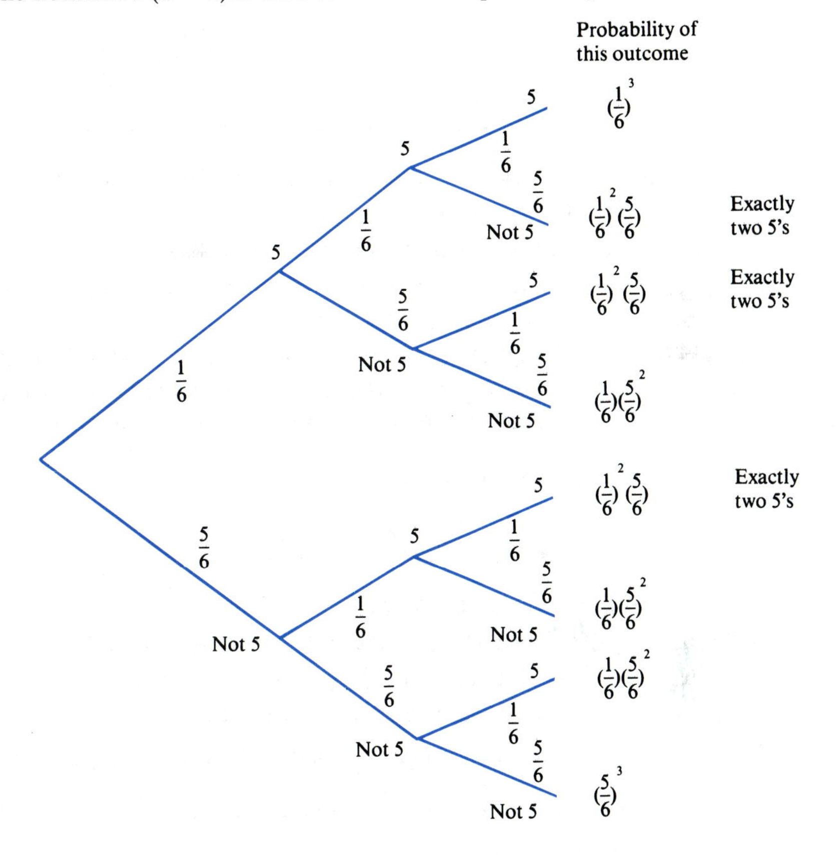

Figure 6–15 Exactly two 5's occurs three times in three tosses of a die.

Example 5 (*Compare Exercise 24*)
A coin is tossed ten times. What is the probability that heads occurs six times?

Solution In this case $n = 10$, $x = 6$, $p = \frac{1}{2}$, and $q = \frac{1}{2}$.

$$P(x = 6) = C(10, 6)\left(\frac{1}{2}\right)^6\left(\frac{1}{2}\right)^4$$

$$= 210\left(\frac{1}{64}\right)\left(\frac{1}{16}\right)$$

$$= .205$$

Justification of the Bernoulli Experiment Formula

Now let's go back to Example 3 and explain how to obtain the expression used to compute the probability of a Bernoulli trial.

Recall the problem: A multiple-choice quiz has five questions with four possible answers to each, one of which is correct. A student guesses the answers. What is the probability that three of the five are correct?

Let's look at how the student succeeds in answering three of the five questions. There are several ways, and the exact number plays a key role in the solution. First, let's list some ways. We will list the sequence of successes (correct answers) and failures (incorrect answers). One sequence is

SSSFF

which indicates that the first three were correct and the last two were incorrect. Some other sequences of three correct and two incorrect are

SSFSF

FSSSF

and so on. Rather than list all the ways three correct and two incorrect answers can be given, let's compute the number.

The basic procedure for forming a sequence of three S's and two F's amounts to selecting the three questions with correct answers. The other two answers are automatically incorrect.

How many different ways can three questions be selected from the five questions? You should recognize this as $C(5, 3)$. Since $C(5, 3) = 10$, there are ten possible sequences of three S's and 2 F's. The student succeeds in passing if his sequence of guesses is any one of the ten sequences.

Recall From compound probability that the probability that his answer is one of the ten is obtained by adding the probabilities of each of the ten sequences.

It helps that the probabilities of all ten sequences are exactly the same. Notice that the probability of SSSFF is

$$\left(\frac{1}{4}\right)\left(\frac{1}{4}\right)\left(\frac{1}{4}\right)\left(\frac{3}{4}\right)\left(\frac{3}{4}\right)$$

by the Multiplication Rule. Also, the probability of SSFSF is

$$\left(\frac{1}{4}\right)\left(\frac{1}{4}\right)\left(\frac{3}{4}\right)\left(\frac{1}{4}\right)\left(\frac{3}{4}\right)$$

Both of these are simply

$$\left(\frac{1}{4}\right)^3\left(\frac{3}{4}\right)^2$$

each written in a different order. In fact, the probability of any sequence of three S's and two F's will contain $\frac{1}{4}$ three times and $\frac{3}{4}$ twice, which gives $(\frac{1}{4})^3(\frac{3}{4})^2$.

When we add up the probabilities of the ten sequences, we are adding $(\frac{1}{4})^3(\frac{3}{4})^2$ ten times, which is

$$10\left(\frac{1}{4}\right)^3\left(\frac{3}{4}\right)^2$$

This is the probability obtained in Example 3. A similar situation holds in general for other values of n, x, p, and q.

Example 6 (*Compare Exercise 38*)
The cycle of a traffic light on Main Street is green, 50 seconds; yellow, 5 seconds; and red, 25 seconds. What is the probability of getting a green three out of four times you approach the intersection?

Solution This is an experiment with four repeated trials. A green light is considered a success. A complete cycle requires 80 seconds, so the probability of green is $\frac{50}{80} = .625$.
Then,

$$\begin{aligned} P(x = 3) &= C(4, 3)(.625)^3(.375)^1 \\ &= 4(.244)(.375) \\ &= .366 \end{aligned}$$

Example 7 (*Compare Exercise 35*)
A multiple-choice quiz with five questions is given. Each question has four possible answers. A student guesses all answers. What is the probability that she passes the test if at least three correct answers are needed to pass.

Solution In this case, $n = 5$, $p = \frac{1}{4}$, and $q = \frac{3}{4}$. The student passes if she gets three, four, or five correct answers. We must find the probability of each of these outcomes and add them.

$$\begin{aligned} P(x = 3) &= C(5, 3)(.25)^3(.75)^2 \\ &= 10(.015625)(.5625) \\ &= .0879 \end{aligned}$$

$$\begin{aligned} P(x = 4) &= C(5, 4)(.25)^4(.75)^1 \\ &= 5(.0039062)(.75) \\ &= .0146 \end{aligned}$$

$$\begin{aligned} P(x = 5) &= C(5, 5)(.25)^5(.75)^0 \\ &= 1(.0009765)(1) \\ &= .00098 \qquad \text{(Rounded)} \end{aligned}$$

Then, the probability of three or more correct, which we indicate by $p(x \geq 3)$, is

$$P(x \geq 3) = .0879 + .0146 + .00098 = .10348$$

6–6 Exercises

I.

Compute the following Bernoulli trial probabilities:

1. $C(6, 4)(.20)^4(.80)^2$
2. $C(7, 3)(.95)^3(.05)^4$
3. $C(4, 2)(.1)^2(.9)^2$
4. $C(5, 2)(.3)^2(.7)^3$
5. $C(8, 7)(.3)^7(.7)$
6. $C(9, 2)(.8)^2(.2)^7$
7. $C(11, 4)(.6)^4(.4)^7$
8. $C(12, 3)(.5)^3(.5)^9$
9. $n = 6, p = 0.5, x = 2$
10. $n = 7, p = \frac{1}{3}, x = 4$
11. $n = 5, p = 0.1, x = 4$
12. $n = 10, p = \frac{1}{5}, x = 6$
13. $P(x = 3)$ for $n = 5, p = \frac{1}{4}$
14. $P(x = 4)$ for $n = 8, p = \frac{2}{3}$
15. (*See Example 2*) The probability that a marksman will hit a moving target is 0.7. Use a tree diagram to determine the probability that he will hit the target in two out of three attempts.
16. The probability that an instructor will give a quiz is 0.4. Use a tree diagram to determine the probability that the instructor will give a quiz in two out of four class meetings.
17. A check of autos passing an intersection reveals that 80% of the drivers are wearing seat belts. Find the probability that three of the next five drivers are wearing seat belts.
18. (*See Example 3*) A multiple-choice quiz of four questions is given. Each question has five possible answers. If a student guesses at all the answers, find the probability that three out of four are correct.
19. If a couple, each with genes for both brown and blue eyes, parent a child, the probability that the child has blue eyes is $\frac{1}{4}$. Find the probability that two of the couple's three children will have blue eyes.
20. A certain drug was developed, tested, and found to be effective 70% of the time. Find the probability of successfully administering the drug to nine out of ten patients.
21. (*See Example 4*) A single die is rolled six times. Find the probability of a 3 turning up four times.
22. A die is tossed three times. Find the probability of throwing two 6's.
23. A die is thrown four times. Determine the probability of throwing (a) zero 6's. (b) one 6. (c) two 6's. (d) three 6's. (e) four 6's.
24. (*See Example 5*) A coin is tossed seven times. Find the probability of it turning up tails five times.
25. A coin is tossed six times. Determine the probability of its landing heads up four times.

26. A coin is tossed four times. Find the probability of its landing on tails

(a) zero times.
(b) once.
(c) twice.
(d) three times.
(e) four times.

II.

27. A certain antibiotic developed in a research lab was tested on patients. It was found successful with 80% of the patients. A hospital has six patients with the disease that the antibiotic treats. Find the probability of curing four of these patients with the antibiotic.

28. Five door prizes are to be awarded at a club meeting. There are 100 tickets in the box, one of them yours. A ticket is drawn, the prize is awarded, and the ticket is placed back in the box. This continues until all five prizes are awarded.

(a) Find the probability that your ticket is drawn once.
(b) Find the probability that your ticket is drawn twice.

29. At State U, 40% of the students are from out of state. Six students are selected at random. Find the probability that half of them are from out of state.

30. The probability of success in drilling for oil is .2. If a company sinks ten wells, find the probability that five wells will strike oil.

31. A certain telephone line is busy 60% of the time. Find the probability of getting through seven times if ten calls are made.

32. A drug was developed, tested, and found to be effective 80% of the time. It is administered to 12 patients. Find the probability that the drug is effective on 8 out of 12 patients.

33. The pass-completion statistic of a certain quarterback is .6. Find the probability that he will complete eight out of the next ten passes that he attempts.

34. A certain tennis player who has a devastating first serve has been getting 70% of her first serves in. She is due to serve. Find the probability that she will get all of her next four first serves in.

III.

35. (*See Example 7*) A single die is rolled five times. Find the probability that a four turns up at least three times.

36. It is estimated that the probability is 0.8 that an oak tree 10 to 12 feet tall can be transplanted and will survive, given that it is well cared for. Find the probability of transplanting five such oak trees and having at least three live.

37. A coin is tossed eight times. Find the probability that it will land on tails at least five times.

38. (*See Example 6*) A certain traffic light is green 60% of the time. What is the probability of getting the green light nine out of ten times one approaches the intersection?

39. A true-false quiz has ten questions. A student guesses all the answers.

(a) Find the probability of getting eight correct.
(b) Find the probability of getting at least eight correct.
(c) Find the probability of getting no more than two correct.

40. The manufacturer of flash bulbs observes that defective bulbs are produced with a probability of .05. A package contains eight flash bulbs.

(a) Find the probability that the package contains no defective bulbs.
(b) Find the probability that the package contains at most one defective bulb.

41. A box contains two red balls and four green balls. Four draws are made, with the ball replaced after each draw. Find the probability that

(a) a red ball is never drawn.
(b) a red ball is drawn once.
(c) a red ball is drawn twice.
(d) a red ball is drawn three times.
(e) a red ball is drawn all four times.

42. A coin is tossed four times. Is it true or false that the probability that it turns up heads twice is $\frac{1}{2}$?

43. A psychology student conducts an ESP experiment in which the subject attempts to identify the number that turns up when an unseen die is rolled. The die is rolled six times. Based on chance alone, find the probability that the subject identifies the number correctly four of the six times.

44. Find the probability that in a family of eight children

(a) three are girls.
(b) there are at least two girls.

45. A baseball player has a 0.360 batting average. Find the probability that he will have at least two hits in five times at bat.

46. It is estimated that a certain brand automobile tire has a 0.8 probability of lasting 25,000 miles. Out of four such tires put on a car, find the probability that one will have to be replaced before 25,000 miles.

47. Suppose that 60% of voters were against the repeal of a certain local law. If ten people were chosen at random from the community, what would be the probability that six or more of these people were in favor of the repeal? (Such a selection would make it appear that the majority favored the repeal.)

6–7 Markov Chains

Transition Matrix

Markov chains provide a means to analyze certain kinds of problems. The techniques used use matrix operations and systems of equations. You may wish to review those topics.

The alumni office of a university knows that generally 80% of the alumni who contribute one year will contribute the next year. They also know that 30% of those who do not contribute one year will contribute the next. We want to answer the following kinds of questions:

If 40% of a graduating class contributes the first year, how many will contribute the second year? The fifth year? The tenth year?

Before we solve this problem, let's define some terminology and basic concepts.

First, the alumni can be placed in exactly one of two possible categories; they either contribute or they do not contribute. These categories are called **states**.

Definition

> A **state** is a category, situation, outcome, or position that a process can occupy at any given time. The states are disjoint and cover all possible outcomes.

For example, the alumni are either in the state of contributors or in the state of noncontributors. A patient is ill or well. A person's emotional state may be happy, angry, or sad. In a Markov process, the system may move from one state to another at any given time. When the process moves from one state to the next, we say a transition is made from the **present state** to the **next state**. An alumnus can make a transition from the noncontributor state to the contributor state.

The information on the proportion of alumni who do or do not contribute can be represented by a **transition matrix**. Let C represent those who contribute and NC those who do not.

		Next State	
		C	NC
Present	C	.8	.2
State	NC	.3	.7

The entries in the transition matrix are probabilities that a person will move from one state (present) to another state (next) the following year. For example, the probability that a person who now contributes will contribute again next year is 0.8. We get this from the statement "80% of the alumni who contribute one year will contribute the next year."

The headings to the left of the matrix identify the present state, and the headings above the matrix identify the next state. Each entry in the matrix is interpreted as follows:

0.8 is the probability that a person passes from present state C to next state C; that is, a contributor remains a contributor.

0.2 is the probability that a person passes from present state C to next state NC; a contributor becomes a noncontributor.

0.3 is the probability that a person passes from present state NC to next state C; that is, a noncontributor becomes a contributor.

0.7 is the probability that a person passes from present state NC to next state NC, that is, a noncontributor remains a noncontributor. Because each row lists the probabilities of going from that state to each of all possible states, the entries in a row always add to 1.

Transition Matrix

A transition matrix is a square matrix with each entry a number from the interval 0 through 1. The entries in each row add to 1.

In the alumni example, a row matrix may be used to represent the proportion of people in each state. For example, the matrix [.40 .60] indicates that 40% are in state C and 60% are in state NC.

The same row matrix may also be interpreted to indicate that the **probability** of a person being in state C is .40, and the probability of being in state NC is .60.

These row matrices are called **state matrices**, or **state probability matrices**. For each state, it shows the probability that a person is in that state.

Why do we put the information in this form? Because we can use matrix operations to provide useful information. Here's how.

Multiply the state matrix and the transition matrix

$$\begin{matrix} C & NC \\ [.40 & .60] \end{matrix} \quad \begin{bmatrix} .8 & .2 \\ .3 & .7 \end{bmatrix}$$

and the result is

$$[(.40)(.8) + (.60)(.3) \quad (.40)(.2) + (.60)(.7)]$$

$$\begin{matrix} & C & NC \\ = & [.50 & .50] \end{matrix} \quad \text{(The next-state matrix)}$$

You should interpret this as

In one year the alumni moved from 40% in C and 60% in NC to 50% in each.

Let's look at the tree diagram in Figure 6–16 to help justify this.

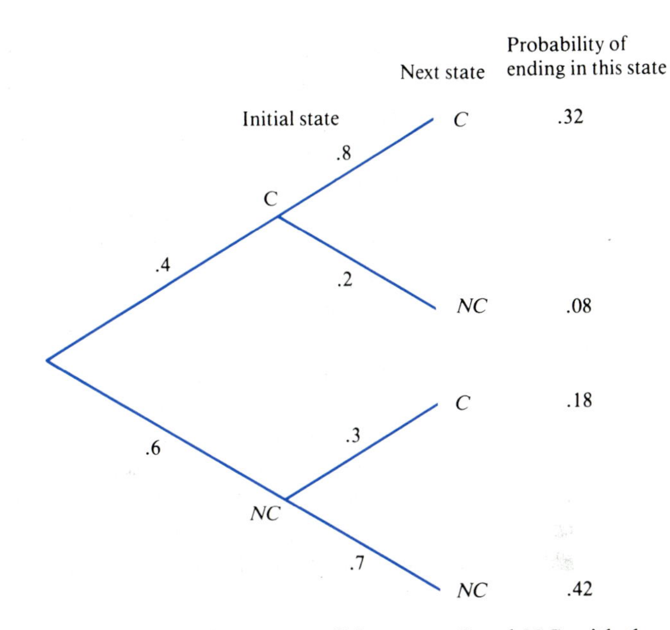

Figure 6–16

The first stage shows the two possible states, C and NC, with the probability of each. The second stage shows the next states and the probability of entering those states. At the end of each branch is the probability of terminating there. There are two branches that terminate in C. We simply add the two probabilities there to obtain the probability of ending in state C. That probability is

$$(.40)(.8) + (.60)(.3) = .32 + .18 = .50$$

Notice that this is exactly the computation used in the matrix product that gave the first entry in the next-state matrix. The second entry in the next-state matrix is

$$(.40)(.2) + (.60)(.7)$$

which is the probability for entering state NC, as obtained from the tree diagram.

This illustrates how the state matrix and transition matrix can be used to get the next-state matrix.

Example 1 (*Compare Exercise 9*)

Using [.40 .60] as the first year's state matrix (sometimes called the **initial**-state matrix) and

$$\begin{bmatrix} .8 & .2 \\ .3 & .7 \end{bmatrix}$$

as the transition matrix for the alumni problem, find the percent of alumni in each category for the second, third, and fourth years.

Solution As illustrated above, the second-year breakdown is

$$\underset{\textbf{1st Year}}{\overset{\begin{matrix} C & NC \end{matrix}}{[.40 \quad .60]}} \begin{bmatrix} .8 & .2 \\ .3 & .7 \end{bmatrix} = \underset{\textbf{2nd Year}}{\overset{\begin{matrix} C & NC \end{matrix}}{[.50 \quad .50]}}$$

The third-year state matrix is obtained by multiplying the second-year state matrix (it is now the present matrix) by the transition matrix:

$$[.50 \quad .50]\begin{bmatrix} .8 & .2 \\ .3 & .7 \end{bmatrix} = [(.50)(.8) + (.50)(.3) \qquad (.50)(.2) + (.50)(.7)]$$

$$= [.40 + .15 \qquad .10 + .35]$$

$$= [.55 \quad .45]$$

The fourth-year state matrix is

$$[.55 \quad .45]\begin{bmatrix} .8 & .2 \\ .3 & .7 \end{bmatrix} = [(.55)(.8) + (.45)(.3) \qquad (.55)(.2) + (.45)(.7)]$$

$$= [.575 \quad .425]$$

This process may be continued for years five, six, and so on.

Example 2 (*Compare Exercise 22*)
A rental firm has three locations. A truck rented at one location may be returned to any of the locations. The company's records show the probability of a truck rented at one location being returned to another. From these records the transition matrix is formed. Each location is considered a state.

		Returned to		
		I	*II*	*III*
Rented from	*I*	.8	.1	.1
	II	.3	.6	.1
	III	.1	.2	.7

Assume that all the trucks are rented daily.

(a) If the trucks are initially distributed with 40% at location I, 25% at location II, and 35% at location III, find the distribution on the second and third days.

(b) If a truck is rented at location II, for each location, find the probability that it will be at that location after 3 days.

Solution **(a)** The initial-state matrix is [.40 .25 .35]. The distribution for day 2 is

$$\underset{\textbf{Day 1}}{\overset{\quad I \qquad II \qquad III}{[.40 \quad .25 \quad .35]}}\begin{bmatrix} .8 & .1 & .1 \\ .3 & .6 & .1 \\ .1 & .2 & .7 \end{bmatrix} = \underset{\textbf{Day 2}}{\overset{\quad I \qquad II \qquad III}{[.43 \quad .26 \quad .31]}}$$

The distribution for day 3 is

$$\underset{\textbf{Day 2}}{[.43 \quad .26 \quad .31]}\begin{bmatrix} .8 & .1 & .1 \\ .3 & .6 & .1 \\ .1 & .2 & .7 \end{bmatrix} = \underset{\textbf{Day 3}}{[.453 \quad .261 \quad .286]}$$

Thus, 45.3% are at location I, 26.1% are at location II, and 28.6% are at location III on the third day.

(b) Since we are interested in the fate of only those trucks originating at location II, we ignore the trucks at the other locations and use [0 1 0] as the initial-state probability matrix.

The probability that a truck is at each location on the second day is given by

$$[0 \quad 1 \quad 0]\begin{bmatrix} .8 & .1 & .1 \\ .3 & .6 & .1 \\ .1 & .2 & .7 \end{bmatrix} = [.3 \quad .6 \quad .1]$$

that is, the probability of being at location I is .3; at location II the probability is .6; and at location III the probability is .1.

For the third day, the state probability matrix is

$$[.3 \quad .6 \quad .1]\begin{bmatrix} .8 & .1 & .1 \\ .3 & .6 & .1 \\ .1 & .2 & .7 \end{bmatrix} = [.43 \quad .41 \quad .16]$$

The probability that a truck is at location II on the third day is 0.41.

Markov Chain

Let's summarize the ideas of a Markov chain.

A **Markov chain**, or **Markov process**, is a sequence of experiments with the following properties.

1. An experiment has a finite number of discrete outcomes, called **states**. The process, or experiment, is always in one of these states.
2. With each additional trial, the experiment can move from its present state to any other state or remain in the same state.
3. The probability of going from one state to another on the next trial depends only on the present state, and not on past states.
4. The probability of moving from any one state to another in one step is represented in a transition matrix.
   - **(a)** The transition matrix is square, since all possible states are used for rows and columns.
   - **(b)** Each entry is between 0 and 1, inclusive.
   - **(c)** The entries in each row add to 1.
5. The state matrix times the transition matrix gives the next-state matrix.

Steady State

A study of Markov processes enables us to determine the probability state matrix for a sequence of trials. It is sometimes helpful to know the long-term trends of a population, of the market of a product, or of political processes. A Markov chain may provide some useful long-term information because some Markov processes will tend toward a **steady state**, or **equilibrium**. Here is a simple example.

Example 3 (*Compare Exercise 23*)
The transition matrix of a Markov process is

$$T = \begin{bmatrix} .6 & .4 \\ .1 & .9 \end{bmatrix}$$

and an initial-state matrix is [.50 .50].

If we compute a sequence of next-state matrices, we obtain the following:

Step	State Matrix
Initial	$[.50 \quad .50]$
1	$[.35 \quad .65] = [.50 \quad .50]T$
2	$[.275 \quad .725] = [.35 \quad .65]T = [.50 \quad .50]T^2$
3	$[.238 \quad .762] = [.275 \quad .725]T = [.50 \quad .50]T^3$
4	$[.219 \quad .781] = [.238 \quad .762]T = [.50 \quad .50]T^4$
5	$[.209 \quad .791] = [.219 \quad .781]T = [.50 \quad .50]T^5$
6	$[.204 \quad .796] = [.209 \quad .791]T = [.50 \quad .50]T^6$
7	$[.202 \quad .798] = [.204 \quad .796]T = [.50 \quad .50]T^7$
8	$[.201 \quad .799] = [.202 \quad .798]T = [.50 \quad .50]T^8$

It appears that the state matrix is approaching [.20 .80] as the sequence of trials progresses. In fact, that is the case. Furthermore, the state matrix [.20 .80] has an interesting property, which we can observe when we find the next-state matrix.

$$\begin{aligned}[.20 \quad .80]\begin{bmatrix} .6 & .4 \\ .1 & .9 \end{bmatrix} &= [(.20)(.6) + (.80)(.1) \qquad (.20)(.4) + (.80)(.9)] \\ &= [.12 + .08 \quad .08 + .72] \\ &= [.20 \quad .80]\end{aligned}$$

There is no change in the next state. The proccess has reached a **steady**, or **equilibrium**, state.

Definition

A state matrix $X = [p_1 \quad p_2 \quad \ldots \quad p_n]$ is a **steady-state** or **equilibrium** matrix for a transition matrix T if $XT = X$.

Example 4 The steady-state matrix for the alumni problem is $[.6 \quad .4]$ because an initial-state matrix will eventually approach $[.6 \quad .4]$ and

$$[.6 \quad .4]\begin{bmatrix} .8 & .2 \\ .3 & .7 \end{bmatrix} = [.48 + .12 \qquad .12 + .28]$$
$$= [.60 \quad .40]$$

This indicates that as long as the transition matrix represents the giving practices of the alumni, they will stablize at 60% contributing and 40% not contributing.

Finding the Steady-State Matrix

Let's show how to find the steady-state matrix for the alumni problem. Let $X = [x \quad y]$ be the desired, but unknown, steady-state matrix. We want to find x and y so that

$$[x \quad y]\begin{bmatrix} .8 & .2 \\ .3 & .7 \end{bmatrix} = [x \quad y]$$

The matrix product on the left gives

$$[.8x + .3y \quad .2x + .7y] = [x \quad y]$$

so

$$\begin{aligned} .8x + .3y &= x \\ .2x + .7y &= y \end{aligned}$$

which is equivalent to

$$\begin{aligned} -.2x + .3y &= 0 \\ .2x - .3y &= 0 \end{aligned}$$

Since $[x \quad y]$ is a probability matrix, we must have $x + y = 1$. This, with the two equations above, gives the system

$$\begin{aligned} x + y &= 1 \\ -.2x + .3y &= 0 \\ .2x - .3y &= 0 \end{aligned}$$

If we use an augmented matrix to solve this system, we have

$$\left[\begin{array}{rr|r} 1 & 1 & 1 \\ -.2 & .3 & 0 \\ .2 & -.3 & 0 \end{array}\right]$$

which eventually reduces to

$$\left[\begin{array}{rr|r} 1 & 0 & .6 \\ 0 & 1 & .4 \\ 0 & 0 & 0 \end{array}\right]$$

so $x = .6$, $y = .4$, and $[.6 \quad .4]$ is the steady-state matrix. This approach will work in general.

Example 5 (*Compare Exercise 25*)
Find the steady-state matrix of the transition matrix

$$T = \begin{bmatrix} .3 & .2 & .5 \\ .1 & .4 & .5 \\ .4 & 0 & .6 \end{bmatrix}$$

Solution Solve the equation

$$[x \quad y \quad z]\begin{bmatrix} .3 & .2 & .5 \\ .1 & .4 & .5 \\ .4 & 0 & .6 \end{bmatrix} = [x \quad y \quad z]$$

for a probability matrix $[x \quad y \quad z]$, which is the system

$$\begin{aligned} .3x + .1y + .4z &= x \\ .2x + .4y \qquad &= y \\ .5x + .5y + .6z &= z \end{aligned}$$

plus the equation $x + y + z + 1$. Write this as

$$\begin{aligned} x + \quad y + \quad z &= 1 \\ -.7x + .1y + .4z &= 0 \\ .2x - .6y \qquad &= 0 \\ .5x + .5y - .4z &= 0 \end{aligned}$$

Use the Gauss-Jordan method to solve. It gives the following sequence of augmented matrices.

$$\left[\begin{array}{rrr|r} 1 & 1 & 1 & 1 \\ -.7 & .1 & .4 & 0 \\ .2 & -.6 & 0 & 0 \\ .5 & .5 & -.4 & 0 \end{array}\right]$$

Multiply the last three equations through by 10, and use row operations to obtain 0's in column 1 below row 1.

$$\left[\begin{array}{rrr|r} 1 & 1 & 1 & 1 \\ 0 & 8 & 11 & 7 \\ 0 & -8 & -2 & -2 \\ 0 & 0 & -9 & -5 \end{array}\right]$$

$$\left[\begin{array}{rrr|r} 1 & 1 & 1 & 1 \\ 0 & 8 & 11 & 7 \\ 0 & 0 & 9 & 5 \\ 0 & 0 & -9 & -5 \end{array}\right]$$

$$\left[\begin{array}{ccc|c} 1 & 1 & 1 & 1 \\ 0 & 8 & 11 & 7 \\ 0 & 0 & 9 & 5 \\ 0 & 0 & 0 & 0 \end{array}\right]$$

$$\left[\begin{array}{ccc|c} 1 & 1 & 1 & 1 \\ 0 & 1 & \frac{11}{8} & \frac{7}{8} \\ 0 & 0 & 1 & \frac{5}{9} \\ 0 & 0 & 0 & 0 \end{array}\right]$$

$$\left[\begin{array}{ccc|c} 1 & 1 & 0 & \frac{4}{9} \\ 0 & 1 & 0 & \frac{1}{9} \\ 0 & 0 & 1 & \frac{5}{9} \\ 0 & 0 & 0 & 0 \end{array}\right]$$

$$\left[\begin{array}{ccc|c} 1 & 0 & 0 & \frac{3}{9} \\ 0 & 1 & 0 & \frac{1}{9} \\ 0 & 0 & 1 & \frac{5}{9} \\ 0 & 0 & 0 & 0 \end{array}\right]$$

This gives $x = \frac{3}{9}$, $y = \frac{1}{9}$, $z = \frac{5}{9}$, and the steady-state matrix $[\frac{3}{9} \quad \frac{1}{9} \quad \frac{5}{9}]$.

Example 6 (*Compare Exercise 26*)
A sociologist made a regional study of the shift of population between rural and urban areas. The transition matrix of the annual shift from one area to another was found to be

		To	
		R	U
From	R	.76	.24
	U	.08	.92

indicating that 76% of rural residents remain in rural areas, 24% move from rural to urban areas; 8% of urban residents move from urban to rural areas, and 92% remain in the urban areas. Find the percentage of the population in rural and urban areas when the population stabilizes.

Solution Let $[x \quad y]$ be the state matrix of the population, with x the proportion in rural areas and y the proportion in urban areas. We want to find the steady-state matrix, that is, the solution to

$$[x \quad y]\begin{bmatrix} .76 & .24 \\ .08 & .92 \end{bmatrix} = [x \quad y]$$

This condition, with $x + y = 1$, gives the system

$$\begin{aligned} x + \quad y &= 1 \\ -.24x + .08y &= 0 \\ .24x - .08y &= 0 \end{aligned}$$

The solution to the system is $x = .25$ and $y = .75$, so the steady-state matrix is $[.25 \quad .75]$, indicating that the population will stabilize at 25% in rural areas and 75% in urban areas.

It is sometimes important to know whether a Markov process will eventually reach equilibrium. In Examples 4 and 5, we found the steady-state matrix. It happens to be true in those cases that we will eventually reach the steady-state matrix after a sequence of trials, regardless of the initial-state matrix. While this is not true for all transition matrices, there is a rather reasonable property that ensures that a Markov process will reach equilibrium. We call transition matrices with this property **regular**. A regular Markov process will eventually reach a steady state and its transition matrix has the following property.

Definition A transition matrix T of a Markov process is called **regular** if some power of T has only positive entries.

Example 7 (*Compare Exercise 34*)

$$T = \begin{bmatrix} .3 & .7 \\ .25 & .75 \end{bmatrix}$$

is regular because its first power contains all positive entries.

$$T = \begin{bmatrix} 0 & 1 \\ .6 & .4 \end{bmatrix} \text{ is regular because } \begin{bmatrix} 0 & 1 \\ .6 & .4 \end{bmatrix}^2 = \begin{bmatrix} .6 & .4 \\ .24 & .76 \end{bmatrix}$$

has all positive entries.

Example 8 (*Compare Exercise 30*)
Find the steady-state matrix of the regular transition matrix

$$T = \begin{bmatrix} 0 & .5 & .5 \\ .5 & .5 & 0 \\ .5 & 0 & .5 \end{bmatrix}$$

(it is regular).

Solution The condition

$$[x \quad y \quad z]\begin{bmatrix} 0 & .5 & .5 \\ .5 & .5 & 0 \\ .5 & 0 & .5 \end{bmatrix} = [x \quad y \quad z]$$

with $x + y + z = 1$ yields a system of four equations whose augmented matrix is

$$\left[\begin{array}{ccc|c} 1 & 1 & 1 & 1 \\ -1 & .5 & .5 & 0 \\ .5 & -.5 & 0 & 0 \\ .5 & 0 & -.5 & 0 \end{array}\right]$$

(Be sure that you can get this matrix.)

We will not show all the row operations that lead to the solution, but the final matrix is

$$\left[\begin{array}{ccc|c} 1 & 0 & 0 & \frac{1}{3} \\ 0 & 1 & 0 & \frac{1}{3} \\ 0 & 0 & 1 & \frac{1}{3} \\ 0 & 0 & 0 & 0 \end{array}\right]$$

so the steady-state matrix is $[\frac{1}{3} \quad \frac{1}{3} \quad \frac{1}{3}]$.

6–7 Exercises

I.

1. Which of the following are transition matrices?

(a) $\begin{bmatrix} .6 & .4 \\ .3 & .7 \end{bmatrix}$

(b) $\begin{bmatrix} .6 & 0 & .4 \\ .2 & .1 & .5 \\ .3 & .4 & .3 \end{bmatrix}$

(c) $\begin{bmatrix} .5 & .5 \\ .3 & .7 \\ .2 & .8 \end{bmatrix}$

(d) $\begin{bmatrix} .1 & 0 & .3 & .6 \\ .2 & .3 & 0 & .5 \\ .4 & .2 & .1 & .3 \\ 0 & .25 & .35 & .4 \end{bmatrix}$

2. Which of the following are probability state matrices?

(a) $[.4 \quad .3 \quad .3]$
(b) $[.3 \quad .3 \quad .3]$
(c) $[.1 \quad .2 \quad .3 \quad .4]$
(d) $[.6 \quad .7]$
(e) $[.1 \quad .2 \quad 0 \quad .3 \quad .4]$

3. The matrix T represents the transition of college students between dorms (D) and apartments (A) at the end of a semester.

		To	
		A	D
From	A	.9	.1
	D	.4	.6

(a) What percent of those living in apartments move to a dorm?

(b) What is the probability that a student remains in a dorm the next semester?

(c) What is the probability that an apartment-dwelling student will remain in an apartment?

(d) What percent of dorm residents move to an apartment?

4. An investment firm invests in stocks, bonds, and mortgages for its clients. One of the partners in the firm analyzed the investment patterns of her clients. She found that during a year they change between types of investments according to the following transition matrix.

		To		
		S	B	M
From	S	.88	.09	.03
	B	.15	.75	.10
	M	.19	.17	.64

(a) What percent of those investing in bonds move to mortgages?

(b) What percent move their investments from stocks to bonds?

(c) What is the probability that a bond investor will leave his investment in bonds?

(d) What is the probability that a mortgage investor will change to stocks or bonds?

5.

	A	B	C
A	.3	.2	.5
B	.4	.6	0
C	.1	.8	.1

$= T$

From the transition matrix T, find

(a) the probability of moving from state B to state A.

(b) the probability of moving from state C to state B.

(c) the probability of remaining in state A.

For Exercises 6 through 8, find the next-state matrix from the given present-state and transition matrices.

6. $S = [.2 \quad .8] \quad T = \begin{bmatrix} .5 & .5 \\ .8 & .2 \end{bmatrix}$

7. $S = [.45 \quad .55] \quad T = \begin{bmatrix} .4 & .6 \\ .9 & .1 \end{bmatrix}$

8. $S = [.2 \quad .5 \quad .3] \quad T = \begin{bmatrix} .3 & .5 & .2 \\ .2 & .2 & .6 \\ .1 & .8 & .1 \end{bmatrix}$

Initial-state and transition matrices are given in Exercises 9 through 11. Find the following two next-state matrices. (*See Example 1*)

9. $M_0 = [.65 \quad .35] \quad T = \begin{bmatrix} .24 & .76 \\ .36 & .64 \end{bmatrix}$

10. $M_0 = [.3 \quad .7] \quad T = \begin{bmatrix} .8 & .2 \\ .2 & .8 \end{bmatrix}$

11. $M_0 = [.25 \quad .50 \quad .25] \quad T = \begin{bmatrix} .3 & .4 & .3 \\ .1 & .3 & .6 \\ .2 & .5 & .3 \end{bmatrix}$

For Exercises 12 through 15, show that the given state matrix S is the steady-state matrix for the transition matrix T.

12. (*See Example 3*)

$S = [.375 \quad .625] \quad T = \begin{bmatrix} .5 & .5 \\ .3 & .7 \end{bmatrix}$

13. $S = \begin{bmatrix} \frac{2}{3} & \frac{1}{3} \end{bmatrix} \quad T = \begin{bmatrix} .6 & .4 \\ .8 & .2 \end{bmatrix}$

14. $S = [.625 \quad .375] \quad T = \begin{bmatrix} .58 & .42 \\ .7 & .3 \end{bmatrix}$

15. $S = \begin{bmatrix} \frac{13}{28} & \frac{15}{28} \end{bmatrix} \quad T = \begin{bmatrix} .25 & .75 \\ .65 & .35 \end{bmatrix}$

16. Let T be the transition matrix

$$T = \begin{bmatrix} .25 & .75 \\ .40 & .60 \end{bmatrix}$$

and $M_0 = [.2 \quad .8]$, an initial-state matrix.

(a) Compute the next-state matrix $M_1 = M_0T$. Then complete the next-state matrix from M_1, that is, $M_2 = M_1T$.

(b) Compute T^2.

(c) Compute M_0T^2. Verify that it equals M_1T.

17. For $M = [.3 \quad .3 \quad .4]$ and

$$T = \begin{bmatrix} .1 & .5 & .4 \\ .2 & .6 & .2 \\ .4 & .0 & .6 \end{bmatrix}$$

compute MT, $(MT)T$, $((MT)T)T$, and MT^3.

II.

18. A department store's charge accounts are either currently paid up or in arrears. Their records show that

    (i) 90% of those who are paid up this month will be paid up next month also.

    (ii) 40% of those in arrears this month will be in arrears next month also.

    (iii) Their current accounts are 85% paid up and 15% in arrears.

    (a) Represent this information with a tree diagram.

    (b) Represent this information with a state matrix and a transition matrix.

19. Find x so that $[.2 \quad x \quad .4]$ is a probability state matrix.

20. Find x, y, z so that

$$\begin{bmatrix} .1 & x & 0 \\ y & .3 & .5 \\ .2 & .7 & z \end{bmatrix}$$

is a transition matrix.

21. The alumni of State University generally contribute (C) or do not contribute (NC) according to the following pattern: 75% of those who contribute one year will contribute the next year; 15% of those who do not contribute one year will contribute the next. The transition matrix is

		Next Year	
		C	NC
Present Year	C	.75	.25
	NC	.15	.85

Forty-five percent of last year's graduating class contributed this year. What percent will contribute next year? In two years?

22. (*See Example 2*) A rental firm has three locations. A truck rented at one location may be returned to any of the three locations. The transition matrix of where the trucks are returned is

		Returned to		
		I	II	III
Rented from	I	.6	.2	.2
	II	.1	.8	.1
	III	.2	.3	.5

(a) If all the trucks are rented daily, and they are initially distributed with 50% at location I, 25% at location II, and 25% at location III, how will they be distributed the next day? Two days later?

(b) A truck is rented at location I. Find, for each location, the probability that the truck will be at that location after 3 days.

23. (*See Example 3*) For the initial-state matrix $[.6 \quad .4]$ and the transition matrix

$$T = \begin{bmatrix} .5 & .5 \\ .2 & .8 \end{bmatrix}$$

find the sequence of the following six next-state matrices. What appears to be the steady-state matrix?

24. Find the steady-state matrix of

$$\begin{bmatrix} .6 & .4 \\ .2 & .8 \end{bmatrix}$$

25. (*See Example 5*) Find the steady-state matrix of the transition matrix

$$T = \begin{bmatrix} .6 & .2 & .2 \\ .1 & .8 & .1 \\ .2 & .4 & .4 \end{bmatrix}$$

26. (*See Example 6*) The transition of college students between dorms and apartments at the end of a semester is given by

		To	
		D	A
From	D	.9	.1
	A	.4	.6

Find the percent of the population in dorms and apartments when the population stabilizes.

27. Find the steady-state matrix of the transition matrix given in Exercise 5.

28. Use the transition matrix of Exercise 21 to find the steady-state distribution of State University Alumni who contribute and who do not contribute.

29. Use the transition matrix of Exercise 22 to find the proportion of trucks at each rental location when equilibrium is reached.

Find the steady-state matrix of the following regular matrices in Exercises 30 through 33.

(*See Example 8*)

30. $\begin{bmatrix} \frac{1}{3} & \frac{2}{3} \\ \frac{3}{4} & \frac{1}{4} \end{bmatrix}$

31. $\begin{bmatrix} .9 & .1 \\ 1 & 0 \end{bmatrix}$

32. $\begin{bmatrix} .3 & .2 & .5 \\ 0 & .5 & .5 \\ .7 & .2 & .1 \end{bmatrix}$

33. $\begin{bmatrix} \frac{1}{3} & \frac{1}{3} & \frac{1}{3} \\ \frac{1}{2} & \frac{1}{2} & 0 \\ 0 & \frac{1}{4} & \frac{3}{4} \end{bmatrix}$

34. (*See Example 7*) Show that

$$T = \begin{bmatrix} 0 & 0 & 1 \\ .2 & .3 & .5 \\ 0 & .3 & .7 \end{bmatrix}$$

is regular.

35. There are five points on a circle as shown in Figure 6–17. A particle is on one of the points and moves to an adjacent point in either direction. The probability is $\frac{1}{2}$ that it will move clockwise and $\frac{1}{2}$ that it will move counterclockwise. Write the transition matrix for this process.

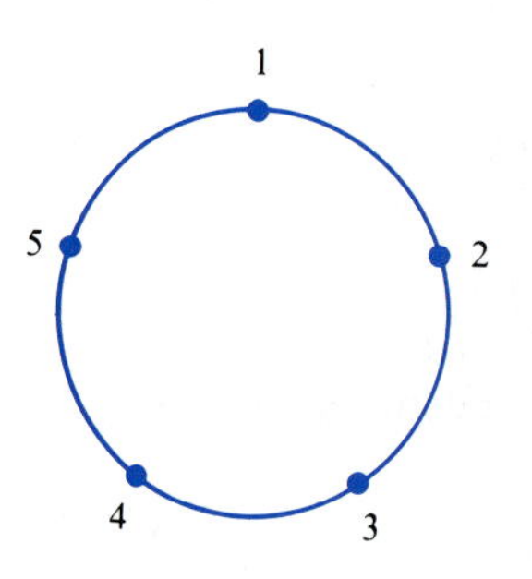

Figure 6–17

36. A nightwatchman must check three locations periodically. They are located in a triangle. To relieve the monotony, after arriving at one location, he tosses a coin to determine which of the other two he will check next.

(a) Form the transition matrix of this process.

(b) Find the steady-state matrix.

37. A plant with genotype *RW* can produce red (*R*), pink (*P*), or white (*W*) flowers. When two plants of this genotype are crossed, they produce the three colors according to the following transition matrix.

		Flowers of Offspring		
		R	*P*	*W*
Flowers of parent	*R*	.50	.50	.0
	P	.25	.50	.25
	W	.0	.50	.50

Flowers of this genotype are crossed for successive generations. When the process reaches a steady state, what percent of the flowers will be red, pink, and white.

38. Assume a person's profession can be classified as professional, skilled, or unskilled. Sociology studies give the following information about a child's profession as related to his or her parents.

Professional: Of their children, 80% are professional, 10% are skilled, and 10% are unskilled.

Skilled: Of their children, 60% are skilled, 20% are unskilled, and 20% are professional.

Unskilled: Of their children, 35% are skilled, 15% are professional, and 50% are unskilled.

(a) Set up the transition matrix of this process.

(b) Find the probability of an unskilled parent having a grandchild who is a professional.

(c) The population will eventually stabilize into fixed proportions of these professions. Find those proportions, that is, find the steady-state population distribution.

39. A country is divided into three geographic regions, I, II, and III. The matrix T is the transition matrix that shows the probabilities of moving from one region to another.

		To		
		I	*II*	*III*
From	*I*	.9	.06	.04
	II	.15	.8	.05
	III	.3	.1	.6

What is the probability of

(a) moving from region I to region II?

(b) moving from region II to region III?

(c) remaining in region II?

(d) moving from region I to region III?

40. A country is divided into three geographic regions. It is found that each year 5% of the residents move from region I to region II, and 5% move from region I to region III. In region II, 15% move to region I and 10% move to region III. In region III, 10% move to region I and 5% to region II.

Find the steady-state population distribution.

IMPORTANT TERMS

6–1	Experiment	Trials	Outcomes
	Sample space	Event	Simple outcome
	Probability assignment	Empirical probability	
6–2	Equally likely	Successes	Multiplication principal
6–3	Compound events	Union	Intersection
	Complement	Mutually exclusive events	Disjoint events
6–4	Conditional probability	Reduced sample space	Multiplication rule
	Independent events		
6–5	Bayes' rule		
6–6	Bernoulli trials	Repeated trials	
6–7	Markov chain	State	Present state
	Next state	Transition matrix	Probability-state matrix
	Initial-state matrix	Steady state	Equilibrium
	Regular matrix		

REVIEW EXERCISES

1. An experiment has six possible outcomes with the following probabilities.

$$P_1 = 0.2, P_2 = 0.3, P_3 = 0.1, P_4 = 0.0, P_5 = 0.2, P_6 = 0.3$$

Is this a valid probability assignment?

2. An experiment has the sample space $S = \{a, b, c, d\}$. Find the probability of each simple outcome in S if

$$P(a) = P(b), P(c) = 2P(b), P(d) = 3P(c)$$

3. A refreshment stand kept a tally of the number of soft drinks sold. One day their records showed the following:

Soft Drinks

Size	Number Sold
Small	94
Medium	146
Large	120

Find the probability that a person selected at random will buy a medium-sized soft drink.

4. Three people are selected at random from a group of five men and two women.

(a) Find the probability that all three selected are men.
(b) Find the probability that two men and one woman are selected.
(c) Find the probability that all three selected are women.

5. In a class of 30 students, ten participate in sports, 12 participate in band, and five participate in both. If a student is selected at random, find the probability that the student participate in sports or band.

6. In a group of 30 school children, 15 are eight-year-olds, 12 are nine-year-olds, and three are ten-year-olds. Of the eight-year-olds, ten are boys; of the nine-year-olds, five are boys; and of the ten-year-olds, two are boys. One child is selected at random from the group. Find the probability that the child is

(a) an eight-year-old.
(b) a boy.
(c) a nine-year-old.
(d) a twelve-year-old.
(e) a nine-year-old girl.
(f) a ten-year-old girl.

7. Three delegates for a conference are to be selected from 12 people. Find the probability that Mrs. Thomas and Ms. Roberts, two of the group, will be selected.

8. A die is rolled. Find the probability that an even number or a number greater than 4 will be rolled.

9. A single card is picked from a deck of 52 playing cards. Find the probability that it will be a king or a spade.

10. A coin and a die are tossed. Find the probability of throwing

(a) a head and a number less than 3.
(b) a tail and an even number.
(c) a head and a 6.

11. A coin is tossed five times. Find the probability that all five tosses will land heads up.

12. A card is selected from a deck of 52 playing cards. Find the probability that it will be

(a) a red card or a 10.
(b) a face card or a spade.
(c) a face card or a 10.

13. A bag contains five red, four white, and three black balls. Find the probability of drawing

(a) a red or a white ball, if one ball is drawn.
(b) two black or two red, if two balls are drawn without replacement.

14. A card is selected at random from a deck of bridge cards. Find the probability that

(a) it is not an ace.
(b) it is not a face card.

15. A bargain table has 40 books; ten are romance, ten are biographies, ten are crafts, and ten are historical fiction. If two books are selected at random, what is the probability that

(a) they are the same kind?
(b) they are different kinds?

16. A load of lumber contains 40 pieces of birch and 50 pieces of pine. Of the lumber, five pieces of birch and three pieces of pine are warped. Let F, G, and H be the events of selecting a birch, pine, and a warped piece of wood, respectively. Compute and interpret the following probabilities.

(a) $P(F)$, $P(G)$, $P(H)$

(b) $P(F \cap H)$

(c) $P(F \cup H)$

(d) $P(F' \cup H)$

(e) $P((F \cup H)')$

17. Compute $C(4, 2)(.4)^2(.6)^2$.

18. Find $P(X = 5)$ for a Bernoulli trial in which $n = 6$ and $p = 0.2$.

19. A card is picked from a deck of 52 playing cards. Let F be the event of selecting an even number, and let G be that of selecting a ten. Are G and F independent events?

20. A card is picked from a deck of 52 playing cards. Let R be the event of selecting a red card, and let Q be the event of selecting a queen. Are R and Q independent?

21. A die is tossed four times. Find the probability of obtaining

(a) one, two, three, four, in that order.

(b) one, two, three, four, in any order.

(c) two even numbers, then a 5, then a number less than 3.

22. Two dice are rolled. Find the probability that

(a) the sum of the numbers on the dice is 6.

(b) the same number is obtained on each die.

23. A student has four examinations to take. He has determined that the probability of his passing the mathematics examination is 0.8, that of passing English 0.5, that of passing history 0.3, and that of passing chemistry 0.7. Assuming independence of examinations, find the probability of his

(a) passing mathematics, history, and English, but failing chemistry.

(b) passing mathematics and chemistry, but failing history and English.

(c) passing all four subjects.

24. A study of juvenile delinquents shows that 60% come from low-income families (LI), 45% come from broken homes (BH), and 35% come from both ($LI \cap BH$). A juvenile delinquent is selected at random.

(a) Find the probability that the juvenile is not from a low-income family.

(b) Find the probability that the juvenile comes from a broken home or a low-income family.

(c) Find the probability that the juvenile comes from a low-income family, given the juvenile comes from a broken home.

(d) Are LI and BH independent?

(e) Are LI and BH mutually exclusive?

25. A study of the adult population in a midwestern state found the following information on drinking habits.

	Men	Women
Abstain	20%	40%
Infrequent	10%	20%
Moderate	50%	35%
Heavy	20%	5%

The adult population of the state is 55% female and 45% male.

(a) An individual selected at random is found to be an abstainer. Find the probability that the person is male.

(b) An individual selected at random is found to be a heavy drinker. Find the probability that the individual is female.

26. A stock analyst classifies stocks as either blue chip (BC) or not (NBC). The analyst also classifies stock by whether it goes up (UP), remains unchanged (UC), or goes down (D) at the end of a day's trading. One percent of the stocks are blue chip. The analyst summarizes the performance of stocks as follows:

	Probability of		
	UP	UC	D
BC	.45	.35	.20
NBC	.35	.25	.40

(a) Show this information with a tree diagram.

(b) A customer selects a stock at random and asks the analyst to buy. Find the probability that the stock is a blue chip stock that goes up the next day.

27. A two-digit number is to be constructed at random from the numbers 1, 2, 3, 4, 5, and 6, repetition not allowed. Find the probability of getting

(a) the number 33.

(b) the number 35.

(c) a number the sum of whose digits is 10.

(d) a number whose first digit is greater than its second.

(e) a number less than 24.

28. A finite mathematics professor observed that 90% of the students who do the homework regularly pass the course. He also observed that only 20% of those who do not do the homework regularly pass the course. One semester, he estimated that 70% of the students did the homework regularly. Given a student who passed the course, find the probability that the students did the homework regularly.

29. A sociology class is composed of ten juniors, 34 seniors, and six graduate students. Two juniors, eight seniors, and three graduate students received an A in the course. A student is selected at random and is found to have received an A. What is the probability that the student is a junior?

30. In a certain population, 5% of the men are colorblind and 3% of the women are colorblind. The population is made up of 55% men and 45% women. If a person chosen at random is colorblind, what is the probability that the person is a man?

31. The probability of having blood type A^+ is $\frac{1}{3}$. In a group of eight people, what is the probability that six of them will have this blood type?

7

Topics in this chapter can be applied to:

Opinion Polls ● Population Studies ● Advertising Analyses ● Sales Comparisons ● Pricing ● Customer Services ● Quality Control ● Revenue Projections ● Risk Analyses ● Product Performance Evaluations ● Decision Making ● Marketing Strategies

Statistics

- **Frequency Distributions**
- **Measures of Central Tendency**
- **Measures of Dispersion: Range, Variance, and Standard Deviation**
- **Random Variables and Probability Distribution**
- **Expected Value**
- **Normal Distribution**
- **Approximating the Binomial Distribution**
- **Estimating Bounds on a Proportion**

When the President of the United States submits an annual budget to Congress in the trillion dollar range, many taxpayers will ask "Where is all that money going?" But few of them wish to be handed a detailed budget a foot thick. They want the information summarized in a few broad categories such as defense, social security, education, agriculture, interest on the debt, and so on. Meanwhile, the President may want to know how the voters react to specific budget items such as Defense and Social Security budgets. To poll all voters regarding their opinion of, for instance, the defense budget is impractical. However, the President can obtain valuable, although incomplete, information on voter opinion by a **sample** opinion poll.

The President's budget summary is an example of applying **descriptive statistics**. Descriptive statistics summarize a mass of data and describe its more prominent features.

The sample poll, on the other hand, falls into the category of **inferential statistics**. Inferential statistics make generalizations or draw conclusions from incomplete information. This chapter presents some methods used in descriptive and inferential statistics. Sections 1 through 3 deal with descriptive statistics, whereas Sections 4 through 8 deal with inferential statistics.

7–1 Frequency Distributions

Frequency Table
Construction of a Frequency Table
Visual Representations of Frequency Distributions
Histogram
Frequency Polygon
Pie Chart

Frequency Table

Opinion polls and population studies use random samples to obtain information. Usually, the data forms a mass of unorganized information. Useful information may be obtained by grouping similar observations into categories and reporting the number of observations in each category. For example, a complete list of the size of each class at State University may be too unorganized to help a building-use committee determine classroom needs. A summary like the following table provides more useful information.

Size of Classes	Number of Classes
Fewer than 5	9
5–10	37
11–20	52
21–50	213
51–100	91
101–200	5
201–500	2
Total	409

We call this kind of summary a **frequency table** or **frequency distribution**. We call the number of observations in a category the **frequency** of that category. It gives the number of classes for each class size category at the university. A range, or interval, of numbers like 5 to 10 or 11 to 20 determines each category.

In some summaries, the categories may not be numerical. For example, we may summarize the students' majors by subject category at the university with a frequency table like the following:

Major	Number of Students
Science	429
Arts	132
Languages	41
Social Sciences	631
Engineering	344
Total	1577

Notice that a frequency table does not present all available information. The frequency table of the size of university courses does not tell the number of students in the smallest or the largest classes. In fact, it does not tell the number of students in any class. It does give general information from which a decision may be made about the number and size of classrooms needed.

Example 1 A mathematics quiz consists of five questions. The professor summarizes the performance of the class of 75 students with the following frequency distribution. Each quiz question determines a different category.

Question	Number of Correct Answers (frequency)
1	36
2	41
3	22
4	54
5	30

Example 2 *(Compare Exercise 1)*
A survey of students reveals that they spent the following amounts of money on books for three courses during a semester:

$ 78	$123	$136	$162	$ 96	$145
$115	$183	$150	$110	$191	$ 88
$157	$137	$122	$172	$165	$119
$105	$127	$148	$170	$131	$118

Make a frequency table to summarize the students' book expense.

Solution We form five categories with $25 intervals, starting at $75, and obtain the following:

Book Expenses	Number of Students (Frequency)
$75–99	3
$100–124	7
$125–149	6
$150–174	6
$175–199	2

Construction of a Frequency Table

Construct a frequency table in three steps, as in the following.

Step 1. Choose the categories by which the information will be grouped—for example, size of classes at State University.

Step 2. Place each piece of information in the appropriate category—for example, sort the classes by size and place them in the appropriate category.

Step 3. Count the data in each category. This is the **frequency** of the category—for example, State University has two classes in the 201 to 500 range.

The category decisions in Step 1 make the other two steps mechanical. The determination of categories can be a two-step decision. First, determine the number of categories (e.g., 7) and then the **range** of values each category covers (e.g., fewer than 5, 5–10, etc.). No magic formula exists to make these decisions. They depend on the nature of the data and the message you wish to convey. However, some generally accepted rules of thumb may help. Just remember, exceptions are appropriate at times.

Hints on Setting Up Category Intervals

1. Number of categories.

Generally, we use 5 to 15 categories. More categories may be unwieldy, and fewer categories may not distinguish between important features. Be sensible about applying this rule. Don't use five categories to summarize a two-category situation, such as male-female interest in football. Fifteen categories are not appropriate when you have only a few observations.

2. Range of values covered by the categories.

   (a) Choose each category so that each piece of information is in some category. Be sure to include the largest and smallest values in some category, and be sure you leave no gaps between categories when they might include some of the data. Don't use intervals like $2.00 to 4.00 and $5.00 to 7.00 when $4.75 is a valid data point.

   (b) Assign each observation to only one category. If one category has a range of values 300 to 400, and the next category has a range of 400 to 500, it is not clear where to place 400. The category designations 300 to 399 and 400 to 499 clearly indicate where to place 400.

   (c) Try to make the category intervals the same length, and use intervals that make sense for the situation. For example, if you summarize traffic speeds for the city traffic department, they probably will not be impressed with category intervals such as 33 to 38 and 39 to 42. It is more natural and sensible to use intervals such as 31 to 35 and 36 to 40. On the other hand, don't force equal intervals when different lengths make sense. We summarize class performance on a test with intervals of 0 to 59, 60 to 69, 70 to 79, 80 to 89, 90 to 100 because these intervals represent letter grades. Notice that the example of class sizes also uses intervals of different lengths.

Visual Representations of Frequency Distributions

Because a picture carries a more forceful message than a column of numbers, a visual presentation of a frequency table sometimes provides a better understanding of the data. We will study three common visual methods, the **histogram**, the **frequency polygon**, and the **pie chart**. In each case, the graph shows information obtained from a frequency table.

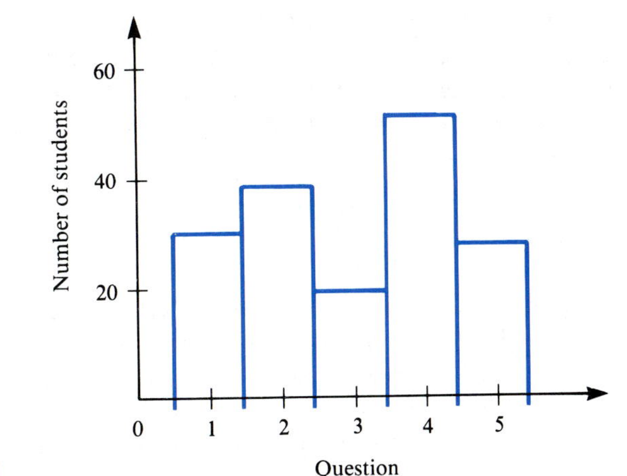

Figure 7–1

Histogram

A **histogram** is a bar chart where each bar represents a category, and its height represents the frequency of the category. Figure 7–1 is the histogram of Example 1.

Mark the categories on a horizontal scale and the frequencies on a vertical scale. The bars are of equal width and are centered at the midpoint of the category interval. In this case, a single number forms the category interval. The bars should be of equal width because two bars with the same height and different widths have different areas. This gives an impression of different frequencies. Thus, the area, not just the height of the bar, customarily represents the frequency when different bar widths are used.

Frequency Polygon

A second visual representation, a **frequency polygon**, is a line chart of a frequency distribution. You construct it in much the same way as the histogram. Again, the horizontal scale represents the categories, and the vertical scale represents the frequencies. Plot the frequency of each category above the midpoint of the category and connect the points with straight lines. Customarily, the polygon is closed by placing two points on the baseline at either end. Locate these two points as if there were another class interval at each end.

The frequency polygon shown in Figure 7–2 represents the frequency distribution of Example 1.

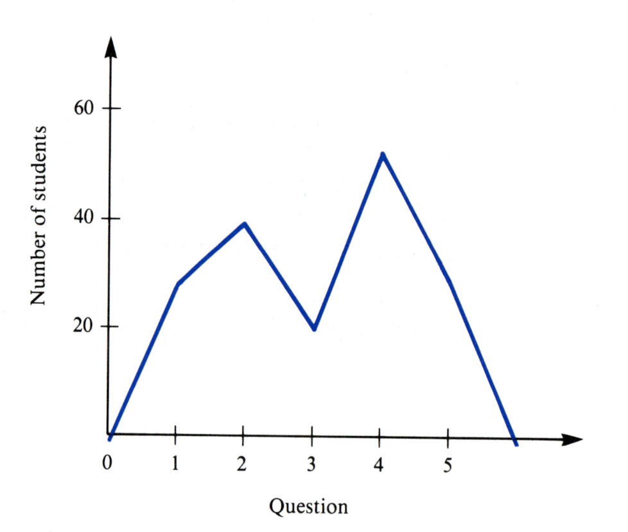

Figure 7–2

Example 3 (*Compare Exercise 8*)
The university registrar selects 100 transcripts at random and records the GPA for each. The frequency distribution follows:

GPA	Frequency	GPA	Frequency
0–0.49	5	2.00–2.49	18
0.5–0.99	9	2.50–2.99	22
1.00–1.49	17	3.00–3.49	11
1.50–1.99	10	3.50–4.00	8

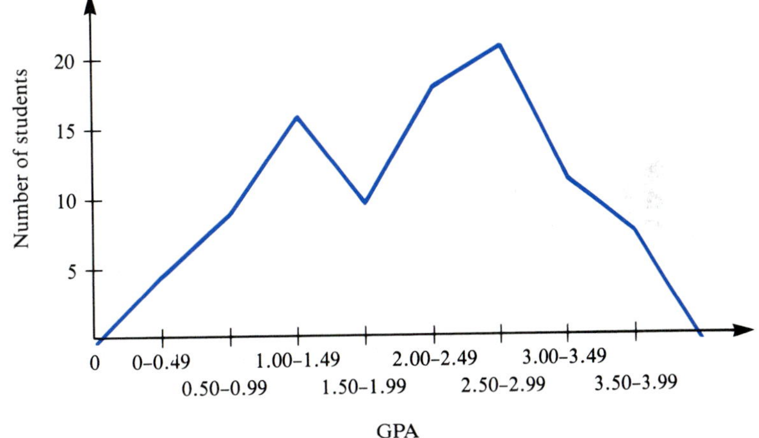

Figure 7–3

The frequency polygon representing this information appears in Figure 7–3.

Pie Chart

The third visual representation of data, the **pie chart**, emphasizes the proportion of data that falls into each category. You sometimes see a pie chart in the newspaper that represents the division of a budget into parts. The parts are frequently reported as percentages of the total. You may obtain the percentage of the data that falls into each category from a frequency table.

Example 4 (*Compare Exercise 13*)
Jim Dandy has $200 for spending money this month. He carefully prepares this budget.

Category	Amount	Category	Amount
Dates	$70	Bicycle Repairs	$48
Books and Records	$20	Miscellaneous	$52
Laundry	$10		

Construct a pie chart that represents this budget.

Solution First, compute the amount in each category as a percentage of the total by dividing the amount in the category by the total, \$200.

Category	Percentage of Total Amount
Dates	35
Books and Records	10
Laundry	5
Bicycle Repair	24
Miscellaneous	26

We "cut the pie" into pieces that have areas in the same proportion as the percentages representing the categories. See Figure 7–4.

A glance at this pie chart tells us the relative share of each category. The angle at the center of each slice determines the size of the slice. The simplest way to determine the size of the angle is to multiply each percentage by 360°. Then use a protractor to mark off the required angle.

For example, the bicycle repair category in the above example accounts for 24% of the budget. The angle used for this category is $.24 \times 360° = 86°$.

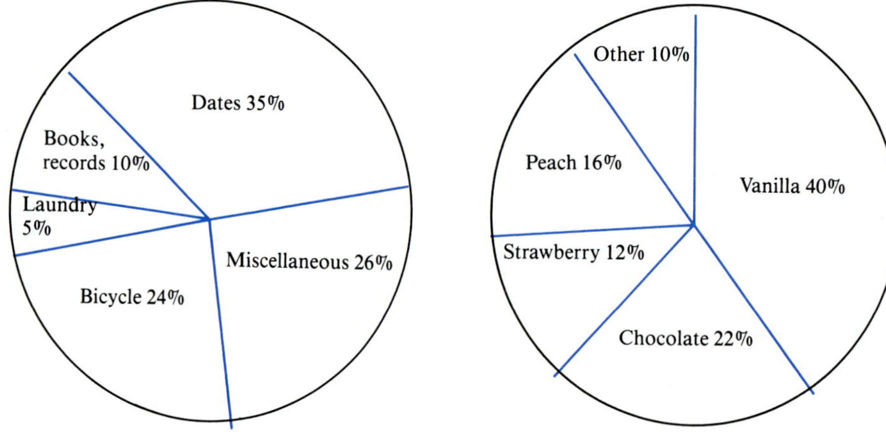

Figure 7–4

Figure 7–5

Example 5 An advertising firm asked 150 children their favorite kind of ice cream. Here is a frequency table of their findings:

Kind of Ice Cream	Number Who Favor
Vanilla	60
Chocolate	33
Strawberry	18
Peach	24
Other	15
Total	150

Summarize this information with a pie chart.

Solution Compute the percentages of each category and the size of the angle to be used.

Kind of Ice Cream	Percentage	Angle in Degrees
Vanilla	40 $\left(\frac{60}{150}\right)$	144 (0.40×360)
Chocolate	22 $\left(\frac{33}{150}\right)$	79 (0.22×360)
Strawberry	12 $\left(\frac{18}{150}\right)$	43 (0.12×360)
Peach	16 $\left(\frac{24}{150}\right)$	58 (0.16×360)
Other	10 $\left(\frac{15}{150}\right)$	36 (0.10×360)

(See Figure 7–5.)

You may wonder why we have several ways to represent a frequency distribution. The histogram, frequency polygon, and pie chart all give visual representations of the same information. The frequency polygon and the histogram show the size of each category in reports on monthly sales, annual gross national product growth for several years, or enrollment in accounting courses.

Use the pie chart when you wish to show the proportion of data that falls in each category in cases such as a breakdown of students by home state, a summary of family incomes, or a percentage breakdown of letter grades in a course.

The histogram is preferable when the categories are **discrete**, that is, when the values in one category are isolated from the values in other categories. When a traffic study summarizes the number of passengers per automobile, the categories such as 1 to 2, 3 to 4, and 5 or more are separated by a "gap." A value of 2.35 passengers in an automobile is not valid. On the other hand, the GPA example does not have discrete categories, any value from 0 through 4.0 is valid for a GPA. There is no jump from one category to another. The frequency polygon works well here. The frequency polygon better represents data when it varies continuously from one category to the next. The frequency polygon is simpler and more suited for making comparisons between distributions, such as quarterly sales comparisons for two consecutive years (see Figure 7–6).

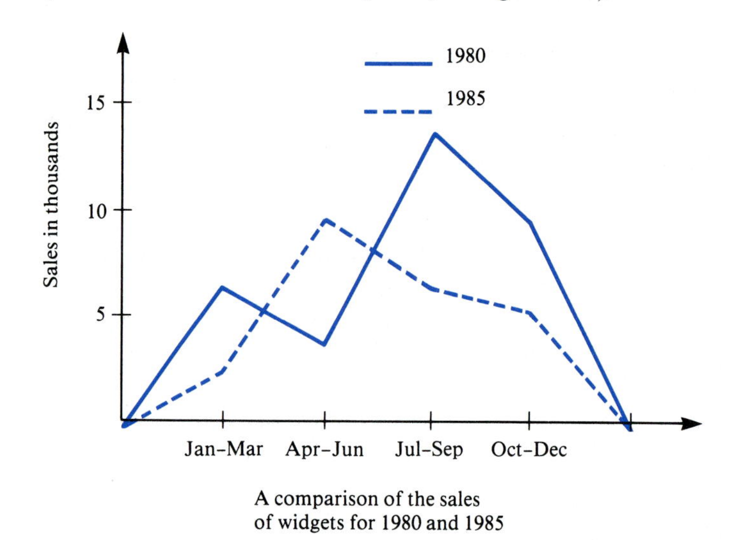

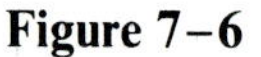

Figure 7–6 A comparison of the sales of widgets for 1980 and 1985

Choice of a Graph of a Frequency Distribution

Histogram: use for discrete data.
Frequency Polygon: use for continuous data and for comparing two distributions.
Pie Chart: use for representing the proportion of data in each category.

7–1 Exercises

1. (*See Example 2*) For the given set of data,

$$-1, 2, 2, 4, 6, 2, -1, 4, 4, 4, 6, 8$$

find the frequency of the following numbers:

(a) -1 **(b)** 2 **(c)** 4 **(d)** 6 **(e)** 8

2. An experiment is repeated 12 times. The results are

$$-3, 2, 4, -3, 4, 8, -3, 2, 8, 8, 8, 9$$

Give the frequency of each result.

3. The daily total number of students who used the state university swimming pool on 40 days during the summer are

90	98	137	108	128	115	152	122
110	132	149	131	102	109	118	126
121	145	89	149	86	120	97	118
142	139	128	110	105	104	131	159
93	119	107	129	132	129	98	116

Form a frequency table with the classes 85 to 99, 100 to 114, 115 to 129, 130 to 144, and 145 to 159.

4. A city police department radar unit recorded the following speeds one afternoon on 17th street:

30	46	53	28	52	39	34	29
42	27	48	33	37	29	44	42
38	47	31	51	40	31	36	49
41	26	50	39	35	30	45	43
38	41	36	28	52	34	37	43
35	44	49					

Form a frequency table representing this data. Use category intervals of 5 miles per hour.

5. The following is the distribution of grades on a math exam:

Grade	Number of Students
0–59	7
60–69	12
70–79	26
80–89	14
90–100	8

Which of the following quantities can be determined from this distribution? If the quantity can be determined, find it.

(a) the number of students who took the test
(b) the number of students who scored below 70
(c) the number of students who scored at least 80
(d) the number of students who scored below 75
(e) the number of students who scored between 69 and 90
(f) the number of students who scored 95

6. The daily totals of students who rode the campus shuttle bus are as follows:

166	172	184	176	181	84	170	198
182	203	210	141	77	93	147	205
164	122	211	137				

Summarize these totals with a frequency table with five categories.

7. A quiz consists of six questions. The tabulation shows the number of students who received each possible score of zero through six.

Score	Number of Students	Score	Number of Students
0	3	4	15
1	5	5	8
2	12	6	5
3	21		

Draw a histogram of the data.

8. (*See Example 3*) The student employment office surveyed 150 working students to determine how many hours they worked each week. They found the following:

Hours per Week	Number of Students
0–4.9	21
5–9.9	34
10–14.9	29
15–19.9	17
20–24.9	27
25–30	22

Draw the frequency polygon of this summary.

9. Professor X polled his students to determine the number of hours per week they spent on homework for his course. Here are his findings:

3.50	7.50	9.00	2.25	3.50	5.25
4.00	2.75	4.75	1.50	4.50	3.50
3.25	8.50	2.75	5.50	0.25	5.75
2.25	6.25	5.00	6.00	3.50	4.00
0.75	2.25	3.50	3.75	4.00	6.50

Using five category intervals,

(a) draw a histogram.
(b) draw a frequency polygon.

10. A sampling of State University freshmen reveals the following data for mathematics ACT scores:

ACT Score	Number of Students
18	10
19	8
20	15
21	21
22	26
23	30
24	44
25	38
26	50
27	28
28	16
29	12

Using two consecutive scores to form category intervals (18–19, 20–21, and so forth),

(a) which graph is more appropriate to represent this data?
(b) draw the graph.

11. The testing center predicts grades in freshman English based on a student's high school grades. They also follow up to see what grades are received. Here is a summary for a fall semester:

Grade	Number Predicted to Receive Grade	Number Who Received Grade
A	114	126
B	138	120
C	252	233
D	60	80
F	36	41

On the same scale, draw two frequency polygons comparing the number predicted to receive each grade and the number who actually did.

12. The Soft Drink Corporation reported sales on their Super Cola and Diet Super Cola for the first six months as follows:

	Number of Drinks Sold	
Month	Super Cola	Diet Super Cola
Jan.	125,000	75,000
Feb.	141,000	79,000
Mar.	145,000	85,000
Apr.	146,000	93,000
May	130,000	104,000
June	120,000	109,000

On the same scale, draw two graphs comparing the number of each drink sold.

13. (*See Example 4*) Draw the pie chart that represents this frequency distribution.

Brand of Coffee	Number Who Prefer This Brand
Brand X	28
Brand Y	34
Brand Z	18

14. Graph the following information with the most appropriate graph.

Items in Family Budget	Percentage of Income
Food	25
Housing	35
Utilities	22
Clothing	12
Recreation	6

Draw the pie charts for the frequency distributions in Exercises 15 through 21.

15.

Concentration of Ozone in Air of Large City (In Parts per Billion)	Number of Days
0–40	8
41–80	22
81–120	18
121–160	12

16.

Budget Items of Old Main University	Amount in Budget (Dollars)
Instructional	10,500,000
Administrative	1,500,000
Buildings and Grounds	2,000,000
Student Services	10,100,000
Other	5,700,000

17.

Income of Old Main University	Amount (Dollars)
Tuition and Fees	11,800,000
Endowment	2,200,000
Gifts	1,500,000
Auxiliary Enterprises	10,100,000
Other	2,400,000

18.

Education Level of Acme Manufacturing Employees	Number of Employees
Less than High School	45
High School Graduate	180
College Graduate	60
Graduate Work	15

19.

GPA of Tech Students	Number of Students
0–0.99	21
1.00–1.99	72
2.00–2.99	98
3.00–4.00	46

20.

Favorite Vacation Spot of State University Students	Number of Students
Beach	340
Mountains	420
A Big City	220
Visiting Relatives	130

21. The Census Bureau tabulates the income of Dogpatch families as follows:

Income (Dollars)	Number of Families
0–4999	7
5000–9999	32
10,000–16,999	19
17,000–30,000	4

7–2 Measures of Central Tendency

We have used histograms, frequency polygons, and pie charts to summarize a set of data. These devices sometimes make it easier to understand the data.

There are times, however, when we want to be more concise in reporting information, so comparisons can be easily made or so the information can be reduced to a single number. You often hear questions like,

"What was the class average on the exam?"
"What kind of gas mileage do you get on your car?"
"What happened to the price of homes from 1977 to 1984?"

You generally expect a response to questions like these to be a single number that is somehow "typical" or at the "center" of the exam grades, distance a car is driven on a certain amount of gas, or price of homes. We will study three ways to obtain this "central number," which is called a **measure of central tendency**.

The Mean

When we talk about "averages" like test averages, the average price of gasoline, or a basketball player's scoring average, we usually refer to one particular measure of central tendency, the arithmetic **mean**. To compute the mean of a set of numbers, simply add the numbers and divide by how many numbers were used.

Definition

The **mean** of the n numbers $x_1, x_2, \ldots, x_n$ is sometimes denoted by $\bar{x}$ and is computed

$$\bar{x} = \frac{x_1 + x_2 + \cdots + x_n}{n}$$

The Convenience Chain wants to compare sales in their 56 stores during July with sales a year ago when they had 49 stores. A comparison of total sales for each July may be misleading because the number of stores differs. Sales can be down from a year ago in each of the 49 stores, but total sales up because there are 7 additional stores. The mean sales of all stores in each year should better indicate if sales are improving.

Example 1 (*Compare Exercise 1*)
The mean of the test grades 82, 75, 96, 74 is

$$\frac{82 + 75 + 96 + 74}{4} = \frac{327}{4} = 81.75$$

Example 2 Find the mean of the annual salaries \$25,000, \$14,000, \$18,000, \$14,000, \$20,000, \$14,000, 18,000, 14,000.

Solution Add the salaries:

$$\begin{array}{r} 25{,}000 \\ 14{,}000 \\ 18{,}000 \\ 14{,}000 \\ 20{,}000 \\ 14{,}000 \\ 18{,}000 \\ \underline{14{,}000} \\ 137{,}000 \end{array}$$

Divide this total by eight, the number of salaries, to obtain the mean.

$$\text{Mean} = \frac{137000}{8} = \$17{,}125$$

This mean can be written in this more compact form:

$$\frac{25{,}000 + 4 \times 14{,}000 + 2 \times 18{,}000 + 20{,}000}{8}$$

where 4 is the **frequency** of the value 14,000, 2 is the frequency of 18,000, and 25,000 and 20,000 each have a frequency of one. The divisor, 8, is the sum of the frequencies. This form is useful in cases like the following where the scores are summarized in a frequency table. The data is **ungrouped** because different scores are not combined into one category.

Example 3 (*Compare Exercise 13*)
Scores are summarized in the frequency table:

Score (X)	Frequency (f)
3	2
4	1
5	8
6	4
	15

The mean is given by

$$\text{Mean} = \frac{2 \times 3 + 1 \times 4 + 8 \times 5 + 4 \times 6}{15} = \frac{74}{15} = 4.93$$

The general formula for the mean of a frequency distribution is as follows:

Formula for the Mean of an Ungrouped Frequency Distribution

Given the scores $x_1, x_2, \ldots, x_k$, which occur with frequency $f_1, f_2, \ldots, f_k$, respectively, the mean is

$$\text{Mean} = \frac{f_1x_1 + f_2x_2 + \cdots + f_kx_k}{n}$$

where $n = f_1 + f_2 + \cdots + f_k$, the sum of the frequencies.

The next example deals with **grouped data**. Scores are combined into categories and the frequency of that category is given.

Example 4 (*Compare Exercise 14*)
Professor Tuff gave a twenty-question quiz. He summarized the class performance with this frequency table.

Number of Correct Answers	Number of Students
0–5	8
6–10	14
11–15	23
16–20	10

Determine the class mean for this grouped data.

Solution We do not have specific values of the data, only the number in the indicated interval. To obtain an estimate of the mean, we use the midpoint of each category as the representative value and compute the mean by

$$\text{Mean} = \frac{8 \times 2.5 + 14 \times 8.0 + 23 \times 13.0 + 10 \times 18.0}{55} = 11.11$$

where 55 is the sum of the frequencies.

Note: The midpoint of an interval is the mean of the first and last numbers in the interval.

Formula for the Mean of Grouped Data

Let $x_1, x_2, \ldots, x_k$ be the **midpoints** of each category interval, and let $f_1, f_2, \ldots, f_k$ be the frequency of each interval, respectively.

$$\text{Mean} = \frac{f_1x_1 + f_2x_2 + \cdots + f_kx_k}{n}$$

where n is the sum of frequencies, $f_1 + f_2 + \cdots + f_k$.

The mean is a useful measure of central tendency because

1. it is familiar to most people;
2. it is easy to compute;
3. it can be computed for any set of numerical data; and
4. each set has just one mean.

The following example illustrates a disadvantage of the mean.

Example 5 The salaries of the Acme Manufacturing Company are

President,	\$200,000
Vice President,	25,000
Production Workers,	18,000

The company has 15 production workers. They complain to the president that company salaries are too low. Mr. President responds that the average company salary is about \$29,117. He maintains that that is a good salary. The production workers and the vice president remain unimpressed with this information because not a single one of them makes this much money. Although the President computed the mean correctly, he failed to mention that a single salary, his, was so large that the mean was in no way typical of all the salaries. This illustrates one of the disadvantages of the mean as a number that summarizes data. One or two extreme values can shift the mean making it a poor representative of the data.

The Median

Another measure of central tendency, this one not easily affected by one or two extreme values, is the **median**. When data from an experiment are listed according to size, people tend to focus on the middle of the list. Thus, the median is a useful measure of central tendency.

Definition

> The **median** is the value of the middle item after the data have been arranged in order. When there is an even number of items, there is no middle term and the median is the mean of the two middle items.

Basically, the median divides the data into two parts. One part contains the lower half of the data and the other part contains the upper half of the data.

Example 6 (*Compare Exercise 17*)
The median of the numbers 3, 5, 8, 13, 19, 22, and 37 is the middle number, 13. Three terms lie below 13, and three lie above.

Example 7 (*Compare Exercise 19*)
The median of the numbers 1, 5, 8, 11, 14, and 27 is the mean of the two middle numbers, 8 and 11. The median is

$$\frac{8 + 11}{2} = 9.5$$

Four terms lie below 9.5, and four lie above.

Example 8 (*Compare Exercise 20*)
Find the median of the set 8, 5, 2, 17, 28, 4, 3, and 2.

Solution First, the numbers must be placed in ascending order: 2, 2, 3, 4, 5, 8, 17, 28. Since there are eight numbers, an even number, there is no middle number. We find the mean of the two middle numbers, 4 and 5, to obtain 4.5 as the median.

Example 9 The set of numbers 2, 5, 9, 10, and 15 has a mean of 8.2 and a median of 9. If 15 is replaced by 140, the mean changes to 33.2, but the median remains 9. A change in one score of a set may make a significant change in the mean and yet leave the median unchanged.

The median is often used to report income, price of homes, and SAT scores.

The Mode

The **mode** is used less often as a measure of central tendency. It is used to indicate which observation or observations dominate the data because of the frequency of their occurrence.

Definition

> The **mode** of a set of data is the value that occurs most often. A distribution may have one mode, several modes, or no mode.

Example 10 (*Compare Exercise 23*)
The mode of the numbers 1, 3, 2, 5, 4, 3, 2, 6, 8, 2, and 9 is 2 because 2 appears more often than any other value.

Example 11 (*Compare Exercise 25*)
The set of numbers 2, 4, 8, 3, 2, 5, 3, 6, 4, 3, and 2 has two modes, 2 and 3, because they both appear three times, more than any other value. The set of numbers 2, 5, 17, 3, and 4 has no mode because each number appears just once.

The mode provides useful information about the most frequently occurring categories. The clothing store does not want to know that the mean size of men's shirts is 15.289. However, they like to know that they sell more size $15\frac{1}{2}$ shirts than any other size. The mode best represents summaries where the most frequent response is desired such as the most popular brand of coffee.

Which Measure of Central Tendency Is Best?

With three different measures of central tendency, you probably are curious about which is best. The answer: "It all depends." It all depends on the nature of the data and the information you wish to summarize.

The mean is a good summary for values that represent magnitudes, like test scores and price of shoes, if there are not one or two extreme values to distort the mean. The mean is the best measure when equal distances between scores represent equal differences between the things being measured. For example, the difference between $15 and $20 is $5, the same amount of money is the difference between $85 and $90.

The median is a positional average. It is better used when ranking people or things. In a ranking, an increase or decrease by a fixed amount may not represent the same amount of change at one end of the scale as it does at the other. The difference between the number one ranked tennis player and the number two ranked player at Wimbleton may be small indeed. The difference between the tenth and eleventh ranked players may be significantly greater. In contests, in student standing in class, and in taste tests numbers are assigned for ranking purposes. However, this does not imply that the people or things ranked all differ by equal amounts. In such cases, the median better measures central tendency.

The mode is better when summarizing dress sizes, or the brands of bread preferred by families. The information desired is the most typical category; the one that occurs most frequently.

Example 12 What is the most appropriate measure of central tendency for the price of hamburgers in town? Since the price is a quantity, not a rank, and since prices tend to occur in a fairly limited price range, the mean is the best measure to use. However, if Joe's Greasy Spoon has a super $39.95 hamburger, this extreme value could distort the mean. In such a case, the median may be more appropriate.

Example 13 A marketing class runs a taste test on hamburgers in town to determine the typical hamburger. What measure of central tendency is most appropriate? The students are evaluating taste, a quality, not a quantity. The mean is not appropriate in this case. It makes sense to use the median because half the hamburgers taste better and half taste worse, or use the mode to indicate the most popular hamburger in town.

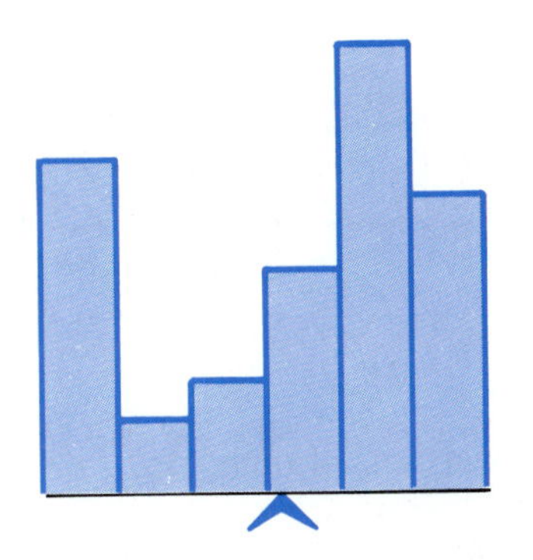

The mean is where the histogram balances.

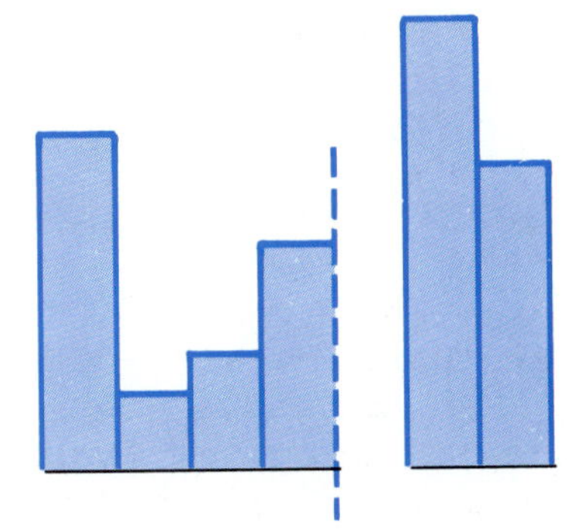

The median divides the histogram into equal parts.

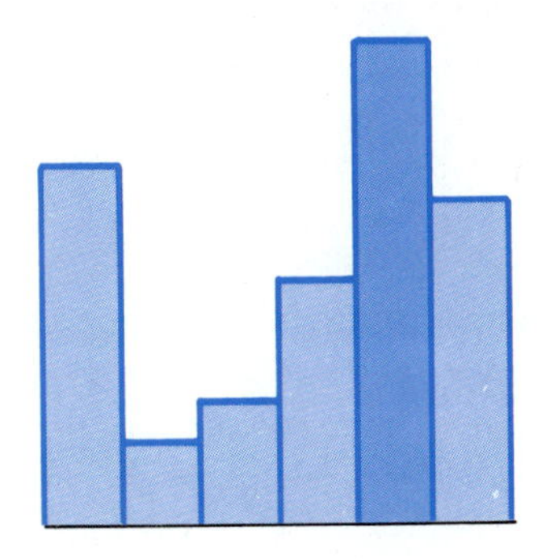

Figure 7–7 The mode is the tallest bar, or bars, in the histogram.

When you summarize a set of data with a histogram you can visualize the mean, median, and mode in the following way. Think of the histogram as being constructed from material of uniform weight, a thickness of plastic or cardboard. If you try to balance the histogram on a point, you will need to position it at the **mean.** If you cut the histogram with a vertical cut so that the area is the same on either side of the cut, then the cut occurs at the **median.** If there is a mode (or modes) it is represented by the tallest rectangle (or rectangles) in the histogram. See Figure 7–7 on previous page.

7–2 Exercises

I.

Find the mean of the following sets of data:

1. (*See Example 1*) 2, 4, 6, 8, 10
2. 3, 8, 2, 14, 21
3. 2.1, 3.7, 5.9
4. 150, 225, 345, 86, 176, 410, 330
5. 6, −4, 3, 5, −8, 2
6. 3, 3, 3, 3, 5, 5, 5, 5
7. 5.9, 2.1, 6.6, 4.7
8. 1525, 1640, 1776, 1492, 2000
9. Find the mean of the following exam grades: 80, 76, 92, 64, 93, 81, 57, and 77.
10. Six women have the following weights: 106 lb, 115 lb, 130 lb, 110 lb, 120 lb, and 118 lb. What is their mean weight?
11. The lap maximum speeds (in mph) during trials for eight cars in the Daytona 500 were

    180, 175, 190, 182, 191, 193, 177, 194

    Determine the average of these speeds.
12. The daily maximum temperatures (in °F) during a certain week in January in Denver were

    65, 60, 54, 32, 34, 40, 41

    Find the mean daily maximum temperature during that week.

II.

13. (*See Example 3*) Professor Y had the following grade distribution in her class:

Grade	Frequency
96	2
91	3
85	7
80	13
75	12
70	10
60	5
50	2

What is the class average (mean)?

14. (*See Example 4*) Find the mean for the grouped data:

Score	Frequency
0–5	6
6–10	4
11–19	12
20–30	7

15. Professor Z posted the following summary of test grades:

Grade	Number of Grades
90–100	8
80–89	15
70–79	22
60–69	11
45–59	5

Find the class average.

16. The Advertising Department of the *Food Today* magazine ran a survey to determine the number of ads read by their subscribers. They found the following:

Number of Ads Read	Frequency
0–2	45
3–5	75
6–10	35
10–20	15

Find the average number of ads read by a subscriber.

In Exercises 17 through 22, find the median of the given set of numbers.

17. (*See Example 6*) 1, 3, 9, 17, 22

18. **(a)** 36, 41, 55, 88, 121, 140, 162

    (b) 12, 4, 8, 3, 1, 10, 6

19. (*See Example 7*) 12, 14, 21, 25, 30, 37

20. (*See Example 8*) 101, 59, 216, 448, 92, 31

21. 72, 86, 65, 90, 72, 98, 81, 72, 68

22. 379, 421, 202, 598, 148

In Exercises 23 through 28, find the mode(s) of the given set of numbers.

23. (*See Example 10*) 1, 5, 8, 3, 2, 5, 6, 11, 5

24. 3, 2, 4, 6, 5, 4, 1, 6, 8, 4, 1, 6, 8, 4, 4, 13, 6

25. (*See Example 11*) 1, 5, 9, 1, 5, 9, 1, 5, 9, 1, 5

26. 10, 14, 10, 16, 10, 14, 16, 14, 21

27. 5, 4, 9, 13, 12, 1, 2

28. 2, 3, 2, 3, 2, 3, 2, 5, 9

29. The daily stock prices of Acme Corporation for one week were $139.50, $141.25, $140.75, $138.50, and $132.00. What was the average price (mean) for the week?

30. Henry scored 80, 72, 84, and 68 on four exams. What must he make on the fifth exam to have an average of 80?

31. Throughout the 1986 wheat harvest, a farmer sold his wheat each day as it was harvested. The prices he received per bushel were $3.60, $3.57, $3.90, $3.85, $4.00, $4.15, $4.25, and $4.40. Find the mean and median prices.

32. The family income in a depressed area is reported as follows:

Family Income (Dollars)	Number of Families
0–1999	3
2000–3999	8
4000–7999	15
8000–11,999	6
12,000–14,000	4

Estimate the average (mean) family income by using the midpoint of each category as the actual income.

III.

33. Professor *X* gave a quiz to five students. He remembered four of the grades, 72, 88, 81, and 67, and the mean, 78. What was the other grade?

34. Seven cash registers at a supermarket averaged sales of $2946.38 one day. What were the total sales?

35. A certain company employs 120 workers. The wages are as follows:

4 people at $23,000
10 people at $18,000
20 people at $16,500
42 people at $12,300
44 people at $11,000

Determine the average wage.

36. The women taking a physical education course were weighed. The results are summarized below:

Weight (Pounds)	Number of Women	Weight (Pounds)	Number of Women
105–115	9	156–165	14
116–125	13	166–175	11
126–135	15	176–185	2
136–145	21	185–195	1
146–155	14		

In which category does the median fall?

37. A travelling salesman lives in Orlando and has to make regular trips to Daytona, Miami, Tallahasee, and Jacksonville. The one-way distances from Orlando to Daytona, Miami, Tallahassee, and Jacksonville are 55 miles, 238 miles, 243 miles, and 136 miles, respectively. During a certain period, the salesman makes five trips to Daytona, three to Miami, six to Tallahassee, and eight to Jacksonville, returning to Orlando each time. Determine the distance of the average (mean) journey.

38. An apartment complex has 110 rooms, 20 of which rent for \$275 a month, 30 for \$250 a month, and the remainder for \$210 a month. Determine the average (mean) rent per room.

39. The median of five test scores is 82. If four of the grades are 65, 93, 77, and 82, what can you determine about the fifth score?

40. If the mean of 10 scores is 13.4, and the mean of 5 other scores is 6.2, find the mean of the 15 scores.

41. The mean price of three sugar coated cereals is \$1.27, and the mean price of two sugar-free cereals is \$1.34. Find the mean price of all five cereals.

42. Honest Joe's used car lot sold an average of 23 cars per month over a twelve month period. Sales for the last seven months of this period averaged 26 cars per month. What was the average for the first five months?

43. The mean of the numbers 6, -2, 4, 9, and 5 is 4.4.

(a) Add 3 to each of the given numbers and find their mean. How is it related to 4.4?

(b) Add 10 to each of the given numbers and find their mean. How is it related to 4.4?

(c) Multiply each of the given numbers by 2 and find their mean. How is it related to 4.4?

(d) Multiply each of the given numbers by 6 and find their mean. How is it related to 4.4?

(e) Make a general statement about the effect on the mean when a constant is added to each number or when each number is multiplied by a constant.

44. Which measure of central tendency best represents each of the following sets of data:

(a) the weight of lab rats used in a diet experiment

(b) SAT scores

(c) the favorite movie at a video rental shop

(d) rank on the state bar exam

(e) number of books read per month by female adults in McLennan County

(f) number of pages in finite mathematics books

(g) the salaries of janitors at State University

(h) the salaries of 75 janitors, the president, and the vice president of Ford Tractor Co.

(i) the waist sizes of jeans sold to junior high boys

(j) the age of college students

7–3 Measures of Dispersion: Range, Variance, and Standard Deviation

Range
Variance and Standard Deviation
Measurements of Position

A score often has little meaning unless it is compared with other scores. We have used the mean as one comparison. If you know your test score and the class average, you can compare how far you are above or below the class average. The average alone does not tell you how many in the class had grades closer to the average than you.

If two bowlers have the same average, do they have the same ability? If two students have the same average in a course, did they learn the same amount? While the mean is a rather simple representation of a set of data, sometimes more information is needed about how the scores are clustered about the mean in order to make valid comparisons. Let's use students' class averages to illustrate **measures of dispersion.**

Student X	Student Y
80	65
87	92
82	95
92	75
84	98

The mean for each student is 85. However, student X is more consistent. Student Y's scores vary more widely. The mean does not distinguish between these two sets of data. As this example shows, two data sets may have the same mean, but in one set values are clustered close to the mean, and in the other set the values are widely scattered. In order to say something about the amount of cluster as well as the average, we need more information.

Range

One way to measure the dispersion of a set of scores is to find the **range**, the distance between the largest and smallest scores. The range is 12 for student X and 33 for student Y, so the range suggests that X's scores are clustered closer to the mean, and the mean, thereby, is a better representation of X's scores than of Y's scores. For grouped data, like a frequency table, or histogram, the range is the difference between the smaller boundary of the lowest category and the larger boundary of the highest category.

Here are two frequency tables.

A.

Category	Frequency
1–2	10
3–4	5
5–6	3
7–8	6
9–10	16

B.

Category	Frequency
1–2	1
3–4	8
5–6	11
7–8	17
9–10	3

The mean for each is 6.15 (using the grouped data mean), and the range for each is 9. Clearly, the scores in table B are clustered near the mean, whereas they tend to be more remote in table A. The range only gives information about the extreme scores and gives no information on their cluster near the mean.

Variance and Standard Deviation

The range is a simple measure of dispersion, but it is of little value. Two other numbers, the **variance** and the **standard deviation**, are useful and widely used measures of dispersion from the mean.

When a score is moved away from, or closer to, the mean, the variance and standard deviation will reflect a change even though the range remains unchanged. While more useful, the variance and standard deviation are more complicated to compute.

Because the computation of the standard deviation is more complicated than the mean and the range, we will use the example of student X and student Y to go through the steps to compute the variation and standard deviation:

Step 1. Determine the **deviation** of each score from the mean. Compute these deviations by subtracting the mean from each score. The computations below show the deviations for student X and student Y using the mean of 85 in each case.

Computation of Standard Deviation

Student X			**Student Y**		
Grade	**Deviation**	**Squared Deviation**	**Grade**	**Deviation**	**Squared Deviation**
80	$80 - 85 = -5$	25	65	$65 - 85 = -20$	400
87	$87 - 85 = 2$	4	92	$92 - 85 = 7$	49
82	$82 - 85 = -3$	9	95	$95 - 85 = 10$	100
92	$92 - 85 = 7$	49	75	$75 - 85 = -10$	100
84	$84 - 85 = -1$	1	98	$98 - 85 = 13$	169
425	0	88	425	0	818

Step 2. Square each of the deviations. These are shown under the heading **Squared Deviation**.

Step 3. Find the mean of the squared deviations. For student X, the sum of the squared deviations is 88, and their mean is $\frac{88}{5}$ which equals 17.6. For student Y, the sum of the squared deviations is 818, and their mean is $\frac{818}{5}$ which equals 163.6. The number 17.6 is the **variance** for student X, and 163.6 is the **variance** for student Y.

Step 4. Find the square root of the means just obtained. For student X, we have $\sqrt{17.6} = 4.20$. For student Y, we have $\sqrt{163.6} = 12.79$. These numbers are **standard deviations**. By tradition, the symbol σ (sigma) denotes standard deviation. Thus, for student X, $\sigma = 4.20$, and for student Y, $\sigma = 12.79$. The standard deviation measures the spread of the values about their mean. Student Y has the larger standard deviation, so her grades are more widely scattered. The grades of student X cluster closer to the mean.

A formal statement of the formula for variance and standard deviation is the following:

Formula for Variance and Standard Deviation

Given the n numbers $x_1, x_2, \cdots x_n$ whose mean is $\bar{x}$. The variance, denoted var, and standard deviation, σ, of these numbers is given by

$$\text{var} = \frac{(x_1 - \bar{x})^2 + (x_2 - \bar{x})^2 + \cdots + (x_n - \bar{x})^2}{n}$$

$$\sigma = \sqrt{\text{var}}$$

Both the variance and the standard deviation measure the dispersion of data. The variance is measured in the **square** of the units of the original data. The standard deviation is measured in the units of the data, so it is usually preferred as a measure of dispersion.

Example 1 (*Compare Exercise 1*)
Compute the mean and standard deviation of the numbers 8, 18, 7, and 10.

Solution The mean is $\frac{8 + 18 + 7 + 10}{4} = \frac{43}{4} = 10.75$.

Scores	Deviation	Squared Deviation
8	$8 - 10.75 = -2.75$	7.56 (rounded)
18	$18 - 10.75 = 7.25$	52.56
7	$7 - 10.75 = -3.75$	14.06
10	$10 - 10.75 = -0.75$	0.56
	0	74.74

$$\sigma = \sqrt{\frac{74.74}{4}} = \sqrt{18.69} = 4.32$$

You may have noticed that the sum of the deviations is zero in each of the cases shown. This is no accident. The deviations will always add to zero. Thus, the sum of deviations gives no information about the dispersion of scores. We need to use something more complicated, like standard deviation, to determine dispersion.

Let's summarize the procedure for obtaining standard deviation:

Procedure for Computing Standard Deviation

Step 1. Compute the mean of the scores.

Step 2. Subtract the mean from each value to obtain the deviation.

Step 3. Square each deviation.

Step 4. Find the mean of the squared deviations. This is the variance.

Step 5. Take the square root of the variance. This is the standard deviation.

Example 2 Find the standard deviation of the scores 8, 10, 19, 23, 28, 31, 32, and 41.

Solution

STEP 1 Find the mean of the scores.

$$\frac{8 + 10 + 19 + 23 + 28 + 31 + 32 + 41}{8} = \frac{192}{8} = 24$$

STEP 2 Compute the deviation from the mean.

$$8 - 24 = -16$$
$$10 - 24 = -14$$
$$19 - 24 = -5$$
$$23 - 24 = -1$$
$$28 - 24 = 4$$
$$31 - 24 = 7$$
$$32 - 24 = 8$$
$$41 - 24 = 17$$

STEP 3 Square each deviation.

$$(-16)^2 = 256$$
$$(-14)^2 = 196$$
$$(-5)^2 = 25$$
$$(-1)^2 = 1$$
$$4^2 = 16$$
$$7^2 = 49$$
$$8^2 = 64$$
$$17^2 = 289$$

STEP 4 Find the mean of the squared deviations.

$$\frac{256 + 196 + 25 + 1 + 16 + 49 + 64 + 289}{8} = 112$$

STEP 5 Take the square root of the result in Step 4. (Use a calculator.)

$$\sigma = \sqrt{112} = 10.6$$

Example 3 *(Compare Exercise 11)*
Find the standard deviation for the **grouped** data:

Score	Frequency
0–5	6
6–10	3
11–19	13
20–25	8

Solution Use the midpoint of each category to compute the mean.

$$\text{Mean} = \frac{2.5 \times 6 + 8.0 \times 3 + 15.0 \times 13 + 22.5 \times 8}{30} = 13.8$$

The deviations are computed using the mean and the midpoints of each category.

Deviation	Squared Deviation
$2.5 - 13.8 = -11.3$	$(-11.3)^2 = 127.69$
$8.0 - 13.8 = -5.8$	$(-5.8)^2 = 33.64$
$15.0 - 13.8 = 1.2$	$(1.2)^2 = 1.44$
$22.5 - 13.8 = 8.7$	$(8.7)^2 = 75.69$

To obtain the variance, we need to use each squared deviation multiplied by the frequency of the corresponding category.

$$\text{Variance} = \frac{127.69 \times 6 + 33.64 \times 3 + 1.44 \times 13 + 75.69 \times 8}{30}$$

$$= \frac{1491.30}{30}$$

$$= 49.71$$

$$\text{Standard deviation} = \sqrt{49.71} = 7.05.$$

You are not expected to compute the standard deviation during your daily activities. Yet, on an intuitive level, all of us are interested in standard deviation. For example, you can recall a trip to the grocery store when you selected the shortest checkout line only to find that it took longer than your friend who got in a longer line. Let's look at the situation from the store's viewpoint. They are interested in customer satisfaction so they work to reduce the waiting time. Proudly they announce that, on the average, they check out a customer in 3.5 minutes. The customer may agree that 3.5 minutes is good service, but an occasional long wait may make the customer question the claim of the store. Basically, the customer wants the store to have a small standard deviation. A small standard deviation indicates that you expect the waiting time to be near 3.5 minutes per customer.

A restaurant uses a different system of waiting lines. The customers do not choose a table and line up there. All customers wait in a single line and are directed to a table as it becomes available—and for a good reason. The diner doesn't want someone standing over him waiting for his place. Ignoring the psychology of lining up at someone's table, which of these two systems is better? Is it better to let the customer choose one of several lines or to have a single line and allow a customer to go to a station as it becomes available? Both systems are reasonable in a bank. One bank may use the system where the customer chooses the teller, and a line forms at each teller. Another bank may use the system where the customers wait in a single line, and the one at the head of the line goes to the next available teller. It turns out that the average waiting time is the same under both systems. Thus, from the bank's viewpoint, both systems allow them to be equally efficient. But it also turns out that the single-line concept tends to reduce the extremes in waiting times, so the standard deviation is decreased. Thus, from the customer's viewpoint, the single-line concept is better because it helps reduce an occasional long delay. Since the bank's efficiency is the same in both cases, management would do well to adopt the single-line concept.

Measurements of Position

Scores are generally meaningless by themselves. "We scored ten runs in our softball game" gives no information about the outcome of the game unless the score of the opponent is known. A grade of 85 is less impressive when it becomes known that everyone else scored 100.

The mean, median, and mode give a central point of reference of a score. You can compare your score to the mean or median and determine on which side of the central point your score lies.

One familiar measurement of position is the **rank**. A student ranks 15th in a class of 119; a runner places 38th in a field of 420 in the ten-kilometer run; a girl is the 4th runner-up in a beauty contest. Your rank simply gives the position of youı score relative to all other scores. It gives no information on the location of the central point and how closely scores are clustered around it.

If you took the SAT test, you received a numerical score and a **percentile** score, another positional score. A score at the 84th percentile states that 84% of those taking the test scored lower and 16% scored higher. A score at the 50th percentile is a median score, 50% score higher and 50% score lower.

The percentile score is used as a positional score when a large number of scores is involved.

Another standard score is the *z-score*. It is more sophisticated than rank or percentile because it uses a central point of the scores (the mean) and a measure of dispersion (the standard deviation). It is useful in comparing scores when two different reference groups are involved. The z-score is

$$z = \frac{\text{score} - \text{mean}}{\text{standard deviation}}$$

Example 4 (*Compare Exercise 18*)
If the mean = 85 and $\sigma = 5$, then the z-score for a raw score of 92 is

$$z = \frac{92 - 85}{5} = \frac{7}{5} = 1.4$$

For a score of 70,

$$z = \frac{70 - 85}{5} = \frac{-15}{5} = -3$$

A negative z-score indicates that the score is below the mean. $z = 1.4$ indicates that the score is 1.4 standard deviations above the mean. A z-score of 0 corresponds to the mean.

Example 5 (*Compare Exercise 26*)
A student took both the SAT and ACT tests and made a 480 on the SAT verbal portion and a 22 on the ACT verbal. Which was the better score? *Note:* You need to know that for the SAT verbal test, mean = 431 and $\sigma = 111$; and for the ACT verbal, mean = 17.8 and $\sigma = 5.5$.

Solution The z-score for each test is:

$$\text{SAT } z = \frac{480 - 431}{111} = \frac{49}{111} = .44$$

$$\text{ACT } z = \frac{22 - 17.8}{5.5} = \frac{4.2}{5.5} = .76$$

The higher z-score, .76, indicates that the student taking the ACT performed better because the ACT score was higher above the mean.

7–3 Exercises

I.

Determine the mean, variance, and standard deviation for each of the following sets of data:

1. (*See Example 1*) 19, 10, 15, 20
2. 0, 3, 5, 9, 10, 12
3. 4, 8, 9, 10, 14
4. 10, 14, 20, 24
5. 17, 39, 54, 22, 16, 46, 25, 19, 62, 50
6. 1.1, 1.3, 1.5, 1.7
7. −8, −4, −3, 0, 1, 2
8. −4, 3, 6, 7, 6, 9
9. −5, −1, 0, 0, 1, 2, 0, 3, 7, 10, 41, 8
10. −20, −7, 0, 1, 23, 47, 81, 276, 93, 512
11. (*See Example 3*) Find the standard deviation for the grouped data:

Score	Frequency
0–10	5
11–20	12
21–30	8

12. The following is a summary of test grades for a class. Find the mean and standard deviation.

Grades	Frequency
91–100	3
81–90	4
71–80	10
61–70	4
51–60	3

13. An experiment is repeated 55 times, and the outcomes are 2, 8, and 12. They occur with frequencies 10, 19, and 26, respectively. Determine the mean, variance, and standard deviation of these results.
14. An experiment is repeated 32 times with outcomes -7, -3, 0, 1, and 2 occurring with frequencies 6, 9, 1, 7, and 9, respectively. Determine the mean, variance, and standard deviation of these results.

II.

15. The following grades were made on a test by a class of ten students:

83, 90, 76, 52, 67, 75, 84, 88, 72, 64

Determine the mean, variance, and standard deviation of these scores.

16. **(a)** Calculate the standard deviation of the set 6, 6, 6, 6, 6, and 6.
    (b) What are the mean and the standard deviation of a set where all scores are the same?
    (c) What is the median and mode of a set where all scores are the same?
17. The following statements give a single-number model of situations. Discuss how our view of the situation might change if we also knew the standard deviation.
    (a) The mean monthly rainfall in Waco is 2.51 inches.
    (b) The average grade on the last test was 82.
    (c) The mean of the temperature on July 11, 1987 was 80 degrees.
    (d) "But, Dad, I averaged 50 miles per hour."
18. (*See Example 4*) Let the mean = 160 and $\sigma = 16$.
    (a) Find z for the score 180.
    (b) Find z for the score 150.
    (c) Find z for the score 160.
    (d) Find the score x if $z = 1$.
    (e) Find the score x if $z = -0.875$.

III.

19. A machine shop uses two lathes to make shafts for electric motors. A sample of five shafts from each lathe are checked for quality control purposes. The diameter of each shaft (in inches) is given as follows:

Lathe I	Lathe II
0.501	0.502
0.503	0.497
0.495	0.498
0.504	0.501
0.497	0.502

The mean diameter in each case is .500.

(a) Find the standard deviation for each lathe.

(b) Which lathe is more consistent?

20. **(a)** Calculate the mean and the standard deviation of the scores

1, 3, 8, and 12.

(b) Add 2 to each of the scores in part (a) and compute the mean and the standard deviation. How are the mean and the standard deviation related to those in part (a)? Would the same kind of relationship exist if you added another constant instead of 2?

21. A set has 8 numbers. Each one is either 2, 3, or 4. The mean of the set is 3.

(a) If $\sigma = 0.5$, what is the set of numbers?

(b) If $\sigma = 1$, what is the set of numbers?

(c) Can σ be larger than 1?

22. A golfer finished 15th in a field of 90 golfers in a tournament. What was his percentile?

23. A student scored at the 70th percentile on a test taken by 110 students. What was his rank?

24. A student learned that 105 students scored higher and 481 scored lower on a standardized test.

(a) What is the student's rank on the test?

(b) What is the student's percentile on the test?

25. Suppose that the mean = 140 and $\sigma = 15$. Find the score 2.2 standard deviations above the mean.

26. (*See Example 5*) A student took an admission's test at two different universities. At one, where the mean was 72 and the standard deviation 12, he made 86. At the other, with mean 62 and standard deviation 8, he scored 82. Which one was the better performance?

27. Joe scored 114 on a standardized test which had a mean of 120 and a standard deviation of 8. Josephine scored 230 on a test which had a mean of 250 and a standard deviation of 14. Which one performed better?

28. A runner ran a 3-kilometer race in 19 minutes. The average time of all who finished was 18 minutes with $\sigma = 1$ minute.

A cyclist rode a 25-mile race in 64 minutes. The average time of all who finished was 59 minutes with $\sigma = 3$. Which athlete had the better performance?

7–4 Random Variables and Probability Distribution

Random Variables
Probability Distribution

Random Variables

A driver approaching a traffic signal expects to see one of three colors: red, yellow, or green. Think of observing the colors on a traffic light as an experiment that has the sample space {red, yellow, green}, because these three colors give all possible outcomes. On a major street, a light is timed so each cycle shows green for 85 seconds, yellow for 5 seconds, and red for 35 seconds. In this manner, we associate a number with each color, the number of seconds the color shows.

Noting the time each color shows per cycle may seem like a minor observation, but this illustrates an important statistical concept, that of assigning a number to each outcome of an experiment. This process of assigning a number to each outcome is called a **random variable**.

Definition

A **random variable** is a rule that assigns a number to each outcome of an experiment.

It is customary to denote the random variable by a capital letter, such as X or Y.

Example 1 (*Compare Exercise 1*)

If a coin is tossed twice, there are four possible outcomes: HH, HT, TH, and TT. We can assign a number to each outcome by simply giving the number of heads that appear in each case. The values of X are assigned as follows.

Outcome	X
HH	2
HT	1
TH	1
TT	0

Example 2 (*Compare Exercise 3*)
A true-false quiz consists of 15 questions. A random variable may be defined by assigning the total number of correct answers to each quiz.

Example 3 (*Compare Exercise 5*)
The Hardware Store has a box of Super-Special items. Selecting an item from the box is an experiment with each of the items as an outcome. Two possible ways to define a random variable are to

(a) assign the original price to each item (outcome); or,
(b) assign the sale price to each item.

In one sense, you may be quite arbitrary in the way you define a random variable for an experiment. It depends on what is most useful for the problem at hand. Here are several examples:

Experiment	Random Variable, X
A survey of cars entering mall parking lot.	Number of passengers in a car: $1, 2, \ldots, 8$.
Rolling a pair of dice.	Sum of numbers that turn up: $2, 3, \ldots, 12$.
Tossing a coin three times.	Number of times tails occurs.
Selecting a sample of five widgets from an assembly line.	Number of defective widgets in the sample.
Measuring the height of a student selected at random.	Observed height.
Finding the average life of Brand X tires.	Number of miles driven.
Selecting a box of Crunchies cereal.	Weight of the box of cereal.
Checking the fuel economy of a compact car.	Distance X the car travels on a gallon of gas.

The first four random variables are **discrete** variables because the number assigned comes from a finite set of distinct numbers with no values between. For example, the number of auto passengers can take on only selected values 1, 2, 3, and so on. The number 2.63 is not a valid assignment for the number of passengers. On the other hand, the last four examples are **continuous variables**. Assuming an accurate measuring device, 5 feet 4.274 inches is a valid height. There are no distinct gaps that must be excluded as a valid height. Similarly, even though one may expect his car to travel about 23 miles on a gallon of gas, 22.64

miles cannot be excluded as a valid distance. Any distance in a reasonable interval is a valid possibility.

Probability Distribution

We introduce another concept associated with a random variable. Look at the traffic signal example again. How many times have you asked, "What are my chances of getting a green light?" You are involved in an "experiment" with three outcomes—green, yellow, and red. We defined a random variable by assigning a number, the number of seconds a color is on during a cycle. Another concept, that of a **probability distribution** assigns a probability to each value of the random variable.

We determine the probabilities of getting, upon arrival at the traffic signal, a green light, a yellow light, or a red light as follows. The duration of the cycle is (85 + 5 + 35) seconds, or 125 seconds.

The green light is on $\frac{85}{125}$ of the time; the yellow is on $\frac{5}{125}$ of the time; and the red is on $\frac{35}{125}$ of the time. This suggests that we define the probability of each outcome as:

$$P(G) = \frac{85}{125} = 0.68$$

$$P(Y) = \frac{5}{125} = 0.04$$

$$P(R) = \frac{35}{125} = 0.28$$

We summarize the traffic signal experiment by:

Outcome	X	$P(X)$
Green	85	0.68
Yellow	5	0.04
Red	35	0.28

Definition

If a random value has the values

$$X_1, X_2, \ldots, X_n$$

then a probability distribution $P(X)$ is a rule that assigns a probability $P(X_i)$ to each value X_i. More specifically,

(a) $0 \leq P(X_i) \leq 1$ for each X_i

(b) $P(X_1) + P(X_2) + \cdots + P(X_n) = 1$

Example 4 (*Compare Exercise 17*)
A coin is tossed twice. Define a random variable as the number of times heads appears. Compute the probability of zero, one, or two heads in the usual manner to get

$$P(0) = P(\text{TT}) = \frac{1}{4}$$

$$P(1) = P(\text{HT or TH}) = \frac{1}{2}$$

$$P(2) = P(\text{HH}) = \frac{1}{4}$$

We can represent a probability distribution graphically using a histogram. Form a category for each value of the random variable. The probability of each value of X determines the height of that bar.

The graph of this probability distribution is Figure 7–8.

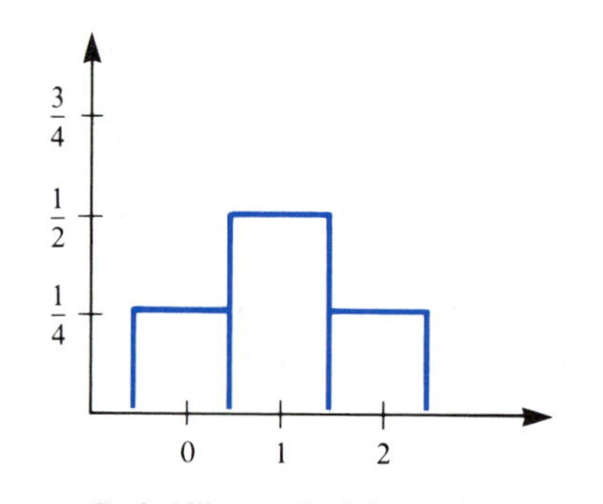

Figure 7–8 Probability graph of the number of heads in two tosses of a coin

Example 5 (*Compare Exercise 19*)
Each work of the phrase "Now is the time for all good men" is written on a card. An experiment is conducted by selecting a card at random. The sample space is the set of words in the phrase. Define a random variable X as the number of letters in the word selected. The values of X range over the number of letters possible: 2, 3, and 4. What is the probability of each value of X?

Solution $P(2)$ is the probability of selecting a two-letter word. Since the selection is from eight words, and one is a two letter word, $P(2) = \frac{1}{8}$. $P(3)$ is the probability of selecting a three-letter word, and $P(3) = \frac{5}{8}$. $P(4)$, the probability of a four-letter word, is $\frac{2}{8}$.

Thus, a probability distribution of X and its graph are shown in Figure 7–9.

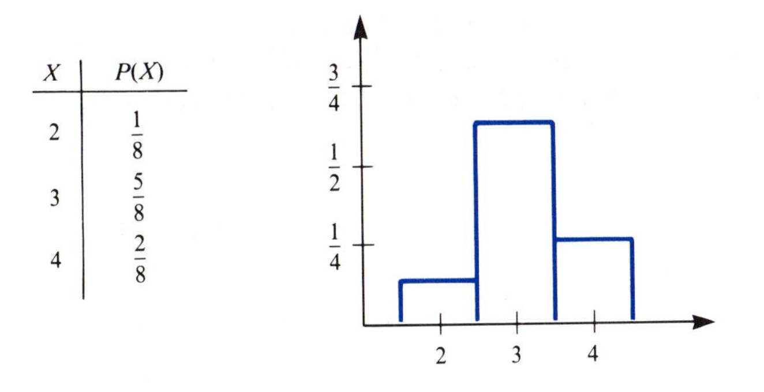

X	$P(X)$
2	$\frac{1}{8}$
3	$\frac{5}{8}$
4	$\frac{2}{8}$

Figure 7–9

Example 6 (*Compare Exercise 22*)
An experiment randomly selects two people from a group of five men and four women. A random variable X is the number of women selected. Find the probability distribution of X.

Solution The values of X range over the set $\{0, 1, 2\}$ since 0, 1, or 2 women can be selected. The probability of each value is computed in the usual manner.

$$P(0) = \frac{C(5, 2)}{C(9, 2)} = \frac{10}{36} \qquad \text{(Probability both are men)}$$

$$P(1) = \frac{C(5, 1) \times C(4, 1)}{C(9, 2)} = \frac{20}{36} \qquad \text{(Probability of 1 man and 1 woman)}$$

$$P(2) = \frac{C(4, 2)}{C(9, 2)} = \frac{6}{36} \qquad \text{(Probability of 2 women)}$$

The probability distribution and its graph are shown in Figure 7–10.

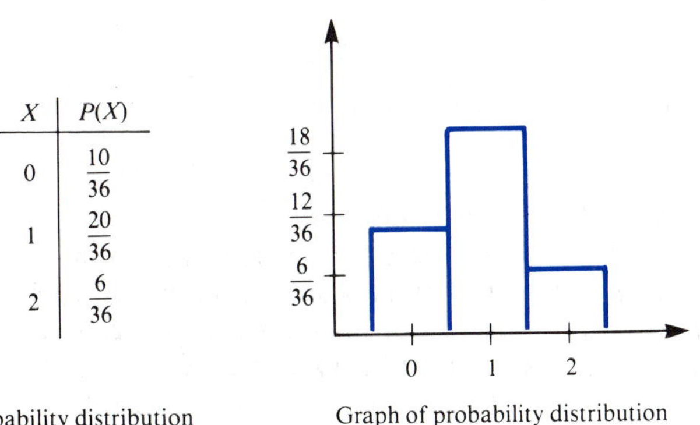

X	$P(X)$
0	$\frac{10}{36}$
1	$\frac{20}{36}$
2	$\frac{6}{36}$

Figure 7–10 Probability distribution — Graph of probability distribution

Example 7 (*Compare Exercise 27*)
The probability of guessing a correct answer on a multiple-choice question is $\frac{1}{4}$. Let X be the number of answers guessed correctly on a four-question quiz. Find the probability distribution of X and draw its graph.

Solution This is a Bernoulli trial problem with $p = \frac{1}{4}$ and the number of trials, $n = 4$. X takes on the values 0, 1, 2, 3, and 4.

$$P(X = 0) = C(4, 0)\left(\frac{1}{4}\right)^0\left(\frac{3}{4}\right)^4 = .316 \quad \text{(Rounded)}$$

$$P(X = 1) = C(4, 1)\left(\frac{1}{4}\right)^1\left(\frac{3}{4}\right)^3 = .422$$

$$P(X = 2) = C(4, 2)\left(\frac{1}{4}\right)^2\left(\frac{3}{4}\right)^2 = .211$$

$$P(X = 3) = C(4, 3)\left(\frac{1}{4}\right)^3\left(\frac{3}{4}\right) = .047$$

$$P(X = 4) = C(4, 4)\left(\frac{1}{4}\right)^4\left(\frac{3}{4}\right)^0 = .004$$

The probability distribution table and graph are given in Figure 7–11.

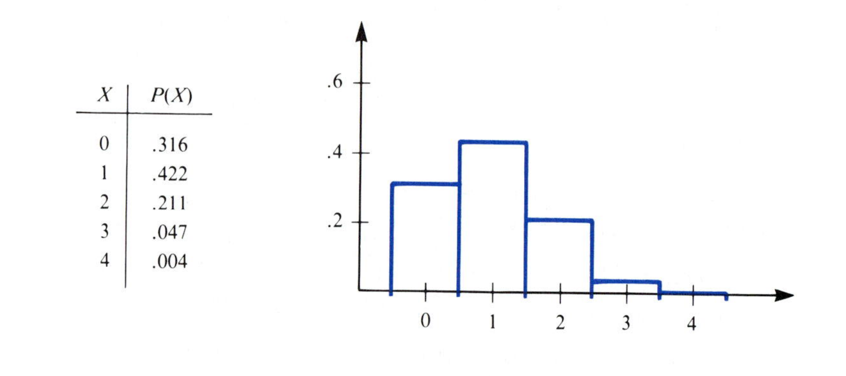

X	$P(X)$
0	.316
1	.422
2	.211
3	.047
4	.004

Figure 7–11

7–4 Exercises

1. (*See Example 1*) A coin is tossed three times, and the number of heads that occurs, X, is assigned to each outcome. List all possible outcomes and the corresponding value of X.

2. A word is selected from the phrase "A stitch in time saves nine." A random variable counts the number of letters in the word selected. List all possible outcomes and the corresponding value of X.

Give all possible values of the random variable in each of Exercises 3 through 5.

3. (*See Example 2*) Three people are selected at random from eight men and six women. The random variable is the number of men selected.

4. Four people are selected at random from five women and three men. The random variable is the number of men selected.

5. (*See Example 3*) An urn contains four red, six yellow, and three green balls. Four balls are drawn.

   (a) The random variable is the number of yellow balls drawn.
   (b) The random variable is the number of green balls drawn.

6. An experiment consists of selecting a whole number and finding the remainder after division by three. What are the possible outcomes?

7. A word is selected from a newspaper article. A random variable X is defined as the number of letters in that word.

   (a) What are the possible values of X?
   (b) Is X discrete or continuous?

Identify the random variables in Exercises 8 through 14 as discrete or continuous.

8. The diameter X of a golf ball.

9. From a random selection of ten students, the number X which have type O blood.

10. Sixty families live in a remote village. The random variable X is

    (a) the number of children in a family.
    (b) the amount of electricity used in a month by a family.
    (c) the monthly income of a family.

11. The number of building permits issued each year in Augusta, Georgia.

12. The length of time to play a tennis match.

13. The number of cartons of Blue Bell ice cream sold each week in the Food Mart.

14. The number of pounds of apples sold each day at Joe's Roadside Fruit Stand.

15. Fifty students are polled on the courses they are taking, and X is the number of courses a student takes. The number of courses range from two to six. Is the following a probability distribution of X?

X (Number of Courses)	**Fraction Taking X Courses**
2	0.08
3	0.12
4	0.48
5	0.30
6	0.02

16. Customers at Lake Air Mall were asked to state which of three brands of coffee tasted the best. The brands were numbered 1, 2, and 3. The results are summarized as follows:

Brand	Fraction Preferring Brand
1	.25
2	.40
3	.20

If the brand number is a random variable, does this define a probability distribution?

17. (*See Example 4*) A coin is tossed three times. X is the number of heads that appear. Find the probability distribution of X and its graph.

18. A pair of dice is tossed. The random variable X is the sum of the numbers that turn up. Find the probability distribution of X.

19. (*See Example 5*) Each word of the slogan "Get ahead, learn finite math" is written on a card, and a card is drawn at random. The random variable X is the number of letters in the word drawn. Find the probability distribution of X.

20. A class of 30 students took a five-point quiz. The results are summarized as follows.

Score	Frequency
0	2
1	4
2	8
3	3
4	12
5	1

Let the random variable X be the score on the quiz. Find the probability distribution of X, and draw its graph.

21. A survey of 75 households revealed the following information on the number of television sets they owned.

Number of Television Sets	Frequency
0	8
1	49
2	13
3	4
4	1

Let the random variable X be the number of television sets owned. Find the probability distribution of X, and sketch its graph.

22. (*See Example 6*) Two people are selected from a group of four men and four women. A random variable X is the number of men selected. Find the probability distribution for X and its graph.

23. In a collection of ten electronic components, three are defective. Two are selected at random and the number of defective components noted. Let X be the number of defective components, and draw the graph of the probability distribution of X.

24. A number is selected at random from the numbers 1, 2, 3, 4, 5, 6, 7, 8, 9, and 10. A random variable X is the remainder when the selected number is divided by 3. Find the probability distribution of X.

25. An urn contains three red, two green, and one yellow ball. Two balls are selected at random.

(a) A random variable X is the number of green balls drawn. Find the probability distribution of X.

(b) A random variable X is the number of yellow balls drawn. Find the probability distribution of X.

26. At each meeting of a club, one person is selected to draw a "lucky number." That person gets the amount in dollars of the number drawn. The box contains 30 cards with the number 1, 14 cards with the number 5, 3 cards with the number 10, 2 cards with the number 20, and 1 card with the number 50. Let the random variable X be the numbers on the cards. Determine the probability distribution of X, and draw its graph.

27. (*See Example 7*) The probability of getting the correct answer on a multiple-choice question is $\frac{1}{5}$. Let X be the number of answers guessed correctly on a four-question quiz. Find the probability distribution of X and its graph.

28. The probability of a baseball player getting a hit in his turn at bat is .300. Let X be the number of hits in three times at bat. Find the probability distribution of X and its graph.

7–5 Expected Value

Expected Value
Variance and Standard Deviation of a Random Variable

Expected Value

Each day a student puts 50¢ in a candy machine to buy a 35¢ candy bar. She observes three possible outcomes.

1. She gets a candy bar and 15¢ change. This happens 80% of the time.
2. She gets a candy bar and no change. This happens 16% of the time.
3. She gets a candy bar and the machine returns her 50¢. This happens 4% of the time.

Over a period of time what is the average cost of a candy bar?

The three possible costs, 35¢, 50¢, and 0¢, have a mean of 28.3¢. This is not the average cost because it assumes that 35¢, 50¢, and 0¢ occur equally often; whereas, in fact, 35¢ occurs more often than the other two. If the machine behaved this way for 500 purchases, we would expect the following.

Cost (X)	Frequency (per 500)	Probability ($P(X)$)
35¢	400	0.80
50¢	80	0.16
0¢	20	0.04
	500	1.00

This presents the information as a frequency table, so we can compute the mean as we have before with a frequency table (Section 7–3). The mean is

$$\frac{400 \times 35 + 80 \times 50 + 20 \times 0}{500} = \frac{14{,}000 + 4{,}000 + 0}{500} = 36$$

so the average cost is 36¢.

Let's write the mean in another way:

$$\begin{aligned}\frac{400 \times 35 + 80 \times 50 + 20 \times 0}{500} &= \frac{400}{500} \times 35 + \frac{80}{500} \times 50 + \frac{20}{500} \times 0 \\ &= .80 \times 35 + .16 \times 50 + .04 \times 0\end{aligned}$$

This last expression is simply the sum obtained by adding each cost of a candy bar times the probability that cost occurred, that is,

$$P(35) \times 35 + P(50) \times 50 + P(0) \times 0$$

This illustrates the procedure used to compute the mean when each value occurs with a specified probability. This mean is called the **expected value**; it represents the long term mean of numerous trials. While a few trials likely would not average to the expected value, a larger and larger number of trials will tend to give a mean closer to the expected value.

We now make a formal definition of expected value in terms of a random variable and probability distribution.

Definition

If X is a random variable with values $x_1, x_2, \ldots, x_n$ and corresponding probabilities $p_1, p_2, \ldots, p_n$, then the **expected value** of X, $E(X)$, is

$$E(X) = p_1 x_1 + \cdots + p_n x_n$$

Example 1 (*Compare Exercise 1*)

Find the expected value of X, where the values of X and their corresponding probabilities are given by the table.

x_i	2	5	9	24
p_i	.4	.2	.3	.1

Solution

$$E(X) = .4 \times 2 + .2 \times 5 + .3 \times 9 + .1 \times 24$$
$$= .8 + 1.0 + 2.7 + 2.4$$
$$= 6.9$$

Example 2 (*Compare Exercise 5*)

(a) A contestant tosses a coin and receives \$5 if heads appears and \$1 if tails appears. What is the expected value of a trial?

(b) A contestant receives \$4.00 if a coin turns up heads and pays \$3.00 if it turns up tails. What is the expected value?

Solution

(a) The probability of heads is $\frac{1}{2}$, and the probability of tails is $\frac{1}{2}$. Then,

$$E(X) = \frac{1}{2}(5) + \frac{1}{2}(1) = \$3.00$$

(b) Again, the probability of heads is $\frac{1}{2}$, and the probability of tails is $\frac{1}{2}$.

$$E(X) = \frac{1}{2}(4) + \frac{1}{2}(-3) = \$2.00 - \$1.50 = \$.50$$

Notice that a negative \$3.00 value indicates that the contestant pays \$3.00. The \$.50 indicates a long term gain that averages \$.50 each trial.

Example 3 (*Compare Exercise 9*)

An IRS study shows that 60% of all income tax returns audited have no errors; 6% have errors that cause overpayments averaging \$25; 20% have minor errors that cause underpayments averaging \$35; 13% have more serious errors averaging \$500 underpayments; and 1% have flagrant errors averaging \$7000 underpayment. If the IRS selects returns at random,

(a) What is the average amount per return that is owed to the IRS, that is, what is the expected value of a return selected at random?

(b) How much should the IRS expect to collect if one million returns are audited at random?

(c) If the budget for the Audit Department is \$15 million, how many returns must they examine to collect enough to cover their budget expenses?

Solution **(a)** Since an underpayment error eventually results in additional money paid to the IRS, and a taxpayer overpayment is money paid back by the IRS, we use positive values for underpayment and negative for overpayment. The probability in each case is the fraction of returns with that type of error.

$$E(X) = 0.60(0) + 0.06(-25) + 0.20(35) + 0.13(500) + 0.01(7000) = 140.5$$

Thus, the IRS expects to collect \$140.50 for each return selected.

(b) If 1 million returns are selected at random, they may expect to collect 140.50×1 million $=$ \$140.5 million.

(c) Since they expect to collect an average of \$140.50 for each return audited, they must examine

$$15{,}000{,}000/140.50 = 106{,}762 \text{ returns}$$

The next example illustrates that expected value need not be an amount of money. It can be any "payoff" associated with each outcome.

Example 4 (*Compare Exercise 14*)
A tray of electronic components contains nine good components and three defective components. It two components are selected at random, what is the expected number of defective components?

Solution Let a random variable X be the number of defective components selected. X can have the values 0, 1, or 2. We need the probability of each of those numbers.

$$P(0) = \text{probability of no defective (both good)}$$

$$= \frac{C(9, 2)}{C(12, 2)} = \frac{36}{66} = \frac{12}{22}$$

$$P(1) = \text{probability of 1 good and 1 defective}$$

$$= \frac{C(9, 1)\,C(3, 1)}{C(12, 2)} = \frac{27}{66} = \frac{9}{22}$$

$$P(2) = \text{probability of two defective}$$

$$= \frac{C(3, 2)}{C(12, 2)} = \frac{3}{66} = \frac{1}{22}$$

The expected value is

$$E(X) = \frac{12}{22}(0) + \frac{9}{22}(1) + \frac{1}{22}(2) = \frac{11}{22} = \frac{1}{2}$$

so the expected number of components is $\frac{1}{2}$. Clearly, you don't expect to get half of a component. The value $\frac{1}{2}$ simply says that if a large number of selections are made, you will *average* one-half each time. You expect to get no defectives a little less than half the time and either one or two the rest of the time, but the average will be one half.

Variance and Standard Deviation of a Random Variable

The expected value is the "central tendency," or mean, of the values of a random variable, taking the probability of their occurrence into consideration. Think back on the mean and median; we shouldn't be surprised that sometimes the dispersion, or spread, of a random variable is needed to give more information. We can find variance of a random variable, and it too measures the dispersion, or spread, of a random variable from the mean (expected value). The greater dispersion gives a greater variance. The histogram of a probability distribution shown in Figure 7–12(a) has a smaller variance than that shown in (b).

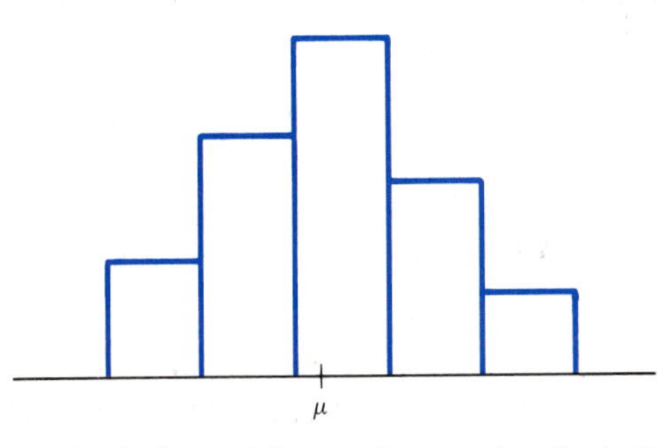

This data is clustered closer to the mean than that in (b).

(a)

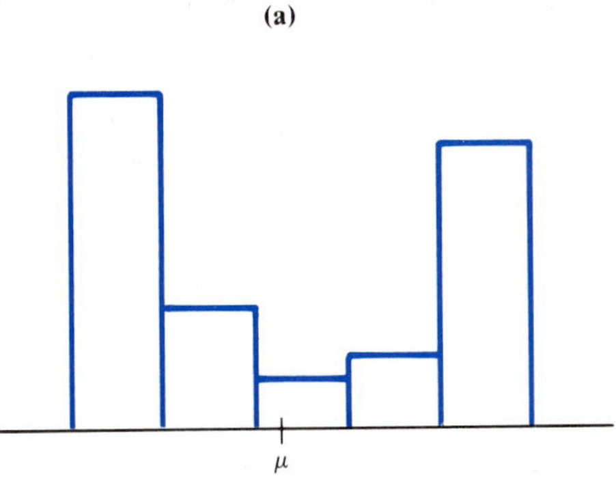

This data has greater dispersion than that in (a).

Figure 7–12 **(b)**

Let's give the formal definition of variance and standard deviation of a random variable, and then give examples showing the computation. We denote the variance by var(X) and let the Greek letter μ (mu) represent the mean (expected value).

Definition

If X is a random variable with values $x_1, x_2, \ldots, x_n$, corresponding probabilities $p_1, p_2, \ldots, p_n$, and expected value $E(X) = \mu$, then

$$\text{var}(X) = p_1(x_1 - \mu)^2 + p_2(x_2 - \mu)^2 + \cdots + p_n(x_n - \mu)^2$$

The standard deviation, $\sigma(X) = \sqrt{\text{var}(X)}$

Example 5 (*Compare Exercise 10*)
Find the variance and standard deviation for the random variable defined by the following table.

x_i	4	7	10	8
p_i	.2	.2	.5	.1

Solution Set up the computations in the following way:

x_i	p_i	$p_i x_i$	$x_i - \mu$	$(x_i - \mu)^2$	$p_i(x_i - \mu)^2$
4	.2	.8	−4	16	3.2
7	.2	1.4	−1	1	.2
10	.5	5.0	2	4	2.0
8	.1	.8	0	0	0
		$\mu = 8.0$			var(X) = 5.4

The variance var(X) = 5.4. So, the standard deviation is

$$\sigma(X) = \sqrt{5.4} = 2.32$$

7–5 Exercises

1. (*See Example 1*) Find the expected value of X.

x_i	3	8	15	22
p_i	.3	.2	.1	.4

2. Find $E(X)$ for the following probability distribution.

x_i	2	4	−1	6	10
p_i	.15	.08	.17	.35	.25

3. Find $E(X)$.

x_i	150	235	350	410	480
p_i	.2	.2	.2	.2	.2

4. Find $E(X)$.

x_i	0	10
p_i	.7	.3

5. (*See Example 2*) A contestant receives \$8 if a coin turns up heads and \$2 if it turns up tails. What is the expected value of a trial?

6. A contestant rolls a die and receives \$5 if it is a one or two, receives \$15 if it is a three, and pays \$2 if it is a four, five, or six. What is the expected value of each trial?

7. A game consists of tossing a coin twice. The player will win \$1 if he throws the same face, heads or tails, twice. How much should the organizer charge to enter the game

(a) if he wants to break even?

(b) if he wants to average \$1 per person profit?

8. A game involves throwing a pair of dice. The player will win, in dollars, the sum of the numbers thrown. How much should the organizer charge to enter the game if he wants to break even?

9. (*See Example 3*) An auto repair shop's records show that 15% of the autos serviced need minor repairs averaging \$20, 65% need moderate repairs averaging \$130, and 20% need major repairs averaging \$700.

(a) What is the expected cost of repair of a car selected at random?

(b) The shop has 125 cars scheduled for repair. What is the total expected repair cost of these cars?

10. (*See Example 5*) Compute the expected value, the variance ($\text{var}(X)$), and standard deviation ($\sigma(X)$) for the random variable defined by the following table.

x_i	5	10	14	20
p_i	.1	.1	.5	.3

11. Computer Var (X) and $\sigma(X)$ for the random variable defined as follows.

x_i	100	140	210
p_i	.4	.5	.1

12. Compute μ and $\sigma(X)$ for the following random variable.

x_i	-1	2	3	4	10
p_i	.1	.2	.1	.3	.3

13. Compute μ and $\sigma(X)$ for the following random variable.

x_i	10	30	50	90
p_i	.4	.2	.3	.1

14. (*See Example 4*) A lawyer finds that 60% of the wills he prepares for clients are routine and require an average of one hour each. The moderately complex wills require an average of three hours each and account for 30% of his clients. The other 10% are complex and require an average of ten hours each.

(a) What is the expected time of each will?

(b) How much time should he schedule to prepare 45 wills?

15. A company considers two business ventures. The executives believe that venture A has a 0.60 probability of a \$50,000 profit, a 0.30 probability of breaking even, and a 0.10 probability of a \$70,000 loss. Venture B has a 0.55 probability of \$100,000 profit and a 0.45 probability of a \$60,000 loss. Determine the expected value of each venture. Which is the better risk?

16. A football team can expect 20,000 fans to attend a game if it rains. If it is fair, 50,000 attend. The weatherman says that there is a 70% chance of rain at kickoff time. What is the expected attendance?

17. Of 1000 animals of a certain species studied, it was found that 50 of the animals had litters consisting of one animal, 100 had litters consisting of two animals, 250 had litters consisting of three animals, 450 had litters consisting of four animals, and 150 had litters consisting of five animals. On the basis of these statistics, what is the expected litter size (expected value of litter size)?

18. A businessman has evaluated a proposed business venture as follows: He stands to make a profit of \$8000 with probability 0.2, to make a profit of \$5000 with probability 0.4, to break even with probability 0.3, and to lose \$4000 with probability 0.1. Find the expected profit.

19. A plant manufactures microchips, 5% of which are defective. They make a profit of \$18 on each good microchip and lose \$23 on each defective one.

(a) What is the expected profit on a microchip?

(b) They plan a production run of 150,000 microchips. What is the expected profit for the run?

20. As Jack Nicklaus was preparing to tee off on the 17th hole of the final round of the Heritage Golf Classic in 1975, the statistics for the hole (a par 3) were

39 birdies (score of 2),
180 pars (score of 3),
50 bogeys (score of 4),
9 double bogeys (score of 5),
1 triple bogey (score of 6).

The tournament took place at the Harbour Town Golf Links, a treacherous 6555-yard course on Hilton Head Island, South Carolina.

(a) What was the *mathematical* probability that Nicklaus would birdie the hole?

(b) What was the mean score on the hole up to that time? (Nicklaus parred the hole, winning the tournament with a record score of 271, or 3 under par.)

Note that the mathematical average score on the hole does not predict an individual's score. You expect a golfer of Nicklaus's caliber to score better than average.

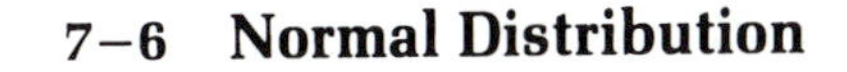

7–6 Normal Distribution

Normal Curve
Area Under a Normal Curve
The *z* Score
Probability and a Normal Distribution

Normal Curve

One of the more important representations of data is the **normal curve**. It is a "smoothed" frequency polygon for a large set of data that has definite characteristics. When a set of data exhibits these characteristics, we say that it is a **normal distribution**. Data sets such as heights of adult males, IQ of 18-year-olds, and scores on SAT tests are normal distributions, and the normal curve is a valid representation of the data.

Normal curves vary in size and shape depending on the mean and standard deviation of the data. However, all normal curves have "bell shapes" similar to Figure 7–13. When the values of the scores are represented by a horizontal line, the normal curve peaks at the mean and is perfectly symmetric about the mean. One half the area under the curve lies to the left of the mean, and the other half

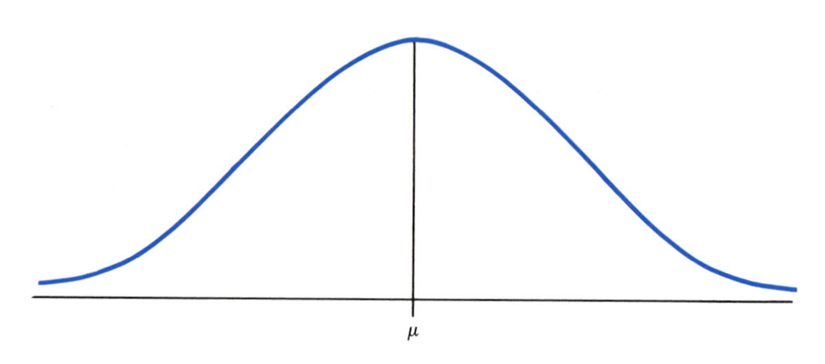

Figure 7–13 A normal curve with mean μ

lies to the right. The normal curve has the property that the area under the bell is always 1.

The normal curve represents **continuous** data, that for which the values could be arbitrarily close together. For example, when we measure the heights of male college freshmen, two men could be of different, but nearly the same height. When we have **discrete** data, like counting the number of heads in five tosses of a coin, two different values must be at least one unit apart. When dealing with continuous data, we usually do not ask how many men are exactly 5′10″ tall. We may ask how many are between 5′9″ and 5′11″ or within one-fourth an inch of 5′10″, and so on.

Area Under a Normal Curve

The ability to find area under a portion of the normal curve is important because it gives the frequency of scores in a specified category or interval. For example, suppose that a normal distribution has a mean of 85. The area under the curve between 90 and 95 represents the fraction of scores, not the number, that lie between 90 and 95. (Since the total area under the curve is 1, the area also represents the probability that the score is between 90 and 95.)

Notice that the area under the curve between 85 and 90 is larger than the area between 90 and 95. This indicates a proportionally larger frequency of values between 85 and 90. Since the curve is symmetric about the mean, it approaches, but never quite touches, the base line in either direction.

The normal curve has the unusual property of being completely determined by the mean and standard deviation. The normal curve can be standardized in a

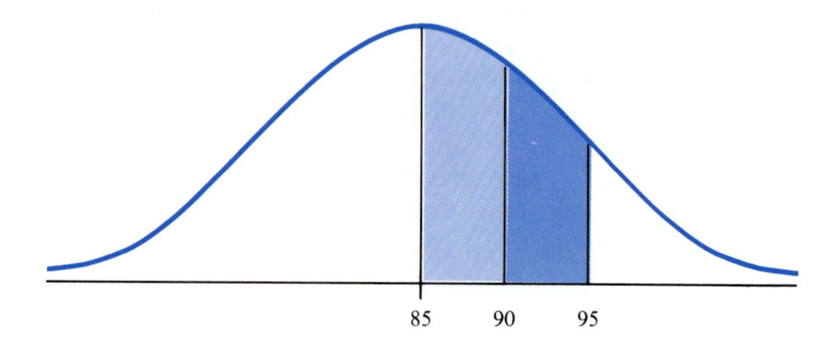

Figure 7–14

way that allows us to find the area under a portion of a normal curve from one table of values. The distance from the mean, measured in standard deviations, determines the area under the curve in that interval. For example, for all normal curves, about 68% of the scores will lie within one standard deviation of the mean, 34% on one side and 34% on the other. Approximately 95% of the scores will lie within two standard deviations, and over 99% will lie within three standard deviations. (See Figure 7–15.)

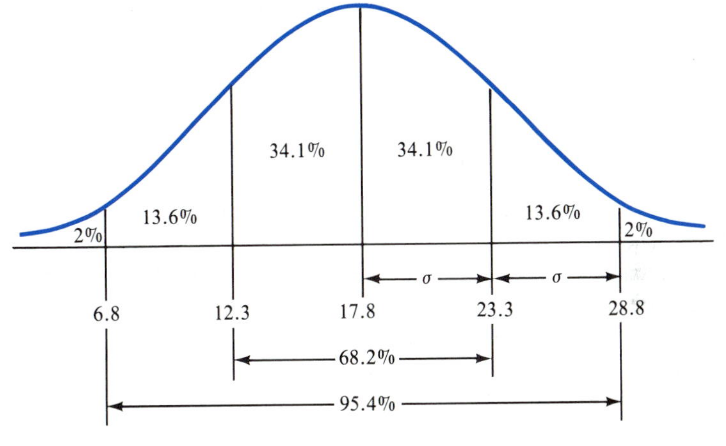

Figure 7–15 Normal curve with $\mu = 17.8$ and $\sigma = 5.5$

We can use the **standard normal curve** to answer questions about normal curves in general. The standard normal curve has mean $\mu = 0$ and $\sigma = 1$. In the standard normal curve, a score X is also the number of standard deviations from the mean. Several examples will help you to understand the use of a normal curve

Let's apply this idea to the following.

Example 1 One year the ACT English test had a mean $\mu = 17.8$ and a standard deviation $\sigma = 5.5$.

(a) The scores 12.3 and 23.3 are both one standard deviation from the mean, so 34% of the ACT English scores lie between 12.3 and 17.8, and 34% lie between 17.8 and 23.3.

(b) About 95% of the scores lie between 6.8 and 28.8, that is, within two standard deviations of the mean.

(c) Because over 99% of the scores lie within three standard deviations of the mean (a distance of 16.5 or less from the mean), you expect less than 1% to lie *more* than three standard deviations away. So, less than 1% of the ACT English scores lie above 34.3 or below 1.3.

Since the mean and standard deviation of a normal distribution completely determine the shape of a normal curve, two normal distributions that have the same mean but different standard deviations will have different shapes. The one with the smaller standard deviation will have a sharper peak, and the one with the larger standard deviation will have a flatter normal curve. For a

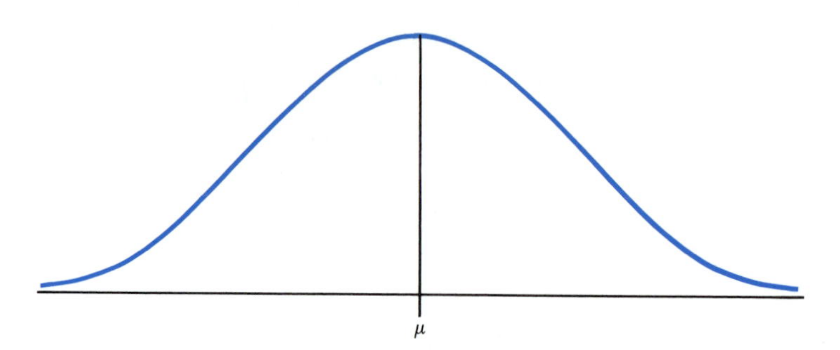

Normal curve with larger standard deviation

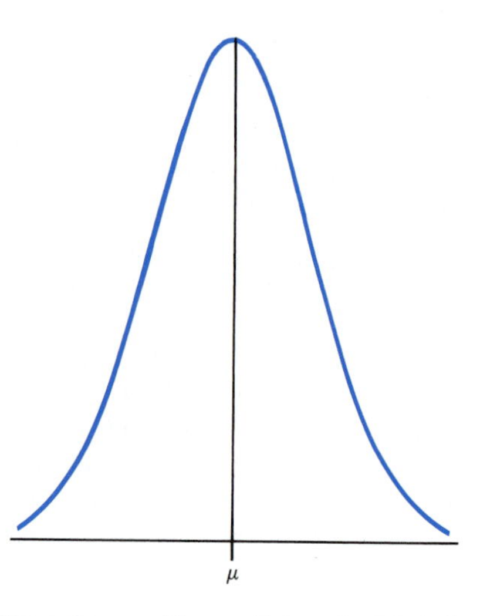

Figure 7–16 Normal curve with a smaller standard deviation

normal distribution, the mean, the median, and the mode are all the same. See Figure 7–16.

The *z* Score

We have already discussed the fraction of test scores that lie within 1, 2, or 3 standard deviations of the mean in a normal distribution. How do we determine the number of scores that lie within 1.25, 0.63, or 2.50 standard deviations of the mean? Tables such as Table 7–1 (a table for the standard normal curve) enable you to find the area between the mean and a score. The key is to locate the score according to the number of standard deviations it lies from the mean. Traditionally, Z represents the number of standard deviations. For selected values of Z, Table 7–1 gives the area under the curve between the mean and the Z score.

Suppose that the score is 1.40 standard deviations above the mean. To determine the area, we look in Table 7–1 at $Z = 1.40$ and find 0.4192 for A (area)

Table 7–1. Area Under the Normal Curve Between the Mean and z, $z \geq 1$

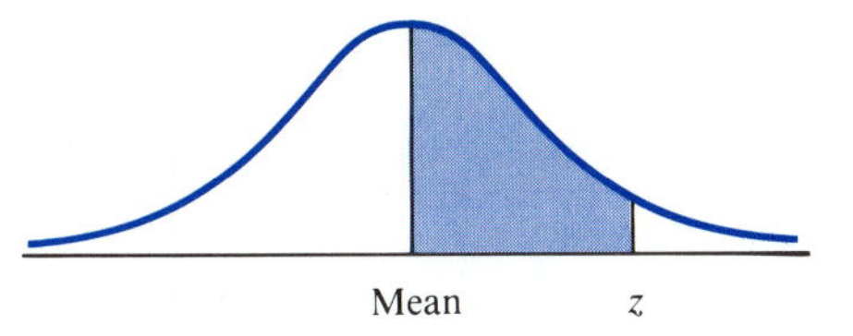

z	A	z	A	z	A	z	A
0.00	0.0000	0.40	0.1554	0.80	0.2881	1.20	0.3849
0.01	0.0040	0.41	0.1591	0.81	0.2910	1.21	0.3869
0.02	0.0080	0.42	0.1628	0.82	0.2939	1.22	0.3888
0.03	0.0120	0.43	0.1664	0.83	0.2967	1.23	0.3907
0.04	0.0160	0.44	0.1700	0.84	0.2996	1.24	0.3925
0.05	0.0199	0.45	0.1736	0.85	0.3023	1.25	0.3944
0.06	0.0239	0.46	0.1772	0.86	0.3051	1.26	0.3962
0.07	0.0279	0.47	0.1808	0.87	0.3079	1.27	0.3980
0.08	0.0319	0.48	0.1844	0.88	0.3106	1.28	0.3997
0.09	0.0359	0.49	0.1879	0.89	0.3133	1.29	0.4015
0.10	0.0398	0.50	0.1915	0.90	0.3159	1.30	0.4032
0.11	0.0438	0.51	0.1950	0.91	0.3186	1.31	0.4049
0.12	0.0478	0.52	0.1985	0.92	0.3212	1.32	0.4066
0.13	0.0517	0.53	0.2019	0.93	0.3238	1.33	0.4082
0.14	0.0557	0.54	0.2054	0.94	0.3264	1.34	0.4099
0.15	0.0596	0.55	0.2088	0.95	0.3289	1.35	0.4115
0.16	0.0636	0.56	0.2123	0.96	0.3315	1.36	0.4131
0.17	0.0675	0.57	0.2157	0.97	0.3340	1.37	0.4147
0.18	0.0714	0.58	0.2190	0.98	0.3365	1.38	0.4162
0.19	0.0754	0.59	0.2224	0.99	0.3389	1.39	0.4177
0.20	0.0793	0.60	0.2258	1.00	0.3413	1.40	0.4192
0.21	0.0832	0.61	0.2291	1.01	0.3438	1.41	0.4207
0.22	0.0871	0.62	0.2324	1.02	0.3461	1.42	0.4222
0.23	0.0910	0.63	0.2357	1.03	0.3485	1.43	0.4236
0.24	0.0948	0.64	0.2389	1.04	0.3508	1.44	0.4251
0.25	0.0987	0.65	0.2422	1.05	0.3531	1.45	0.4265
0.26	0.1026	0.66	0.2454	1.06	0.3554	1.46	0.4279
0.27	0.1064	0.67	0.2486	1.07	0.3577	1.47	0.4292
0.28	0.1103	0.68	0.2518	1.08	0.3599	1.48	0.4306
0.29	0.1141	0.69	0.2549	1.09	0.3621	1.49	0.4319
0.30	0.1179	0.70	0.2580	1.10	0.3643	1.50	0.4332
0.31	0.1217	0.71	0.2612	1.11	0.3665	1.51	0.4345
0.32	0.1255	0.72	0.2642	1.12	0.3686	1.52	0.4357
0.33	0.1293	0.73	0.2673	1.13	0.3708	1.53	0.4370
0.34	0.1331	0.74	0.2704	1.14	0.3729	1.54	0.4382
0.35	0.1368	0.75	0.2734	1.15	0.3749	1.55	0.4394
0.36	0.1406	0.76	0.2764	1.16	0.3770	1.56	0.4406
0.37	0.1443	0.77	0.2794	1.17	0.3790	1.57	0.4418
0.38	0.1480	0.78	0.2823	1.18	0.3810	1.58	0.4430
0.39	0.1517	0.79	0.2852	1.19	0.3830	1.59	0.4441

Table 7–1. (*continued*)

z	A	z	A	z	A	z	A
1.60	0.4452	2.00	0.4773	2.40	0.4918	2.80	0.4974
1.61	0.4463	2.01	0.4778	2.41	0.4920	2.81	0.4975
1.62	0.4474	2.02	0.4783	2.42	0.4922	2.82	0.4976
1.63	0.4485	2.03	0.4788	2.43	0.4925	2.83	0.4977
1.64	0.4495	2.04	0.4793	2.44	0.4927	2.84	0.4977
1.65	0.4505	2.05	0.4798	2.45	0.4929	2.85	043978
1.66	0.4515	2.06	0.4803	2.46	0.4931	2.86	0.4979
1.67	0.4525	2.07	0.4808	2.47	0.4932	2.87	0.4980
1.68	0.4535	2.08	0.4812	2.48	0.4934	2.88	0.4980
1.69	0.4545	2.09	0.4817	2.49	0.4936	2.89	0.4981
1.70	0.4554	2.10	0.4821	2.50	0.4938	2.90	0.4981
1.71	0.4564	2.11	0.4826	2.51	0.4940	2.91	0.4982
1.72	0.4573	2.12	0.4830	2.52	0.4941	2.92	0.4983
1.73	0.4582	2.13	0.4834	2.53	0.4943	2.93	0.4983
1.74	0.4591	2.14	0.4838	2.54	0.4945	2.94	0.4984
1.75	0.4599	2.15	0.4842	2.55	0.4946	2.95	0.4984
1.76	0.4608	2.16	0.4846	2.56	0.4948	2.96	0.4985
1.77	0.4616	2.17	0.4850	2.57	0.4949	2.97	0.4985
1.78	0.4625	2.18	0.4854	2.58	0.4951	2.98	0.4986
1.79	0.4633	2.19	0.4857	2.59	0.4952	2.99	0.4986
1.80	0.4641	2.20	0.4861	2.60	0.4953	3.00	0.4987
1.81	0.4649	2.21	0.4865	2.61	0.4955	3.01	0.4987
1.82	0.4656	2.22	0.4868	2.62	0.4956	3.02	0.4987
1.83	0.4664	2.23	0.4871	2.63	0.4957	3.03	0.4988
1.84	0.4671	2.24	0.4875	2.64	0.4959	3.04	0.4988
1.85	0.4678	2.25	0.4878	2.65	0.4960	3.05	0.4989
1.86	0.4686	2.26	0.4881	2.66	0.4961	3.06	0.4989
1.87	0.4693	2.27	0.4884	2.67	0.4962	3.07	0.4989
1.88	0.4700	2.28	0.4887	2.68	0.4963	3.08	0.4990
1.89	0.4706	2.29	0.4890	2.69	0.4964	3.09	0.4990
1.90	0.4713	2.30	0.4893	2.70	0.4965	3.10	0.4990
1.91	0.4719	2.31	0.4896	2.71	0.4966	3.11	0.4991
1.92	0.4726	2.32	0.4898	2.72	0.4967	3.12	0.4991
1.93	0.4732	2.33	0.4901	2.73	0.4968	3.13	0.4991
1.94	0.4738	2.34	0.4904	2.74	0.4969	3.14	0.4992
1.95	0.4744	2.35	0.4906	2.75	0.4970	3.15	0.4992
1.96	0.4750	2.36	0.4909	2.76	0.4971	3.16	0.4992
1.97	0.4756	2.37	0.4911	2.77	0.4972	3.17	0.4992
1.98	0.4762	2.38	0.4913	2.78	0.4973	3.18	0.4993
1.99	0.4767	2.39	0.4916	2.79	0.4974	3.19	0.4993

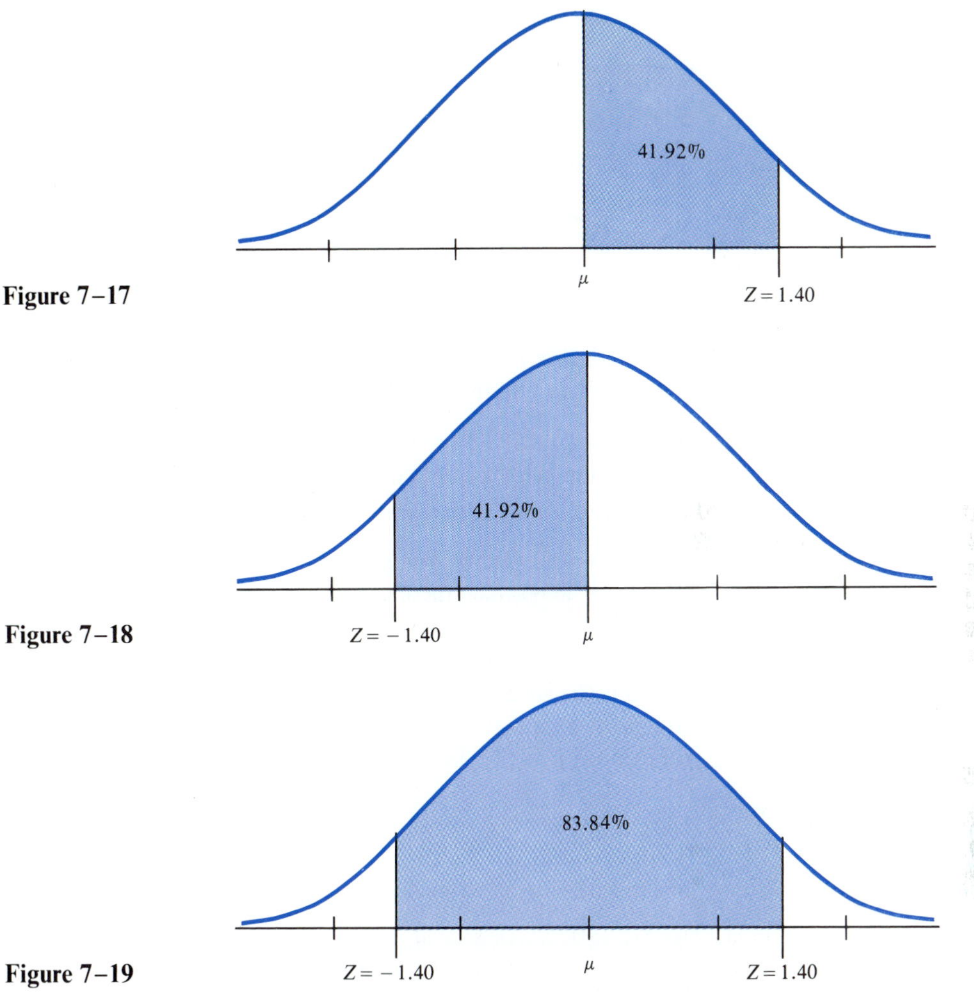

Figure 7–17

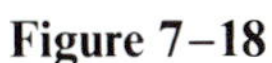
Figure 7–18

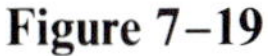
Figure 7–19

(see Figure 7–17). This indicates that the shaded area is 0.4192, or 41.92%, of the area under the curve. Stated another way, it indicates that 41.92% of the scores lie between the mean and a point 1.40 standard deviations above the mean. Because the normal curve is symmetric, we also know that 41.92% of the scores lie between the mean and a point that is 1.40 standard deviations *below* the mean. A *negative* value for Z indicates that the score lies to the left of the mean ($Z = -1.40$). (See Figure 7–18.) Combining the two areas on each side of the mean, we see that 83.84% of the scores lie *within* 1.40 standard deviations of the mean (see Figure 7–19).

We can also find the fraction of scores that are *more* than 1.40 standard deviations away from the mean. Since 83.84% are within 1.40 standard deviations of the mean, then the remainder, 16.16%, lie more than 1.40 standard deviations away. Half of these scores lie above $Z = 1.40$. The other half lie below $Z = -1.40$. See Figure 7–20.

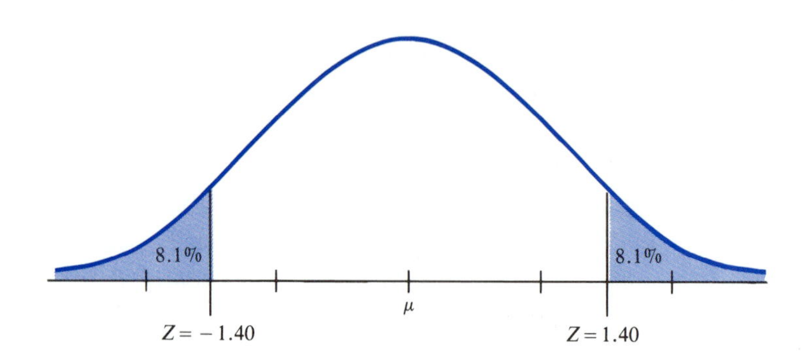

Figure 7–20

Example 2 (*Compare Exercise 1*)
Use Table 7–1 to find the fraction of scores that lie

(a) between the mean and a score 0.80 standard deviations above the mean.
(b) between the mean and a score 0.80 standard deviations below the mean.
(c) within 0.80 standard deviations of the mean.

Solution **(a)** Here $Z = 0.80$. Table 7–1 gives $A = 0.2881$. Thus, 0.2881, or 28.81%, of the scores lie between the mean and a point 0.80 standard deviations above the mean.

(b) Since the normal curve is symmetric with respect to the mean, the area between the mean and a point 0.80 standard deviations *below* the mean equals the area between the mean and a point 0.80 standard deviation *above* the mean. So, 28.81% lie between the mean and the point 0.80 below the mean ($Z = -0.80$).

(c) Here $Z = 0.80$, but we want scores on both sides of the mean. Thus, $2 \times 0.2881 = 0.5762$, or 57.62%, of the scores lie within 0.80 standard deviations of the mean (or between $Z = -0.80$ and $Z = 0.80$).

We find the number of standard deviations a score lies from the mean the same way we did in Section 7–4. The formula is

$$Z = \frac{\text{distance between score and mean}}{\text{standard deviation}} = \frac{\text{score} - \text{mean}}{\text{standard deviation}} = \frac{x - \mu}{\sigma}$$

Example 3 (*Compare Exercise 18*)
Compute Z for each of the following values of the mean, the standard deviation, and a given score:

(a) mean = 25, standard deviation = 2, score = 31
(b) mean = 25, standard deviation = 2, score = 19
(c) mean = 7.5, standard deviation = 1.2, score = 10.5
(d) mean = 16.85, standard deviation = 2.1, score = 14.12

Solution **(a)** Using the formula

$$Z = \frac{\text{score} - \text{mean}}{\text{standard deviation}}$$

we obtain

$$Z = \frac{31 - 25}{2} = \frac{6}{2} = 3$$

(b) $Z = \dfrac{19 - 25}{2} = \dfrac{-6}{2} = -3$

(c) $Z = \dfrac{10.5 - 7.5}{1.2} = \dfrac{3}{1.2} = 2.5$

(d) $Z = \dfrac{14.12 - 16.85}{2.1} = \dfrac{-2.73}{2.1} = -1.3$

Example 4 (*Compare Exercise 28*)
A normal distribution has a mean of 30 and a standard deviation of 7. Find the fraction of scores between 30 and 42.

Solution Here,

$$Z = \frac{42 - 30}{7} = \frac{12}{7} = 1.71$$

From Table 7–1, we obtain $A = 0.4564$ when $Z = 1.71$. Thus, 45.64% of the scores lie between 30 and 42. See Figure 7–21.

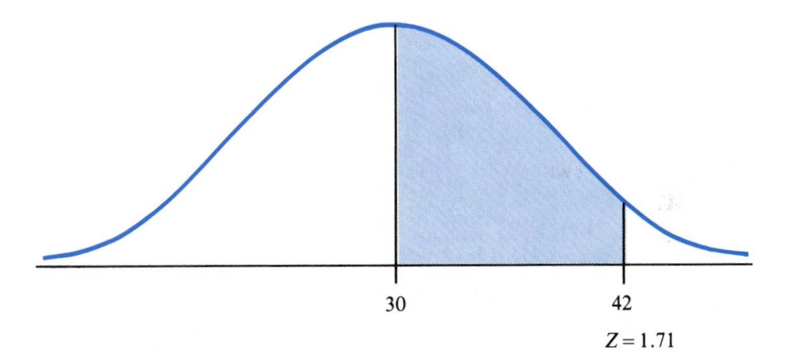

Figure 7–21

When two scores are on opposite sides of the mean and are different distances from the mean, we add areas to find the area between the scores.

Example 5 (*Compare Exercise 35*)
A normal distribution has a mean $\mu = 50$ and a standard deviation $\sigma = 6$. Find the fraction of scores between 47 and 58.

Solution Since the mean is between 47 and 58, we need to find the area under the curve in two steps; that is, we need to find the area from the mean to each score.

1. The area between 47 and 50.

$$Z_1 = \frac{47 - 50}{6} = \frac{-3}{6} = -.50$$

From Table 7–1, this area is $A = .1915$.

2. The area between 50 and 58.

$$Z_2 = \frac{58 - 50}{6} = \frac{8}{6} = 1.33$$

$$A = .4082$$

The total area between 47 and 58 is .1915 + .4082 = .5997, so .5997, or 59.97%, of the scores lie between 47 and 58. See Figure 7–22.

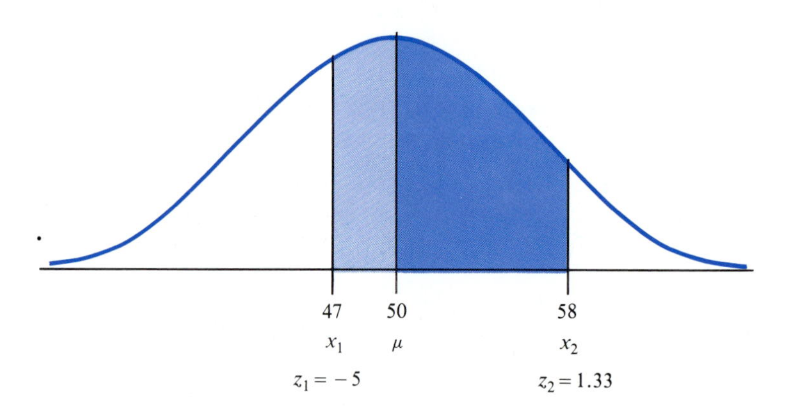

Figure 7–22

Sometimes, we want to find an area between two scores that lie on the same side of the mean.

Example 6 (*Compare Exercise 36*)
A normal distribution has a mean $\mu = 100$ and a standard deviation $\sigma = 8$. Find the fraction of scores that lie between 110 and 120.

Solution Since all areas are measured from the mean to a score, we can find the area between 100 and 110 and the area between 100 and 120. To find the area between 110 and 120, subtract the area between 100 and 110 from the area between 110 and 120.

For $X_1 = 110$,

$$Z_1 = \frac{110 - 100}{8} = \frac{10}{8} = 1.25$$

and $A_1 = .3944$

For $X_2 = 120$,

$$Z_2 = \frac{120 - 100}{8} = \frac{20}{8} = 2.5$$

and $A_2 = .4938$.

The area between 110 and 120 is then .4938 − .3944 = .0994, so 9.94% of the scores lie between 110 and 120. See Figure 7–23.

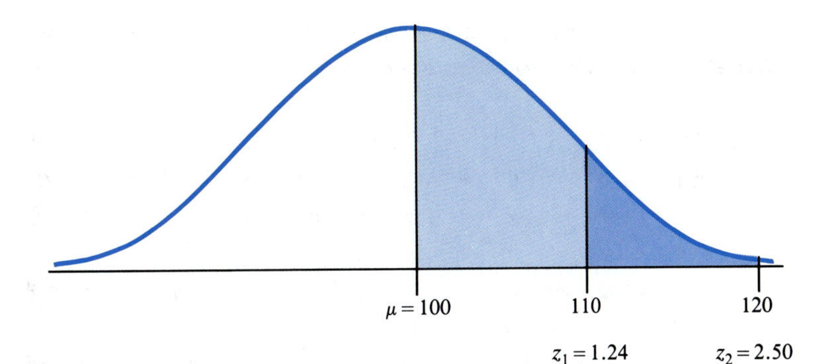

Figure 7–23

Probability and a Normal Distribution

Because the total area under the normal curve is 1, and because the area between two scores gives the fraction of the data that lie between the two scores, an experiment whose data forms a normal distribution is connected to its related normal curve in the following way.

> The probability that an outcome is between two scores X_1 and X_2 is just the area under the normal curve between X_1 and X_2.

Example 7 (*Compare Exercise 39*)
Students at Flatland University spend an average of 24.3 hours per week on homework, with a standard deviation of 1.4 hours.

(a) What percentage of the students spend more than 28 hours per week on homework?

(b) What is the probability that a student spends more than 28 hours per week on homework?

Solution **(a)** The value of Z corresponding to 28 hours is

$$Z = \frac{28 - 24.3}{1.4} = \frac{3.7}{1.4} = 2.64$$

From Table 7–1, we have $A = 0.4959$ when $Z = 2.64$. The value $A = 0.4959$ represents the area from the mean, 24.3, to 28 ($Z = 2.64$). All of the area under the curve to the right of the mean is one half of the total area. The area to the right of $Z = 2.64$ is $0.5000 - 0.4959 = 0.0041$. Thus, 0.41% of the students study more than 28 hours.

(b) The probability that a student studies more than 28 hours is the fraction of the area that is to the right of 28, that is, 0.0041.

Example 8 (*Compare Exercise 41*)

(a) Find a value of Z such that 4% of the scores are to the right of Z.

(b) Find the value of Z such that 0.04 is the probability that a score lies to the right of Z.

Solution **(a)** If 4% of the scores are to the right of Z, then the other 46% of the scores to the right of the mean are between the mean and Z. (**Remember** that 50% of the scores are to the right of the mean.) Look for $A = 0.4600$ in Table 7–1. It occurs at $Z = 1.75$. This is the desired value of Z.

(b) This also occurs at $Z = 1.75$.

Since the normal curve is symmetric, Table 7–1 gives A only for positive values of Z. For negative values of Z, when the score is to the left of the mean, simply use the value of A for the corresponding positive Z. Keep in mind that each Z value determines an area *from the mean* to the Z position.

We can find a variety of areas by adding or subtracting areas given from the table. (See Figure 7–24.)

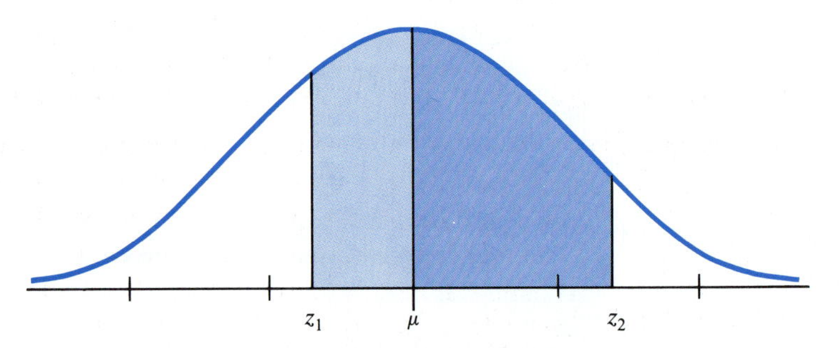

(a) Add the areas for z_1 and z_2 to get the area between z_1 and z_2.

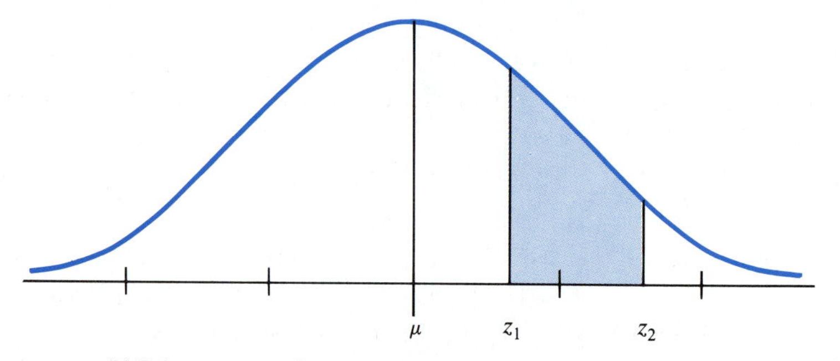

(b) Subtract z_1 area from z_2 area to get the area between z_1 and z_2.

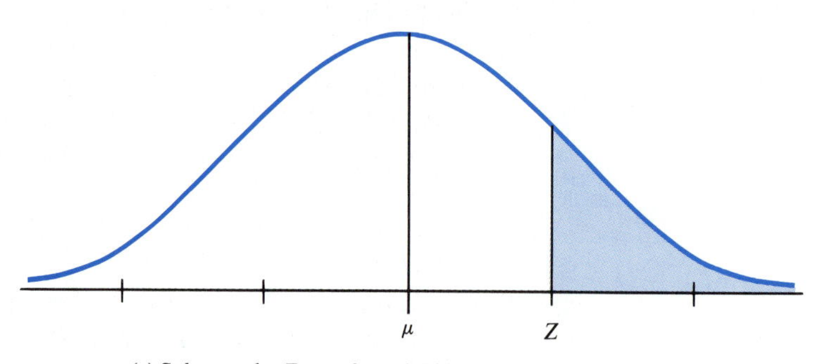

Figure 7–24

(c) Subtract the Z area from 0.500 to get the area beyond Z.

Procedure to Determine the Fraction of Scores Between Two Scores of a Normal Distribution

Step 1. Determine the Z value for each score.

Step 2. From Table 7–1, determine A corresponding to Z.

Step 3.

(a) If the two scores are on opposite sides of the mean (Z_1 and Z_2 have opposite signs), add the values of A corresponding to Z_1 and Z_2.

(b) If the two scores are on the same side of the mean (Z_1 and Z_2 have the same signs), subtract the smaller value of A from the larger value.

Example 9 (*Compare Exercise 44*)
A standardized test is given to several hundred thousand junior high students. The mean is 100, and the standard deviation is 10. If a student is selected at random, what is the probability that he scores in the 114 to 120 interval?

Solution This question may be answered by using the properties of the normal curve since it represents standardized test scores well. Because the area under the normal curve is 1, we can use the area between 114 and 120 as the probability that the student scores in the 114 to 120 interval.

The values of Z corresponding to 114 and 120 are $Z = 1.4$ and $Z = 2.0$, respectively. For $Z = 1.4$, $A = 0.4192$, and for $Z = 2.0$, $A = 0.4773$. So, the area between 114 and 120 is $0.4773 - 0.4192 = 0.0581$. The probability that the student's score is in the interval 114 to 120 is 0.0581.

Summary of Properties of a Normal Curve

1. All normal curves have the same general bell shape.
2. It is symmetric with respect to a vertical line that passes through the peak of the curve.
3. The vertical line through the peak occurs where the mean, median, and mode coincide.
4. The area under any normal curve is always 1.
5. The mean and standard deviation completely determine a normal curve. For the same mean, a smaller standard deviation gives a taller and narrower peak. A larger standard deviation gives a flatter curve.
6. The area to the right of the mean is 0.5; the area to the left of the mean is 0.5.
7. About 68.26% of the area under a normal curve is enclosed in the interval formed by the score one standard deviation to the left of the mean and the score one standard deviation to the right of the mean.
8. If a random variable X has a normal probability distribution, the probability that a score lies between X_1 and X_2 is the area under the normal curve between X_1 and X_2.
9. The probability that a score is less than X_1 equals the probability that a score is less than, or equal to X_1, that is, $P(X < X_1) = P(X \leq X_1)$.

7–6 Exercises

I.

In Exercises 1 through 9, find the percentage of the area under the normal curve that lies between the mean and the given number of standard deviations above the mean.

1. (*See Example 2*) 0.50
2. 1.90
3. 0.25
4. 2.50
5. 1.10
6. 0.10
7. 2.90
8. 2.25
9. 0.75

In Exercises 10 through 13, find the fraction of scores that lie between the mean and the given number of standard deviations below the mean.

10. 1.30
11. 0.46
12. 2.17
13. 1.00

In Exercises 14 through 17, find the percentage of the area under the normal curve that lies within the given number of standard deviations of the mean.

14. 2.73
15. 0.65
16. 1.34
17. 0.38

Compute Z in Exercises 18 through 23.

18. (*See Example 3*) score $= 192$, mean $= 150$, $\sigma = 7$
19. score $= 3.1$, mean $= 4.0$, $\sigma = 0.3$
20. score $= 38.0$, mean $= 22.5$, $\sigma = 6.2$
21. score $= 10.1$, mean $= 10.0$, $\sigma = 2.0$
22. score $= 31$, mean $= 31$, $\sigma = 5$
23. score $= 2.65$, mean $= 0$, $\sigma = 1.0$

Find the percentage of the area under the normal curve that lies more than the given number of standard deviations away from the mean in Exercises 24 through 27.

24. (*See Example 1*) 2.50
25. 0.86
26. 1.70
27. 1.50

A normal distribution has a mean of 75 and a standard deviation of 5. Find the fraction of scores in the intervals indicated in Exercises 28 through 33.

28. (*See Example 4*) lie between 75 and 82
29. lie between 80 and 85
30. lie between 77 and 83
31. lie above 76
32. lie between 68 and 79
33. are either below 70 or above 80

A normal distribution has a mean of 168 and a standard deviation of 10. Find the percent of scores in the intervals indicated in Exercises 34 through 39.

34. are between 168 and 175

35. (*See Example 5*) are between 155 and 169

36. (*See Example 6*) are between 170 and 180

37. are less than 172

38. are less than 150

39. (*See Example 7*) are larger than 173

40. Find the fraction of scores from a normal distribution that lie between $Z = -0.60$ and $Z = 1.20$.

Find the value of Z in Exercises 41 through 43.

41. (*See Example 8*) 8% of the scores are to the right of Z.

42. 96% of the scores are to the left of Z.

43. 91% of the scores are between Z and $-Z$.

II.

44. (*See Example 9*) A standardized test has a mean of 80 and a standard deviation of 5. A test is selected at random. Find the probability that

(a) the score is between 84 and 90.
(b) the score is above 88.
(c) the score is below 74.
(d) the score is between 75 and 83.

45. One year the freshmen of all U.S. colleges had a mean IQ of 110 and a standard deviation of 12. The IQ scores form a normal distribution. If a student is selected at random, find the probability that

(a) the student has an IQ between 120 and 125.
(b) the student has an IQ below 100.
(c) the student has an IQ between 105 and 115.

46. The IQ of individuals forms a normal distribution with mean = 100 and $\sigma = 15$.

(a) What percentage of the population have IQs below 85?
(b) What percentage have IQs over 130?
(c) A college requires an IQ of 120 or more for entrance. From what percentage of the population must they draw their students?
(d) In a state with 400,000 high school seniors, how many meet the IQ requirement?
(e) What is the probability that a student has an IQ below 90?

III.

47. The weights of college students at a particular college is a normal distribution with a mean of 126 lb and a standard deviation of 6 lb. The college has 700 students.

(a) How many students weigh between 120 and 132 lb?

(b) How many students weigh less than 114 lb?

(c) How many students weigh more than 134 lb?

48. The batteries used for a calculator have an average life of 2000 hours and a standard deviation of 200 hours. The normal distribution closely represents the life of the battery. Find the fraction of batteries that can be expected to last the following lengths of time:

(a) between 1800 and 2200 hours

(b) between 1900 and 2100 hours

(c) at least 2500 hours

(d) no more than 2200 hours

(e) less than 1500 hours

(f) between 1600 and 2300 hours

(g) less than 1900 or more than 2100 hours

(h) between 2200 and 2400 hours

(i) What is the probability that a battery lasts longer than 2244 hours?

49. The scores of students on a standardized test form a normal distribution with a mean of 300 and a standard deviation of 40.

(a) What fraction of the students scored between 270 and 330?

(b) What fraction of the students scored between 312 and 330?

(c) What fraction of the students scored less than 300? Less than 326?

(d) What must a student score to be in the upper 10%?

(e) A student is selected at random. What is the probability that the student scored between 312 and 324?

50. An automatic lathe makes shafts for a high-speed machine. The specifications call for a shaft with a diameter of 1.800 inches. An inspector closely monitors production for a week. He finds the actual diameters form a normal distribution with a mean of 1.800 inches and a standard deviation of 0.00033 inches. The design engineer will accept a shaft that is within 0.001 inches of the specified diameter of 1.800. If a shaft is selected at random, what is the probability that it is within the specified tolerance; that is, the diameter is in the interval 1.799 to 1.801 inches?

51. A standardized test has a mean of 120 and a standard deviation of 10. If two students are selected at random, what is the probability that both score below 128?

52. The mean of the weights of a random sample of students on a college campus is 140 pounds, and the standard deviation is 20 pounds. Assuming that the weights of the students are normally distributed, find the probability that a student selected at random will weigh between 120 and 130 pounds.

53. Applicants for certain jobs must pass a qualifying examination. Those with scores between 85 and 100 are suitable for the positions. The scores on the examination are normally distributed with a mean of 75 and a standard deviation of 11. What percentage of those taking the examination pass?

54. A company manufactures car tires. In an analysis of tire wear under normal driving conditions, it was found that a random sample had an average life of 30,000 miles. The standard deviation of the sample was 2000 miles. What percentage of tires can be expected to have a life of 33,800 miles or more?

55. The heights of individuals in a certain society are normally distributed with a mean of 66.4 inches and a standard deviation of 4.3 inches. What is the probability that an individual selected at random will be at least 6 feet tall?

56. A set of data has a mean of 150 and a standard deviation of 8. Find the two numbers, both the same distance from the mean but on opposite sides, that enclose

    (a) 90% of the area under the normal curve.
    (b) 95% of the area under the normal curve.
    (c) 98% of the area under the normal curve.

57. In a mathematics test, grades were awarded on a scale of 0 through 100 points. The mean grade was 80, and the standard deviation was 15. Assuming that the grades were normally distributed, determine the percentage of students who scored below 90.

58. In an examination, the mean score is 75, the standard deviation is 15, and the grades follow a normal distribution. The instructor wants to assign A's to the top 10%, B's to the next 20%, C's to the next 40%, and D's to the remainder of the class. Where should the cutoff points be?

7–7 Approximating the Binomial Distribution

Binomial Distribution

Recall from Section 6–6 that a Bernoulli experiment is one that meets the following requirements:

(a) It consists of a sequence of repeated trials.
(b) It has two possible outcomes—success and failure.
(c) The probability of success remains fixed for each trial.

(d) The trials are independent.

(e) The probability of x successes in n trials is

$$C(n, x)p^x q^{n-x}$$

We can use Bernoulli trials to form an important probability distribution. This distribution is called a **binomial distribution**. Define it in the following way.

Binomial Distribution

For a Bernoulli experiment of n repeated trials, define a random variable X as the number of successes in n trials. For each value of X, $0 \leq x \leq n$, find the probability of x successes in n trials. The probability distribution obtained is the **binomial distribution**.

Example 1 (*Compare Exercise 5*)

Form the binomial distribution for the experiment of rolling a die five times and counting the times a 4 appears.

Solution The random variable X takes on the values 0, 1, 2, 3, 4, and 5, the possible number of successes in five trials. The probability of each value occurring is computed using Bernoulli trials with $p = \frac{1}{6}$ and $q = \frac{5}{6}$. ($\frac{1}{6}$ is the probability of rolling a 4 in a single trial.)

X	$P(X)$
0	$C(5, 0)(\frac{5}{6})^5 = 0.4019$
1	$C(5, 1)(\frac{1}{6})^1(\frac{5}{6})^4 = 0.4019$
2	$C(5, 2)(\frac{1}{6})^2(\frac{5}{6})^3 = 0.1608$
3	$C(5, 3)(\frac{1}{6})^3(\frac{5}{6})^2 = 0.0322$
4	$C(5, 4)(\frac{1}{6})^4(\frac{5}{6}) = 0.0032$
5	$C(5, 5)(\frac{1}{6})^5 = 0.0001$

The binomial distribution got its name from the **binomial** $(p + q)^n$. Each term of the expansion of the binomial gives one of the probabilities in the binomial distribution. Notice that

$$(p + q)^3 = p^3 + 3p^2q + 3pq^2 + q^3$$

which can be written

$$C(3, 3)p^3q^0 + C(3, 2)p^2q + C(3, 1)pq^2 + C(3, 0)q^3$$

where each term represents the probability of 3, 2, 1, or 0 successes, respectively, in a binomial experiment with three trials.

Example 2 (*Compare Exercise 7*)
Form the binomial distribution of the experiment of tossing a coin six times and counting the number of heads.

Solution The random variable X takes on the values 0, 1, 2, 3, 4, 5, and 6, the possible number of successes in six tosses. Both p and q are $\frac{1}{2}$. The values of X and the corresponding probabilities computed using Bernoulli trials are the following:

X	$P(X)$
0	$C(6, 0)(\frac{1}{2})^6 = 0.0156$
1	$C(6, 1)(\frac{1}{2})^1(\frac{1}{2})^5 = 0.0938$
2	$C(6, 2)(\frac{1}{2})^2(\frac{1}{2})^4 = 0.2344$
3	$C(6, 3)(\frac{1}{2})^3(\frac{1}{2})^3 = 0.3125$
4	$C(6, 4)(\frac{1}{2})^4(\frac{1}{2})^2 = 0.2344$
5	$C(6, 5)(\frac{1}{2})^5(\frac{1}{2}) = 0.0938$
6	$C(6, 6)(\frac{1}{2})^6 = 0.0156$

The histogram of the distribution is given in Figure 7–25.

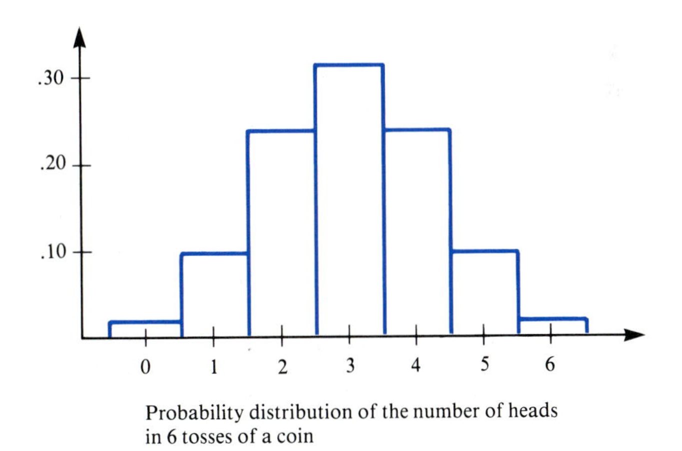

Figure 7–25 Probability distribution of the number of heads in 6 tosses of a coin

Example 3 (*Compare Exercise 1*)
Find the binomial distribution for $n = 4$ and $p = 0.3$.

Solution The random variable X may take on the values 0, 1, 2, 3, and 4. The probability distribution is the following.

X	$P(X)$
0	$C(4, 0)(0.7)^4 = 0.2401$
1	$C(4, 1)(0.3)(0.7)^3 = 0.4116$
2	$C(4, 2)(0.3)^2(0.7)^2 = 0.2646$
3	$C(4, 3)(0.3)^3(0.7) = .0756$
4	$C(4, 4)(0.3)^4 = .0081$

Example 4 (*Compare Exercise 10*)
The probability that a new drug will cure a certain blood disease in 0.7. If it is administered to 100 patients with the disease, what is the probability that 60 of them will be cured?

Solution You should set up

$$P(X = 60) = C(100, 60)(0.7)^{60}(0.3)^{40}$$

with little difficulty. You then should find it extremely tedious to compute the probability, and even more tedious to form the binomial distribution of this experiment.

The normal distribution provides a means to avoid this wearisome computation. It can be used to estimate the binomial distribution for large values of n. Let's see how to do this.

Normal Approximation to the Binomial Distribution

You recall from Section 7–5 that the computation of the mean (or expected value), the variance, and the standard deviation of a probability distribution can be time consuming for a random variable with a number of values.

In Example 3, $n = 4$ and $p = 0.3$. The expected value of the random variable X is

$$\begin{aligned} E(X) &= .2401(0) + .4116(1) + .2646(2) + .0756(3) + .0081(4) \\ &= 0 + .4116 + .5292 + .2268 + .0324 \\ &= 1.20 \end{aligned}$$

The significance of this result is that $1.20 = 4(.3)$, which is np for this example. This result holds for all binomial distributions, $E(X) = np$. Likewise, the variance and standard deviation can be expressed in rather simple terms of n, p, and q, as follows.

Mean, Variance, and Standard Deviation of a Binomial Distribution

Let X be the random variable for a binomial distribution with n repeated trials, with p the probability of success, q the probability of failure, and

$$P(X = x) = C(n, x)\, p^x(1 - p)^{n-x}$$

Then, the mean (expected value), variance, and standard deviation of X are given by

Mean $\mu = np$
Variance: $\mathrm{var}(X) = np(1 - p) = npq$
Standard Deviation: $\sigma(X) = \sqrt{np(1 - p)} = \sqrt{npq}$

Example 5 (*Compare Exercise 15*)

(a) For the binomial distribution with $n = 20$, $p = 0.35$:

$$\text{Mean} = \mu = 20(.35) = 7$$
$$\text{Var}(X) = 20(.35)(.65) = 4.55$$
$$\sigma(X) = \sqrt{4.55} = 2.133$$

(b) If $n = 160$, $p = .21$,

$$\mu = 160(.21) = 33.6$$
$$\text{Var}(X) = 160(.21)(.79) = 26.544$$
$$\sigma(X) = \sqrt{26.544} = 5.152$$

To illustrate how the normal curve is used to estimate binomial probabilities, we use the following simple example. Let me warn you that this example violates the following rule.

Rule

The normal distribution provides a good estimate of the binomial distribution when both np and nq are greater than five.

The next example violates this rule to illustrate the basic ideas without lengthy computations.

Example 6 (*Compare Exercise 27*)
Use the normal curve to estimate the probability of

(a) three heads in six tosses of a coin.

(b) three or four heads in six tosses of a coin.

Solution **(a)** We first point out that the binomial distribution of this problem, and its histogram, are given in Example 2. From it, for example, we see that $P(X = 3) = 0.3125$.

For the binomial distribution with $n = 6$ and $p = 0.5$, the mean and standard deviation are

$$\mu = 6(.5) = 3$$
$$\sigma = \sqrt{6(.5)(.5)} = \sqrt{1.5} = 1.225$$

Figure 7–26 shows a normal curve with $\mu = 3$ and $\sigma = 1.225$ superimposed on the binomial distribution for $n = 6$ and $p = 0.5$. Notice that a portion of the

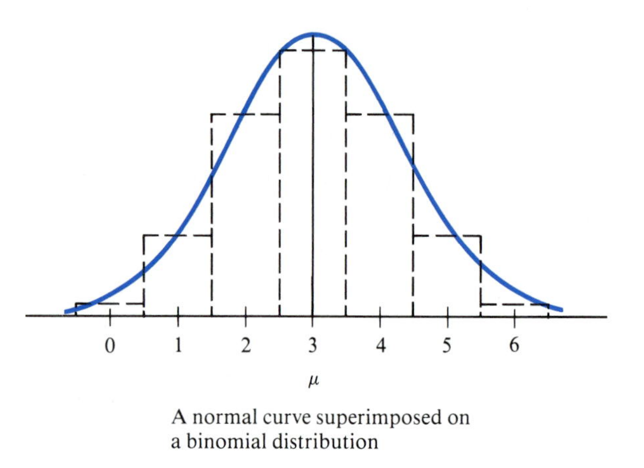

Figure 7–26

A normal curve superimposed on a binomial distribution

histogram lies above the curve, while some space under the curve is not filled by the rectangles. It appears that if the portions of the rectangle outside were moved into the empty spaces below the curve, then the area under the curve would "pretty well" be filled. This states that the total area enclosed by the histogram is "close" to the area under the normal curve.

Let's make another observation. Each bar in the histogram is of width 1, and its height is equal to the probability it represents. The bar for $P(X = 3)$ is located between 2.5 and 3.5, with height 0.3125. The area of that bar $(1 \times .3125)$ represents the probability that $X = 3$. We mention this because it holds the key to using the normal distribution. To find $P(X = 3)$, find the area under the normal curve between 2.5 and 3.5. Here's how:

Use $\mu = 3$, $\sigma = 1.225$, $X_1 = 2.5$, and $X_2 = 3.5$. Then,

$$Z_1 = \frac{2.5 - 3}{1.225} = \frac{-.5}{1.225} = -0.41$$

$$Z_2 = \frac{3.5 - 3}{1.225} = \frac{.5}{1.225} = 0.41$$

Table 7–1 shows that the area under the normal curve between the mean and $Z = 0.41$ is $A = 0.1591$. Then, the area between $Z_1 = -0.41$ and $Z_2 = 0.41$ is $0.1591 + 0.1591 = 0.3182$. This estimates $P(X = 3)$ as 0.3182. This compares to the actual probability of 0.3125.

(b) To find the probability of three or four heads in six tosses of a coin, $P(3 \leq X \leq 4)$, we find the total area of the bars for three and four. This amounts to finding the area in the histogram between 2.5 and 4.5. The normal approximation is the area between

$$Z_1 = \frac{2.5 - 3}{1.225} \qquad \text{and} \qquad Z_2 = \frac{4.5 - 3}{1.225}$$

that is,

$$Z_1 = -0.41 \qquad \text{and} \qquad Z_2 = 1.22$$

The corresponding areas are

$$A_1 = 0.1591 \qquad \text{and} \qquad A_2 = 0.3888$$

so the desired probability estimate is $0.1591 + 0.3888 = 0.5479$. The actual probability, from Example 2, is $0.3125 + 0.2344 = .5469$.

In summary, estimate a binomial probability with a normal distribution as follows.

Procedure for Estimating a Binomial Probability

1. If np and nq are both greater than 5, you may assume that the normal distribution provides a good estimate.
2. Compute $\mu = np$ and $\sigma = \sqrt{npq}$.
3. To estimate $P(X = c)$, find the area under the normal curve between $c - .5$ and $c + .5$. To do so, find the area between the Z scores

$$Z_1 = \frac{c - .5 - \mu}{\sigma} \qquad \text{and} \qquad Z_2 = \frac{c + .5 - \mu}{\sigma}$$

4. To estimate $P(c \leq X \leq d)$, $c < d$, find the area under the normal curve between $c - .5$ and $d + .5$. To do so, find the area between

$$Z_1 = \frac{c - .5 - \mu}{\sigma} \qquad \text{and} \qquad Z_2 = \frac{d + .5 - \mu}{\sigma}$$

5. To estimate $P(c < X < d)$, $c < d$, find the area under the normal curve between $c + .5$ and $d - .5$. To do so, find the area between

$$Z_1 = \frac{c + .5 - \mu}{\sigma} \qquad \text{and} \qquad Z_2 = \frac{d - .5 - \mu}{\sigma}$$

Example 7 (*Compare Exercise 29*)
The probability that a new drug will cure a certain blood disease is 0.7. If it is administered to 100 patients,

(a) what is the probability that 60 of them will be cured?
(b) what is the probability that 60 to 75 of them will be cured?
(c) what is the probability that more than 75 will be cured?

Solution Since $n = 100$, $p = 0.7$, and $q = 0.3$, we know that $np = 70$ and $nq = 30$. Since both values are greater than 5, a normal curve provides a good estimate.

$$\mu = 100(.7) = 70$$
$$\sigma = \sqrt{100(.7)(.3)} = \sqrt{21} = 4.583$$

(a) Find the area under the normal curve between 59.5 and 60.5 that is between

$$Z_1 = \frac{59.5 - 70}{4.583} = \frac{-10.5}{4.583} = -2.29$$

and

$$Z_2 = \frac{60.5 - 70}{4.583} = \frac{-9.5}{4.583} = -2.07$$

The corresponding areas are $A_1 = 0.4890$ and $A_2 = 0.4808$. Since the Z scores are on the same side of the mean, we must subtract areas, $0.4890 - 0.4808 = .0082$. So, $P(X = 60) = .0082$.

(b) To find $P(60 \leq X \leq 75)$, we find the area between 59.5 and 75.5. The corresponding Z values and areas are

$$Z_1 = \frac{59.5 - 70}{4.583} = 2.29 \qquad \text{and} \qquad Z_2 = \frac{75.5 - 70}{4.583} = 1.20$$

$A_1 = 0.4890$ and $A_2 = 0.3849$. Since the scores are on opposite sides of the mean, we add areas to obtain the probability.

$$P(60 \leq x \leq 75) = 0.4890 + 0.3849 = 0.8739$$

(c) To find the probability that more than 75 will be cured, $P(X > 75)$, find the area under the normal curve that lies to the right of 75.5 (*Note:* To find the probability of 75 or more, $P(X \geq 75)$, find the area to the right of 74.5.)

For a score of 75.5, $Z = 1.20$ and $A = 0.3849$. (See part [*b*].) Since we want the area above $Z = 1.20$, we need to subtract $0.5000 - 0.3849 = 0.1151$ to get $P(X > 75) = 0.1151$.

7–7 Exercises

1. (*See Example 3*) Find the binomial distribution for $n = 5$ and $p = 0.3$.
2. Find the binomial distribution for $n = 4$ and $p = 0.6$.
3. Form the binomial distribution of an experiment with $n = 5$ and $p = .4$.
4. Draw the probability histogram for the binomial distribution for $n = 4$ and $p = 0.2$.
5. (*See Example 1*) Form the binomial distribution for the experiment of rolling a die four times and counting the number of times a two appears.
6. Form the binomial distribution of rolling a die three times and counting the number of times a six appears.
7. (*See Example 2*) Form the binomial distribution of the experiment of tossing a coin four times and counting the number of heads. Draw the probability histogram.
8. A coin is tossed seven times, and the number of heads is counted. Determine the binomial distribution of the experiment.
9. Five cards are numbered 1 through 5. The cards are shuffled and a card drawn. The number drawn is noted and replaced. This is done four times, and

a count is made of the number of times an odd number appears. Find the probability distribution and draw its histogram.

10. (*See Example 4*) The probability that a new drug will cure a skin rash is 0.85. If it is administered to 500 patients with the skin rash, what is the probability that it will cure 400 of them. Set up, do not compute, this probability.

Set up, do not compute, the probability for each of the problems in Exercises 11 through 14.

11. the probability of 40 successes, where $n = 90$ and $p = 0.4$
12. the probability of 100 successes, where $n = 500$ and $p = 0.7$
13. $P(X = 35)$, where $n = 75$ and $p = 0.25$
14. $P(40 \leq x \leq 42)$, where $n = 100$ and $p = 0.35$

For each of the following binomial distributions, compute the mean, variance, and standard deviation.

15. (*See Example 5*) $n = 50$, $p = 0.4$
16. $n = 210$, $p = 0.3$
17. $n = 600$, $p = .52$
18. $n = 1850$, $p = .24$
19. $n = 470$, $p = .08$

The normal distribution is a good estimate for which of the following binomial distributions in Exercises 20 through 26?

20. $n = 30$, $p = 0.4$
21. $n = 50$, $p = 0.7$
22. $n = 40$, $p = 0.1$
23. $n = 40$, $p = 0.9$
24. $n = 200$, $p = 0.08$
25. $n = 25$, $p = 0.5$
26. $n = 15$, $p = 0.3$
27. (*See Example 6*) Given the binomial experiment with $n = 50$ and $p = 0.7$, use the normal distribution to estimate

    (a) $P(X = 40)$.

    (b) $P(28 \leq X \leq 40)$.

    (c) $P(32 < X < 40)$.

28. Given the binomial experiment with $n = 12$ and $p = 0.5$, use the normal distribution to estimate

    (a) $P(X = 6)$.

    (b) $P(6 \leq X \leq 7)$.

    (c) $P(x > 6)$.

29. (*See Example 7*) The probability that a new drug will cure a certain disease is 0.6. It is administered to 100 patients.

(a) Find the probability that it will cure 50 to 75 of them.
(b) Find the probability that it will cure more than 75.
(c) Find the probability that it will cure less than 50.

30. A coin is tossed 30 times. Find the probability that it will land heads up at least 20 times.

31. A die is rolled 20 times. What is the probability that it turns up a 1 or a 3 six, seven, or eight times?

32. A coin is tossed 100 times. Find the probability of 55 to 60 heads.

33. A coin is tossed 100 times. Find the probability of tossing 50 to 55 heads, inclusive.

34. A die is tossed 36 times. Find the probability of throwing six to eight 4's, inclusive.

35. A certain Bernoulli trial is repeated 64 times. The probability of success in a single trial is $\frac{1}{4}$. Find the probability of success in 20 to 24 of the times, inclusive.

36. In a certain city of 30,000 people, the probability of a person being involved in a motor accident in any one year is .01. Find the probability of more than 250 people having accidents in a year.

37. An airline has 10% of its reservations result in no-shows. It books 270 passengers on a flight that has 250 seats. Find the probability that all passengers who show will have a seat.

38. A drug company has developed a new drug that it believes is 90% effective. It tests the drug on 500 people. Find the probability that at least 90% of the people respond favorably to the drug.

39. A college report states that 30% of its students commute to school. If a random sample of 250 students is taken, find the probability that at least 65 commute.

40. An insurance company estimates that 5% of automobile owners do not carry liability insurance. If 60 cars are stopped at random, what is the probability that less than 5% of them have no liability insurance?

41. A true-false test consists of 90 questions, three points are given for each correct answer, and one point is deducted for each incorrect answer. A student must score at least 98 points to pass.

(a) How many correct answers are required to pass if a student answers all questions?
(b) If a student guesses all answers, find the probability of a passing grade.

7–8 Estimating Bounds on a Proportion

Standard Error
Computing Error Bounds for a Proportion

Mr. Alexander Quality, President of Quality Cola Company, grew tired of Quality Cola. He wanted a cola with more zest and a new taste. Like his father before him, he had vowed never to drink his competitor's cola. Thus, he must develop a new quality cola or resign himself to the traditional taste. He discussed the problem with department heads. The head of the research division agreed they could develop a new formula, but it would cost thousands of dollars. The chief accountant insisted that they should recover the development cost and make a profit. The marketing manager hesitated to put a new product on the market unless he was confident that it would succeed.

Mr Quality, an astute executive, agreed his managers had valid points, so he instructed them to develop a new formula, find out if the public preferred it to the old cola, and if so, pour money into advertising it. After weeks of work, the research division developed a formula that both they and Mr Quality liked. Now the marketing manager wanted to know if the public liked it. He quickly determined that it was quite unrealistic and prohibitive in cost to give everyone in the country a taste test. So he asked the company statistician to help him. He told the statistician he was confident that the new cola would be successful if 40% or more of the population liked it. The statistician outlined his plan:

1. Select, by a random means, 500 people over the country.
2. Give each one a taste test.
3. Find the proportion of the sample that like the new cola.
4. Use the sample proportion as an estimate of the proportion of the total population that like the new cola.

It took the statistician several weeks to select and survey his sample. When the information was in and tabulated, they found that 43% of the people in the sample test liked the new quality cola. At first the marketing manager was elated. Enough people liked the new product to make it successful. Then he had second thoughts. What about the millions of people who did not participate in the taste test? They were the ones who would determine the success of the new cola, so he called the statistician.

"Can I depend on 43% of everyone liking the new cola? Perhaps you just happened to pick the few people who like it."

"I cannot guarantee that precisely 43% of the general public will like it. I told you this was an estimate."

"How good is the estimate? If the estimate is off by two or three percentage points, we are O.K. If the proportion for the entire population is actually only 20%, we are in real trouble. Can you put some bounds on how much the estimate might be in error?"

"We can compute what is called the **standard error** and use it to put error bounds on the proportion. It is similar to the standard deviation used in describing the dispersion of data around the mean."

Here is how the statistician determined the standard error.

Standard Error

Let n be the number of people in the sample. In our sample, $n = 500$.

Let p be the proportion of the sample who like the cola. In our sample, $p = 0.43$. The **standard error** is defined as

$$SE = \sqrt{\frac{p(1-p)}{n}}$$

In our case,

$$\begin{aligned} SE &= \sqrt{\frac{(.43)(.57)}{500}} \\ &= \sqrt{\frac{(.2451)}{500}} \\ &= \sqrt{0.000490} = 0.0221 \end{aligned}$$

We must emphasize that we cannot find absolute bounds. We determine bounds based on a level of confidence. Some concepts of a normal distribution can be used. For example, suppose that we are willing to accept bounds which will enclose the true proportion 90% of the time. Think of $p = .43$ as the mean in a normal distribution, and find the scores positioned equal distances above and below $p = .43$ that enclose 90% of the area under the curve. We first find the Z score from the normal distribution that corresponds to $A = 0.45$ (the 90% is divided equally, 45% on each side of p). Find Z in Table 7–1 that corresponds to $A = 0.45$. In this case, $Z = 1.65$. The number $E = Z \times SE = (1.65)(.0221) = 0.036$ is the maximum error and gives the distance above and below p that encloses the true proportion at the 90% confidence level:

$$p + Z \times SE = 0.43 + 1.65 \times 0.0221 = 0.43 + 0.036 = 0.466$$
$$p - Z \times SE = 0.43 = 1.65 \times 0.0221 = 0.43 - 0.036 = 0.394$$

Thus, when we say that about 40% to 47% of the total population like new quality cola, we will be correct 90% of the time. If we want the error bounds at a higher confidence level, say 98%, we use Z for $A = .49$ ($.49 = \frac{.98}{2}$) and get $p + Z \times SE$ and $P - Z \times SE$ as the error bounds at the .98 confidence level. The bounds are

$$.43 + 2.33(.0221) = .43 + .051 = .481$$

and

$$.43 - 2.33(.0221) = .43 - .051 = .379$$

Thus, we state that from 38% to 48% of the general public like the new quality cola, and we are 98% confident of the bounds. If we take a number of samples, we will be in error about 2% of the time.

Computing Error Bounds for a Proportion

Notice that the higher confidence level gives a wider spread of the error bounds. In order to obtain a higher level of confidence, we must widen the bounds on the error.

Procedure for Computing Error Bounds for a Proportion

Let n be the sample size and p the proportion that respond favorably.

1. Decide on the confidence level to be used, and write it as a decimal c.
2. Compute $A = c/2$. This corresponds to the area under the normal curve between the mean and a score yet unknown. (It is the same as A found in the normal distribution table.)
3. Find the value of Z in Table 7–1 that corresponds to A.
4. Compute the standard error.

$$SE = \sqrt{\frac{p(1-p)}{n}}$$

5. Compute the maximum error.

$$E = Z \times SE = Z \times \sqrt{\frac{p(1-p)}{n}}$$

6. Compute the upper and lower limits.

$$p + E \qquad \text{and} \qquad p - E$$

7. Let p' be the proportion of the total population that respond favorably (p' is the proportion being estimated). Then, the probability that the interval

$$p - E < x < p + E$$

contains p' is c, the confidence level.

Example 1 (*Compare Exercise 1*)
Compute the error bounds for the proportion $p = 0.55$, obtained from a sample of size $n = 120$. Use the 95% confidence level.

Solution The steps for the procedure give

1. $n = 120$, $p = 0.55$, and $c = 0.95$;
2. $A = \dfrac{0.95}{2} = 0.475$;
3. the value of Z that corresponds to $A = 0.475$ is $Z = 1.96$;
4. $SE = \sqrt{\dfrac{(.55)(.45)}{120}} = \sqrt{.0020625} = .045$;

5. $E = Z \times SE = 1.96 \times .045 = 0.0882$;

6. the upper and lower bounds of the proportion are

$$0.55 + 0.0882 = 0.6382$$
$$0.55 - 0.0882 = 0.4618$$

and the confidence interval is $.4618 < x < .6382$.

Example 2 (*Compare Exercise 9*)
A marketing class made a random selection of 150 shoppers at a shopping mall to participate in a taste test of different brands of coffee. They found that 54 shoppers preferred Brand X. Find, at the 95% confidence level, the error bounds of the proportions of shoppers who prefer Brand X.

Solution For this problem, $n = 150$, $p = \frac{54}{150} = 0.36$, and $c = .95$.
Then,

$$A = \frac{.95}{2} = 0.475;$$

$Z = 1.96$ corresponds to $A = .475$ in Table 7–1;

$$SE = \sqrt{\frac{(.36)(.64)}{150}} = \sqrt{.001536} = .0392;$$

$$E = 1.96(.0392) = 0.0768;$$

The bounds are $0.36 + 0.0768 = 0.4368$ and $0.36 - 0.0768 + 0.2832$; the confidence interval is $.2832 < x < .4368$.

The marketing class is 95% confident that between 28.32% and 43.68% of all shoppers prefer Brand X. (Or we may say that the probability is 0.95 that the proportion of the population who prefer Brand X is in the interval $.2832 < x < .4368$.)

Example 3 (*Compare Exercise 11*)
Suppose that the sample in Example 2 was $n = 500$ in size, but the proportion remained the same. Compute the error bounds of the proportion.

Solution Now, $n = 500$, $p = 0.36$, and $c = 0.95$. We still have $Z = 1.96$, but

$$SE = \sqrt{\frac{(.36)(.64)}{500}} = \sqrt{.0004608} = 0.0215$$

Then, $E = 1.96(.0215) = .0421$, and the upper and lower bounds are

$$0.36 + 0.0421 = 0.4021 \qquad \text{and} \qquad 0.36 - 0.0421 = 0.3179$$

and the confidence interval is

$$.3179 < x < .4021$$

Notice that this confidence interval is smaller, so the sample proportion is a better estimate of the total population. This is generally true, a larger sample size reduces the maximum error. The sample size that will keep the maximum error to a specified level can be determined (see Example 6).

Example 4 (*Compare Exercise 13*)
A random sample of 200 people shows that 46 of them use No-Plaque toothpaste. Based on a 98% confidence level, estimate the proportion of the general population that uses the toothpaste.

Solution For this sample,

$$n = 200,$$

$$p = \frac{46}{200} = .23,$$

$$c = .98,$$

$$A = .49,$$

$$Z = 2.33,$$

$$SE = \sqrt{\frac{(.23)(.77)}{200}} = .02976,$$

$$E = 2.33(.02976) = .0693.$$

Then, the interval that contains the proportion of the general population is from $.23 - .0693 = .1607$ to $.23 + .0693 = .2993$. We conclude, with 98% confidence, that about 16% to 30% of the population use No-Plaque toothpaste.

Example 5 (*Compare Exercise 15*)
A random sample of 25 shoppers showed that 24% shopped at the department store. Find a 90% confidence interval of the proportion of the general population that shop there.

Solution $n = 25$, $p = .24$, $c = .90$, $A = .45$, $Z = 1.65$, $SE = \sqrt{.007296} = .0854$, and $E = .1409$. (Be sure you check these computations.) So, the upper and lower bounds are

$$.24 + .1409 = .3809$$

and

$$.24 - .1409 = .0991$$

So, the 90% confidence interval is $.0991 < x < .3809$. Notice that this small sample yields a wide confidence interval.

Example 6 (*Compare Exercise 16*)
A television station wants an estimate of the proportion of the population that watches their late movie. They want the estimate correct within 5% at the 95% confidence level. How big a sample should they select?

Solution Basically, the television station is requesting a maximum error of 5%, written .05 in our computations, at the 95% confidence level.

Look at the computations to obtain E. They are

$$E = Z\sqrt{\frac{p(1-p)}{n}}$$

We are given $E = .05$ and know that $Z = 1.96$ for the 95% confidence level. We need to find n so that

$$.05 = 1.96\sqrt{\frac{p(1-p)}{n}}$$

We face a dilemma. We need to know the value p so we can solve for n. However, we find p from the sample. Thus, it appears that we need p before we know the size of the sample, and we need the sample to find p. There is a way out of this vicious circle. Thanks to techniques of calculus, it can be shown (we won't) that the largest possible value of $p(1-p)$ is 0.25 and occurs when $p = 0.5$. So if we use 0.25 for $p(1-p)$, the value of E may be a little large and the resulting confidence interval a little larger than necessary, but we have erred on the safe side. We proceed using 0.25. Then,

$$.05 = 1.96\sqrt{\frac{.25}{n}}$$

Squaring both sides, we get

$$.0025 = (1.96)^2\frac{(.25)}{n} = \frac{.9604}{n}$$

Then,

$$.0025n = .9604$$

$$n = \frac{.9604}{.0025} = 384.16$$

A sample size of 385 will be sufficient to provide the desired maximum error.

7–8 Exercises

Find the error bounds for the proportion given in each of the following.

1. (*See Example 1*) $p = .64$, $n = 140$, 95% confidence level
2. $p = .22$, sample size = 60, 90% confidence level
3. $p = .30$, sample size = 50, 90% confidence level
4. $p = .30$, sample size = 50, 95% confidence level
5. $p = .30$, sample size = 50, 98% confidence level
6. $p = .40$, $n = 300$, 95% confidence level
7. $p = .5$, $n = 400$, 95% confidence level
8. $p = .1$, $n = 1000$, 95% confidence level
9. (*See Example 2*) A survey of 300 people showed that 45% preferred Brand X cola. Find error bounds on the proportion of the population who prefer

Brand X

(a) using a 90% confidence level.
(b) using a 95% confidence level.

10. A survey of 450 people showed that 35% thought that inflation would decrease the next year. Find error bounds on the proportion of the population who think inflation will decrease

(a) using a 95% confidence level.
(b) using a 98% confidence level.

11. (*See Example 3*) A radar unit on an interstate found that 28% of the vehicles exceeded 60 mph. Find the error bounds on the proportion at the 95% confidence level if

(a) the number of vehicles checked was 200.
(b) the number of vehicles checked was 400.

12. A random sample of freshmen transcripts revealed that 21% made lower than a C average their first semester. Find error bounds on the proportion at the 95% confidence level if

(a) the sample consisted of 275 freshmen.
(b) the sample consisted of 500 freshmen.

13. (*See Example 4*) A high school paper reported that 243 students out of 300 surveyed cruise Main Street on Saturday nights. Based on a 98% confidence level, estimate the proportion of the high school population that cruise Main Street.

14. A health class found that 145 of 320 people surveyed were overweight by 5 lb or more. Based on a 95% confidence level, estimate the proportion of the population that is overweight by 5 lb or more.

15. (*See Example 5*) A random sample of 60 shoppers showed that 54% shopped at Nate Chadrow's Department Store. Find a 90% confidence interval of the proportion of the general population that shop there.

16. (*See Example 6*) A university food service wants to estimate the proportion of students who like their food. They want the estimate correct within 6% at the 95% confidence level. How big a sample should they select?

17. A candidate for public office wants an estimate of the proportion of voters who will vote for her. She wants the estimate to be accurate within 1% (.01) at the 95% confidence level.

(a) How big a sample should be used?
(b) If a 99% confidence level is desired, how big should the sample be?

18. A state highway department needs an estimate of the percent of vehicles that exceed 60 mph on an interstate. How large a sample is needed if the maximum error of the estimate is no more than 2% at the 95% confidence level?

19. How big a sample should be taken to ensure a maximum error of .0785 in the estimate of the proportion of adults who smoke? Use a 95% confidence level.

20. A random check of 90 students at Classic College revealed that 40 of them did not know the location of the Reserve Room in the Library. Find error bounds on the proportion of Classic students who do not know where the Reserve Room is located. Use a 98% confidence level.

21. A survey of 100 adults showed that 60 of them drank coffee daily. Estimate the proportion of the adult population that drink coffee daily, and find error bounds at the 95% confidence level.

22. The manufacturer of Brand Y coffee wants to estimate the proportion of adults who prefer their coffee. How large a sample should they survey to estimate the proportion with an error of 5% or less. They want a 95% confidence level.

23. A Harris Poll taken in November 1986 showed that confidence in President Reagan slid sharply in the wake of an arms deal with Iran. Based on 1252 telephone interviews, 43% gave Reagan a positive rating. Using a 95% confidence level, find the error bounds of this proportion.

24. A 1986 survey of 2000 families showed that 18% owned at least one handgun. Find the error bounds of this proportion using a 95% confidence level.

IMPORTANT TERMS

7–1	Frequency table	Frequency distribution histogram	Frequency polygon
	Pie chart	Discrete data	
7–2	Measure of central tendency	Mean	Grouped data
	Median	Mode	
7–3	Variation	Deviation	Squared deviation
	Standard deviation		
7–4	Random variable	Discrete variable	Continuous variable
	Probability distribution		
7–5	Expected value	Variance of a random variable	Standard deviation of a random variable
7–6	Normal curve	Normal distribution	Continuous data
	Discrete data	z score	Probability
	Normal distribution		
7–7	Binomial distribution	Bernoulli trials	Mean, variance, and standard deviation of a binomial distribution
	Estimating a binomial probability		
7–8	Error bounds	Standard error	Confidence level

REVIEW EXERCISES

1. Draw the histogram, frequency polygon, and pie chart based on this frequency table.

New Accounts Opened	Frequency
Monday	17
Tuesday	31
Wednesday	20
Thursday	14
Friday	8

2. Find the mean of 4, 6, -5, 12, 3, 2, and 9.
3. Find the median of

(a) 8, 12, 3, 5, 6, 3, and 19.

(b) 4, 9, 16, 12, 3, 22, 1, and 95.

(c) 3, -2, 6, 1, 4, and -3.

4. Find the mean for the following quiz data:

Score on Quiz	Frequency
0	2
1	3
2	6
3	9
4	2
5	4

5. Find the mean for the number of passengers in cars arriving at a play.

Number of Passengers in a Car	Frequency
1–2	54
3–4	32
5–6	12

6. Find the mean, median, and mode for the numbers 2, 8, 4, 3, 2, 9, 6, 2, and 7.
7. A shopper paid a total of \$90.22 for her purchases. The mean price was \$3.47. How many items were purchased?
8. Find the variance and standard deviation for the numbers 8, 18, 10, 16, 3, and 11.
9. A professor selects three students each day. Let X be the random variable that represents the number who completed their homework. Give all possible values of X.
10. Professor X asked his students to evaluate his teaching at the end of the semester. Their response to the statement "The tests were a good measure of my knowledge" were marked on a scale of 1 to 5 as follows.

Outcome	Random Variable x	Responses
Strongly agree	1	10
Agree	2	35
Neutral	3	30
Disagree	4	20
Strongly disagree	5	15

Find the probability distribution of X, and sketch its graph.

11. A store has a special on bread with a five-loaf limit. The probability distribution for sales is listed in the following table. Find the average number of loaves per customer.

Number of Loaves per Customer	Probability
0	0.05
1	0.20
2	0.15
3	0.20
4	0.25
5	0.15

12. Find the expected value and variance for the following probability distribution:

X	$P(X)$	X	$P(X)$
1	.14	4	.15
2	.06	5	.36
3	.22	6	.07

13. An instructor summarized his student ratings (scale of 1–5) in the following probability distribution.

Response	X	**Probability**
Excellent	1	.20
Good	2	.32
Average	3	.21
Fair	4	.15
Poor	5	.13

What is his expected average rating?

14. A normal distribution has a mean of 80 and a standard deviation of 6. Find the fraction of scores

(a) between 80 and 88.
(b) between 70 and 84.
(c) greater than 90.

15. A distributor averages sales of 350 mopeds per month, with a standard deviation of 25. Assume that sales follow a normal distribution. What is the probability that sales will exceed 400 during the next month?

16. A construction company contracts to build an apartment complex. The total construction time follows a normal distribution with an average time of 120 days and a standard deviation of 15 days.

(a) The company will suffer a penalty if construction is not completed within 140 days. What is the probability that they will be assessed the penalty?
(b) The company will be given a nice bonus if they complete construction in less than 112 days. What is the probability that they will receive the bonus?
(c) What is the probability that the construction will be completed in 115 to 130 days?

17. Find the binomial distribution for $n = 4$ and $p = .25$.

18. A binomial distribution has $n = 22$ and $p = .35$. Find the mean and standard deviation of the distribution.

19. The probability that a new drug will cure a certain disease is 0.65. If it is administered to 80 patients, estimate

(a) the probability that it will cure more than 50 patients.
(b) the probability that it will cure 65 patients.
(c) the probability that it will cure between 55 and 60 patients.

20. Compute the standard error for the given sample sizes and proportions.

(a) $n = 50$, $p = .45$
(b) $n = 100$, $p = .32$
(c) $n = 400$, $p = .20$

21. Compute the error bounds for the following.

(a) $n = 50$, $p = .45$, 95% confidence level
(b) $n = 100$, $p = .30$, 98% confidence level

22. A survey of 30 individuals revealed that 22 watched *Monday Night Football*. Find, at the 95% confidence level, the error bounds of the proportion who watch *Monday Night Football.*

23. A manufacturer wants an estimate of the proportion who will respond favorably to their new product. They want the estimate correct within 3% at the 95% confidence level. How big a sample should they select?

24. For a set of scores, the mean $= 240$ and $\sigma = 10$.

(a) Find Z for the score 252.
(b) Find Z for the score 230.
(c) Find the score corresponding to $Z = -2.3$.

25. In a cross-country ski race, a skier came in 43rd in a field of 216 skiers. What was her percentile?

26. A brother and sister took two different standarized tests. He scored 114 on a test that had a mean of 100 and a standard deviation of 18. She scored 85 on a test that had a mean of 72 and a standard deviation of 12. Which one had the better score?

8

Topics in this chapter can be applied to:

Loans and Credit • Interest • Discounts • Securities • Present Value • Investments • Investment Funds • Business Finance

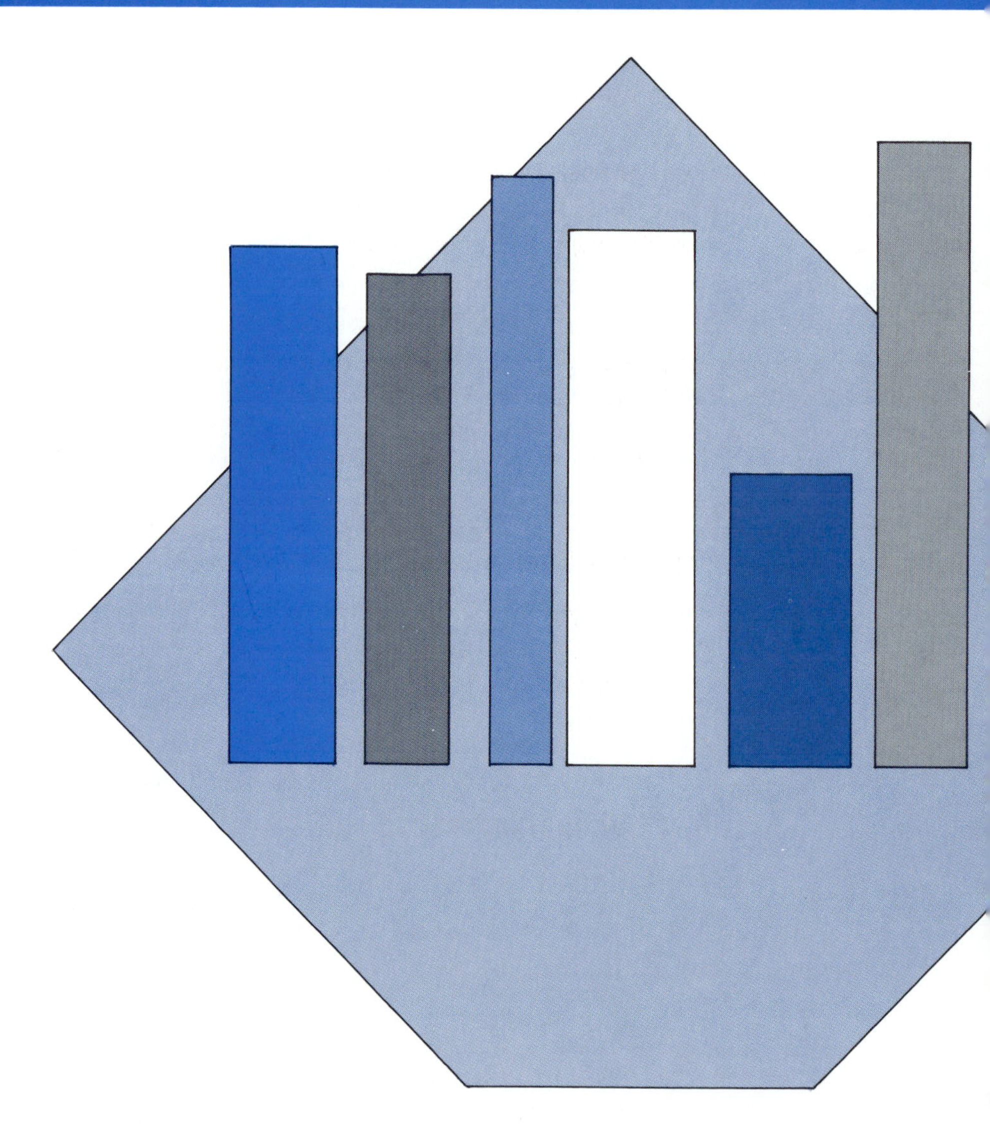

Mathematics of Finance

- **Simple Interest**
- **Compound Interest**
- **Annuities and Sinking Funds**
- **Present Value of an Annuity and Amortization**

Our modern economy depends on borrowed money. Very few families would own a house, and many would not have a car, if it were not for credit. Business depends on borrowed money for day-to-day operations and for long-term expansion. Government at all levels borrows money. Banks depend on loans for a major source of their income.

"Rented money" is more accurate than "borrowed money" because you pay **interest** on borrowed money. Interest is the fee you pay to use, or "rent" money for a period of time.

When you place money in a savings account, you are in effect renting, or loaning, your money to the bank, and they pay you interest for the use of the money. In turn, they loan the money to someone who pays them interest for the use of the money.

8–1 Simple Interest

Simple Interest
Future Value
Simple Discount
Treasury Bills

Simple Interest

The fee charged for use of money may be determined a number of ways. We first look at **simple** interest. It is most often used for loans of shorter duration.

The money borrowed in a loan is called the **principal**. The number of dollars received by the borrower is the **present value**. In a simple interest loan, the principal and present value are the same. The fee for a simple interest loan is usually expressed as a percent of the principal per year and is called the **interest**

rate. If the interest rate is 10% per year, then the borrower pays 10% of the amount borrowed each year. The total fee, interest, paid for a simple interest loan is given by the following formula.

Simple Interest

$$I = Prt$$

where

P = principal (amount borrowed)
r = interest rate per year (expressed in decimal form)
t = time in years.

Example 1 (*Compare Exercise 1*)
Compute the interest paid on a loan of \$1400 at 9% interest rate for 18 months.

Solution

$$P = \$1400$$

$$r = 9\% = .09 \text{ in decimal form}$$

$$t = 18 \text{ months} = \frac{18}{12} \text{ years} = 1.5 \text{ years}$$

$$I = 1400 \times .09 \times 1.5 = 189.$$

So the interest paid is \$189.

Warning! The time units for r and t must be consistent, so months was converted to years.

Example 2 (*Compare Exercise 11*)
An individual borrows \$300 for six months at 1% simple interest per month. How much interest is paid?

Solution Notice that the interest rate is given as 1% **per month**. We can still use the $I = Prt$ formula provided r and t are consistent in time units, in this case months.

$$I = 300 \times .01 \times 6 = \$18$$

If the interest rate is stated as 8%, 10%, and so on, it is understood to be an annual rate. The time period will be stated when it is not annual.

Example 3 (*Compare Exercise 14*)
Joe borrows money at 8% for two years. He paid \$124 interest. How much did he borrow?

Solution In this case, $I = 124$, $r = .08$, and $t = 2$, so

$$124 = P(.08)(2) = .16P$$

$$P = \frac{124}{.16} = 775$$

The loan was \$775.

Example 4 (*Compare Exercise 18*)
Jane borrowed \$950 for 15 months. The interest was \$83.13. Find the interest rate.

Solution We are given $P = 950$, $t = \frac{15}{12} = 1.25$ years, and $I = 83.13$, so

$$\begin{aligned} 83.13 &= 950r(1.25) \\ &= 1187.5r \\ r &= \frac{83.13}{1187.5} = .07 \end{aligned}$$

so the interest rate was 7% per year.

Future Value

A loan made at simple interest requires the borrower to pay back the sum borrowed (principal) plus the interest. This total, $P + I$, is called the **future value** or **amount** of the loan.

Amount or Future Value of a Loan

$$\begin{aligned} A &= P + I \\ &= P + Prt \\ &= P(1 + rt) \end{aligned}$$

where

P = principal, or present value
r = annual interest rate
t = time in years
A = amount, or future value

Example 5 (*Compare Exercise 22*)
Find the amount (future value) of a \$2400 loan for nine months at 11% interest rate.

Solution We know that $P = 2400$, $r = .11$, and $t = \frac{9}{12} = .75$ years, then

$$I = 2400(.11)(.75) = 198$$

and

$$A = 2400 + 198 = 2598$$

We can also use the formula for $A = P(1 + rt)$ and compute A as

$$\begin{aligned} A &= 2400(1 + .11(.75)) \\ &= 2400(1 + .0825) \\ &= 2400(1.0825) \\ &= 2598 \end{aligned}$$

The total of principal and interest is \$2598.

Example 6 (*Compare Exercise 26*)
How much should you invest at 12% for 21 months to have \$3000 at the end of the 21 months?

Solution In this example you are given the future value, \$3000, and are asked to find the present value, P. We know that $r = .12$, $t = \frac{21}{12} = 1.75$ years, and $A = 3000$. Then,

$$\begin{aligned} 3000 &= P(1 + .12(1.75)) \\ &= P(1 + .21) \\ &= 1.21P \end{aligned}$$

so $P = \dfrac{3000}{1.21} = \$2479.34$ (rounded to nearest cent).

Example 7 (*Compare Exercise 47*)
Your friend loaned some money. The debtor is scheduled to pay him \$560 in four months. (The future value is \$560.) Your friend needs the money now, so you agree to pay him \$525 for the note, and the debtor will pay you the \$560 in four months. What annual interest rate will you earn?

Solution For the purposes of this problem, $A = 560$, $P = 525$, and $t = 4$ months $= \frac{1}{3}$ year. Using the formula $A = P + Prt$, we have

$$\begin{aligned} 560 &= 525 + 525r\left(\frac{1}{3}\right) \\ &= 525 + 175r \\ 35 &= 175r \end{aligned}$$

so
$$r = \frac{35}{175} = .20$$

You will earn 20% annual interest.

Simple Discount

The borrower of a simple interest loan receives the full amount of the principal and pays the interest at the end of the loan. Some loans deduct the interest from the principal at the time the loan is made, so the borrower receives less than the principal. This type of loan is a **simple discount** note. For example, if someone borrows \$600 for one year at 9% interest, the lender will retain 9% of \$600, \$54 as the **discount** and give the borrower \$546, the **proceeds**. The borrower is required to repay \$600, the **maturity value**, at the end of the year.

This differs from simple interest in that the principal is the future value (maturity value) and the present value (proceeds) is less than the principal. The **discount rate** plays the same role as the interest rate except that the discount rate is applied to the maturity value. Let's compare a 12% simple interest loan and a 12% discount loan.

On a \$1000 simple interest loan at 12%, the borrower receives \$1000 and pays \$1120 at the end of one year (\$1000 principal plus \$120 interest).

On a \$1000 simple discount loan at 12%, the borrower receives \$880 and pays \$1000 at the end of one year. The \$120 interest is deducted from \$1000 before the borrower receives the money. The interest is computed on the principal (what the borrower receives) on a simple interest loan. For a simple discount loan, the interest is computed on the maturity value (what the borrower pays back).

Simple Discount

$$\begin{aligned} D &= Mdt \\ PR &= M - D \\ &= M - Mdt \\ &= M(1 - dt) \end{aligned}$$

where

M = maturity value (principal)
d = annual discount rate, written in decimal form
t = time, in years
D = discount
PR = proceeds, the amount the borrower receives

Example 8 (*Compare Exercise 29*)
Find the discount and the amount a borrower receives (proceeds) on a \$1500 simple discount loan at 8% discount rate for 1.5 years.

Solution In this case, $M = 1500$, $d = .08$, and $t = 1.5$ years, then

$$\begin{aligned} D &= 1500(.08)(1.5) \\ &= 180 \\ PR &= 1500 - 180 \\ &= 1320 \end{aligned}$$

so the bank keeps the discount \$180, and the borrower receives \$1320.

Example 9 (*Compare Exercise 49*)
A customer borrows money from a bank that gives simple discount loans. The customer needs \$4500. How much should he borrow (maturity value) at 11% for two years so that the proceeds will be \$4500?

Solution The information gives $PR = 4500$, $d = .11$, and $t = 2$, so using the formula

$$PR = M(1 - dt)$$

we get

$$\begin{aligned} 4500 &= M(1 - .11(2)) \\ &= M(1 - .22) \\ &= .78M \end{aligned}$$

so $$M = \frac{4500}{.78} = 5769.23 \qquad \text{(Rounded to 2 decimals)}$$

The customer should borrow \$5769.23.

If a customer has the choice between a simple interest and a simple discount loan, which is better? The following example gives an answer.

Example 10 A discount note for \$800 is made at a 10% discount rate for three months. The discount is

$$D = 800(.10)\frac{3}{12}$$

$$= 20$$

so the borrower gets \$780.

If the borrower had obtained \$780 at simple interest, how much interest would be paid?

Solution The simple interest is

$$I = 780(.10)\left(\frac{3}{12}\right) = 19.50$$

This illustrates a general fact that, for the same amount received by the borrower, the simple interest paid is less than the discount paid.

Treasury Bills

Treasury bills are short-term securities issued by the federal government. The bills do not specify a rate of interest. They are sold at weekly public auctions with financial institutions making competitive bids. For example, a bank may bid \$978,300 for a 90-day \$1 million treasury bill. At the end of 90 days, the bank receives \$1 million, which covers the cost of the bill and interest earned on the bill. This is basically a simple discount transaction.

Example 11 (*Compare Exercise 65*)
A bank wants to earn 7.5% simple discount interest on a 90-day \$1 million treasury bill. How much should they bid?

Solution We are given the maturity value, $M = 1{,}000{,}000$, the discount rate, $d = .075$, and the time, $t = \frac{90}{360}$. (Banks often use 360 days per year when computing daily interest.) We want to find the proceeds, PR.

Substituting these in the simple discount formula,

$$PR = 1{,}000{,}000\left(1 - (.075)\frac{90}{360}\right)$$

$$= 1{,}000{,}000(1 - 0.01875)$$

$$= 1{,}000{,}000(.98125)$$

$$= 981{,}250$$

So, the bank should bid \$981,250.

Example 12 (*Compare Exercise 67*)
A bank paid \$983,000 for a 90-day \$1 million treasury bill. What is their simple discount rate?

Solution We are given $M = 1{,}000{,}000$, $PR = 983{,}000$, and $t = \frac{90}{360}$. We want to find d.

$$983{,}000 = 1{,}000{,}000\left(1 - d\left(\frac{90}{360}\right)\right)$$
$$= 1{,}000{,}000 - 250{,}000d$$
$$-17{,}000 = -250{,}000d$$
$$d = \frac{17{,}000}{250{,}000} = .068$$

So, the annual discount rate was 6.8%.

8–1 Exercises

I.

1. (*See Example 1*) Compute the interest on a loan of \$1100 at 8% interest rate for nine months.

Compute I for Exercises 2 through 5.

2. $P = \$2300$, $r = 9\%$, $t = 1.25$ years
3. $P = \$600$, $r = 10\%$, $t = 18$ months
4. $P = \$4200$, $r = 8\%$, $t = 16$ months
5. $P = \$745$, $r = 8.5\%$, $t = 6$ months
6. Compute the simple interest on \$500 for one year at 7%.
7. Compute the simple interest and the amount for \$300 at 6% for one year.
8. Compute the simple interest on \$400 at 6% for three months.
9. Jones borrows \$500 for three years at 8% simple interest (annual rate). How much interest will be paid?
10. Smith borrows \$250 at 9% (annual rate) for three months. How much interest will be paid?
11. (*See Example 2*) Mrs. Smith borrows \$700 for five months at 1.5% simple interest per month. How much interest is paid?
12. \$1450 is borrowed for three months at a 1.5% monthly rate. How much interest is paid?
13. How much interest is paid on a \$950 loan for seven months at 1.75% monthly rate?

How much was borrowed for the loans in Exercises 14 through 17?

14. (*See Example 3*) $I = \$24.19$, $r = 7.5\%$, $t = 9$ months
15. $I = \$42.90$, $r = 13\%$, $t = 6$ months

16. $I = \$38.00$, $r = 10\%$, $t = 10$ months
17. Mary paid \$116.10 interest on a loan at 9% simple interest for 1.5 years. How much did she borrow?

Find the interest rate in Exercises 18 through 21.

18. (*See Example 4*) $P = \$650$, $I = \$91.00$, $t = 2$ years
19. $P = \$1140$, $I = \$49.40$, $t = 8$ months
20. $P = \$75$, $I = \$1.50$, $t = 2$ months
21. $P = \$5800$, $I = \$616.25$, $t = 1.25$ years

Find the amount (future value) of the loans in Exercises 22 through 25.

22. (*See Example 5*) $P = \$1800$, $r = 10.5\%$, $t = 7$ months
23. $P = \$2700$, $r = 8\%$, $t = 1.5$ years
24. $P = \$240$, $r = 12\%$, $t = 10$ months
25. $P = \$6500$, $r = 8.6\%$, $t = 1.75$ years
26. (*See Example 6*) How much should you invest at 11% for 16 months to have \$3000 at the end of that period?
27. How much should be invested at 8% to have \$1800 at the end of 1.5 years?
28. Sandi wants to buy a \$400 television set in six months. How much should she invest at 7% simple interest now to have the money then?

Find the discount and the amount the borrower receives (proceeds) for the following discount loans.

29. (*See Example 8*) $M = \$1850$, $d = 7.5\%$, $t = 1$ year
30. $M = \$960$, $d = 10.25\%$, $t = 10$ months
31. $M = \$485$, $d = 13\%$, $t = 1.5$ years
32. $M = \$9540$, $d = 16.5\%$, $t = 8$ months

II.

33. A sum of \$5000 is borrowed for a period of three years at simple interest of 7%. Determine the total interest paid over this period.
34. A sum of \$2000 is borrowed for a period of two years at simple interest of $5\frac{3}{4}\%$. Determine the total interest paid.
35. Determine the total simple interest paid on \$50,000 over six years at the rate of 7.5%. The interest is to be paid back annually. Find the amount of each annual payment.
36. A sum of \$4000 is borrowed for a period of three years at simple interest. The rate is 8% per annum to be paid semiannually (twice a year).
    (a) Determine the interest that is to be paid semiannually.
    (b) Compute the total interest that will be paid over the three-year period.
37. A sum of \$100,000 is borrowed for a period of six years at 9% simple interest. The interest is to be paid monthly.

(a) Determine the monthly interest payments.
(b) Compute the total interest to be paid over the six-year period.

38. Joe borrowed $150 from a loan company. At the end of one month he paid off the loan with $163.50. What annual interest rate did he pay?

39. Kathy borrowed $800 at 6% interest. The amount of interest paid was $144. What was the length of the loan?

40. For how long did Amy borrow $1200 if she paid $112 in interest at a 7% interest rate?

41. How long will it take $800 to earn $18 at 9% simple interest?

42. A student borrows $200 at 6% simple interest (annual rate). When the loan is repaid, the amount of the principal and interest is $218. What was the length of the loan?

43. A girl borrowed some money from the bank at 7.5% simple interest for a year. She paid $48.75 in interest. How much money was loaned?

44. The Smith family borrowed some money for 15 months (1.25 years). The interest rate was 9%, and they paid $9.90 in interest. How much did they borrow?

45. A family borrows $20,000 to buy a home. The interest rate is 9%, and each monthly payment is $179.95. How much of the first payment goes for interest and how much for principal?

46. Mr. Jones borrows $750 for four months at 12% interest. How much interest does he pay?

47. (*See Example 7*) You pay $860 for a loan that has a future value of $900 in six months. What annual interest rate will you earn?

48. How much should you pay for a loan that has a future value of $320 in three months to earn 9% simple interest?

49. (*See Example 9*) A grocer needs $6500 to stock her store. What should the amount of a simple discount loan be for her to receive proceeds of $6500 at 14% discount rate for one month?

50. A contractor needs to buy $16,000 worth of materials. He receives a simple discount loan for two months at 15% discount rate. What should be the principal of the loan so that he will receive the amount he needs?

51. A car dealer goes to the bank to obtain $38,000 to purchase some autos. The bank will give her a simple discount loan at 12% discount rate for three months. What maturity value should be used so that the dealer will receive $38,000?

52. (*See Example 10*) A discount note for $1000 is made at a 12% discount rate for six months.

(a) What are the discount and proceeds?
(b) How much interest would be paid on a simple interest loan for the amount of the proceeds for the same rate and time?

53. A woman goes to the bank to borrow $650 to pay some obligations. She may get either a simple interest or a simple discount loan at 8% for two months. What is the fee (interest and discount) paid in each case?

54. A mechanic borrowed $320 from a loan company and at the end of one month paid off the loan with $324.80. What simple interest rate did he pay?

55. A man borrowed $950. Six months later, he repaid the loan (principal and interest) with $1000. What simple interest rate did he pay?

56. Find the future value of a $1300 loan for 14 months at 9.5% simple interest.

III.

57. A jeweler wants to obtain $3000 for four months. She has the choice of a simple interest loan at 10.3% or a discount loan at 10.1%. Which one will result in the lower fee for use of the money?

58. The low point in gross public debt was $38,000 in 1836. If the government paid 6% simple interest, how much were the annual interest payments?

59. The total government debt (federal, state, and local) in 1970 was $450 billion. How much interest was paid for one year if the interest rate averaged 11%?

60. In 1984, the estimated debts of the United States were

    Government, $1,780.7 billion
    Business and corporations, $3,206.5 billion
    Individuals, $2,133.8 billion

    Find the interest paid for one year

    (a) for government loans, assuming 9% interest rate.
    (b) for business and corporations, assuming 10.5% interest rate.
    (c) for individuals, assuming 12% interest rate.
    (d) for the total of all the above.

61. A city has raised $500,000 for the purpose of putting in a new sewer system. This money was raised by issuing bonds which mature after five years. The interest on the bonds is 9% per annum. Determine the total interest paid.

62. A city has sold $750,000 in bonds for the purpose of raising money to develop a transportation system. The interest on the bonds is to be paid annually at 7% simple interest over a period of five years. Determine the annual interest payments that will be made to the bondholders and the total interest that will have been paid over the five-year period.

63. A company has issued $400,000 in ten-year bonds that pay 8% simple interest. The interest is payable semiannually. Determine the amount of each semiannual payment and the total interest that will have been paid on the bonds.

64. An investment company bought $50,000 of three-year bonds at 7% in simple interest payable quarterly. Find the quarterly interest payment and the total interest paid over the three-year period.

65. (*See Example 11*) A bank wants to earn 7% simple discount interest on a 90-day \$1 million treasury bill. How much should they pay?

66. How much should a bank pay for a 90-day \$1 million treasury bill to earn 6.5% simple discount?

67. (*See Example 12*) A bank paid \$982,000 for a 90-day \$1 million treasury bill. What was the simple discount rate?

68. A bank paid \$987,410 for a 90-day \$1 million treasury bill. What was the simple discount rate?

69. How much should a bank bid on a 30-day \$2 million treasury bill to receive 7.125% interest?

8–2 Compound Interest

Compound Interest

When you deposit money into a savings account, the bank will pay for the use of your money. Normally, they pay interest at specified periods of time, such as every three months. Unless instructed otherwise, the bank credits your account with the interest, and for the next time period, the bank pays interest on the new total. This is called **compound interest**. Let's look at a simple example.

Example 1 (*Compare Exercise 10*)
You put \$1000 into an account that pays 8% annual interest. The bank will compute interest and add it to your account at the end of each year. This is called **compounding interest annually**. Here's how your account builds up. We use the formula $A = P(1 + rt)$, where $r = .08$ and $t = 1$.

End of Year	Balance in Your Account
0 (start)	\$1000.00
1	$1000(1.08) = 1080.00$
2	$1080(1.08) = 1166.40 = 1000(1.08)^2$
3	$1166.40(1.08) = 1259.71 = 1000(1.08)^3$
4	$1259.71(1.08) = 1360.49 = 1000(1.08)^4$
5	$1360.49(1.08) = 1469.33 = 1000(1.08)^5$

So at the end of five years, the account has grown to \$1469.33.

Notice how the pattern of growth involves powers of 1.08. This same pattern can be used to obtain the amount in the account after 10 years, 15 years, and so on. For example, the amount in the account at the end of ten years is $1000(1.08)^{10} = 1000(2.15892) = 2158.92$. In general,

When P dollars is invested at an annual interest rate r, and the interest is compounded annually, the amount at the end of t years is

$$A = P(1 + r)^t$$

Example 2 (*Compare Exercise 12*)
$800 is invested at 6%, and it is compounded annually. What is the amount in the account at the end of four years?

Solution Here $P = 800$, $r = .06$, and $t = 4$, so

$$A = 800(1.06)^4 = 800(1.26248) = 1009.98$$

In the preceding examples, interest was compounded annually. Interest is often calculated and added to the principal at other regular intervals. The most common intervals are semiannually, quarterly, monthly, and daily.

The **interest rate per compound period** is the annual rate divided by the number of compound periods per year. An 8% annual rate becomes 8%/4 = 2% quarterly rate. Let's repeat Example 1, but compound interest quarterly.

Example 3 $1000 is invested at 8% annual rate. The interest is compounded quarterly. Find the amount in the account at the end of five years.

Solution Since interest is compounded quarterly, it must be computed every three months and added to the principal. The quarterly interest rate is 2%, so the computations are the following.

End of Quarter	Amount in Account
0	1000.00
1	$1000(1.02) = 1020.00$
2	$1020(1.02) = 1040.40 = 1000(1.02)^2$
3	$1040.40(1.02) = 1061.21 = 1000(1.02)^3$
4	$1061.21(1.02) = 1082.43 = 1000(1.02)^4$

At this stage, we have the amount at the end of one year. We will not continue for another 16 computations, but, observe the pattern. At the end of 10 quarters, we correctly expect the amount to be $A = 1000(1.02)^{10}$. Since five years is 20 quarters, the amount at the end of five years is

$$A = 1000(1.02)^{20} = 1000(1.48595) = 1485.95$$

The number $(1.02)^{20} = 1.48595$ may be computed on a scientific calculator or obtained from a table. Notice that compounding quarterly gives an amount of \$1485.95 **compared** to an amount of \$1469.33 from compounding annually. The general formula for finding the amount after a specified number of compound periods is the following:

Compound Interest—Amount

$$A = P(1 + i)^n$$

where

r = annual interest rate.
m = number of times compounded per year.
$i = r/m$ = the interest rate per period.
n = the number of periods; $n = mt$, where t is the number of years.
A = amount (future value) at the end of n compound periods.

Example 4 (*Compare Exercise 14*)
\$800 is invested at 12% for two years. Find the amount at the end of two years if the interest is compounded

(a) annually. **(b)** semiannually. **(c)** quarterly.

Solution In this example, $P = 800$, $r = .12$, and $t = 2$ years.

(a) $m = 1$, so $i = .12$, $n = 2$

$$A = 800(1 + .12)^2 = 800(1.12)^2 = 800(1.2544) = 1003.52$$

(b) $m = 2$, so $i = \frac{.12}{2} = .06$, $n = 4$

$$A = 800(1 + .06)^4 = 800(1.06)^4 = 800(1.26248) = 1009.98$$

(c) $m = 4$, so $i = \frac{.12}{4} = .03$, $n = 8$

$$A = 800(1 + .03)^8 = 800(1.03)^8 = 800(1.26677) = 1013.42$$

The computation of numbers like $(1.03)^8$ and $(1.02)^{20}$ can be tedious if you don't have a scientific calculator. Table 8–1 is provided to ease the arithmetic. It lists the answers to the computations needed for a limited number of interest rates. Here's how you use it.

To compute $(1 + i)^n$, you need the values of i and n. Different values of i and n give different results. Table 8–1 has columns of answers for various values of i and n. The value of $(1.03)^8$ is found in the column headed $i = .03$. Look under the n heading until you find 8. Adjacent to 8 you will find 1.26677. This number is the answer to $(1.03)^8$. Find $(1.02)^{20}$ by looking under the $i = .02$ column to the entry where n is 20. There you should see 1.48595, the value of $(1.02)^{20}$.

Check the table to make sure that you find $(1.01)^{15} = 1.16097$, $(1.04)^{12} = 1.60103$, and $(1.025)^{24} = 1.80873$.

Table 8–1. Compound Interest $A = (1 + i)^n$

Number of Periods n	Interest per period i									
	.01	.015	.02	.025	.03	.04	.05	.06	.07	.075
1	1.01000	1.01500	1.02000	1.02500	1.03000	1.04000	1.05000	1.06000	1.07000	1.07500
2	1.02010	1.03022	1.04040	1.05063	1.06090	1.08160	1.10250	1.12360	1.14490	1.15563
3	1.03030	1.04568	1.06121	1.07689	1.09273	1.12486	1.15762	1.19102	1.22504	1.24230
4	1.04060	1.06136	1.08243	1.10381	1.12551	1.16986	1.21551	1.26248	1.31080	1.33547
5	1.05101	1.07728	1.10408	1.13141	1.15927	1.21665	1.27628	1.33823	1.40255	1.43563
6	1.06152	1.09344	1.12616	1.15969	1.19405	1.26532	1.34010	1.41852	1.50073	1.54330
7	1.07214	1.10984	1.14869	1.18869	1.22987	1.31593	1.40710	1.50363	1.60578	1.65905
8	1.08286	1.12649	1.17166	1.21840	1.26677	1.36857	1.47746	1.59385	1.71819	1.78348
9	1.09369	1.14339	1.19509	1.24886	1.30477	1.42331	1.55133	1.68948	1.83846	1.91724
10	1.10462	1.16054	1.21899	1.28008	1.34392	1.48024	1.62889	1.79085	1.96715	2.06103
11	1.11567	1.17795	1.24337	1.31209	1.38423	1.53945	1.71034	1.89830	2.10485	2.21561
12	1.12683	1.19562	1.26824	1.34489	1.42576	1.60103	1.79586	2.01220	2.25219	2.38178
13	1.13809	1.21355	1.29361	1.37851	1.46853	1.66507	1.88565	2.13293	2.40985	2.56041
14	1.14947	1.23176	1.31948	1.41297	1.51259	1.73168	1.97993	2.26090	2.57853	2.75244
15	1.16097	1.25023	1.34587	1.44830	1.55797	1.80094	2.07893	2.39656	2.75903	2.95888
16	1.17258	1.26899	1.37279	1.48451	1.60471	1.87298	2.18287	2.54035	2.95216	3.18079
17	1.18430	1.28802	1.40024	1.52162	1.65285	1.94790	2.29202	2.69277	3.15882	3.41935
18	1.19615	1.30734	1.42825	1.55966	1.70243	2.02582	2.40662	2.85434	3.37993	3.67580
19	1.20811	1.32695	1.45681	1.59865	1.75351	2.10685	2.52695	3.02560	3.61653	3.95149
20	1.22019	1.34686	1.48595	1.63862	1.80611	2.19112	2.65330	3.20714	3.86968	4.24785

21	1.23239	1.36706	1.51567	1.67958	1.86029	2.27877	2.78596	3.39956	4.14056	4.56644
22	1.24472	1.38756	1.54598	1.72157	1.91610	2.36992	2.92526	3.60354	4.43040	4.90892
23	1.25716	1.40838	1.57690	1.76461	1.97359	2.46472	3.07152	3.81975	4.74053	5.27709
24	1.26973	1.42950	1.60844	1.80873	2.03279	2.56330	3.22510	4.04893	5.07237	5.67287
25	1.28243	1.45095	1.64061	1.85394	2.09378	2.66584	3.38635	4.29187	5.42743	6.09834
26	1.29526	1.47271	1.67342	1.90029	2.15659	2.77247	3.55567	4.54938	5.80735	6.55571
27	1.30821	1.49480	1.70689	1.94780	2.22129	2.88337	3.73346	4.82234	6.21387	7.04739
28	1.32129	1.51722	1.74102	1.99650	2.28793	2.99870	3.92013	5.11168	6.64884	7.57594
29	1.33450	1.53998	1.77584	2.04641	2.35657	3.11865	4.11614	5.41839	7.11426	8.14414
30	1.34785	1.56308	1.81136	2.09757	2.42726	3.24340	4.32194	5.74349	7.61225	8.75495
31	1.36133	1.58653	1.84759	2.15001	2.50008	3.37313	4.53804	6.08810	8.14511	9.41157
32	1.37494	1.61032	1.88454	2.20376	2.57508	3.50806	4.76494	6.45338	8.71527	10.11744
33	1.38869	1.63448	1.92223	2.25885	2.65234	3.64838	5.00319	6.84059	9.32534	10.87625
34	1.40258	1.65900	1.96068	2.31532	2.73191	3.79432	5.25335	7.25102	9.97811	11.69196
35	1.41660	1.68388	1.99989	2.37321	2.81386	3.94609	5.51602	7.68608	10.67658	12.56886
36	1.43077	1.70914	2.03989	2.43254	2.89828	4.10393	5.79182	8.14725	11.42394	13.51153
37	1.44508	1.73478	2.08068	2.49335	2.98523	4.26809	6.08141	8.63608	12.22361	14.52489
38	1.45953	1.76080	2.12230	2.55568	3.07478	4.43881	6.33548	9.15425	13.07927	15.61426
39	1.47412	1.78721	2.16474	2.61958	3.16703	4.61637	6.70475	9.70350	13.99481	16.78532
40	1.48886	1.81402	2.20804	2.68506	3.26204	4.80102	7.03999	10.28571	14.97445	18.04422

Example 5 (*Compare Exercise 16*)
\$2500 is invested at 10%, compounded quarterly. Find the amount at the end of eight years.

Solution $P = 2500, m = 4, r = .10, i = \frac{.10}{4} = .025, t = 8,$ and $n = 4 \times 8 = 32$

$$\begin{aligned} A &= 2500(1.025)^{32} \\ &= 2500(2.20376) \quad \text{From Table 8-1} \\ &= 5509.40 \end{aligned}$$

Example 6 (*Compare Exercise 21*)
A department store charges 1% per month on the unpaid balance of a charge account. This means that 1% of the bill is added to the account each month it is unpaid. This makes it compound interest. A customer owes \$135.00 and the bill is unpaid for four months. What is the amount of the bill at the end of four months?

Solution The interest rate is 1% per month and is compounded monthly. The value of i is then .01, and the number of periods is four. The amount of the bill is

$$\begin{aligned} A &= 135(1.01)^4 \\ &= 135(1.04060) \\ &= 140.48 \end{aligned}$$

Notice that i and n were given in this example so they can be substituted directly into the formula.

Present Value

The **present value** of an amount due at a specified time is the principal that must be invested now to accumulate the amount due.

Example 7 (*Compare Exercise 25*)
How much should Jack invest at 8%, compounded quarterly, so that he will have \$5000 at the end of seven years?

Solution In this case, we are given $A = 5000, r = .08, m = 4,$ and $t = 7$. Then, $i = \frac{.08}{4} = .02$ and $n = 4(7) = 28$. Using the compound interest formula,

$$\begin{aligned} 5000 &= P(1.02)^{28} \\ &= 1.74102\,P \end{aligned}$$

Solve the equation for P.

$$P = \frac{5000}{1.74102} = 2871.88$$

Jack should invest \$2871.88 to have \$5000 in seven years. In other words, the present value of \$5000 due in seven years is \$2871.88.

Doubling an Investment

If you invest \$1000 at 8% interest, compounded quarterly, how long will it be before your investment has doubled in value?

This is a compound interest problem with $P = 1000$, $A = 2000$, $r = .08$, and $m = 4$. The value of n is unknown. Use the compound interest formula with $i = .08/4 = .02$. We know all parts of the formula except the exponent n, so we can write

$$2000 = 1000(1.02)^n$$
$$2 = (1.02)^n$$

We want the value of n that makes this true; the power of 1.02 that gives 2. We can find this in Table 8–1. Since we know that the answer to $(1.02)^n$ is 2, look for 2 in the *answer part* under the $i = .02$ column. The nearest numbers you find are 1.99989 and 2.03989. These correspond to $n = 35$ and $n = 36$. Thus, the value will certainly double when $n = 36$ time periods. Since the interest is compounded quarterly, $n = 36$ represents 36 quarters, or nine years.

Actually, the amount invested is irrelevant. If we use P and $2P$ in the formula, we have

$$2P = P(1.02)^n$$

and

$$2 = (1.02)^n$$

the same as for $P = 1000$. So, any investment will double in value in nine years if the interest rate is 8% and is compounded quarterly.

Example 8 (*Compare Exercise 35*)
An investor has visions of doubling his money in six years. What interest rate is required for him to do so if the investment draws interest compounded quarterly?

Solution If P dollars are invested, $2P$ dollars are wanted in six years. The future value formula for interest compounded quarterly becomes

$$2P = P(1 + i)^{24}$$

where we wish to find i. We do not need to know the value of P because we can divide both sides by P and have $2 = (1 + i)^{24}$.

Again, we know the answer, 2, in Table 8–1, and we know that we look in the row for $n = 24$. We don't know which column, but we can look for 2 in the row for $n = 24$. The column for $i = .03$ gives 2.03279, so we conclude that 3% quarterly interest (12% annual interest) will double an investor's money in six years.

Effective Rate

For a given annual rate, a more frequent compounding of interest gives a larger value of an investment at the end of the year. A 10% annual rate, compounded monthly, gives a larger amount at the end of the year than does 10%, compounded semiannually. However, a lower rate, compounded more frequently, may or may not give a larger return. For example, which yields the better return, 9% compounded semiannually or 8.9% compounded quarterly? To answer this, let's assume that $1000 is invested in each case and compute the amount at the end of one year.

1. $r = 9\%$ compounded semiannually.

Let $P = 1000$, $t = 1$ year, $m = 2$, and $i = \frac{.09}{2} = .045$. Then,

$$A = 1000(1.045)^2 = 1000(1.09203) = 1092.03$$

(Use your calculator to find $(1.045)^2$, it isn't in the table.)

2. $r = 8.9\%$ compounded quarterly.

Let $P = 1000$, $t = 1$ year, $m = 4$, and $i = \frac{.089}{4} = .02225$. Then,

$$A = 1000(1.02225)^4 \quad \text{(Use calculator for } (1.02225)^4\text{)}$$
$$= 1000(1.09201)$$
$$= 1092.01$$

Thus, 9% compounded semiannually is a slightly better investment, but there is no significant difference.

Another way to put different rates and frequency of compounding on a comparable basis is to find the **effective rate**.

Definition

The **effective rate** of an annual interest rate r compounded m times per year is the simple interest rate that produces the same amount of interest per year as the compound interest (or the same total value of the investment).

Example 9 (*Compare Exercise 28*)

Find the effective rate of 8% compounded quarterly.

Solution If we invest P dollars, the amount of the investment at the end of the year is

$$A = P(1.02)^4$$

If we invest the same amount, P, at a simple interest rate, r, the amount of the investment at the end of the year is $A = P + Pr(1)$.

Now, r is the unknown simple interest rate that gives the same amount A as does compound interest. So,

$$P + Pr = P(1.02)^4$$

We can divide throughout by P to get

$$1 + r = (1.02)^4$$

and so,

$$r = (1.02)^4 - 1$$
$$= 1.08243 - 1$$
$$= .08243$$

In percent form, $r = 8.243\%$ is the effective rate of 8% compounded quarterly.

This method works generally, so we can make the following statement.

Effective Rate

If money is invested at an annual rate r and compounded m times per year, the effective rate x, in decimal form, is

$$x = (1 + i)^m - 1 \qquad \text{where} \qquad i = \frac{r}{m}$$

8–2 Exercises

I.

1. Use a calculator to find the following.

   (a) $(1.07)^3$ **(b)** $(1.02)^5$ **(c)** $(1.015)^4$
   (d) $(1.035)^2$ **(e)** $(1.045)^4$ **(f)** $(1.025)^5$

2. Use Table 8–1 to find the following.

   (a) $(1.02)^{15}$ **(b)** $(1.03)^{25}$ **(c)** $(1.04)^8$
   (d) $(1.01)^{30}$ **(e)** $(1.025)^{18}$ **(f)** $(1.02)^5$

3. The annual interest rate is 15%. Find

   (a) the semiannual interest rate.
   (b) the quarterly interest rate.
   (c) the monthly interest rate.

4. The annual interest rate is 9%. Find

   (a) the semiannual interest rate.
   (b) the quarterly interest rate.
   (c) the monthly interest rate.

In Exercises 5 through 9, find (a) the final amounts and (b) the total interest earned on the amounts.

5. \$10,000 at 9%, compounded annually, for five years
6. \$12,000 at 5%, compounded semiannually, for three years
7. \$20,000 at 8%, compounded annually, for five years

8. \$7500 at 10%, compounded semiannually, for eight years
9. \$2000 at 6%, compounded annually, for three years
10. (*See Example 1*) A family deposits \$1500 into an account that pays 6%, compounded quarterly. Find the amount in their account at the end of each year for four years.
11. \$800 is invested at 12%, compounded quarterly. Find the amount at the end of the first four quarters.
12. (*See Example 2*) \$2600 is invested at 7%, compounded annually. What is the amount of the account at the end of three years?
13. \$1800 is invested at 9%, compounded semiannually. How much is in the account at the end of one year?
14. (*See Example 4*) \$4500 is invested at 8% annual rate. Find the amount at the end of two years if the interest is compounded

    (a) annually. **(b)** semiannually. **(c)** quarterly.
15. \$12,000 is invested at 10% interest. Find the amount at the end of three years if the interest is compounded

    (a) annually. **(b)** semiannually. **(c)** quarterly.

Find the amount at the end of the specified time in Exercises 16 through 19.

16. (*See Example 5*) $P = \$7000$, $r = 10\%$, $t = 4$ years, $m = 4$ (compounded quarterly)
17. $P = \$10{,}000$, $r = 12\%$, $t = 5$ years, $m = 4$
18. $P = \$550$, $r = 8\%$, $t = 4$ years, $m = 2$ (compounded semiannually)
19. $P = \$460$, $r = 10\%$, $t = 6$ months, $m = 12$

II.

20. \$3500 is invested at 12%, compounded quarterly. Find the amount at the end of six years.
21. (*See Example 6*) A loan shark charges 2% per month on the unpaid balance of a loan. A student's loan was for \$640. He was unable to pay for six months. What was his loan balance at the end of six months?
22. A store has an interest rate of 1.5% per month on the unpaid balance of charge accounts. (Interest is compounded monthly.) A customer charges \$60, but allows it to become four months overdue. What is the bill at that time?
23. Your bookstore charges 1% per month on your unpaid balance. You charge \$32.75 for books. You do not pay the bill until four months after the first billing. What is the amount of interest you owe after four months of accumulated interest?
24. The Browns fail to pay a bill for items they charged. If the original bill was \$140, and the store charges 1.5% interest per month on the unpaid balance, what does the bill total when it is eight months overdue? (Interest is compounded monthly.)

25. (*See Example 7*) A man buys an investment that pays 6%, compounded semiannually. He wants \$25,000 when he retires in 15 years. How much should he invest? (Find the present value.)
26. Jack and Jill want to put money in a savings account now so that they will have \$1800 in five years. The savings bank pays 6% interest, compounded quarterly. How much should Jack and Jill invest?
27. Alexandra wishes to have \$3000 available to buy a car in four years. How much should she invest in a savings account now so that she will be able to do this? The bank pays 10% interest, compounded quarterly.
28. (*See Example 9*) A savings company pays 8% interest, compounded semiannually. What is the effective rate of interest?
29. What is the effective interest rate of 6% interest compounded semiannually?
30. What is the effective interest rate of 6% interest compounded quarterly?

How long does it take an investment to double at the interest rates given in Exercises 31 through 34?

31. 6% compounded semiannually
32. 10% compounded quarterly
33. 12% compounded quarterly
34. 8% compounded semiannually
35. (*See Example 8*) An investor wants to double her money in seven years. What interest rate, compounded quarterly, will enable her to do so?
36. How long will it take to increase an investment by 50% if it is invested at 1.5% per month, compounded monthly?
37. How long does it take to triple an investment if it draws 10% interest per. annum and is compounded semiannually?
38. What interest rate enables an investor to obtain a 60% increase in an investment in three years if the interest is compounded quarterly?

In Exercises 39 through 41, determine which is the better investment.

39. 8.8% compounded semiannually or 8.6% compounded quarterly?
40. 7.2% compounded semiannually or 6.8% compounded quarterly?
41. 9.4% compounded annually or 9.2% compounded quarterly?
42. A credit union pays interest at 6%, compounded semiannually, on accounts totaling \$275,000. They are considering compounding quarterly. How much would this increase the interest the credit union will pay in the period of one year?
43. Which is the better investment, 10.2% compounded annually or 10% compounded quarterly?
44. A loan company charges 36% a year, compounded monthly, for small loans. How long will it take to triple money?

45. A town increased in population 2% per year for ten years. If the population was 35,000 at the beginning, how much was it at the end of ten years?

46. The Acme Company sales have been increasing 4% per year. If sales in the latest year were $600,000 and the trend continues, how much are they expected to be in five years?

47. On her 58th birthday, a woman invests $15,000 in an account that pays 8%, compounded quarterly. How much will be in her account when she retires on her 65th birthday?

48. An investment company advertises that an investment with them will yield 12% annual rate, compounded monthly. Another firm advertises that its investments will yield 15%, compounded semiannually. Which of these gives the better yield?

49. Susan received a $1000 gift that she deposited in a savings bank that compounded interest quarterly. After five years of accumulating interest, the account had grown to $1485.95. What was the annual interest rate of the bank?

50. The cost of an average family house increased from $32,500 in 1975 to $52,950 in 1984. Determine approximately the average annual inflation in house prices over this ten-year period.

51. Jerry placed $500 in a credit union that compounded interest semiannually. His account had grown to $633.39 after four years. What was the annual interest rate of the credit union?

52. A couple has moved from California to Georgia. Since real estate prices are much higher in California than in Georgia, the family realized a net profit on selling their California house and buying one in Georgia. The husband is due to retire in five years time and would like to invest part of the profit into an account paying 6% per annum, compounded semiannually, that would yield him $20,000 at that time. How much should be deposited into the account?

53. Which should you choose—a savings account that originates with $6000 and earns interest at 10%, compounded semiannually, for 10 years or a lump sum of $16,000 at the end of ten years?

54. A child's grandparents have opened a savings account for that child on the day of her birth. The account contains $5000 and pays interest at 9%, compounded annually. The child will be allowed to withdraw the money when she becomes 21 years old. What will the account be worth at that time? (Use your calculator.)

55. The Anchor Company has $44,850, which will be deposited in a savings account until needed. It is anticipated that approximately $84,000 will be needed in eight years time to expand manufacturing facilities. What rate of interest will be required to accumulate the necessary amount, assuming compounding on a semiannual basis?

56. A student on graduating from college gets a job that pays $15,000 a year. If inflation is averaging 10%, what will he have to be making five years later to keep pace with inflation? What will he have to make five years later to increase his standard of living by 4%?

57. How long will it take $8000 to increase to $18,000 in an account that pays 6%, compounded semiannually?

58. A company acquires an asset and signs an agreement to pay $10,000 for it in three years. The interest rate is to be 10%, compounded semiannually. What is the current market value of the asset to the company?

59. A company will need $240,000 cash to modernize machinery in five years. A financial institution will invest in a fund at 8% interest, compounded semiannually. Determine the cash that must be deposited at present to meet this need.

8–3 Annuities and Sinking Funds

Ordinary Annuity
Sinking Funds

Ordinary Annuity

A series of equal payments like monthly house payments, installment payments, premiums on life insurance, and regular monthly deposits in a credit union are examples of **annuities**. In general, any set of equal payments paid at equal time intervals form an **annuity**. The time between successive payments is the **payment period**, and the amount of each payment is the **periodic payment**. Interest on an annuity is compound interest. Payments may be paid annually, quarterly, monthly, or at any specified time interval. We will study the **ordinary annuity**.

Ordinary Annuity

An ordinary annuity is an annuity with periodic payments made at the *end* of each period.

Example 1 (*Compare Exercise 1*)

A family enters a savings plan whereby they will invest $1000 at the end of each year for five years. The annuity will pay 7% interest, compounded annually. Find the value of the annuity at the end of the five years.

Solution Since $1000 is deposited each year, the first deposit will draw interest longer than subsequent deposits. The value of each deposit at the end of five years is the original $1000 plus the compound interest for the time it draws interest. Use the formula for the amount at compound interest. Here is a summary of payments. $1000 deposited at the end of each year for five years. Interest rate $r = .07$ is compounded annually.

Year Deposited	Amount of Time Deposit Draws Interest	Value of Deposit at End of Five Years
1	4 years	$1000(1.07)^4 = 1310.80$
2	3 years	$1000(1.07)^3 = 1225.04$
3	2 years	$1000(1.07)^2 = 1144.90$
4	1 year	$1000(1.07) = 1070.00$
5	0 year	$1000(1) = 1000.00$
	Final Value	5750.74

The final value is obtained by adding the five payments and the interest accumulated on each one. The final value is called the **amount** or **future value** of the annuity.

Clearly, computing this total period-by-period can be tedious if the annuity runs for a number of periods. A formula for computing the future value of an annuity helps ease the computations.

Future Value (Amount) of an Ordinary Annuity (Payments are made at the end of each period)

$$A = RS$$

where

$$S = \frac{(1+i)^n - 1}{i}$$

i = interest rate per period

n = number of periods

R = amount of each periodic payment

A = future value or amount

Historical note: The number S is traditionally denoted by $s_{\overline{n}|i}$ (read s sub n angle i). For simplicity, we use S. You should be aware that the value of S depends on the interest rate and number of periods. See Appendix A for a derivation of S.

Applying this formula to Example 1,

$$S = \frac{(1+.07)^5 - 1}{.07} = \frac{(1.07)^5 - 1}{.07} = \frac{1.4025517 - 1}{.07}$$

$$= \frac{.4025517}{.07}$$

$$= 5.75074$$

so $A = 1000(5.75074) = \$5750.74$

Table 8–2. Amount of an Ordinary Annuity $S = \frac{(1+i)^n - 1}{i}$

Number of Periods n	Interest per Period i											
	.01	.015	.02	.025	.03	.04	.05	.06	.07	.08	.09	.10
1	1.00000	1.00000	1.00000	1.00000	1.00000	1.00000	1.00000	1.00000	1.00000	1.00000	1.00000	1.00000
2	2.01000	2.01500	2.02000	2.02500	2.03000	2.04000	2.05000	2.06000	2.07000	2.08000	2.09000	2.10000
3	3.03010	3.04523	3.06040	3.07562	3.09090	3.12160	3.15250	3.18360	3.21490	3.24640	3.27810	3.31000
4	4.06040	4.09090	4.12161	4.15252	4.18363	4.24646	4.31012	4.37462	4.43994	4.50611	4.57313	4.64100
5	5.10100	5.15227	5.20404	5.25633	5.30914	5.41632	5.52563	5.63709	5.75074	5.86660	5.98471	6.10510
6	6.15201	6.22955	6.30812	6.38774	6.46841	6.63298	6.80191	6.97532	7.15329	7.33593	7.52333	7.71561
7	7.21354	7.32299	7.43428	7.54743	7.66246	7.89829	8.14201	8.39384	8.65402	8.92280	9.20043	9.48717
8	8.28567	8.43284	8.58297	8.73612	8.89234	9.21423	9.54911	9.89747	10.25980	10.63663	11.02847	11.43589
9	9.36853	9.55933	9.75463	9.95452	10.15911	10.58280	11.02656	11.49132	11.97799	12.48756	13.02104	13.57948
10	10.46221	10.70272	10.94972	11.20338	11.46388	12.00611	12.57789	13.18079	13.81645	14.48656	15.19293	15.93742
11	11.56683	11.86326	12.16872	12.48347	12.80780	13.48635	14.20679	14.97164	15.78360	16.64549	17.56029	18.53117
12	12.68250	13.04121	13.41209	13.79555	14.19203	15.02581	15.91713	16.86994	17.88845	18.97713	20.14072	21.38428
13	13.80933	14.23683	14.68033	15.14044	15.61779	16.62684	17.71298	18.88214	20.14064	21.49530	22.95338	24.52271
14	14.94742	15.45038	15.97394	16.51895	17.08632	18.29191	19.59863	21.01507	22.55049	24.21492	26.01919	27.97498
15	16.09690	16.68214	17.29342	17.93193	18.59891	20.02359	21.57856	23.27597	25.12902	27.15211	29.36092	31.77248
16	17.25786	17.93237	18.63929	19.38022	20.15688	21.82453	23.65749	25.67253	27.88805	30.32428	33.00340	35.94973
17	18.43044	19.20136	20.01207	20.86473	21.76159	23.69751	25.84037	28.21288	30.84022	33.75023	36.97370	40.54470
18	19.61475	20.48938	21.41231	22.38635	23.41444	25.64541	28.13238	30.90565	33.99903	37.45024	41.30134	45.59917
19	20.81090	21.79671	22.84056	23.94601	25.11687	27.67123	30.53900	33.75999	37.37896	41.44626	46.01846	51.15909
20	22.01900	23.12367	24.29737	25.54466	26.87037	29.77808	33.06595	36.78559	40.99549	45.76196	51.16012	57.27500
21	23.23919	24.47052	25.78332	27.18327	28.67649	31.96920	35.71925	39.99273	44.86518	50.42292	56.76453	64.00250
22	24.47159	25.83758	27.29898	28.86286	30.53678	34.24797	38.50521	43.39229	49.00574	55.45676	62.87334	71.40275
23	25.71630	27.22514	28.84496	30.58443	32.45288	36.61789	41.43048	46.99583	53.43614	60.89330	69.53194	79.54302
24	26.97346	28.63352	30.42186	32.34904	34.42647	39.08260	44.50200	50.81558	58.17667	66.76476	76.78981	88.49733
25	28.24320	30.06302	32.03030	34.15776	36.45926	41.64591	47.72710	54.86451	63.24904	73.10594	84.70090	98.34706

Just as we have a table for computing values of compound interest, we have a table for computing future values of annuities. Table 8–2 gives values of S for selected values of n and i.

Example 2 (*Compare Exercise 3*)

(a) To locate S for $i = .04$ and $n = 10$, look down the n column to 10 and then across to the .04 column. There you see 12.00611. This is the desired value of S.

(b) Deposits are made quarterly to an annuity for three years. The annual interest rate is 8%. In this case, the periodic rate is $i = \frac{.08}{4} = .02$, and the number of periods (quarters) is 12. So, $S = 13.41209$ is found in the .02 column across from $n = 12$.

Example 3 (*Compare Exercise 4*)

A family pays \$800 at the end of each six months for seven years into an ordinary annuity paying 10%, compounded semiannually. What is the future value at the end of seven years?

Solution Since the payments are made semiannually, the periodic rate $i = \frac{.10}{2} = .05$, the number of periods is 14, and the periodic payments are $R = 800$. Then, the future value is $A = 800(19.59863) = \$15678.90$, where S is obtained from Table 8–2.

Sinking Funds

When an amount of money will be needed at some future date, you can systematically accumulate a fund that will build to the desired amount at the time needed. Money accumulated in this way is called a **sinking fund**. A sinking fund is an annuity with the future value specified and the periodic payments known. The formula for future value still applies.

Example 4 (*Compare Exercise 14*)

Susie wants to deposit her savings at the end of every three months so that she will have \$7500 available in four years. The account will pay 8% interest per annum, compounded quarterly. How much should she deposit every quarter?

Solution Susie is accumulating a sinking fund with a future value of \$7500, periodic rate $i = \frac{.08}{4} = .02$, and $n = 16$ time periods. The formula for future value of an annuity can be used to find R, the periodic payments.

$$7500 = R(18.63929)$$

$$R = \frac{7500}{18.63929}$$

$$= 402.38$$

Susie should deposit \$402.38 every quarter to accumulate the desired \$7500.

The formula for periodic payments into a sinking fund is a variation of the future value formula for an annuity.

Formula for Periodic Payments of a Sinking Fund

$$R = \frac{A}{S}$$

where

A = value of the annuity after n payments
n = number of payments
i = periodic interest rate
S is obtained from Table 8–2 using i and n
R = amount of each periodic payment

Example 5 (*Compare Exercise 20*)
A company expects to replace a $65,000 piece of equipment in ten years. They establish a sinking fund with annual payments. The fund draws 7% interest, compounded annually. What are these periodic payments?

Solution $A = 65{,}000$, $i = .07$, and $n = 10$, so $S = 13.81645$

$$R = \frac{65000}{13.81645} = \$4704.54$$

8–3 Exercises

1. (*See Example 1*) $600 is deposited into an account at the end of each year for four years. The money earns 10%, compounded annually. Determine the value of each deposit at the end of four years and the total in this account at that time.
2. A boy deposits $500 into an account at the end of each year for three years. The money earns 6%, compounded annually. Determine the value of each deposit at the end of three years and the total amount in the account.
3. (*See Example 2*) Locate the value of S in Table 8–2 for the following.

   (a) $n = 15$, $i = .03$
   (b) $n = 20$, $i = .04$
   (c) 12% compounded quarterly for three years
   (d) 10% compounded semiannually for seven years
   (e) 12% compounded monthly for one year

In Exercises 4 through 13, determine the amount of the ordinary annuities at the end of the given periods.

4. (*See Example 3*) annual payments of $2,000 at 7% for four years
5. annual payments of $10,000 at 8% for five years

6. annual payments of \$20,000 at 10% for 20 years
7. semiannual payments of \$400 at 10% for 12 years
8. quarterly payments of \$100 at 8% for five years
9. monthly payments of \$500 at 24% for two years
10. annual payments of \$5000 at 7% for ten years
11. semiannual payments of \$6000 at 8% for nine years
12. quarterly payments of \$200 at 10% for six years
13. semiannual payments of \$3000 at 6% for eight years

In Exercises 14 through 19, the amount (future value) of an ordinary annuity is given. Find the periodic payments.

14. (*See Example 4*) Amount is \$10,000. Annuity pays 6%, compounded annually, for 10 years.
15. Amount is \$4,000. Annuity pays 7%, compounded annually, for three years.
16. Amount is \$50,000. Annuity pays 10%, compounded quarterly, for five years.
17. Amount is \$40,000. Annuity pays 8%, compounded semiannually, for four years.
18. Amount is \$30,000. Annuity pays 10%, compounded semiannually, for 12 years.
19. Amount is \$2000. Annuity pays 8%, compounded quarterly, for six years.

In Exercises 20 through 23, the amount desired in a sinking fund is given. Find the periodic payments required to obtain the desired amount.

20. (*See Example 5*) $A = \$12{,}000$; interest rate = 8%; payments are made semiannually for five years.
21. $A = \$75{,}000$; interest rate = 10%; payments are made quarterly for four years.
22. $A = \$40{,}000$; interest rate = 6%; payments are made annually for eight years.
23. $A = \$15{,}000$; interest rate = 12%; payments are made monthly for 18 months.
24. A company has decided to prepare for the modernization of its plant by setting up an ordinary annuity consisting of five annual deposits of \$40,000 that will earn 9% per annum. What will be the amount of the annuity?
25. Parents decide to prepare for their son's college education by putting \$500 semiannually into an account that pays 6%, compounded semiannually. What will be the amount of the account after ten years?
26. The Radcliff Corporation decides to create a plant expansion fund by making four annual deposits of \$70,000 each on January 1 for four years. The fund will be held by a trustee who will invest the fund at 8% interest, compounded annually. At the end of four years, no further deposits are made, but the investment earns 8% interest, compounded annually, for

the next two years. How much money will be in the fund on January 1 of the sixth year?

27. A young couple pays $400 at the end of each six months for five years into an ordinary annuity paying 8%, compounded semiannually. What is the future value at the end of five years?

28. A woman invests $2000 a year in a mutual fund for 20 years. If the market value of the fund increases 6% per year, what will be the value of her shares when she makes the 20th payment?

29. Sam wants to invest the same amount at the end of every three months so that he will have $4000 in three years. The account will pay 6%, compounded quarterly. How much should he deposit each quarter?

30. How much should a family deposit at the end of every six months in order to have $8000 at the end of five years? The account pays 6% interest, compounded semiannually.

31. A 13-year-old child received an inheritance of $5000 per year. This was to be invested and allowed to accumulate until the child reached 21 years of age. The first payment was made on the child's 13th birthday and the last on the 21st birthday. If the money was invested at 7%, compounded annually, what did the child receive at age 21?

32. A condominium association decided to set up a sinking fund to accumulate $50,000 by the end of four years to build a new sauna and swimming pool. What quarterly deposits are required if the annual interest rate is 8%, and it is compounded quarterly?

33. A company projects that it will need to expand its plant in six years. They expect the expansion to cost $150,000. How much should they put into a sinking fund each year at 7%, compounded annually?

34. A university professor, anticipating a well-earned around-the-world cruise on retirement, has opened an ordinary annuity that gives interest at a rate of 8%, compounded quarterly. The quarterly payment of $70 is made for a period of six years. What will be the amount of the annuity?

35. A city has issued bonds to finance a new convention center. The bonds have a total face value of $750,000 and are payable in eight years. A sinking fund has been opened to meet this obligation. If the interest rate on the fund is 8%, compounded semiannually, what will the semiannually payments be?

36. A person plans to start a business of her own in six years. She plans to have $10,000 cash available at the time for this purpose. To raise the $10,000, a fund has been started that earns interest at 8%, compounded quarterly. What would the quarterly payments into this fund have to be to raise the $10,000?

37. Electronic Instruments plans to establish a debt retirement fund. The company wants at least $23,800 in five years. Deposits of $4000 are to be made to a trustee each year. Compute an approximate interest rate for the fund on an annual compounding basis to meet these requirements.

38. A company requires each of two subsidiaries to make deposits into a debt retirement fund over the next three years. The one subsidiary is to contribute \$2000 quarterly and the other \$6000 semiannually. Interest at 8%, compounded quarterly and semiannually, will be paid on the fund. Determine the amount of the fund at the end of three years.

39. A savings institution advertised, "Invest \$1000 a year for ten years with us and we will pay you \$1000 a year forever." (They pay 8% interest rate.)

(a) What is the value of the investment at the end of ten years?

(b) The investor ceases payments at the end of ten years, and the investment firm pays him 8% of his investment at the end of each year. How much interest will he receive at the end of each year?

40. **(a)** A young woman invested \$2000 per year in an IRA each year for 9 years. The interest rate was 7%, compounded annually. At the end of 9 years, she ceased the IRA payments but invested the total of her investment at 7%, compounded annually, for the next 25 years.

(i) What was the value of her IRA investment at the end of 9 years?

(ii) What was the value of her investment at the end of the next 25 years? *Note*: $(1.07)^{25} = 5.42743$.

(b) A friend of the young woman (Part a) started his IRA investment in the 10th year and invested \$2000 per year for the next 25 years at 7%, compounded annually. What was the value of his investment at the end of the 25 years?

(c) Which of the two investments had the greater value at the end of the period?

8–4 Present Value of an Annuity and Amortization

Present Value
Amortization

Present Value

We have presented an annuity as a sequence of payments made at regular intervals into an account. The object of an annuity is to accumulate an amount of money at a future date.

Let's look at two variations of this problem.

1. How much should be put into a savings account at compound interest so that the amount accumulated at the end of five years is the same as the amount accumulated by putting in \$25 per month into a savings account?

   This lump sum payment that yields the same total amount as equal periodic payments made over the same period of time is called the **present value** of the annuity.

2. How much should be invested so that \$100 can be withdrawn each month for five years, with the account depeleted at the end of five years?

Similar problems seek to find how much should be placed in a college fund so that a certain amount will be available each year, or how much should be in a retirement fund to provide a monthly retirement benefit for 15 years.

The first problem, finding the present value of an annuity, is solved as follows. Recall that the amount of an annuity is

$$A = RS = \frac{R((1+i)^n - 1)}{i}$$

and the amount of compound interest is,

$$A = P(1+i)^n$$

where i is the periodic interest rate, n is the number of periods, R is the periodic payment for the annuity, and P is the lump sum invested at compound interest. We want to find P so that the amount of compound interest equals the amount of the annuity, that is

$$P(1+i)^n = \frac{R((1+i)^n - 1)}{i}$$

Solving for P gives

$$P = \frac{R((1+i)^n - 1)}{i(1+i)^n}$$

We let

$$K = \frac{(1+i)^n - 1}{i(1+i)^n}$$

to obtain the formula for present value.

Present Value of an Annuity

$$P = RK$$

where

$$K = \frac{(1+i)^n - 1}{i(1+i)^n}$$

i = periodic rate
n = number of periods
R = periodic payments
P = present value of the annuity

Historical note: A common notation for K is $a_{\overline{n}|\,i}$; we use K for simplicity. In any given problem, K depends on the value of n and i. You may find the value K with a scientific calculator or from Table 8–3.

Table 8–3. Present Value of an Ordinary Annuity $K = \dfrac{(1 + i)^n - 1}{i(1 + i)^n}$

Number of Periods n	Interest per period i											
	.01	.015	.02	.025	.03	.04	.05	.06	.07	.08	.09	.10
1	.99010	.98522	.98039	.97561	.97087	.96154	.95238	.94340	.93458	.92593	.91743	.90909
2	1.97040	1.95588	1.94156	1.92742	1.91347	1.88609	1.85941	1.83339	1.87802	1.80802	1.75911	1.73554
3	2.94099	2.91220	2.88388	2.85602	2.82861	2.77509	2.72325	2.67301	2.62432	2.57710	2.53129	2.48685
4	3.90197	3.85438	3.80773	3.76197	3.71710	3.62990	3.54595	3.46511	3.38721	3.31213	3.23972	3.16987
5	4.85343	4.78264	4.71346	4.64583	4.57971	4.45182	4.32948	4.21236	4.10020	3.99271	3.88965	3.79079
6	5.79548	5.69719	5.60143	5.50813	5.41719	5.24214	5.07569	4.91732	4.76654	4.62288	4.48592	4.35526
7	6.72819	6.59821	6.47199	6.34939	6.23028	6.00205	5.78637	5.58238	5.38929	5.20637	5.03295	4.86842
8	7.65168	7.48593	7.32548	7.17014	7.01969	6.73274	6.46321	6.20979	5.97130	5.74664	5.53482	5.33493
9	8.56602	8.36052	8.16224	7.97087	7.78611	7.43533	7.10782	6.80169	6.51523	6.24689	5.99525	5.75902
10	9.47130	9.22218	8.98259	8.75206	8.53020	8.11090	7.72173	7.36009	7.02358	6.71008	6.41766	6.14457
11	10.36763	10.07112	9.78685	9.51421	9.25262	8.76048	8.30641	7.88687	7.49867	7.13896	6.80519	6.49508
12	11.25508	10.90750	10.57534	10.25776	9.95400	9.38507	8.86325	8.38384	7.94269	7.53608	7.16073	6.81369
13	12.13374	11.73153	11.34837	10.98318	10.63496	9.98565	9.39357	8.85268	8.35765	7.90378	7.48690	7.10336
14	13.00370	12.54338	12.10625	11.69091	11.29607	10.56312	9.89864	9.29498	8.74547	8.24424	7.78615	7.36669
15	13.86505	13.34323	12.84926	12.38138	11.93794	11.11839	10.37966	9.71225	9.10791	8.55948	8.06069	7.60608
16	14.71787	14.13126	13.57771	13.05500	12.56110	11.65230	10.83777	10.10590	9.44665	8.85137	8.31256	7.82371
17	15.56225	14.90765	14.29187	13.71220	13.16612	12.16567	11.27407	10.47726	9.76322	9.12164	8.54363	8.02155
18	16.39827	15.67256	14.99203	14.35336	13.75351	12.65930	11.68959	10.82760	10.05909	9.37189	8.75563	8.20141
19	17.22601	16.42617	15.67846	14.97889	14.32380	13.13394	12.08532	11.15812	10.33560	9.60360	8.95011	8.36492
20	18.04555	17.16864	16.35143	15.58916	14.87747	13.59033	12.46221	11.46992	10.59401	9.81815	9.12855	8.51356
21	18.85698	17.90014	17.01121	16.18455	15.41502	14.02916	12.82115	11.76408	10.83553	10.01680	9.29224	8.64869
22	19.66038	18.62082	17.65805	16.76541	15.93692	14.45112	13.16300	12.04158	11.06124	10.20074	9.44243	8.77154
23	20.45582	19.33086	18.29220	17.33211	16.44361	14.85684	13.48857	12.30338	11.27219	10.37106	9.58021	8.88322
24	21.24339	20.03041	18.91393	17.88499	16.93554	15.24696	13.79864	12.55036	11.46933	10.52876	9.70661	8.98474
25	22.02316	20.71961	19.52346	18.42438	17.41315	15.62208	14.09394	12.78336	11.65358	10.67478	9.82258	9.07704

Example 1 (*Compare Exercise 1*)
Find K for $n = 7$ and $i = .04$ by looking in the $i = .04$ column across from $n = 7$. You find 6.00205.

For $r = 12\%$ compounded semiannually for eight years, find K. In this case, $i = .12/2 = .06$ and $n = 16$. The table shows that $K = 10.10590$.

Example 2 (*Compare Exercise 5*)
Find the present value of an annuity with periodic payments of \$2000, semiannually, for a period of ten years at an interest rate of 6%, compounded semiannually.

Solution Here $R = 2000$, $i = .06/2 = .03$, and $n = 20$. The formula for present value gives

$$\begin{aligned} P &= 2000(14.87747) \\ &= 29754.94 \end{aligned}$$

where Table 8–3 gives $K = 14.87747$ for $i = .03$ and $n = 20$.

The present value of the annuity is \$29,754.94. This lump sum will accumulate the same amount in ten years as \$2000 semiannually for ten years.

The second problem of finding the amount to be invested so that it will pay out equal amounts periodically is an "annuity in reverse." For example, a lump sum is invested at 8%, compounded quarterly, so that quarterly payments of \$500 can be made for ten years. In such a case, the amount invested is the present value of \$500 invested quarterly at 8% for ten years. So, the formula for present value applies to this problem.

Example 3 (*Compare Exercise 36*)
Find the present value of an annuity that will pay \$1000 per quarter for four years. The interest rate is 10%.

Solution Since $R = 1000$, $i = 2.5\%$, and $n = 16$,

$$P = 1000(13.05500) = 13{,}055.00$$

where 13.05500 is obtained from Table 8–3 at $i = .025$ and $n = 16$. A lump sum investment of \$13,055 will yield \$1000 per quarter for four years.

Amortization

The **amortization** method is a way to repay an interest-bearing debt. The amortization method makes a series of equal periodic payments. Each payment pays all the interest for that time period and repays part of the principal. As the principal is gradually reduced, the interest on the unpaid balance decreases. Thus, a larger portion of each payment becomes available to apply to the debt.

The last payment pays interest for that period and completes the repayment of the loan. At some point, most of us borrow money to buy a car or a home and amortize the debt.

When a debt is amortized by equal payments at equal time intervals, the amount borrowed is just the present value of an annuity. This method usually applies to car payments and house payments. We can use the present value formula for an annuity to find the periodic payments.

Example 4 (*Compare Exercise 17*)
An employee borrows \$8000 from the company credit union to purchase a car. The interest rate is 12%, compounded quarterly, with payments due every quarter. The employee wants to pay off the loan in three years. (The loan is amortized over three years.) How much are the quarterly payments?

Solution We have $P = 8000$, $i = .12/4 = .03$, $n = 12$ quarters, and $K = 9.95400$ (from Table 8–3).

From the present value formula, we have

$$8000 = R(9.95400)$$

Solving for R,

$$R = \frac{8000}{9.95400} = 803.697$$

The quarterly payments on the car are \$803.70 each.

In general, when we solve for R in the present value formula, we have the amortization payment formula.

Amortization Payments

$$R = \frac{P}{K}$$

where

P = amount of the debt (the present value)

K is found in Table 8–3 and depends on the periodic rate, i, and number of time periods, n.

R = amount of each periodic payment

You should be aware that the amount of the periodic payment may involve a fraction of a cent. In that case, the bank rounds up to the next cent. Consequently, the final payment may be a little less than the other payments.

Example 5 (*Compare Exercise 13*)
A student obtained a 24-month loan on a car. The monthly payments are \$395.42 and are based on a 12% interest rate. What was the amount borrowed?

Solution The amount borrowed is just the present value of the annuity. We then have

$R = 395.42$

$i = \dfrac{.12}{12} = .01$ (monthly rate)

$n = 24$ (number of months)

$K = 21.24339$ (from Table 8–3)

so,

$$P = 395.42(21.24339) = 8400.06$$

It is reasonable to round this to \$8400, the amount borrowed.

Table 8–4. $K = \dfrac{(1+i)^n - 1}{i(1+i)^n}$

Number of Periods n	Periodic interest rate i					
	$\frac{2}{3}\%$	$\frac{3}{4}\%$	$\frac{10}{12}\%$	1%	$\frac{14}{12}\%$	1.25%
240	119.55429	111.14495	103.62460	90.81942	80.41683	75.94228
300	129.56452	119.16162	110.04723	94.94655	83.07297	78.07434
360	136.28349	124.28187	113.95082	97.21833	84.39732	79.08614

Example 6 (*Compare Exercise 24*)
A family borrows $75,000 to buy a house. They are to repay the debt in 20 years with equal monthly payments. The interest rate is 14%. How much are their monthly payments?

Solution This is an amortization problem with $P = 75{,}000$, $i = \dfrac{14\%}{12}$, and $n = 12 \times 20 = 240$ months.

Table 8–4 is a special table giving values of K for 20, 25, and 30 years of monthly payments. In it, for $n = 240$ and $i = .14/12$, we find $K = 80.41683$. Then,

$$P = \frac{75{,}000}{80.41683}$$
$$= 932.64$$

The monthly payments are $932.64.

Example 7 (*Compare Exercise 29*)
A family borrowed $60,000 to buy a house. The loan was for 30 years at 12% interest rate. Their monthly payments were $617.17.

(a) How much of the first month's payment was interest and how much was principal?

(b) What total amount did they pay over the 30 years?

Solution **(a)** The monthly interest rate was $1\% = .01$, so the first month's interest was $60{,}000(.01) = 600.00$.

They paid $600 interest the first month. The rest of the payment, $17.17, went to repay part of the principal.

(b) They paid $617.17 each month for 360 months, so the total amount paid was $617.17(360) = \$222{,}181.20$. You may be surprised at this figure, but it is true. Notice that the total amount paid for interest was

$$\$222{,}181.20 - 60{,}000 = \$162{,}181.20.$$

When a family makes monthly payments on a house mortgage, some of each month's payment goes to reduce the loan. In Example 7, the first payment reduced the loan by \$17.17. To find the balance of the loan after a period of time, say five years, you can find the amount repaid each month for five years and deduct these from the loan. For example, here is an amortization schedule for the first 12 months of the loan.

\$60,000 Loan for 30 Years at 12%

Month	Monthly Payment	Interest Paid	Principal Paid	Balance
0				\$60,000.00
1	\$617.17	\$600.00	\$17.17	59,982.83
2	617.17	599.83	17.34	59,965.49
3	617.17	599.65	17.52	59,947.97
4	617.17	599.48	17.69	59,930.28
5	617.17	599.30	17.87	59,912.41
6	617.17	599.12	18.05	59,894.36
7	617.17	598.94	18.23	59,876.13
8	617.17	598.76	18.41	59,857.72
9	617.17	598.58	18.59	59,839.13
10	617.17	598.39	18.78	59,820.35
11	617.17	598.20	18.97	59,801.38
12	617.17	598.01	19.16	59,782.22

This approach is too tedious. There is a formula that helps considerably.

The Balance of an Amortization

$$\text{Bal} = P(1 + i)^n - \frac{M(1 + i)^n - 1}{i}$$

where

P = amount borrowed
i = periodic interest rate
n = number of time periods
M = monthly payments

Note that this formula can also be written

$$\text{Bal} = P(1 + i)^n - MS$$

where S is obtained from Table 8–2. In this form,

Bal = Future value of compound interest − future value of an annuity.

See Appendix *A* for a derivation of the formula for balance.

Example 8 (*Compare Exercise 30*)
What is the balance of the loan in Example 7 after two years?

Solution In Example 7,

$P = 60{,}000$
$i = 1\% = .01$ per month
$M = 617.17$
$n = 24$ months (the balance after 24 months is desired)
$S = 26.97346$ (from Table 8–2 using $i = 0.1$ and $n = 24$)

$$\begin{aligned}\text{Bal} &= 60{,}000(1.01)^{24} - 617.17(26.97346)\\ &= 60{,}000(1.26973) - 617.17(26.97346)\\ &= 76{,}183.80 - 16{,}647.21\\ &= 59{,}536.59\end{aligned}$$

So, the balance owed after two years is \$59,536.59. The part of the loan repaid is the **equity**.

$$\text{Equity} = \text{loan} - \text{balance}$$

In this case, the equity after two years is

$$\text{Equity} = 60{,}000 - 59{,}536.59 = \$463.41$$

The results of these computations may seem wrong. Hardly any principal is repaid each month. In two years the 24 payments total \$14,812.08, but only \$463.41 of the \$60,000 loan has been repaid. However, the amount repaid increases a little each month (about 19¢, see the amortization table for this loan). While a 19¢ per month increase hardly seems worthwhile, it eventually becomes a significant increase and the loan is paid off. Notice from Example 7 that the total interest on the loan of that example is over \$160,000.

Now let's look at a situation that combines the future value and present value of an annuity.

Example 9 (*Compare Exercise 44*)
The parents of a new baby want to provide for the child's college education. How much should be deposited on each of the child's first 17 birthdays to be able to withdraw \$10,000 on each of the next four birthdays? Assume an interest rate of 8%.

Solution First, compute the amount that must be in the account on the child's 17th birthday to withdraw \$10,000 per year for four years. This is the present value of an annuity where

P is to be found
$i = .08$
$n = 4$
$R = 10{,}000$
$K = 3.31213$ (from Table 8–3)

So $P = RK = 10{,}000(3.31213) = \$33{,}121.30$, the total necessary on the 17th birthday.

Next, find the annual payments that will yield a future value of \$33,121.30 in 17 years at 8%. This is $A = RS$ where

$A = 33{,}121.30$
$i = .08$
$n = 17$
$S = 33.75023$ (from Table 8–2 using $i = .08$ and $n = 17$)

$$33{,}121.30 = 33.75023R$$

$$R = \frac{33{,}121.30}{33.75023} = 981.37.$$

So, \$981.37 should be deposited every birthday for 17 years to provide \$10,000 per year for 4 years.

8–4 Exercises

Find K from Table 8–3 for Exercises 1 through 4.

1. (*See Example 1*)
   $n = 15$ and $i = .02$
2. $n = 6$ and $i = .05$
3. 10% compounded semiannually for 8 years
4. 5% compounded annually for 10 years

In Exercises 5 through 12, determine the present values of the annuities that will pay the given periodic payments.

5. (*See Example 2*) \$5000 annually for a period of 10 years. Interest is 8%, compounded annually.
6. \$1000 annually for five years. Interest is 6%, compounded annually.
7. \$500 quarterly for six years. Interest is 8%, compounded quarterly.
8. \$2000 semiannually for ten years. Interest is 10%, compounded semiannually.
9. \$200 monthly for two years. Interest is 24%, compounded monthly.
10. \$10,000 annually for 20 years. Interest is 6%, compounded annually.
11. \$50,000 semiannually for four years. Interest is 8%, compounded semiannually.
12. \$6000 quarterly for five years. Interest is 10%, compounded quarterly.

For Exercises 13 through 16, find the amount borrowed for each loan described.

13. (*See Example 5*) $R = \$226.94$; interest rate = 8%; payments are made quarterly for three years.

14. $R = \$1071.84$; interest rate $= 10\%$; payments are made semiannually for six years.

15. $R = \$2135.66$; interest rate $= 7\%$; payments are made annually for ten years.

16. $R = \$3993.63$; interest rate $= 10\%$; payments are made annually for 20 years.

In Exercises 17 through 23, the present value of an annuity is given, and you are asked to compute the periodic payments.

17. (*See Example 4*) \$5000. Interest is 7%, compounded annually, for four years.

18. \$10,000. Interest is 8%, compounded semiannually, for 12 years.

19. \$50,000. Interest is 10%, compounded quarterly, for five years.

20. \$4000. Interest is 6%, compounded annually, for three years.

21. \$150,000. Interest is 12%, compounded semiannually, for ten years.

22. \$3000. Interest is 12%, compounded quarterly, for four years.

23. \$2000. Interest is 10%, compounded quarterly, for three years.

24. (*See Example 6*) A family borrows \$92,000 to buy a house. They are to repay the debt in 25 years with equal monthly payments at 10% interest. How much are their monthly payments?

25. Determine the monthly payments on a 20-year loan of \$20,000 at 8% interest.

26. The monthly payments on a 20-year loan of \$25,000 at 8.50% are \$216.95.

(a) What is the total amount paid over the 20 years?

(b) What is the total amount of interest paid?

27. A family obtains a \$75,000 house loan for 30 years at 12% interest. The monthly payments are \$825.75 each.

(a) What is the total amount paid over the 30 years?

(b) What is the total interest paid?

28. A house loan of \$30,000 is made for 30 years at 9% interest. For each of the first two monthly payments, find the amount that is paid for interest and the amount that is paid toward the principal.

29. (*See Example 7*) A family borrowed \$68,000 to buy a house. The loan was at 15% for 25 years. Their monthly payments were \$870.96 each.

(a) How much of the first month's payment was interest and how much was principal?

(b) What was the total amount paid over the 25 years?

30. (*See Example 8*) A family obtained a house loan of \$48,000 for 25 years at 12% interest. Their monthly payments are \$505.54. What is the balance of their loan after

(a) one year?

(b) two years?

(c) What is their equity after two years?

31. A company amortizes a $75,000 loan, at 10% interest, with quarterly payments of $2987.72 for ten years. What is the balance of the loan after five years?

32. Find the present value of an annuity that will pay $1500 every six months for nine years from an account paying interest at a rate of 8%, compounded semiannually.

33. Miss Jones obtained an 18-month loan on a car. The monthly payments are $484.92 and are based on an 18% interest rate. How much did Miss Jones borrow?

34. A druggist borrows $4500 from a bank to stock his drugstore. The interest rate is 8%, compounded semiannually, with payments due every six months. He wants to repay the loan in 18 months. How much are the semiannual payments?

35. A mechanic borrows $7500 to expand her garage. The interest rate is 10%, compounded quarterly, with payments due every quarter. What are the quarterly payments if the loan is to be paid off in four years?

36. (*See Example 3*) A businessman wishes to deposit an amount at the present time into an account that will give him $5000 annually for the next ten years. The account earns compound interest at 7% per annum. How much would he have to deposit to obtain this financial arrangement?

37. An accounting department plans to start a fund that will give a $1000 annual scholastic award for 20 years to the graduating senior with the best four-year academic record. The account from which the money would come pays compound interest at 8% per annum. How much money will have to be raised to form this account?

38. An alumnus of a certain university wishes to donate a sum of money that can be used for a chair in the mathematics department. The plans are for the sum of money to yield $25,000 per year for a period of 25 years towards the salary of the professor holding the chair. To accomplish this, how much money would have to be deposited into an account which yields 9% per annum?

39. A company wishes to allocate a lump sum of funds into an account that will cover quarterly payments of $4000 for the leasing of machinery for a period of five years. The fund pays an 8% annual rate. What amount should be deposited into the account to cover these quarterly payments for five years?

40. A car dealer is willing to finance a loan of $5499 for a car, paid back with interest of 12%, to be compounded quarterly over a period of five years. What will the quarterly payments be?

41. A $600 television set can be amortized over two years at 10%, compounded quarterly. What will the quarterly payments be? Determine the total amount that will have been paid for the set in this manner.

42. Williams and Co., Inc. is bidding for a certain contract. If they win the contract, they will have to lease additional equipment for the four-year duration of the contract. The cost of the leasing would be $8000 per quarter,

plus increases due to inflation, for the four years. Taking the current inflation rate as being 10% per year, how much should be budgeted into the bid for the leasing of the equipment?

43. A property owner leases a building for $8000 per year plus a 6% increase each year for inflation. The leasing agreement is for five years. Instead of paying annual rentals, the tenant wishes to pay a lump sum that would contribute the total rent for the five years. What lump sum would be appropriate to cover the annual rent plus the anticipated inflation?

44. (*See Example 9*) How much should parents invest on each of their child's first 18 birthdays to provide $15,000 per year for the next four years if the investment pays 8% interest?

45. How much should be invested each year for 10 years to provide you with $5000 per year for the next 25 years? Assume a 10% interest rate.

IMPORTANT TERMS

8–1	**Simple interest**	**Principal**	**Present value**
	Interest rate	**Future value**	**Amount**
	Simple discount	**Proceeds**	**Maturity value**
	Discount rate		
8–2	**Compound interest**	**Periodic interest rate**	**Future value**
	Amount	**Present value**	**Effective rate**
8–3	**Annuity**	**Payment period**	**Periodic payment**
	Ordinary annuity	**Amount**	**Future value**
	Sinking fund		
8–4	**Present value of an annuity**	**Amortization**	

REVIEW EXERCISES

1. A loan is made for $500 at 9% simple interest for two years. How much interest is paid?
2. How much simple interest was paid on a $1100 loan at 6% for ten months?
3. A loan, principal and interest, was paid with $1190.40. The loan was made at 8% simple interest for three years. How much was borrowed?
4. The interest on a loan was $94.50, the interest rate was 7.5%, and the loan was for 1.5 years. What was the amount borrowed?
5. A sum of $3000 is borrowed for a period of five years at simple interest of 9% per annum. Compute the total interest paid over this period.
6. Find the future value of a $6500 loan at 7.5% for 18 months.
7. A simple discount note at 9% discount rate for two years has a maturity value of $8500. What is the discount and the proceeds?

8. How much should be borrowed on a discount note with 11% discount rate for 1.5 years so that the borrower obtains $900?

9. Compute the interest earned on $5000 in three years if the interest is 7%, compounded annually.

10. Compute the interest and the amount of $500 after two years if the interest is 10%, compounded semiannually.

11. If money is invested at 8% interest, compounded quarterly, how long will it take to double in value?

12. A sum of $1000 is deposited in a savings account that pays interest of 5%, compounded annually. Determine the amount in the account after four years.

13. Find the effective rate of 6% compounded semiannually.

14. Find the effective rate of 5% compounded semiannually.

15. Find the effective rate of 24% compounded monthly.

16. Which account gives the better interest, one that pays 12% compounded quarterly or one that pays 13% compounded annually.

17. Which account gives the better interest, one that gives 18% per annum compounded quarterly or one that gives 19% compounded annually?

18. The price of an automobile is now $5000. What would be the anticipated price of that automobile in two years time if prices are expected to increase at an annual rate of 8%?

Find (a) the future values and (b) the total interest earned on the amounts in Exercises 19 through 21.

19. $8000 at 6%, compounded annually, for four years

20. $3000 at 5%, compounded annually, for two years

21. $5000 at 10%, compounded quarterly, for six years

22. A company sets aside $120,000 cash in a special building fund to be used at the end of five years to construct a new building. The fund will earn 6%, compounded semiannually. How much money will be in the fund at the end of the period?

23. On January 1, 1985 a Chicago firm purchased a new machine to be used in the plant. The list price of the machine was $15,000 payable at the end of two years at interest of 8%, compounded annually. How much was due on the machine on January 1, 1987?

24. Parents, anticipating college tuition for their child in ten years, want to deposit a lump sum of money into an account that will provide $16,000 at the end of that ten-year period. The account selected pays 10%, compounded semiannually. How much should they deposit?

25. A landowner suspects that real estate prices are too high at the present time, so she decides to sell part of her land and deposit the money into an account that will pay 10%, compounded quarterly. She wants this account to yield $60,000 in six years to meet an expense she will have at that time. In dollar terms, how much land should she sell?

26. How long will it take $1000 to increase to $3000 in an account that pays 10%, compounded semiannually?

27. A sum of $2000 is invested into an account that pays 12%, compounded quarterly. It is to be left in the account until it has grown to $3500. How long will it take?

28. Find S for each of the following.

(a) $i = .02, n = 20$
(b) $i = .04, n = 20$
(c) $i = .05, n = 15$

Find the future value of each of the ordinary annuities in Exercises 29 through 32.

29. $R = \$200, i = .02, n = 10$

30. $R = \$500, i = .05, n = 8$

31. $R = \$1000, r = .06, n = 5$

32. $R = \$300, i = .025, n = 20$

33. Find the future value of $600 paid into an annuity at the end of every six months for five years. The annual interest rate is 10%.

34. Find the future value of $1000 paid into an annuity at the end of every year for six years. The interest rate is 8%.

35. $250 is invested in an annuitv at the end of every three months at 8%. Find the value of the annuity at the end of six years.

36. An annuity consists of payments of $1000 at the end of each year for a period of five years. Interest is paid at 6% per year, compounded annually. Determine the amount of the annuity at the end of five years.

37. A company has issued $400,000 in bonds to raise money for expansion. The bonds are payable in five years. A sinking fund earning interest of 7%, compounded annually, has been formed to meet this obligation. What will the annual payments into the sinking fund be?

38. A couple wants to start a fund with annual payments that will give them $20,000 cash on retirement in ten years. The proposed fund will give interest of 7%, compounded annually. What will the annual payments be?

In Exercises 39 through 42, determine the present value of the given amounts using the given interest rate and length of time.

39. $5000 over five years at 6%, compounded annually

40. $4750 over four years at 5%, compounded semiannually

41. $6000 over five years at 8%, compounded quarterly

42. $1000 over eight years at 10%, compounded semiannually

43. A company wants to deposit a certain sum at the present time into an account that pays compound interest at 8%, compounded quarterly, to meet an expected expense of $50,000 in five years time. How much should it deposit?

44. The Corporation is planning plant expansion as soon as adequate funds can be accumulated. The Corporation has estimated that the additions will cost approximately $105,000. At the present time, it has $70,935 cash on hand that will not be needed in the near future. A local savings institution will pay 8%, compounded semiannually. How many years will be required for the $70,935 to accumulate to approximately $105,000?

45. A medical supplier is making plans to issue $200,000 in bonds to finance plant modernization. The interest rate on the bonds will be 8%, compounded quarterly. It estimates that it can pay up to a maximum of $97,000 in total interest on the bonds.

The company wants to stretch out the period of payment as long as possible. What should be the planned term of the bonds?

46. The Corporation owed a $40,000 debt. The creditor agreed to let them pay the debt in five equal annual payments at 8% interest. Compute the annual payments.

47. A student borrowed $3000 from a credit union towards purchasing a car. The interest rate on such loans was 12%, compounded quarterly, with payments due every quarter. The student wants to pay off the loan in three years. Find the quarterly payments.

48. The interest on a house mortgage of $53,000 for 30 years is 14%, compounded monthly, with monthly payments. Compute the monthly payments.

49. A student takes a loan out for $4000. The interest rate is 8%, compounded annually; the loan is to be paid off in four years. Find the annual payments.

50. What would the semiannual payments be on a loan of $5000 at interest rate of 10%, compounded semiannually, for five years? Determine the total amount paid in five years.

51. A house mortgage is for $35,000 at 10%, compounded annually, for a period of 25 years. What would the annual payments be on such a mortgage? How much will have been paid at the end of the 25-year period?

52. A seller is willing to finance the purchase of land costing $15,000 with a loan at 12% interest, compounded quarterly, over a period of six years. What would the quarterly payments be? What will be the total amount paid for the land in this manner?

53. On January 1, a Corporation owed a debt of $160,000. The creditor agreed to let them pay the debt back in eight equal annual installments. Interest of 10%, compounded annually, was to be paid. Compute the annual payments.

54. A university purchases a new computer for $120,000. The payment is to be amortized over a period of five years at 8%, compounded semiannually. Compute the semiannual payments.

55. An investor invests $1000 at 8%, compounded quarterly, for five years. At the end of the five years, the total amount in the account is reinvested at 10%, compounded quarterly, for another five years. How much is in the account at the end of ten years?

56. A family borrows $65,000 at 12% for 30 years. Their house payments are $668.59 per month. How much of the first month's payment is interest? How much of the first month's payment is principal? What is the total amount they will pay over the 30 years?

9

Topics in this chapter can be applied to:

Contract Negotiations • Sales Strategies • Market Competition • Jamaican Fishing • City Planning • Agricultural Planning • Resource Allocations

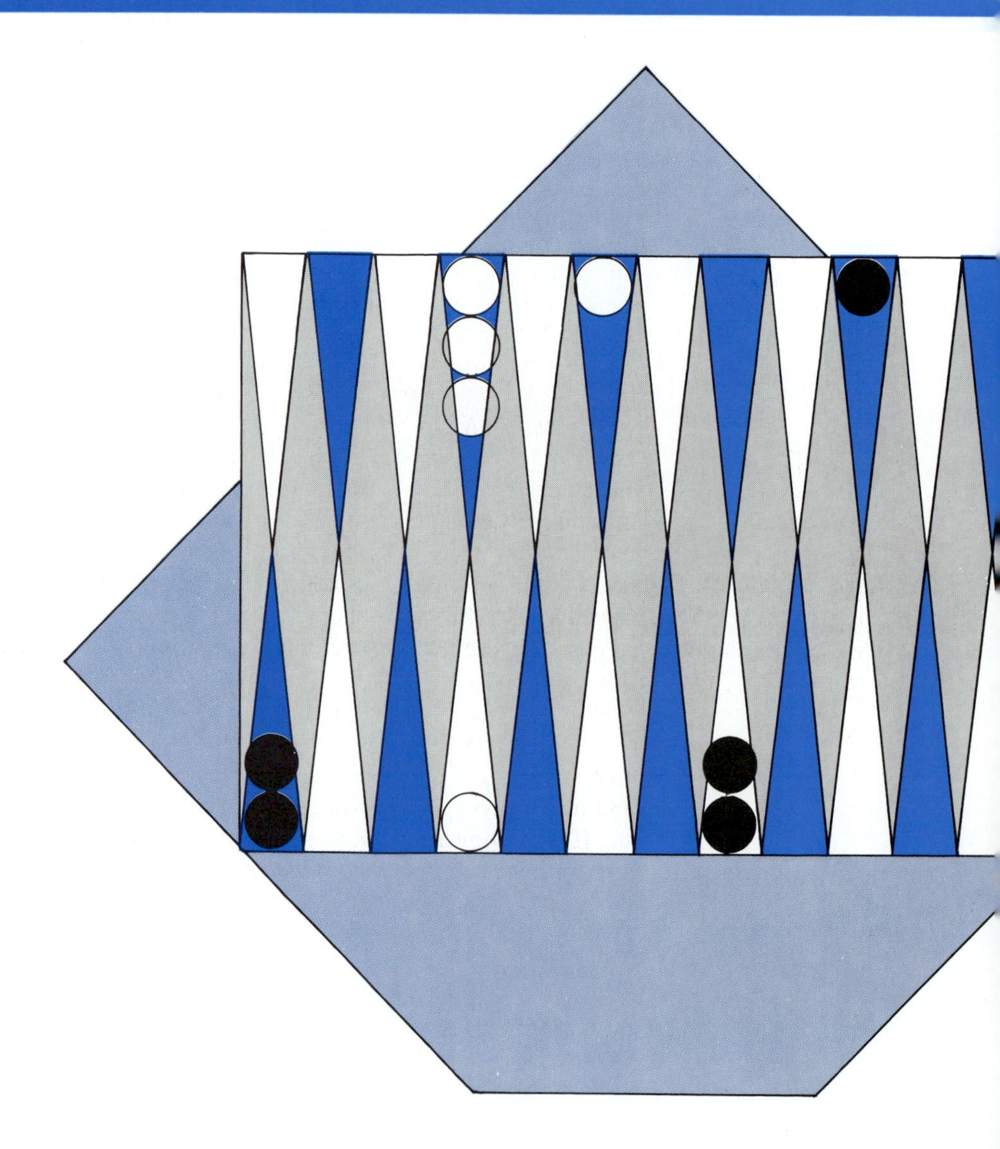

Game Theory

- **Strictly Determined Games**
- **Mixed Strategy Games**

People of all ages like to play games. The games may range from simple children's games (hide and seek), to games of luck (roulette), to games requiring special skills (baseball) or strategies (chess). In some games, each opponent tries to anticipate the other's actions and act accordingly. For example, the defense in a football game may anticipate a pass and set up a blitz while the offense, in turn, tries to anticipate the defense's strategy. The study of game theory is not restricted to our usual concept of a game. It analyzes strategies when the goals of "competing" parties conflict. Such a situation could arise when labor and management meet to discuss a contract, when airlines vie for a certain route, or when supermarkets compete against one another.

This chapter focuses on a simple situation with exactly two sides, or players, involved. These two players compete for a **payoff** that one player pays to the other. Let's begin with a simple example using a coin-matching game.

Two friends, Rob and Chad, play the following game. Each one has a coin and each decides which side he will turn up. They show the coins simultaneously and make payments according to the following.

If both show heads, Rob pays Chad 50¢. If both show tails, Rob pays Chad 25¢. If a head and a tail show, Chad pays Rob 35¢. Figure 9–1 shows these payoffs for each possible outcome.

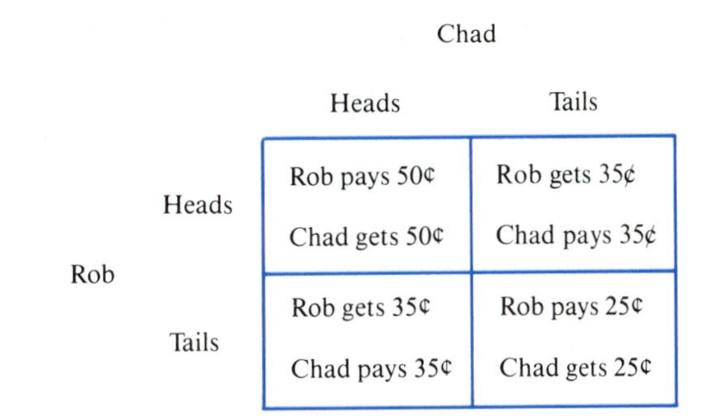

		Chad: Heads	Chad: Tails
Rob	Heads	Rob pays 50¢ Chad gets 50¢	Rob gets 35¢ Chad pays 35¢
	Tails	Rob gets 35¢ Chad pays 35¢	Rob pays 25¢ Chad gets 25¢

Figure 9–1

This game is called a **two-person** game because exactly two persons participate. Notice that the amount won by a player is exactly the same as the amount lost by the opponent. Whenever this is the case, we call the game a **zero-sum** game. We can represent the payoffs in a two-person zero-sum game by a payoff matrix. The payoffs indicated by Figure 9–1 can be simplified to

		Chad:	
		Heads	*Tails*
Rob:	*Heads*	−50¢	35¢
	Tails	35¢	−25¢

The matrix shows all possible payoffs for the way heads and tails are paired. An entry represents the amount Rob receives for that pair. Consequently, a negative entry indicates that Rob pays Chad.

A player's plan of action against his opponent is called a **strategy**. Game theory attempts to determine the best strategy so that each player will maximize his payoff. It is assumed that each player knows all strategies available to himself and to his opponent, but each player selects a strategy without the opponent knowing which strategy is selected. In this coin-matching game, the choice of a strategy simply amounts to selecting heads or tails. Notice that when Rob selects the strategy heads, he has selected the first row of the payoff matrix. Chad's selection of a strategy is equivalent to selecting a column of the matrix.

Example 1 (*Compare Exercise 1*)
Rob and Chad match quarters. When the coins match, Rob receives 25¢. When the coins differ, Chad receives 25¢. The payoff matrix is

		Chad:	
		H	*T*
Rob:	*H*	25¢	−25¢
	T	−25¢	25¢

Example 2 (*Compare Exercise 5*)
The payoff matrix for a game is

		C	
		c_1	c_2
R	r_1	14	−3
	r_2	−6	−5

(a) What are the possible payoffs to R if strategy r_1 is selected?
(b) What are the possible payoffs to C if strategy c_2 is selected?
(c) What is the payoff when R selects r_2, and C selects c_1?

Solution **(a)** R receives 14 if C selects c_1, and R pays 3 if C selects c_2.
(b) C receives 3 or 5, depending on whether R selects r_1 or r_2.
(c) C receives 6, and R pays 6.

9–1 Strictly Determined Games

We begin our analyses of the best strategy with some simple games that have fixed strategies.

Two players, R and C, play a game in which R can take two alternative actions (strategies), called r_1 and r_2, and C can take two actions, c_1 and c_2. Assume that each player chooses a strategy with no foreknowledge of the other's likely strategy. Since there are two possible strategies for R and two for C, a 2×2 payoff matrix represents the outcome, or payoff, corresponding to each pair of strategies selected. Here is an example of a payoff matrix:

$$C:\quad \begin{array}{ccc} & c_1 & c_2 \end{array}$$

$$R:\quad \begin{array}{c} r_1 \\ r_2 \end{array} \begin{bmatrix} 4 & -9 \\ 6 & 8 \end{bmatrix}$$

Following the convention that the entries represent the payoff to the row player, R, we interpret the matrix as follows: If R chooses strategy r_1, and C chooses c_1, then C pays 4 to R. If R chooses strategy r_1, and C chooses c_2, then R pays 9 to C, and so on. Notice that the sum of the amounts won by R and C is zero whatever strategies are selected. For example, if R selects r_2, and C selects c_1, then R wins 6 and C loses 6 (or wins -6), so the total is zero. Therefore, this is a zero-sum game.

R and C each wish to gain as much as they can (or lose as little as possible). Which strategies should they select? Let us analyze the situation. First, observe that for R to gain as much as possible, his strategy attempts to select the **maximum** entry. Since a gain for C is represented by a negative number, C attempts to select the **minimum** entry.

R should select r_2, since he then stands to gain the most, 6 or 8, regardless of C's strategy.

At first glance, it appears that C's strategy should be c_2 since that is his only chance of winning. However, we assume that both parties know all possible strategies, so C knows that R's best strategy is r_2. Thus, C expects R to choose r_2, so C can only choose between losing 8 or losing 6. C should select strategy

c_1 since he then risks giving the least away. Thus, using the strategies r_2 and c_1, c pays 6 to R. This entry of the matrix is called the **value** of the game; its location in the matrix is called a **saddle point**; and the pair of strategies leading to the value is called a **solution**. This game has a value of 6, the (2, 1) location is the saddle point, and the solution is the pair of strategies r_2 and c_1.

Let us use this example and analyze the general approach that determines whether a solution exists in a two-person game and how we can find a solution.

There may be several courses of action available to both players, and there need not be the same number for each.

The strategies available to persons R and C in a two-person game can be represented by a **payoff** matrix A:

$$
\begin{array}{cc}
 & C: \\
 & \begin{array}{cccc} c_1 & c_2 & \cdots & c_n \end{array} \\
R: \begin{array}{c} r_1 \\ r_2 \\ \\ r_m \end{array} & \begin{bmatrix} a_{11} & a_{12} & \cdots & a_{1n} \\ a_{21} & a_{22} & \cdots & a_{2n} \\ \vdots & \vdots & & \vdots \\ a_{m1} & a_{m2} & \cdots & a_{mn} \end{bmatrix} = A
\end{array}
$$

Note that the number of strategies available of R is m, and the number of strategies available to C is n. The entry a_{ij} represents the payoff that R receives when R adopts strategy r_i and C adopts c_j. The following are the conditions of the game:

1. Each player aims to choose the strategy that will enable him to obtain as large a payoff as possible (or to give as little as possible).
2. It is assumed that neither player has any prior knowledge of what strategy the other will adopt.
3. Each player makes his choice of strategy under the assumption that his opponent is an intelligent person adopting an equally rational approach to the game.

Let us first approach the game from the viewpoint of R who is trying to maximize winnings. He scans the rows of A trying to decide which strategy, r_1, $r_2, \ldots, r_m$, to adopt. For any given strategy (row), what is the least possible payoff R can expect? It is the minimum entry of that row. In the example, the minimum row entries are -9 for row 1 and 6 for row 2.

R is aware that he is playing an intelligent opponent who is aiming to hold R to a small payoff. When R chooses a row, he is guaranteed at least the minimum entry of that row. He then selects the row containing the largest of these minima. He is then guaranteed that payoff, and it is the largest payoff he can hope for against C's best counterstrategy.

We now look at the situation from C's viewpoint, who wants the minimum payoff. Recall that a gain for C is represented by a negative number, so C seeks the minimum entry. C scans the columns of A trying to decide which strategy, c_1, $c_2, \ldots c_n$, to adopt. He marks the maximum element in each column, his smallest possible gain (or largest possible loss) for that strategy. In the example given, this is 6 in column 1 and 8 in column 2. He then selects the strategy that has the smallest of those maxima. The smallest of these maxima represents C's greatest gain if it is negative and C's smallest loss if it is positive. This is the least payoff he need make against R's strategy.

Thus, the approaches that R and C should adopt are the following.

Strategy for a Two-Person Zero-Sum Game

R marks off the minimum element in each row and selects the row that has the largest of these minima.

C marks off the maximum element in each column and selects the column that has the smallest of these maxima.

If the largest of the row minima occurs in the same location as the smallest of the column maxima, that payoff is the **value** of the game.

The two strategies leading to the value, the **solution**, are the ones that should be adopted.

When the largest row minimum and the smallest column maximum are the same, the game is said to be **strictly determined**. The location of this element is the **saddle point** of the game. (There are games that are not strictly determined. We shall discuss strategies for such games in the following section.)

In the example, R selects the largest of the row minima, 6. C selects the smallest of the column maxima, 6. Since these are the same, the game is strictly determined, and the (2, 1) location is the saddle point of the game. This game is strictly determined because each player should always play the same strategy. If R wants to maximize his average earnings, he should always play r_2. If C wants to minimize his average payoff, he should always play c_1. It is in this sense that the game is strictly determined.

Example 3 *(Compare Exercise 6)*

The following is a payoff matrix representing a game involving two strategies for player R and three strategies for player C:

		C:		
		c_1	c_2	c_3
R:	r_1	4	3	5
	r_2	6	2	8

Determine the value of the game, if it exists.

Solution Mark a row minimum to the right of the row and a column maximum below the column. Then select the largest of the row minima and the smallest of the column maxima.

$$
\begin{array}{cc}
 & \begin{matrix} & & & \text{row minimum} \end{matrix} \\
 & \begin{bmatrix} 4 & 3 & 5 \\ 6 & 2 & 8 \end{bmatrix} \begin{matrix} \textcircled{3} \text{ maximum} \\ 2 \end{matrix} \\
\text{column maximum} & \begin{matrix} 6 & \textcircled{3} & 8 \end{matrix} \\
 & \text{minimum}
\end{array}
$$

The largest of the minima occurs in the location of the smallest of the maxima, so the game is strictly determined with a value of 3. The saddle point is located at (1, 2). The solution consists of strategies r_1 and c_2.

Thus, if R and C adopt the strategies that they should, namely, r_1 and c_2, then R will always receive 3 from C.

Perhaps you think that this game is unfair; C has no chance of winning. The best he can do is to lose 3 each time the game is played. Unfortunately, a game like this may correspond to a real situation. Sometimes a businessman may have several options in his business, but economic conditions are such that each option results in a financial loss. His best strategy is to minimize losses.

It is instructive to observe the outcome when one player strays from this rule while the other keeps to it. Suppose that R decides to go for strategy r_2 in the hope of winning 6 or 8. C stays with the rule, selecting strategy c_2, and the result is a payoff of 2 rather than 3 for R.

The following example illustrates the possibility of two-person games with more than one saddle point.

Example 4 (*Compare Exercise 12*)
Consider the following game:

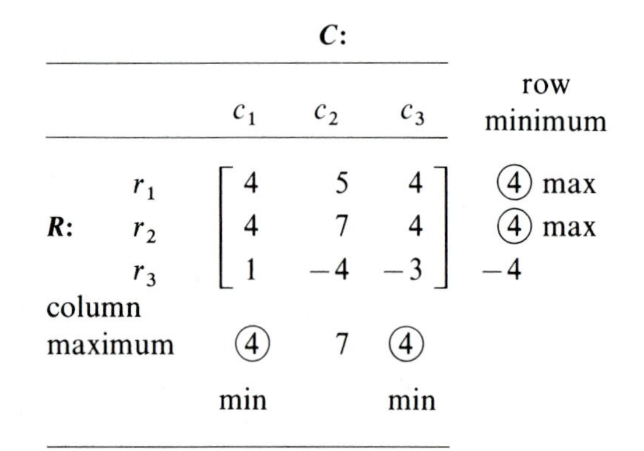

		C: c_1	c_2	c_3	row minimum
R:	r_1	4	5	4	(4) max
	r_2	4	7	4	(4) max
	r_3	1	−4	−3	−4
column maximum		(4) min	7	(4) min	

Each of the 4's in the above matrix is a saddle point. (Check to verify this.) In this game, R has a choice of best strategies r_1 or r_2, and C has a choice of best strategies c_1 or c_3. Note that whatever pair of best strategies is selected, the payoff will be 4.

Whenever there is more than one saddle point to a game, the elements in each of these locations are identical, each being the value of the game.

Now we present a game that has no saddle point to illustrate a game that is not strictly determined.

Example 5 (*Compare Exercise 13*)
A game has the following payoff matrix.

		C:		
		c_1	c_2	row minimum
R:	r_1	4	6	(4) maximum
	r_2	8	3	3
column maximum		8	(6) minimum	

In the rows, the largest of the minima is 4. The smallest of the maxima in the columns is 6. Since these appear in different locations, there is no saddle point, and the game is not strictly determined.

We discuss the approaches that R and C should take to this game in the next section.

Example 6 (*Compare Exercise 18*)
Management representatives of a company meet with union representatives to work out a contract for the following year. The following model enabled the union to come up with a strategy of cooperation rather than aggression:

		Management:	
		Cooperation	*Inflexibility*
Union:	*Cooperation*	10%	12%
	Aggression	8%	15%

The entries in the above matrix correspond to percentage wage increases forecast by an union advisory group based on predictions of various approaches by union and management. The union was aware that the management would also be advised on the approach that they should take in the talks, and that they would come up with a similar model. Their model might vary slightly in the various matrix entries, but the relative sizes of the elements within each matrix would be similar.

According to the above model, what strategy should the union take, and what strategy does it expect management to take?

The largest of the row minima is 10%. This is also the smallest of the column maxima. Thus, the value of the game is 10%. The union should adopt a strategy of cooperation. It can expect that management will do likewise. The result: a 10% wage increase.

Let us examine what would happen if other strategies were selected. Suppose that the union decided to adopt a strategy of aggression, hoping for the 15% increase. If management adopted the cooperative attitude (as it should), the final result of the negotiations would be an 8% wage increase. If the initial attitude of cooperation by management turns into a hardened attitude in the face of aggressiveness by the union, the result, in this case, is a 15% increase.

Game theory can be used in this manner to indicate how participants should behave in certain situations to achieve maximum mutual benefit; it does not, of course, indicate how the participants will behave in reality! If both sides decide to gamble rather than adopt solution strategies, then the whole analysis breaks down, and one side will gain or lose more than the value.

Example 7 (*Compare Exercise 14*)
Consider the strategies of the offense and defense of two opposing football teams. Let the offense have three strategies under consideration, r_1, r_2, r_3; and let the defense have three strategies, c_1, c_2, c_3. The payoff matrix is

		Defense:		
		c_1	c_2	c_3
	r_1	40	−5	20
Offense:	r_2	6	4	5
	r_3	−4	3	15

The elements represent yards gained by the offense. Thus, for example, if the offense chooses strategy r_3, and the defense chooses strategy c_2, the offense will gain 3 yards.

It can be seen that the largest of the minima of the rows occurs in the location of the smallest of the maxima, namely, in the (r_2, c_2) location. Hence, this is the saddle point of the game, so the offense should choose strategy r_2 and the defense, strategy c_2. The offense will then gain 4 yards.

Let us examine what could happen if other strategies were selected. Suppose that the offense gambles on strategy r_1, hoping for a long gain of 20 or 40 yards with "the bomb." If the defense plays safe (as it should) by choosing solution strategy c_2, then the offense will be thrown for a loss of 5 yards.

9–1 Exercises

1. (*See Example 1*) Write the payoff matrix for the coin-matching game between two players. If both coins are heads, R pays C \$1. If both coins are tails, C pays R \$2. If the coins differ, R pays C 50¢.

2. Two players, R and C, each have two cards. R has one black card with the number 5 written on it and one red card with the number 3. C has a black card with a 4 written on it and a red card with a 2. They each select one of their cards and simultaneously show the cards. If the cards are the same color, R gets, in dollars, the difference of the two numbers shown. If the cards are different colors, C gets, in dollars, the smaller of the two numbers shown. Write the payoff matrix of this game.

3. Two-finger Morra is a game in which two players each hold up one or two fingers. The payoff, in dollars, is the total number of fingers shown. R receives the payoff if the total is even, and C receives the payoff if the total is odd. Write the payoff matrix.

4. Player R has a \$1 bill, a \$10 bill, and a \$50 bill. Player C has a \$5 bill and a \$20 bill. They simultaneously select one of their bills at random. The one with the larger bill collects the bill shown by the other. Write the payoff matrix for this game.

5. (*See Example 2*) Given the following payoff matrix

		C:		
		c_1	c_2	c_3
	r_1	5	−3	4
R:	r_2	0	10	−7
	r_3	−8	4	0

(a) What is the payoff when strategies r_2 and c_2 are selected?

(b) What is the largest gain possible for the row player?

(c) What is the greatest loss possible for the column player?

(d) What is the largest gain possible for the column player?

Exercises 6 through 11 are payoff matrices for two-person games. Decide whether the games are strictly determined. Find the saddle point, value, and solution for each strictly determined game. (*See Example 3*)

6. $\begin{bmatrix} 2 & -2 \\ 3 & 4 \end{bmatrix}$

7. $\begin{bmatrix} 0 & 1 \\ -1 & 2 \end{bmatrix}$

8. $\begin{bmatrix} 1 & 2 \\ 4 & -3 \end{bmatrix}$

9. $\begin{bmatrix} 3 & 1 \\ -1 & 2 \end{bmatrix}$

10. $\begin{bmatrix} -2 & 0 \\ 3 & 1 \end{bmatrix}$

11. $\begin{bmatrix} 0 & -3 \\ 3 & 0 \end{bmatrix}$

12. (*See Example 4*) Find the saddle points and values of the payoff matrices.

(a) $\begin{bmatrix} 3 & 3 & 8 \\ 1 & -2 & -3 \\ 3 & 3 & 9 \end{bmatrix}$ (b) $\begin{bmatrix} 6 & 7 & 6 & 8 \\ 3 & 6 & 5 & 12 \\ 6 & 9 & 6 & 11 \end{bmatrix}$

13. (*See Example 5*) The following are 3×3 payoff matrices for games. Determine the saddle points, values, and solutions, if they exist.

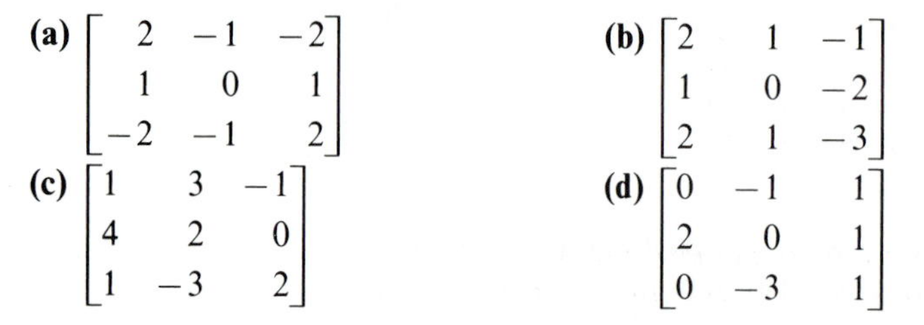

(a) $\begin{bmatrix} 2 & -1 & -2 \\ 1 & 0 & 1 \\ -2 & -1 & 2 \end{bmatrix}$ (b) $\begin{bmatrix} 2 & 1 & -1 \\ 1 & 0 & -2 \\ 2 & 1 & -3 \end{bmatrix}$

(c) $\begin{bmatrix} 1 & 3 & -1 \\ 4 & 2 & 0 \\ 1 & -3 & 2 \end{bmatrix}$ (d) $\begin{bmatrix} 0 & -1 & 1 \\ 2 & 0 & 1 \\ 0 & -3 & 1 \end{bmatrix}$

14. (*See Example 7*) The following are payoff matrices that describe possible strategies of the offense and the defense in a football game. Determine whether solutions exist, and discuss the interpretations of possible conflicting strategies.

Defense

(a) *Offense* $\begin{bmatrix} 5 & 0 & 1 \\ 50 & 20 & -4 \\ 4 & 5 & 3 \end{bmatrix}$

Defense

(b) *Offense* $\begin{bmatrix} 25 & -1 & 2 & 8 \\ 7 & 3 & -2 & 1 \\ 3 & 7 & 7 & -8 \end{bmatrix}$

15. R and C are two companies that market certain competing products. Each has sales options it can use; the results of these strategies are represented by the following payoff matrix:

		C:	
		c_1	c_2
R:	r_1	120	140
	r_2	80	60

The elements of the matrix represent sales (in units of one thousand) that could be gained by R, the major company, at the expense of its rival C. Which strategy should each adopt?

16. A and B are airlines that directly compete with one another on two routes. Because of the different facilities available and other factors, there are various strategies each can adopt in competing for passengers on these routes. The strategies available are represented by the following payoff matrix:

		B:	
		b_1	b_2
A:	a_1	30	−20
	a_2	40	60

The elements in the matrix represent numbers of passengers per year in units of one hundred. A is established on the routes; B is the newcomer attempting to break in. Which strategy should each airline adopt?

17. A and B are opposing sides in armed conflict. A has two strategies open to it, a_1 and a_2. B has three, b_1, b_2, and b_3. The following is the payoff matrix:

		B:		
		b_1	b_2	b_3
A:	a_1	30	−20	40
	a_2	40	10	60

The elements represent hundreds of square miles gained by A. Which strategy should each side adopt?

18. (*See Example 6*) The management of a company and a major union are involved in contract negotiations. The results of various initial approaches by the two sides are described by the following payoff matrix, where the entries are pay increases:

		Management:		
		Accommodating	*Flexible*	*Inflexible*
Union:	*Accommodating*	9%	7%	6%
	Flexible but Tough	10%	9%	8%
	Inflexible	13%	12%	7%

Which strategy should each adopt?

19. Two companies are competing against one another in the purchase of land. In the competition, each has three available options that would result in

various land deals. The following is the payoff matrix:

$$C:\quad \begin{array}{c} \\ R: \end{array}\begin{array}{c} \\ r_1 \\ r_2 \\ r_3 \end{array}\begin{array}{c} \begin{array}{ccc} c_1 & c_2 & c_3 \end{array} \\ \begin{bmatrix} 4 & -2 & 2 \\ -1 & 5 & 2 \\ 5 & 4 & 3 \end{bmatrix} \end{array}$$

The elements represent hundreds of acres of land purchased by R at the expense of C. Which strategy should each company adopt?

9–2 Mixed Strategy Games

Games Not Strictly Determined

We now look at two-person games that are **not strictly determined**. Consider the following game, which was discussed in Example 5 of the previous section:

$$C:\quad R:\ \begin{array}{c} \\ r_1 \\ r_2 \end{array}\begin{array}{c} \begin{array}{cc} c_1 & c_2 \end{array} \\ \begin{bmatrix} 4 & 6 \\ 8 & 3 \end{bmatrix} \end{array}$$

In the rows, the largest of the minima is 4. The smallest of the maxima in the columns is 6. This game has no saddle point.

Two-person games are often of a repetitive nature. Thus, R may select strategy r_1 part of the time and r_2 the remainder of the time; C may select c_1 part of the time and c_2 the remainder of the time. Such games are called **mixed strategy games**. It is important in these games that each person conceal his or her strategy prior to commencing any game; otherwise, the other player can take advantage of the knowledge. In this example, if R decides on strategy r_2, and C becomes aware of this fact beforehand, he will choose strategy c_2 and thereby minimize R's winnings. Thus, R and C should select strategies at random to conceal their intent from the opponent.

The challenge confronting R in this game is to control in some way the randomness of his selection of strategies to maximize his average payoff received from C over the long run against C's counterstrategies. Conversely, C wants to control the randomness of his strategies to minimize long-term payoff received by R. Probability theory is the tool used to control the randomness of both player's strategies.

Consider a general two-person game of this type with the following payoff matrix:

$$
\begin{array}{cc} & C: \\ & \begin{array}{cc} c_1 & c_2 \end{array} \\ R: \begin{array}{c} r_1 \\ r_2 \end{array} & \begin{bmatrix} a_{11} & a_{12} \\ a_{21} & a_{22} \end{bmatrix} \end{array}
$$

Let the probability that player R will adopt strategy r_1 be p_1; and let the probability of his adopting r_2 be p_2. Let the probabilities of the strategies of C be q_1 and q_2. Thus, over many games, R would adopt strategy r_1 the fraction p_1 of the time and r_2 the fraction p_2 of the time, the immediate selection at any time being made in a random manner. The probability of R's winning a_{11} is p_1q_1; the probability of his winning a_{12} is p_1q_2; and so on.

When the strategies are chosen randomly according to these probabilities, we have the **expected payoff** of the game:

$$\text{Expected payoff} = p_1q_1a_{11} + p_1q_2a_{12} + p_2q_1a_{21} + p_2q_2a_{22}$$

(**Notice** that this is the expected value from Section 7–5.) This represents an average payoff that R can expect when the game is played many times, assuming that R and C base their strategies on these probabilities.

Let us write strategy probabilities of R and C as matrices:

$$P = [p_1 \quad p_2] \qquad \text{and} \qquad Q = \begin{bmatrix} q_1 \\ q_2 \end{bmatrix}$$

The expected payoff associated with these probabilities, $E(P, Q)$, can be represented in matrix form as

$$E(P, Q) = PAQ$$

where A is the payoff matrix. P and Q are said to be the **strategies adopted** by R and C.

Example 1 (*Compare Exercise 1*)

Lets' return to the mixed strategy game at the beginning of the section. The payoff matrix is

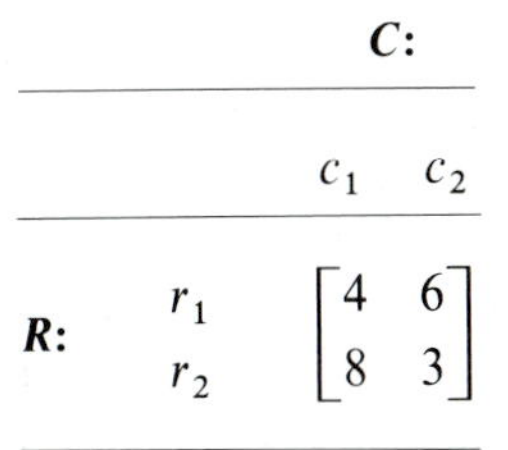

$$
\begin{array}{cc} & C: \\ & \begin{array}{cc} c_1 & c_2 \end{array} \\ R: \begin{array}{c} r_1 \\ r_2 \end{array} & \begin{bmatrix} 4 & 6 \\ 8 & 3 \end{bmatrix} \end{array}
$$

Suppose that the strategies of R is $[\frac{1}{2} \quad \frac{1}{2}]$ (r_1 is chosen $\frac{1}{2}$ the time and r_2 the other half), and the strategy of C is $[\frac{1}{3} \quad \frac{2}{3}]$ (c_1 is chosen $\frac{1}{3}$ of the time and c_2, $\frac{2}{3}$). Then, the expected payoff of the game is

$$\begin{aligned} E &= [\tfrac{1}{2} \quad \tfrac{1}{2}]\begin{bmatrix} 4 & 6 \\ 8 & 3 \end{bmatrix}\begin{bmatrix} \frac{1}{3} \\ \frac{2}{3} \end{bmatrix} \\ &= [6 \quad \tfrac{9}{2}]\begin{bmatrix} \frac{1}{3} \\ \frac{2}{3} \end{bmatrix} \\ &= 5 \end{aligned}$$

Thus, if the game is played many times, and both players select their strategies at random, with R playing evenly and C using strategy c_2 twice as often as c_1, R can expect his winnings to average 5. If different probabilities are used to vary strategies, the expected payoff of the game will naturally vary.

In the case of the strictly determined game, there was a solution that was the optimal strategy for both players. Likewise, in a mixed strategy game, R wants to select the strategy P that will give him the largest possible payoff against C's best counterstrategy. This strategy is called the **optimal strategy for player R.**

C wants to select the strategy Q that will give him the largest possible payoff against R's best strategy; this strategy is called the **optimal strategy for player C.**

These optimal strategies have been determined for two-person games involving 2×2 matrices. It is the following.

If the payoff matrix is

$$\begin{array}{cc} & C \\ \hline R & \begin{bmatrix} a_{11} & a_{12} \\ a_{21} & a_{22} \end{bmatrix} \\ \hline \end{array}$$

then the optimal strategy that should be followed by player R is $P = [p_1 \quad p_2]$, where

$$p_1 = \frac{a_{22} - a_{21}}{a_{11} + a_{22} - a_{12} - a_{21}} \qquad \text{and} \qquad p_2 = 1 - p_1$$

(It can be shown that $a_{11} + a_{22} - a_{12} - a_{21}$ is nonzero in a game that is not strictly determined.)

The optimal strategy of player C is $Q = \begin{bmatrix} q_1 \\ q_2 \end{bmatrix}$, where

$$q_1 = \frac{a_{22} - a_{12}}{a_{11} + a_{22} - a_{12} - a_{21}} \qquad \text{and} \qquad q_2 = 1 - q_1$$

The expected payoff of the game is the value of the game for these optimal strategies:

$$E = PAQ = \frac{a_{11}a_{22} - a_{12}a_{21}}{a_{11} + a_{22} - a_{12} - a_{21}}$$

Example 2 (*Compare Exercise 6*)
Let us determine the optimal strategies and value of the game that we have been discussing in this section. The payoff matrix is

$$\begin{bmatrix} 4 & 6 \\ 8 & 3 \end{bmatrix}$$

Thus,

$$p_1 = \frac{3 - 8}{4 + 3 - 8 - 6} = \frac{5}{7} \qquad \text{and} \qquad p_2 = \frac{2}{7}$$

$$q_1 = \frac{3 - 6}{-7} = \frac{3}{7} \qquad \text{and} \qquad q_2 = \frac{4}{7}$$

The strategy of player R should be $[\frac{5}{7} \quad \frac{2}{7}]$, and that of C should be $[\frac{3}{7} \quad \frac{4}{7}]$. The value of the game is

$$E = \frac{(4 \times 3) - (6 \times 8)}{-7} = \frac{36}{7} = 5.14$$

Thus, over many games, R can expect to win approximately 5.14 per game if both players follow these optimal strategies.

Warning! If a game is strictly determined, the above method of computing the value is not correct.

Example 3 The payoff matrix

$$\begin{bmatrix} 4 & 6 \\ 2 & -5 \end{bmatrix}$$

is strictly determined with value 4. The row player's strategy is $[1 \quad 0]$ (always play r_1), and the column player's strategy is $[1 \quad 0]$. Notice that the formulas for mixed strategy give

$$E = \frac{32}{9}, \qquad R \text{ strategy} = \begin{bmatrix} \frac{7}{9} & \frac{2}{9} \end{bmatrix}, \qquad C \text{ strategy} = \begin{bmatrix} \frac{11}{9} & -\frac{2}{9} \end{bmatrix}$$

These are in error and are given only to emphasize that you must be sure that a game is not strictly determined before applying the formulas.

The next example illustrates a payoff matrix that can be reduced to a simpler matrix.

Example 4 *(Compare Exercise 11)*
The following football situation involves three strategies for both offense and defense:

		Defense:		
		y_1	y_2	y_3
Offense:	x_1	30	10	−3
	x_2	−6	4	−2
	x_3	2	6	5

The numbers again represent yards gained by the offense. The game is not strictly determined.

Notice that every entry in row 3 is larger than the corresponding entry in row 2. This indicates that the offense will always do better with strategy x_3 than with x_2, no matter what counterstrategy the defense adopts.

Also notice that the defense should never adopt strategy y_2 because y_3 will give up fewer yards no matter what offensive strategy is used.

In a situation like this, we say that x_3 **dominates** x_2, and y_3 **dominates** y_2.

Definition

Row i of a payoff matrix **dominates** row j if every entry of row i is greater than or equal to the corresponding entry of row j.

Column i of a payoff matrix **dominates** column j if every entry of column i is less than or equal to the corresponding entry of column j.

Whenever row i dominates row j, then row j may be removed from the payoff matrix, and the analysis is not affected.

When column i dominates column j, then column j may be removed from the payoff matrix, and the analysis is not affected.

The game of the example thus reduces to the following payoff matrix.

		Defense	
		y_1	y_3
Offense:	x_1	30	−3
	x_3	2	5

The strategy of the offense should be $P = [p_1 \quad 0 \quad p_3]$, where

$$p_1 = \frac{5-2}{30+5+3-2} = \frac{3}{36} \qquad \text{and} \qquad p_3 = \frac{33}{36}$$

The zero is in P because x_2 should never be used. The strategy of defense should be $Q = \begin{bmatrix} q_1 \\ 0 \\ q_3 \end{bmatrix}$, where

$$q_1 = \frac{5+3}{36} = \frac{8}{36} \quad \text{and} \quad q_3 = \frac{28}{36}$$

$q_2 = 0$ because strategy y_2 is never used.

The value of the game is

$$\frac{(30 \times 5) - (-3 \times 2)}{36} = \frac{156}{36} = 4.33 \text{ yards}$$

Thus, if the offense and defense follow the above strategies, as they should, the offense will gain, on an average, approximately 4.33 yards per play.

Notice that the deletion of strategies dominated by others allowed us to reduce the payoff matrix to a 2 × 2 payoff matrix. We could then find the optimal solution using the technique for 2 × 2 matrices.

Example 5 (*Compare Exercise 12*)
Reduce the following payoff matrix by deleting dominant rows and/or columns.

$$\begin{bmatrix} 12 & 4 & -2 & 7 \\ 3 & 1 & -3 & 6 \\ -1 & 5 & 0 & 9 \end{bmatrix}$$

Solution Since row 1 dominates row 2, row 2 can be deleted, giving

$$\begin{bmatrix} 12 & 4 & -2 & 7 \\ -1 & 5 & 0 & 9 \end{bmatrix}$$

Column 3 dominates both column 2 and 4, so both can be deleted, giving

$$\begin{bmatrix} 12 & -2 \\ -1 & 0 \end{bmatrix}$$

Fair Games

If a game is not strictly determined and it's value is zero, it is called **fair**. This means that the long-term average gain of each player is zero, that is, the losses of each player equal his gains, so his net gain is zero.

Example 6 (*Compare Exercise 14*)
The game

$$\begin{bmatrix} 3 & -1 \\ 2 & 4 \end{bmatrix}$$

is not strictly determined, and its value is

$$\frac{3(4) - 2(-1)}{3 + 4 + 1 - 2} = \frac{14}{6}$$

so it is not a **fair** game.

The game

$$\begin{bmatrix} 2 & -4 \\ -3 & 6 \end{bmatrix}$$

is not strictly determined, and its value is

$$\frac{2(6) - (-3)(-4)}{2 + 6 + 4 + 3} = \frac{0}{15} = 0$$

so this game is fair.

Generally, a strictly determined game is not fair.

Jamaican Fishing

Example 7 This example from anthropology describes strategies adopted by fishermen in Jamaica. It involves a fishing village of about 200 persons, located on the south shore of Jamaica. The male villagers are all fishermen, and they make their living exclusively by fishing in the local waters and selling their catches to the general marketing system of the island. The fishing grounds that they use extend out from shore to the 100-fathom line, about 22 miles, and comprise about 168 square miles. Twenty-six fishing crews, in sailing-dugout canoes, fish this area by fish pots, which are drawn and reset, weather and sea permitting, on three regular fishing days each week. The fishing crews are composed of a captain, who owns the canoe, and two or three crewmen.

The fishing grounds are divided into inside and outside banks. The inside banks lie from 5 to 15 miles offshore, while the outside banks all lie beyond. However, the distinction between inside and outside banks is made not only on the basis of their distance from shore but, more importantly, on the basis of the strengths of the currents that flow across them. The inside banks are almost fully protected from the currents, and the sea never gets as rough as it does on the outside banks. The currents in the outside banks are not related in any apparent way to weather and sea conditions. The fishermen cannot predict when the current will come or go; nor can they tell from shore whether the current is running. The currents in the outside banks lead to loss of fishing pots and loss of fish. However, higher quality grades of fish, in both variety and size, are caught on the outside banks. Generally, larger catches are obtained on the outside banks. These differences between the current-ridden outside banks and the safe inside banks lead to a choice situation for the captains as to where they will fish. The choices are as follows:

1. all pots set inside,

2. some pots set inside and some outside,
3. all pots set outside.

Actually, no fishermen choose option 3, for it is entirely too risky. We shall ask game theory to predict the proportions of captains that choose each of strategies 1 and 2.

In this game, the village as a whole (represented by their captains) is one team of players, while the other team consists of all the uncontrollable forces of the environment that affect the fishing outcome. The following payoff matrix is based on income estimates of fishing the inside and outside banks.

		Nature:	
		Current	**No Current**
Village Captains:	**Inside**	17.3	11.5
	In-out	5.2	17.0

The entries represent net incomes of captains over an average month, with the units in Jamaican pounds. The matrix is interpreted as follows: The two players are the village captains and nature. The captains select from two strategies: to fish the inside banks only (called **inside**); or to fish both the inside banks and the outside banks (called **in-out**). Nature has two strategies: current or no current. Thus, for example, the above matrix tells us that if a captain elects to fish in-out, and there is no current, his net monthly income will be £17.

Observe that this matrix describes a mixed strategy game. Let us compute the optimal strategy for the fishermen:

$$p_1 = \frac{a_{22} - a_{21}}{a_{11} + a_{22} - a_{12} - a_{21}} = \frac{17 - 5.2}{17 + 17.3 - 11.5 - 5.2} = .67$$

$$p_2 = 1 - .67 = .33$$

Thus, the optimal strategy for the fishermen is [.67 .33]. This strategy tells us that the fishermen should use the inside fishing strategy 67% of the time and the in-out strategy 33% of the time. Observations of actual fishing practices found an average of 18 out of the 26 captains, or 60%, were fishing the inside banks only. The game theory closely describes the actual situation.

Although the proportion of captains choosing each of the alternatives seems to remain fairly constant, the persons choosing each strategy change from time to time. Captains and crewmen fishing the in-out alternative frequently get disheartened by their constant losses from the current and shift their gear to the inside banks. Inside fishermen sometimes decide to gamble a bit, and they shift some of their gear to the in-out strategy in hopes of making more money.

What does this payoff matrix tell about the behavior of nature? No accurate statistics were available for days with and days without current. According to the theory, the optimal strategy of nature is

$$q_1 = \frac{a_{22} - a_{12}}{a_{11} + a_{22} - a_{12} - a_{21}} = \frac{17 - 11.5}{17 + 17.3 - 11.5 - 5.2} = .31$$

$$q_2 = 1 - .31 = .69$$

The optimal strategy is $[.31 \quad .69]$ or $[\frac{31}{100} \quad \frac{69}{100}]$. Thus, the model predicts that current flows in the outside fishing banks 31% of the time.

9–2 Exercises

I.

1. (*See Example 1*) Find the expected payoff using the payoff matrix

$$\begin{array}{c} \quad C \\ R \begin{bmatrix} 4 & 8 \\ 12 & 3 \end{bmatrix} = A \end{array}$$

and $[\frac{1}{2} \quad \frac{1}{2}]$ for the strategy of R, and $[\frac{1}{4} \quad \frac{3}{4}]$ for the strategy of C.

2. Find the expected payoff for the payoff matrix

$$\begin{bmatrix} 3 & -1 \\ -2 & 4 \end{bmatrix} = A$$

using the row strategy $[.3 \quad .7]$ and the column strategy $[.6 \quad .4]$.

3. Find the expected payoff for the payoff matrix

$$\begin{bmatrix} 3 & -6 \\ -2 & 4 \end{bmatrix} = A$$

using the row strategy $[\frac{2}{5} \quad \frac{3}{5}]$ and the column strategy $[\frac{2}{3} \quad \frac{1}{3}]$.

4. If the strategy of R is $[\frac{1}{2} \quad \frac{1}{2} \quad 0]$, that of C is $[\frac{1}{2} \quad 0 \quad \frac{1}{2}]$, and the payoff matrix is

$$\begin{array}{c} \quad C \\ R \begin{bmatrix} 12 & 6 & -4 \\ 5 & 14 & 10 \\ 15 & 0 & -2 \end{bmatrix} \end{array}$$

find the expected payoff.

5. Determine the expected payoff using the following strategies with the matrix game of Exercise 4.

(a) $R\,[1 \quad 0 \quad 0]$, $C\,[0 \quad 0 \quad 1]$
(b) $R\,[\frac{1}{2} \quad \frac{1}{2} \quad 0]$, $C\,[0 \quad \frac{1}{2} \quad \frac{1}{2}]$
(c) $R\,[0 \quad \frac{1}{2} \quad \frac{1}{2}]$, $C\,[\frac{1}{2} \quad \frac{1}{2} \quad 0]$
(d) $R\,[0 \quad 0 \quad 1]$, $C\,[1 \quad 0 \quad 0]$
(e) $R\,[.3 \quad .5 \quad .2]$, $C\,[.2 \quad .2 \quad .6]$
(f) $R\,[.1 \quad .7 \quad .2]$, $C\,[.3 \quad .2 \quad .5]$

The payoff matrices in Exercises 6 through 10 describe mixed strategy games. Determine the optimal strategies for each game and the resulting value. (*See Example 2*)

6. $\begin{bmatrix} 5 & 7 \\ 9 & 2 \end{bmatrix}$ **7.** $\begin{bmatrix} 6 & 4 \\ -3 & 8 \end{bmatrix}$ **8.** $\begin{bmatrix} 8 & -2 \\ -4 & 10 \end{bmatrix}$

9. $\begin{bmatrix} 6 & 4 & 4 \\ 4 & 5 & 2 \\ 5 & 6 & 8 \end{bmatrix}$ **10.** $\begin{bmatrix} 4 & 6 & 5 \\ 6 & -1 & 7 \\ 8 & 2 & 9 \end{bmatrix}$

11. (*See Example 4*) Consider the following football game involving three strategies for both offense and defense:

		Defense:		
		c_1	c_2	c_3
	r_1	4	6	−2
Offense:	r_2	6	8	3
	r_3	−4	2	10

How should the sides mix their strategies? What is the value of the game?

12. (*See Example 5*) Reduce the following matrices by deleting rows or columns that are dominated by other rows or columns.

(a) $\begin{bmatrix} 6 & -5 & 1 & 9 \\ 2 & 0 & 4 & 7 \\ -3 & -6 & 0 & 3 \end{bmatrix}$ **(b)** $\begin{bmatrix} 5 & 2 & -2 & 8 \\ 3 & 1 & 13 & 22 \\ 1 & 3 & 6 & 11 \end{bmatrix}$

13. Reduce the following matrices by deleting rows or columns that are dominated by other rows or columns.

(a) $\begin{bmatrix} 6 & 2 & 3 \\ 2 & 0 & 4 \\ 7 & 2 & 1 \\ -3 & -8 & 3 \end{bmatrix}$ **(b)** $\begin{bmatrix} 6 & 2 & -2 & 1 \\ 4 & 1 & 9 & 2 \\ 2 & 4 & 6 & 8 \\ 1 & 3 & 6 & 1 \end{bmatrix}$

14. (*See Example 6*) Determine whether the following games are fair.

(a) $\begin{bmatrix} 6 & -8 \\ -3 & 4 \end{bmatrix}$ **(b)** $\begin{bmatrix} 5 & 2 \\ 1 & 3 \end{bmatrix}$ **(c)** $\begin{bmatrix} 3 & -9 \\ 2 & 6 \end{bmatrix}$

15. Determine whether the following games are fair.

(a) $\begin{bmatrix} 3 & 2 \\ -1 & 5 \end{bmatrix}$ **(b)** $\begin{bmatrix} 5 & -10 \\ -2 & 4 \end{bmatrix}$ **(c)** $\begin{bmatrix} 7 & 8 \\ 6 & 2 \end{bmatrix}$

II.

16. Farmers in central Ghana can grow maize and hill rice. Crops are affected by the weather: Maize grows best in wet weather, hill rice grows best in dry weather. Variation in crop incomes with weather can be represented by the following payoff matrix:

	Wet Years	*Dry Years*
Maize	61	49
Hill Rice	30	71

where the entries represent hundreds of pounds per acre. Determine the relative percentages of maize and hill rice that the farmers should plant each year.

17. A restaurant owner has two types of clientele, tourists and local people. The size of each group is, to a great extent, determined by the weather. The following payoff matrix describes the number of expected customers:

	Wet Week	*Dry Week*
Tourists	50	200
Locals	150	100

Determine the percentage of tourists and locals that the restauranteur should plan for.

18. A farmer in north central Florida can grow citrus trees and fern plants. The ferns fare better than citrus during a cold winter. The strategies available to the farmer are described by the following matrix

	Cold Winter	*Warm Winter*
Citrus	160	240
Fern	210	180

where the payoff entries are thousands of dollars income. Determine the relative percentages of resources that the farmer should devote to citrus and ferns.

19. A community must decide on the relative emphases to give to industry and tourism. The following payoff matrix describes the strategies available:

	Wet Year	*Dry Year*
Industry	100	90
Tourism	60	150

Determine the relative emphases, in terms of percentages, that should be devoted to industry and tourism.

20. A farmer breeds two types of cattle for meat, A and B. The demand for A is great in times of high inflation, and the demand for B is high during normal inflation. The demands are described by the following matrix:

	High Inflation	*Normal Inflation*
A	300	150
B	180	220

where the payoff entries represent monthly demand in hundreds of pounds of beef. What percentage of each breed should the farmer raise?

21. An automobile company must make long-term plans concerning the relative resources that it is going to devote to economy cars and standard cars. The demands for these cars are related to inflation as follows:

	Normal Inflation	*High Inflation*
Standard Car	2700	2900
Economy Car	3000	2000

where the entries represent the expected number of purchases, in thousands, per year. Determine the percentages of economy cars and standard cars that the company should plan on manufacturing.

22. Suppose that the price of fish changes so the payoff matrix for Jamaican fishermen is the following.

	Current	*No Current*
Inside	16.8	11.0
In-Out	6.4	20.5

What is the optimal strategy for the fishermen?

IMPORTANT TERMS

9–1	**Two-person game**	**Payoff matrix**	**Strategy**
	Value of game	**Saddle point**	**Solution**
	Zero-sum game	**Strictly determined game**	
9–2	**Mixed strategy game**	**Probability of a strategy**	**Expected payoff**
	Optimal strategy	**Fair game**	

REVIEW EXERCISES

1. The following represent payoff matrices for two-person games. For each one determine if the game is strictly determined. If a game is strictly determined, find the saddle point and value.

(a) $\begin{bmatrix} 5 & -1 \\ 2 & 4 \end{bmatrix}$ **(b)** $\begin{bmatrix} 1 & 3 & 9 \\ 7 & 4 & 8 \\ -5 & 3 & 4 \end{bmatrix}$

(c) $\begin{bmatrix} 140 & 210 \\ 300 & 275 \end{bmatrix}$ **(d)** $\begin{bmatrix} -6 & 2 & 9 & 1 \\ 5 & -4 & 0 & 2 \\ 4 & 2 & 8 & 3 \end{bmatrix}$

2. R and C are competing retail stores. Each has two options for an advertising campaign. The results of the campaign are expected to result in the payoff matrix

$$\begin{array}{cc} & C \\ R & \begin{bmatrix} 40 & 35 \\ -25 & 30 \end{bmatrix} \end{array}$$

where the entries represent the amount in sales that R will gain from C. What strategy should each adopt?

3. The following are payoff matrices that describe the payoff (in yards gained) of possible strategies of the offense and defense in a football game. Determine the solutions, if they exist.

(a) $\begin{bmatrix} 3 & 15 & -4 \\ 2 & -1 & 6 \\ 4 & 2 & 7 \end{bmatrix}$ **(b)** $\begin{bmatrix} -5 & 8 & 7 \\ 4 & 5 & 25 \\ 3 & -2 & 2 \end{bmatrix}$

4. Two players X and Y play the following game. X gives Y \$5 before the game starts. They each write 1, 2, or 3 on a slip of paper. Then Y gives X an amount determined by the numbers written. The following payoff matrix gives the amount paid to X in each possible case.

		Y		
		1	*2*	*3*
X	*1*	2	3	7
	2	4	0	6
	3	5	4	9

What strategy should each player use?

5. Determine the expected payoff of the following games.

(a) The strategy of X is $[.3 \quad .7]$; the strategy of Y is $[.6 \quad .4]$; the payoff matrix is

$$X \quad \overset{Y}{\begin{bmatrix} 5 & 9 \\ 11 & 2 \end{bmatrix}}$$

(b) The strategy of X is $[.5 \quad .5]$; the strategy of Y is $[.1 \quad .9]$; the payoff matrix is

$$X \quad \overset{Y}{\begin{bmatrix} -2 & 6 \\ 3 & 9 \end{bmatrix}}$$

(c) The strategy of X is $[.1 \quad .4 \quad .5]$; the strategy of Y is $[.2 \quad .2 \quad .6]$; the payoff matrix is

$$X \quad \overset{Y}{\begin{bmatrix} -3 & 2 & 1 \\ 4 & -2 & 5 \\ 3 & 1 & 2 \end{bmatrix}}$$

6. Determine the optimal strategy and its resulting value for each of the following games.

(a) $\begin{bmatrix} 2 & 9 \\ 6 & 3 \end{bmatrix}$ (b) $\begin{bmatrix} -2 & 5 \\ 8 & 4 \end{bmatrix}$ (c) $\begin{bmatrix} 3 & 5 & 7 \\ 1 & 4 & 6 \\ 6 & 7 & 5 \end{bmatrix}$

7. A truck-garden farmer raises vegetables in the field and in a greenhouse. The weather determines where they should be planted. The payoff matrix describes the strategies available.

	Wet	*Dry*
Field	250	140
Greenhouse	175	210

What percentage of the crop should be planted in the field and what percentage in the greenhouse?

Appendix

Geometric Progressions and Mathematics of Finance

You have encountered problems on I.Q. tests and other standardized tests that ask you to identify the pattern of a sequence of numbers and to determine additional numbers in the sequence. Patterns of numbers have uses beyond standardized tests. Certain patterns of numbers are often useful in mathematics. A sequence that has several applications in mathematics of finance is the **geometric progression**.

The sequence 2, 4, 8, 16, 32,... is a geometric progression because each number, after the first, is obtained by multiplying the preceding number by 2.

The sequence 1329, 132.9, 13.29, 1.329, .1329,... is a geometric progression because each number, after the first, is obtained by multiplying the preceding number by 0.1.

The sequence 1000, 1040, 1081.60, 1124.864,... is a geometric progression because each term, after the first, is obtained by multiplying the preceding term by 1.04.

Definition

A **geometric progression** is a sequence of numbers in which each number, after the first, is obtained by multiplying the preceding term by a fixed number.

A general expression of a geometric progression with first term a and fixed multiplier r is

$$a, ar, ar^2, ar^3, \ldots, ar^n$$

Example 1 An amount P is invested at compound interest, which is compounded periodically at an interest rate of i per period. The compound interest formula gives the amount accumulated in the first, second, third,..., nth periods as, respectively,

$$P(1 + i), P(1 + i)^2, \ldots, P(1 + i)^n$$

These form a geometric progression with $a = P(1 + i)$ and $r = 1 + i$.

Sum of a Geometric Progression

Because we sometimes need the total of terms of a geometric progression, let's find a formula for such a sum.

Suppose that we want to find the sum of the first ten terms of a geometric progression. The general terms are

$$a, ar, ar^2, ar^3, \ldots, ar^8, ar^9$$

Notice that the exponent of r is always one less than the position number of the term. The sum, S, of these ten terms can be written

$$S = a + ar + ar^2 + \cdots + ar^8 + ar^9$$

Now multiply both sides of this equation by r to obtain

$$rS = ar + ar^2 + ar^3 + \cdots + ar^9 + ar^{10}$$

We use these two equations to get a formula for the sum of a geometric progression. Subtract the second from the first:

$$\begin{aligned} S &= a + ar + ar^2 + \cdots + ar^8 + ar^9 \\ -rS &= \phantom{a + {}} ar + ar^2 + ar^3 + \cdots + ar^9 + ar^{10} \\ \hline S - rS &= a \phantom{+ ar + ar^2 + ar^3 + \cdots + ar^9 + {}} - ar^{10} \end{aligned}$$

(Notice how the terms $ar, \ldots, ar^9$ subtract out.)
This equation can be written

$$S(1 - r) = a(1 - r^{10})$$

Dividing by $1 - r$, we have

$$S = \frac{a(1 - r^{10})}{1 - r}, \quad \text{or} \quad \frac{a(r^{10} - 1)}{r - 1}$$

This formula holds in general.

> The sum of the first n terms of a geometric progression $a, ar, \ldots, ar^{n-1}$ is
>
> $$S = \frac{a(1 - r^n)}{1 - r} = \frac{a(r^n - 1)}{r - 1}$$

Example 2 The sum of the first 8 terms of 1, 2, 4, ... is

$$\frac{1(2^8 - 1)}{2 - 1} = \frac{255}{1} = 255$$

where $a = 1$, $r = 2$, and $n = 8$. The sum of the first 12 terms of 1000, 1000(1.02),

$1000(1.02)^2, \ldots$ is

$$\frac{1000[(1.02)^{12} - 1]}{1.02 - 1} = \frac{1000[1.26824 - 1]}{.02}$$

$$= \frac{1000(.26824)}{.02}$$

$$= 13{,}412$$

Note: The value of 1.02^{12} may be obtained by a calculator or from Table 8–1.

Future Value of an Ordinary Annuity

The formula for the future value of an annuity can be obtained by using the ideas of a geometric progression. Let's first look at a specific example.

$1000 is deposited at the end of each year for four years. The interest rate $r = .07$ is compounded annually. Year-by-year values of their annuity are the following.

$1000 Deposited at End of Year:	Draws Interest for:	Value of the Deposit at End of 4 Years:
1	3 years	$1000(1.07)^3 = 1225.04$
2	2 years	$1000(1.07)^2 = 1144.90$
3	1 year	$1000(1.07) = 1070.00$
4	0 years	$1000(1) = 1000.00$
		Final value 4439.94

The final value is obtained by adding the four payments and the interest accumulated on each one. The final value is the future value of the annuity. Notice that the final value can be written

$$1000 + 1000(1.07) + 1000(1.07)^2 + 1000(1.07)^3$$

This is the sum of the first four terms of a geometric progression with $a = 1000$, $r = 1.07$, and $n = 4$, so the sum is

$$\frac{1000[(1.07)^4 - 1]}{.07} = \frac{1000(.310796)}{.07} = 4439.94$$

Thus, the future value may be obtained without finding the values year by year.

In general, if a fixed amount R is invested at the end of each period for n periods at a periodic interest rate i, the value of each deposit by the end of the nth period is given by the following:

R Deposited at End of Period:	Draws Interest for:	Value of Deposit at End of n Periods:
1	$n - 1$ periods	$R(1 + i)^{n-1}$
2	$n - 2$ periods	$R(1 + i)^{n-2}$
3	$n - 3$ periods	$R(1 + i)^{n-3}$
$\vdots$		$\vdots$
$n - 1$	1 period	$R(1 + i)$
n	0 periods	R

The sum of these values gives the future value. Writing the values in reverse order gives

$$\text{Future value} = R + R(1 + i) + \cdots + R(1 + i)^{n-2} + R(1 + i)^{n-1}$$

Using the formula for the sum of a geometric progression with $a = R, r = 1 + i$, for n terms gives

$$\text{Future value} = R\left[\frac{(1 + i)^n - 1)}{1 + i - 1}\right]$$

$$= R\left[\frac{(1 + i)^n - 1}{i}\right]$$

The expression

$$\frac{(1 + i)^n - 1}{i}$$

is the value of S in the formula for future value

$$A = RS$$

Values for selected i and n are given in Table 8–2.

Balance of a Loan Payment

Typically, a loan is repaid with a series of equal payments paid periodically. Each payment pays all the interest for that period and repays some of the principal borrowed. Since the principal is reduced with each payment, the interest for the next period is computed on a new, and somewhat smaller, principal. The principal that remains to be paid is called the **loan balance**. Let's look at the general case to observe the pattern of loan balances for a sequence of payments.

We use the notation

P = original amount of the loan,
n = number of periodic payments made,
i = interest rate per period,
M = periodic payment, the amount of the fixed payment.

Here is what happens for several time periods. Remember that the interest for a specific time period is based on the principal remaining after the previous payment was made. The amount borrowed, the original principal, is P.

First Time Period

Interest paid: Pi
Principal repaid: $M - Pi$
Balance: $P - (M - Pi) = P(1 + i) - M$

Second Time Period

Interest paid: $[P(1 + i) - M]i$
Principal repaid: $M - [P(1 + i) - M]i = M(1 + i) - P(1 + i)i$
Balance: $[P(1 + i) - M] - [M(1 + i) - P(1 + i)i] =$
$P(1 + i)^2 - M(1 + i) - M$

Third Time Period

Interest paid: $[P(1 + i)^2 - M(1 + i) - M]i$
Principal repaid: $M - [P(1 + i)^2 - M(1 + i) - M]i =$
$M(1 + i) - [P(1 + i)^2 - M(1 + i)]i$
Balance: $P(1 + i)^2 - M(1 + i) - M - M(1 + i) + [P(1 + i)^2 -$
$M(1 + i)]i = P(1 + i)^3 - M(1 + i)^2 - M(1 + i) - M$

We see a pattern emerging for the balance. It does, in fact, hold in general. The balance after n payments is

$$\begin{aligned}\text{Balance} &= P(1 + i)^n - M(1 + i)^{n-1} - M(1 + i)^{n-2} - M(1 + i)^{n-3} \ldots \\ &\quad - M(1 + i) - M \\ &= P(1 + i)^n - M[(1 + i)^{n-1} + (1 + i)^{n-2} + \cdots \\ &\quad + (1 + i) + 1]\end{aligned}$$

The sum in brackets to the right is the sum of a geometric progression. If we reverse the order in which it is written, we have $a = 1, r = 1 + i$, and n terms. The sum is

$$S = \frac{(1 + i)^n - 1}{1 + i - 1} = \frac{(1 + i)^n - 1}{i}$$

so the balance may be written

$$\text{Balance} = P(1 + i)^n - M\left[\frac{(1 + i)^n - 1}{i}\right] = P(1 + i)^n - MS$$

where S is the same expression used in the future value of an annuity.

Answers to Selected Problems

CHAPTER 0

Section 0–1, page 4

1. -13 **3.** 23 **5.** -30 **7.** -35 **9.** -5 **11.** -7 **13.** -10 **15.** -8 **17.** 3 **19.** $\frac{9}{5}$

21. $\frac{17}{12}$ **23.** $-\frac{11}{12}$ **25.** $\frac{13}{20}$ **27.** $-\frac{1}{35}$ **29.** $\frac{2}{3}$ **31.** $\frac{5}{14}$ **33.** $\frac{1}{15}$ **35.** $\frac{8}{15}$ **37.** $\frac{12}{35}$ **39.** $\frac{20}{33}$

41. $\frac{4}{3}$ **43.** $\frac{15}{8}$ **45.** $\frac{171}{40}$ **47.** $-6a - 22b$ **49.** $-10a - 50b$

Section 0–2, page 7

1. -3 **3.** 4 **5.** 2 **7.** $-\frac{3}{4}$ **9.** $-\frac{7}{4}$ **11.** $\frac{20}{11}$ **13.** $\frac{15}{23}$ **15.** 6

17. (a) \$242 **(b)** \$412 **(c)** 950 miles **19. (a)** \$7.14 **(b)** \$22.05 **(c)** 61 lbs

Section 0–3, page 11

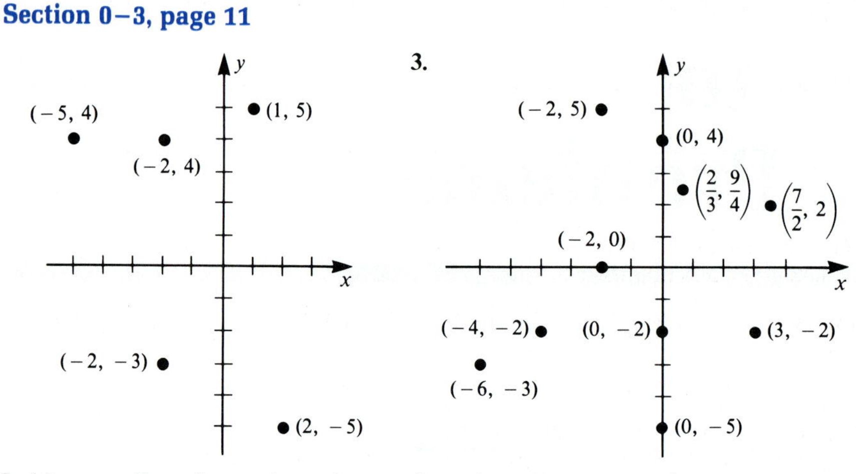

5. **(a)** x coordinate is negative and y coordinate is positive.
(b) x and y coordinates are both negative.
(c) x coordinate is positive and y coordinate is negative.

7.

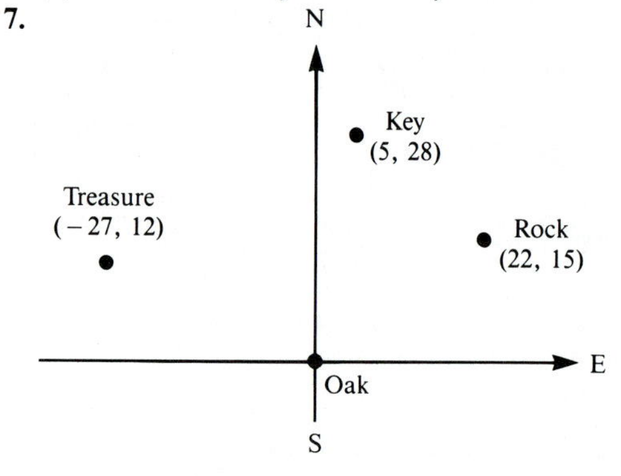

Section 0–4, page 18

1. **(a)** $9 - 3$ is a positive number 6, so true.
(b) $4 - 0$ is a positive number 4, so true.
(c) $-5 - 0$ is not a positive number, so false.
(d) $-3 - (-15)$ is a positive number 12, so true.
(e) $\frac{5}{6} - \frac{2}{3}$ is a positive number $\frac{1}{6}$, so true.

3. $x < \frac{9}{2}$ **5.** $x \geq -16$ **7.** $x < -4$

9. $0 \leq x < 12$ **11.** **13.** **15.** **17.**

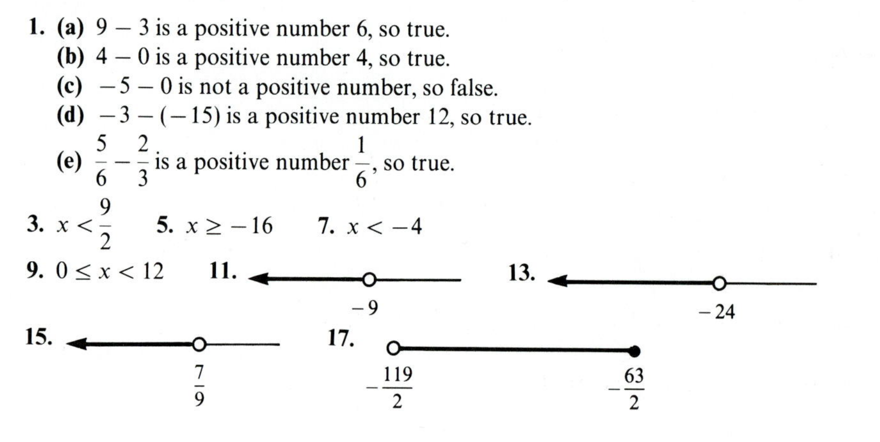

19. $(-\infty, -1]$ **21.** $\left(-\infty, \frac{1}{9}\right]$ **23.** $[7, 13)$ **25.** $x \le -\frac{1}{2}$ **27.** $\left[-11, -\frac{23}{3}\right)$ **29.** 8, 9, or 10

CHAPTER 1

Section 1–1, page 25

1. **(a)** The domain is the set of positive numbers in hours.
(b) The range is the set of numbers, in dollars, greater than 20.

3. **(a)** $\{1, 2, 3, 4, 5, 6, 7\}$ **(b)** 14.25 **5.** **(a)** $f(1) = 1$ **(b)** $f(-2) = -11$ **(c)** $f\left(\frac{1}{2}\right) = -1$ **(d)** $f(a) = 4a - 3$

7. **(a)** $f(5) = \frac{6}{4}$ **(b)** $f(-6) = \frac{5}{7}$ **(c)** $f(0) = -1$ **(d)** $f(2c) = \frac{(2c+1)}{(2c-1)}$

9. **(a)** $f(5) = 390$ calories, $f(2.5) = 195$ calories, $f(6.4) = 499.2$ calories **(b)** 9.5 oz

11. **(a)** 540 calories **(b)** 83.3 minutes **13.** $f(x) = 1.25x + 25$ (x = number of hamburgers)

15. $f(x) = .80x$ (x = regular price) **17.** $f(x) = .60x + 12$ (x = number of miles)

19. $f(x) = 3500x + 5{,}000{,}000$ (x = number of students) **21.** $f(x) = .88x$ (x = list price)

23. **(a)** 13,500 sq feet **(b)** $f(125) = 3750$ **(c)** \$7800 **(d)** 530 ft

Section 1–2, page 38

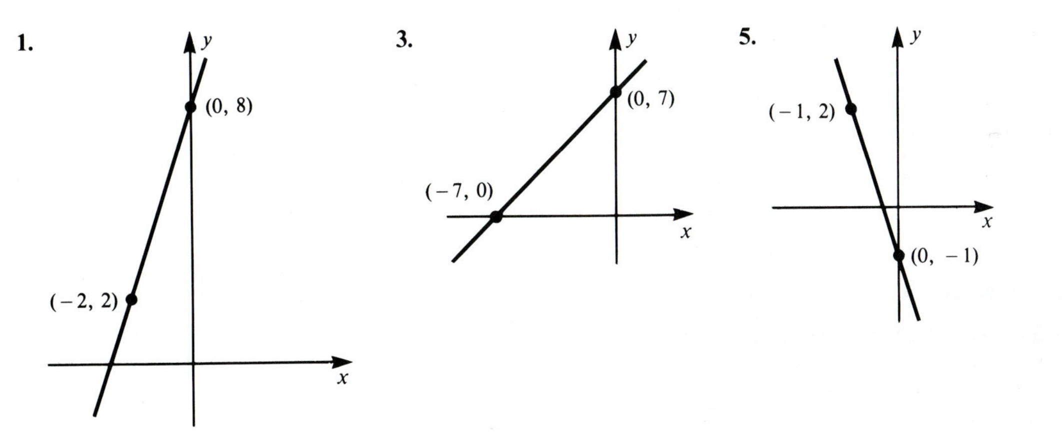

7. $m = 7, b = 22$ **9.** $m = -\frac{2}{5}, b = 6$ **11.** $m = -\frac{2}{5}, b = \frac{3}{5}$ **13.** $m = \frac{1}{3}, b = 2$ **15.** $m = 2$

17. not defined **19.** $m = 0$ **21.** $m = 0$ **23.** $x = 3$ **25.** $x = 10$ **27.** $y = 4x + 3$ **29.** $y = -x + 6$

31. $y = \frac{1}{2}x$ **33.** $y = -4x + 9$ **35.** $x - 2y = -3$ **37.** $y - 5 = 7(x - 1)$ **39.** $y - 6 = \frac{1}{5}(x - 9)$

41. $y = \left(\frac{1}{3}\right)x + \frac{1}{3}$ **43.** $y = 2x$ **45.** $y = 4$ **47.** $m_1 = 1, m_2 = 1$; lines parallel

49. $m_1 = -\frac{9}{7}, m_2 = -\frac{9}{7}$; lines parallel **51.** parallel, both slopes are $-\frac{3}{2}$

53. parallel, both slopes are $\frac{3}{5}$ but there is only one line **55.** $3x + 2y = 18$ **57.** $y = x + 9$

59. $c = .67x + 480$ **61.** $c = \left(\frac{7}{5}\right)x + 640$ **63.** $f(x) = 3500x + 3000$

65. **(a)** 4 **(b)** -3 **(c)** $\frac{2}{3}$ **(d)** $-\frac{1}{2}$ **(e)** $-\frac{2}{3}$ **(f)** 0 **67.** $140y = 15.9x + 3017.5$

Section 1–3, page 48

1. **(a)** \$10,040 **(b)** 223 **3.** **(a)** Fixed cost = \$400, unit cost = \$3 **(b)** $C(600) = \$2200$, $C(1000) = \$3400$
5. **(a)** $R(x) = 32x$ **(b)** $R(78) = \$2496$ **(c)** 21 pairs **7.** **(a)** $R(x) = 3.39x$ **(b)** $R(834) = \$2827.26$
9. 123,000 pages **11.** **(a)** $C = \left(\frac{1}{2}\right)x + 750$ **(b)** Fixed cost = \$750, unit cost = \$.50 **(c)** \$1350
13. $C = 4x + 500$ $C(800) = 3700$
15. **(a)** $C(x) = 649x + 1500$ **(b)** $R(x) = 899x$ **(c)** $C(37) = \$25,513$ **(d)** $R(37) = \$33,263$ **(e)** 6 computers
17. \$70.00 **19.** **(a)** $BV = -50x + 425$ **(b)** \$50 **(c)** $BV(3) = \$275$
21. **(a)** $BV = -1575x + 9750$ **(b)** \$1575 **(c)** $BV(2) = \$6600$, $BV(5) = \$1875$ **23.** \$885 **25.** 3.5 years
27. $C(x) = 4.15x + 1850$ **29.** $R(x) = 38x$
31. **(a)** **(i)** $P(x) = 24x - 465$ **(ii)** \$135 **(b)** **(i)** $P(x) = 80x - 1200$ **(ii)** \$240
(c) **(i)** $P(x) = 80x - 5200$ **(ii)** \$800

Section 1–4, page 58

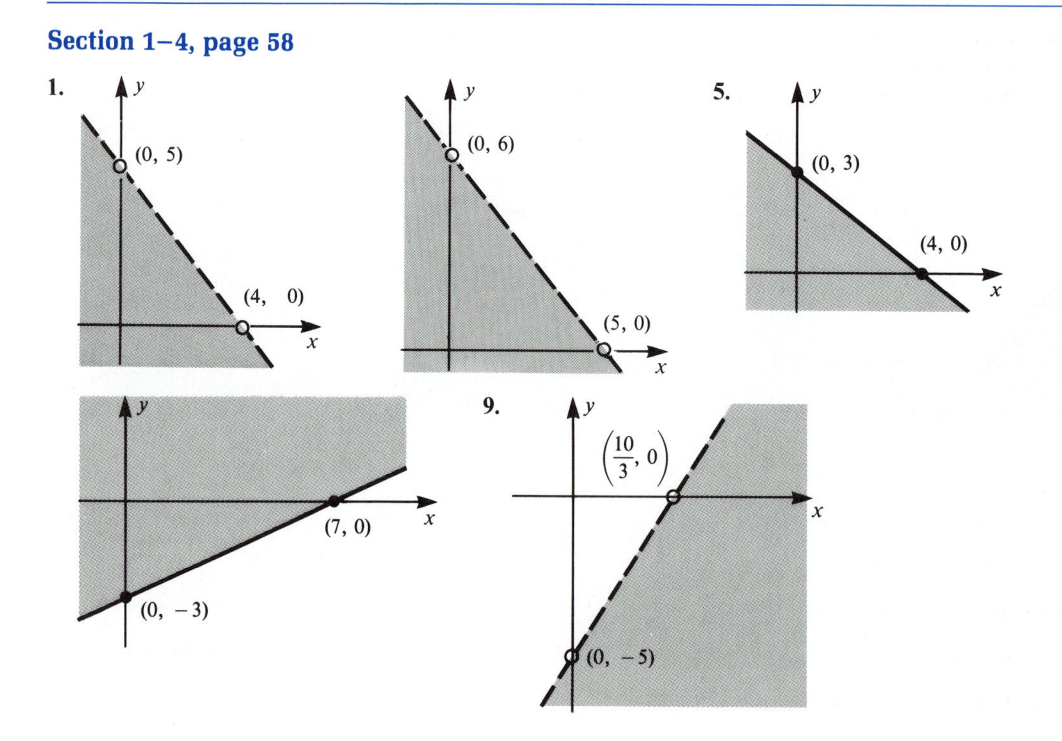

11.

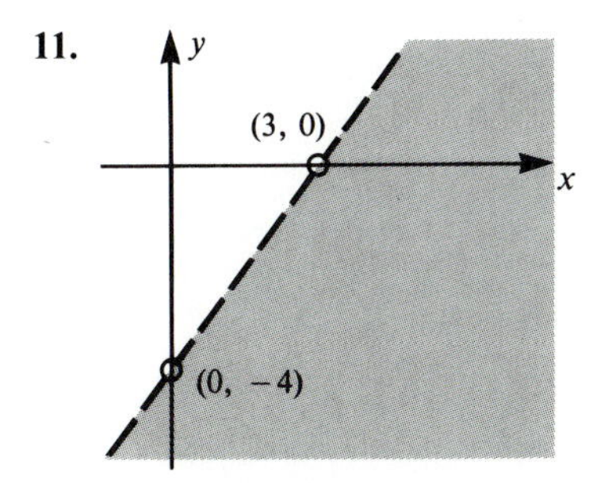

13.

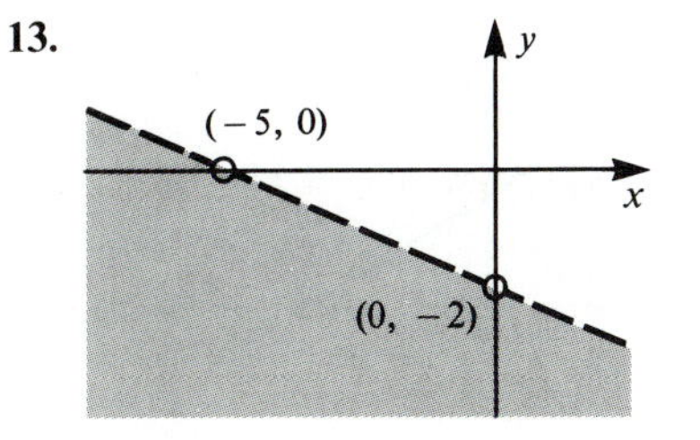

15. x = number of air conditioners
y = number of fans
$3.2x + 1.8y \leq 144$

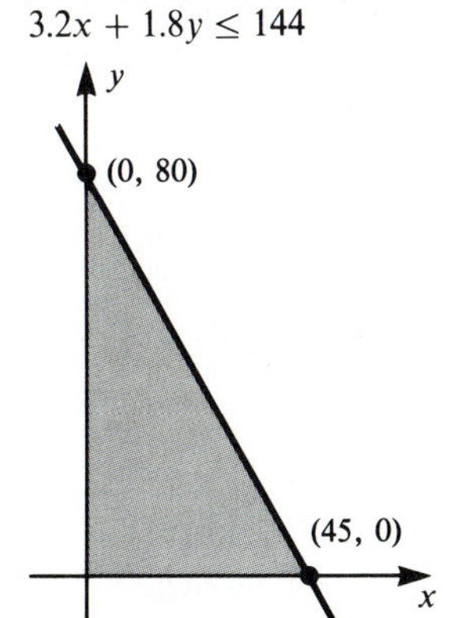

17. **(a)** $4x + 6y \leq 500$
x = number of members, y = number of pledges
(b)

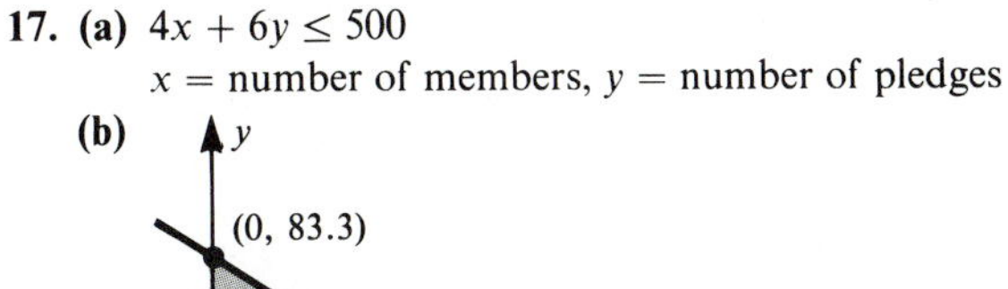

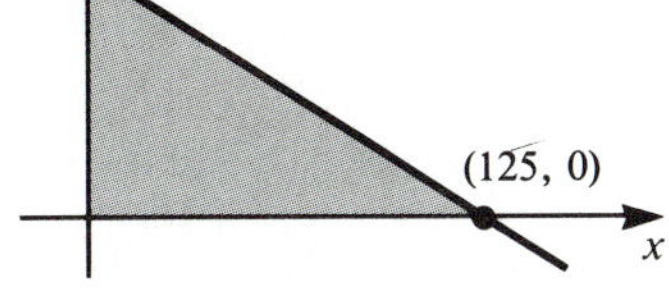

19. x = number of television ads
y = number of newspaper ads
x and y must satisfy $900x + 830y \leq 75{,}000$
The graph is

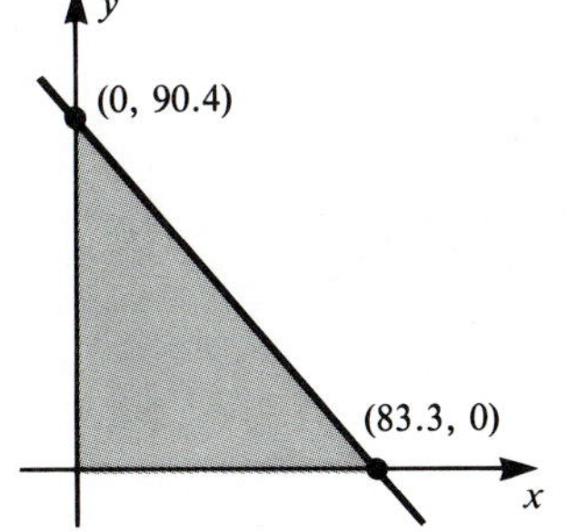

Review Exercises, Chapter 1, page 59

1. **(a)** all nonnegative integers **(b)** all nonnegative numbers **2.** **(a)** 12 **(b)** −28 **(c)** 0 **(d)** $8c - 4$

3. 22 **4.** **(a)** 6 **(b)** −5 **(c)** 0 **(d)** $\dfrac{a}{(2a - 11)}$ **5.** **(a)** \$4.20 **(b)** 2.75 lbs **6.** **(a)** \$2475 **(b)** 13

7. **(a)** $f(x) = 29.95x$ (x = number of pairs bought) **(b)** $f(x) = 1.25x + 40$ (x = number of people)

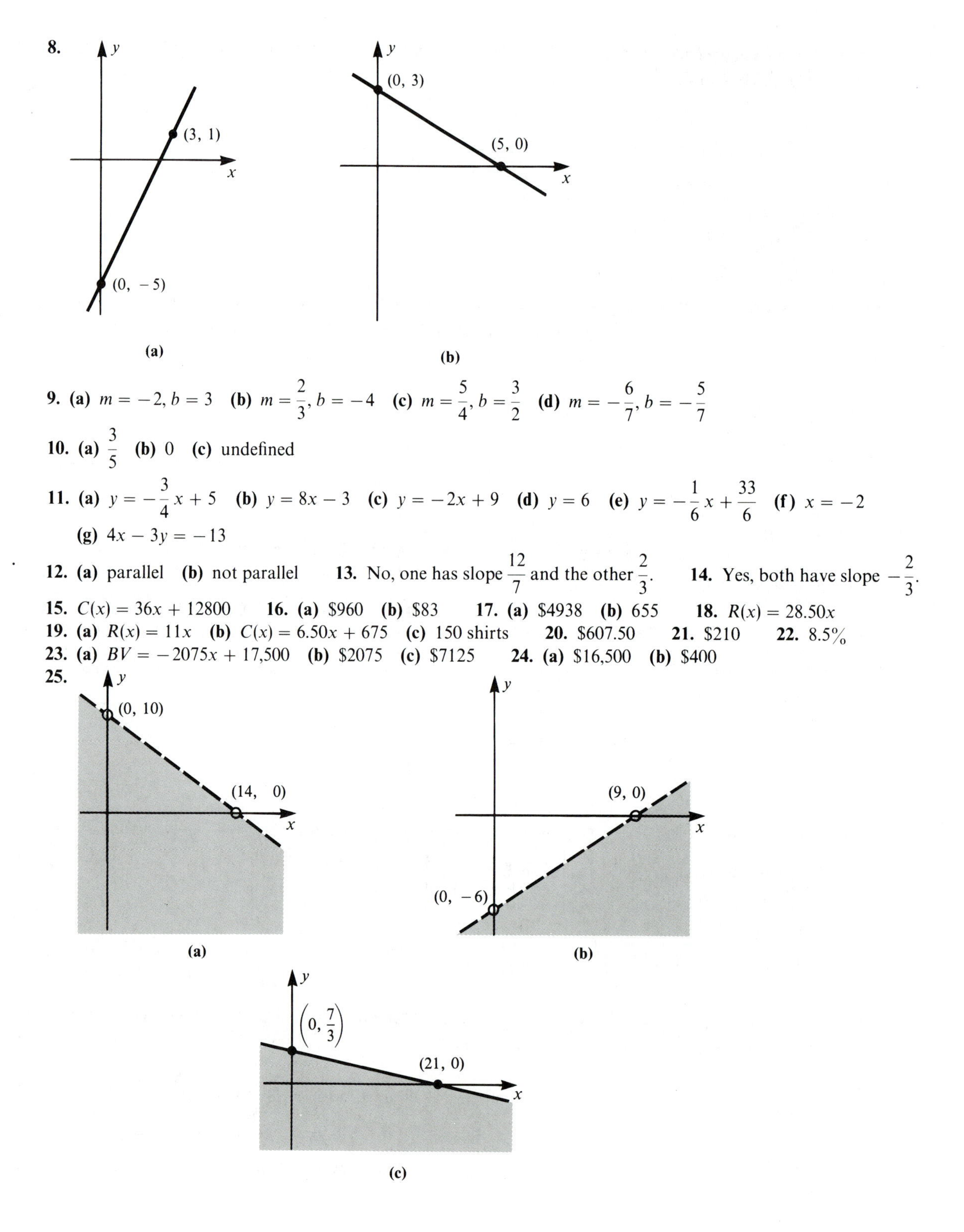

9. **(a)** $m = -2, b = 3$ **(b)** $m = \frac{2}{3}, b = -4$ **(c)** $m = \frac{5}{4}, b = \frac{3}{2}$ **(d)** $m = -\frac{6}{7}, b = -\frac{5}{7}$

10. **(a)** $\frac{3}{5}$ **(b)** 0 **(c)** undefined

11. **(a)** $y = -\frac{3}{4}x + 5$ **(b)** $y = 8x - 3$ **(c)** $y = -2x + 9$ **(d)** $y = 6$ **(e)** $y = -\frac{1}{6}x + \frac{33}{6}$ **(f)** $x = -2$
(g) $4x - 3y = -13$

12. **(a)** parallel **(b)** not parallel **13.** No, one has slope $\frac{12}{7}$ and the other $\frac{2}{3}$. **14.** Yes, both have slope $-\frac{2}{3}$.

15. $C(x) = 36x + 12800$ **16.** **(a)** \$960 **(b)** \$83 **17.** **(a)** \$4938 **(b)** 655 **18.** $R(x) = 28.50x$

19. **(a)** $R(x) = 11x$ **(b)** $C(x) = 6.50x + 675$ **(c)** 150 shirts **20.** \$607.50 **21.** \$210 **22.** 8.5%

23. **(a)** $BV = -2075x + 17{,}500$ **(b)** \$2075 **(c)** \$7125 **24.** **(a)** \$16,500 **(b)** \$400

26. **(a)** x = number of hours of line A
y = number of hours of line B
(b) $65x + 105y \leq 1500$

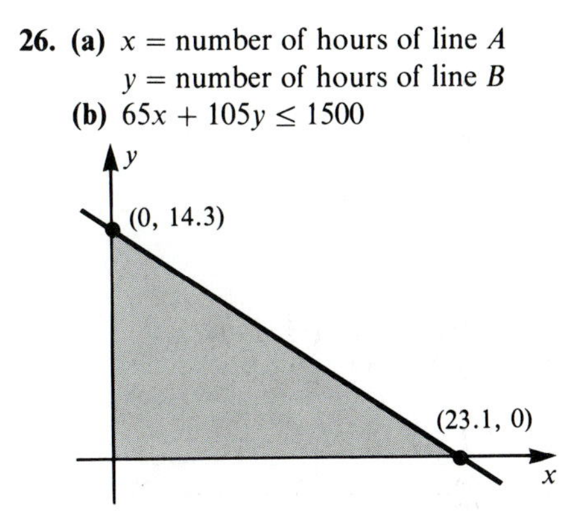

CHAPTER 2

Section 2–1, page 75

1. (2, 3) **3.** (−3, 0) **5.** (−3, −15) **7.** (5, 3) **9.** (6, −8) **11.** $\left(\frac{7}{2}, 4\right)$ **13.** $\left(\frac{2}{3}, -\frac{5}{3}\right)$

15. all points $(k, 3k - 5)$ **17.** (6, −1) **19.** $\left(\frac{5}{2}, -3\right)$ **21.** (−2, 3) **23.** (−14, 19) **25.** (−2.4, 5.1)

27. (.06, .13) **29.** No solution **31.** $\left(k, \frac{1}{5} - \frac{4}{5}k\right)$ **33.** $(4 + 6k, k)$ **35.** (4, 3) **37.** (30, 10)

39. (15, 8) **41.** (25, 85.50) **43.** 43 nickels, 122 dimes **45.** 8 and 6 **47.** 62 chopped beef and 53 sausage
49. 2.5 oranges, 3.25 apples **51.** \$23,500 at 7.4%, \$26,500 at 8.8%

Section 2–2, page 89

1. (2, −1, 2) **3.** (7, 5, 3) **5.** (3, 1, 2) **7.** **(a)** 5, 8, 6, 2, −5 **(b)** (1, 6) and (3, 2)

9. $\begin{bmatrix} 1 & -1 & 1 \\ 1 & 3 & 4 \\ 2 & -2 & 1 \end{bmatrix} \left[\begin{array}{ccc|c} 1 & -1 & 1 & 4 \\ 1 & 3 & 4 & 2 \\ 2 & -2 & 1 & 5 \end{array}\right]$ **11.** $\begin{bmatrix} 6 & 4 & -1 \\ -1 & -1 & -1 \\ 5 & 0 & 0 \end{bmatrix} \left[\begin{array}{ccc|c} 6 & 4 & -1 & 0 \\ -1 & -1 & -1 & 7 \\ 5 & 0 & 0 & 15 \end{array}\right]$

13. $\begin{bmatrix} 7 & -1 & 1 & 2 \\ 0 & 1 & -8 & -1 \\ 1 & 0 & 7 & 1 \end{bmatrix} \left[\begin{array}{cccc|c} 7 & -1 & 1 & 2 & 4 \\ 0 & 1 & -8 & -1 & -1 \\ 1 & 0 & 7 & 1 & -2 \end{array}\right]$ **15.** $\begin{aligned} x_1 + 3x_2 + 2x_3 &= 0 \\ 2x_1 + 4x_2 - 3x_3 &= 2 \\ -x_1 + 6x_2 + x_3 &= 4 \end{aligned}$ **17.** $\begin{aligned} 2x_2 + x_3 &= 3 \\ x_1 + 3x_2 \quad &= 5 \\ 4x_1 + 2x_2 + 3x_3 &= 8 \end{aligned}$

19. −3, 2, −6, and 13. **21.** $\begin{bmatrix} 0 & -1 & 3 & 4 \\ 2 & 1 & 5 & 2 \\ -1 & 2 & 3 & 8 \end{bmatrix}$ **23.** $\begin{bmatrix} 1 & 0 & 0 & -8 \\ 0 & 1 & 0 & 11 \\ 0 & 0 & 1 & 3 \end{bmatrix}$ **25.** $\begin{bmatrix} 1 & 0 & 2 & 3 \\ 0 & 1 & 4 & 1 \\ 0 & 0 & 1 & -3 \end{bmatrix}$

27. (2, −1, 2) **29.** $\left(4, -2, \frac{1}{2}\right)$ **31.** (2, 0, 1, −1) **33.** (−1, 3) **35.** (−1, 1, 2) **37.** (−1, 2, 1)

39. (1, 1, 1) **41.** (3, −2, 4) **43.** $\left(\frac{1}{2}, \frac{3}{2}, -\frac{1}{2}\right)$ **45.** (2, −1, 3, 1) **47.** (4, −1, 2, 3)

49. stock *A*: \$22,000, stock *B*: \$3,000, stock *C*: \$15,000
51. 32 minutes jogging, 12 minutes handball, and 16 minutes cycling
53. 540 high school students, 1230 college students, 680 adults

---

Section 2–3, page 102

1. Reduced echelon form. **3.** Not reduced echelon form. The first nonzero entry in row 2 is not a 1.
5. Not reduced echelon form. The leading 1 in row 3 is to the left of the leading 1 in row 2.
7. Not reduced echelon form. The columns containing the leading 1's of rows 3 and 4 have other nonzero entries.
9. Not reduced echelon form. The first nonzero entry in row 2 is not a 1.

11. $\left[\begin{array}{ccc|c} 1 & 0 & 1 & 4 \\ 0 & 1 & 2 & 3 \\ 0 & 0 & 1 & 2 \end{array}\right]$ **13.** $\left[\begin{array}{ccc|c} 1 & 3 & 4 & -1 \\ 0 & 4 & 3 & 4 \\ 0 & 0 & 6 & 2 \end{array}\right]$ **15.** $\left[\begin{array}{cccc|c} 1 & 4 & 2 & 4 & 5 \\ 0 & 0 & 3 & 1 & 2 \\ 0 & 0 & 0 & 2 & 3 \\ 0 & 0 & 0 & 0 & 0 \end{array}\right]$ **17.** $\left[\begin{array}{ccc|c} 1 & 0 & 2 & 3 \\ 0 & 1 & 4 & 4 \\ 0 & 0 & 0 & 5 \\ 0 & 0 & 0 & 0 \end{array}\right]$

19. $\left[\begin{array}{ccc|c} 1 & 0 & 0 & \frac{5}{4} \\ 0 & 1 & 0 & -\frac{3}{2} \\ 0 & 0 & 1 & -\frac{1}{4} \end{array}\right]$ **21.** $\left[\begin{array}{ccc|c} 1 & 0 & 0 & \frac{22}{13} \\ 0 & 1 & 0 & 0 \\ 0 & 0 & 1 & -\frac{12}{13} \end{array}\right]$

23. $x_1 = 2$, $x_2 + 2x_3 = 3$, $x_4 = -1$; solution $x_1 = 2$, $x_3 = r$, $x_2 = 3 - 2r$, $x_4 = -1$

25. $x_1 + 2x_3 + 2x_5 = -1$, $x_2 + 4x_3 + 3x_5 = 2$, $x_4 + 4x_5 = 1$; solution $x_1 = -1 - 2r - 2s$, $x_2 = 2 - 4r - 3s$, $x_3 = r$, $x_4 = 1 - 4s$, $x_5 = s$

27. $x_1 + 2x_3 = 2$, $x_2 - 3x_3 = 1$, $x_4 = 3$; solution $x_1 = 2 - 2r$, $x_2 = 1 + 3r$, $x_3 = r$, $x_4 = 3$

29. $x_1 + 2x_2 + x_4 = 4$, $x_3 + 3x_4 = -1$, $x_5 = 3$, $x_6 = 2$; solution $x_1 = 4 - 2r - s$, $x_2 = r$, $x_3 = -1 - 3s$, $x_4 = s$, $x_5 = 3$, $x_6 = 2$

31. $x_1 + 2x_2 + 4x_4 + x_6 = 7$, $x_3 + 3x_4 - x_6 = 3$, $x_5 + 4x_6 = 2$; solution $x_1 = 7 - 2r - 4s - t$, $x_2 = r$, $x_3 = 3 - 3s + t$, $x_4 = s$, $x_5 = 2 - 4t$, $x_6 = t$

33. (2, 1, 3) **35.** $x_1 = -32 - 2r$, $x_2 = 45 + r$, $x_3 = r$, $x_4 = 16$ **37.** $x_1 = -59 + 11x_3$, $x_2 = 23 - 5x_3$ **39.** No solution **41.** No solution

43. $x_1 = 3 + x_3$, $x_2 = 6 + x_3 + 2x_4 - x_5$ **45.** $x_1 = 2$, $x_2 = -1 - 2x_3$, $x_4 = 3$ **47.** $x_1 = -2x_2 + 3x_3 - x_5 + 3x_6$, $x_4 = -2x_5 - x_6$

49. $x_1 = \frac{35}{8} - 2x_2 - 3x_3 - \left(\frac{15}{8}\right)x_7$

$x_4 = \frac{6}{8} - \left(\frac{6}{8}\right)x_7$

$x_5 = \frac{29}{8} - \left(\frac{9}{8}\right)x_7$

$x_6 = -\frac{5}{8} + \left(\frac{1}{8}\right)x_7$

51. $15,000 in stocks, $18,000 in bonds, $15,000 in money markets

Section 2–4, page 111

1. 2×2 **3.** 3×3 **5.** 3×1 **7.** 2×3 **9.** 2×1 **11.** 2×2 **13.** Not equal **15.** Equal

17. Not equal **19.** $\begin{bmatrix} 3 & 3 & 2 \\ 7 & 4 & 3 \end{bmatrix}$ **21.** $\begin{bmatrix} 5 \\ 5 \end{bmatrix}$ **23.** Not possible **25.** $\begin{bmatrix} 6 & 5 & 8 \\ 5 & 1 & 3 \\ 5 & -3 & 4 \end{bmatrix}$ **27.** $\begin{bmatrix} 2 & 4 \\ 6 & 8 \end{bmatrix}$

29. $[4 \quad -4 \quad 8 \quad -12]$ **31.** $\begin{bmatrix} .5 \\ -1.0 \\ -.15 \end{bmatrix}$

33. **(a)** $2A = \begin{bmatrix} 2 & 4 \\ 6 & 0 \end{bmatrix}$ $3B = \begin{bmatrix} -3 & 6 \\ 3 & 3 \end{bmatrix}$ $-2C = \begin{bmatrix} 0 & -2 \\ -2 & -8 \end{bmatrix}$

(b) $A + B = \begin{bmatrix} 0 & 4 \\ 4 & 1 \end{bmatrix}$ $B + A = \begin{bmatrix} 0 & 4 \\ 4 & 1 \end{bmatrix}$ $A + C = \begin{bmatrix} 1 & 3 \\ 4 & 4 \end{bmatrix}$ $B + C = \begin{bmatrix} -1 & 3 \\ 2 & 5 \end{bmatrix}$

(c) $A + 2B = \begin{bmatrix} -1 & 6 \\ 5 & 2 \end{bmatrix}$ $3A + C = \begin{bmatrix} 3 & 7 \\ 10 & 4 \end{bmatrix}$ $2A + B - C = \begin{bmatrix} 1 & 5 \\ 6 & -3 \end{bmatrix}$

35.

	I	*II*	*III*
PC	23	17	20
Print	19	22	11
Disk	151	151	105

37. $x = 3$ **39.** $x = \frac{17}{8}$

41.

	Sulphur Dioxide	*Nitric Oxide*	*Part. Matter*
A	2760	1080	1680
B	3120	1380	1992

43.

A	89.7
B	68.0
C	75.3
D	82.0
E	75.7

45. $\begin{bmatrix} 4.96 & 13.57 & 22.24 & 36.74 \\ -206.96 & -287.87 & -360.85 & -464.18 \\ -136.47 & -133.05 & -142.12 & -148.91 \\ -273.55 & -292.55 & -299.63 & -350.47 \end{bmatrix}$

Negative entry means more energy was consumed than produced.

Section 2–5, page 119

1. 14 **3.** 12 **5.** 14 **7.** $6.25 **9.** $\begin{bmatrix} -5 & 11 \\ 0 & 14 \end{bmatrix}$ **11.** $\begin{bmatrix} 30 & 2 \\ 39 & -3 \end{bmatrix}$ **13.** $\begin{bmatrix} 14 & 6 \\ 16 & -1 \end{bmatrix}$ **15.** $\begin{bmatrix} 10 \\ 20 \end{bmatrix}$

17. $\begin{bmatrix} 7 & 4 \\ 0 & 5 \\ 22 & 14 \end{bmatrix}$ **19.** $\begin{bmatrix} -12 & 4 & -3 \\ -14 & 22 & 5 \\ 6 & 14 & 9 \end{bmatrix}$ **21.** $\begin{bmatrix} 2 & 3 \\ 8 & 7 \end{bmatrix}$ **23.** Not possible **25.** Not possible

27. Not possible **29.** $AB = \begin{bmatrix} 1 & 8 \\ -1 & 2 \end{bmatrix}$ $BA = \begin{bmatrix} 3 & 10 \\ -1 & 0 \end{bmatrix}$ **31.** $AB = \begin{bmatrix} -3 & 10 \\ -2 & 5 \end{bmatrix}$ $BA = \begin{bmatrix} -1 & 2 \\ -4 & 3 \end{bmatrix}$

33. $AB = \begin{bmatrix} 6 & 2 & 13 \\ -11 & -6 & -4 \end{bmatrix}$, BA Not possible **35.** $AB = [27 \quad 38]$, BA Not possible **37.** $[39 \quad 91]$

39. $\begin{bmatrix} 5 & 37 & \frac{5}{2} & 29 & 20 \\ 7 & 59 & \frac{31}{2} & 53 & 52 \\ 11 & 31 & -\frac{1}{2} & 9 & 11 \\ 25 & 113 & \frac{17}{2} & 103 & 65 \end{bmatrix}$ **41.** $\begin{bmatrix} 11 & 9 \\ 10 & 19 \end{bmatrix}$ **43.** $\begin{bmatrix} 1 & 2 \\ 3 & 1 \end{bmatrix}$ **45.** $\begin{bmatrix} 1 & 2 & 3 \\ 4 & -1 & 0 \\ 2 & 1 & 7 \end{bmatrix}$

47. $\begin{bmatrix} 4 & 5.5 \\ 1 & 2 \end{bmatrix}\begin{bmatrix} 300 \\ 450 \end{bmatrix} = \begin{bmatrix} 3675 \\ 1200 \end{bmatrix}$
3675 stereos, 1200 televisions

49. $[5 \quad 3 \quad 3 \quad 2]\begin{bmatrix} 64 & 62 & 68 \\ 63 & 68 & 71 \\ 59 & 58 & 57 \\ 80 & 82 & 78 \end{bmatrix} = [8.46 \quad 8.52 \quad 8.80]$ (*Store 1*, *Store 2*, *Store 3*)

51. $\begin{bmatrix} .5 & 1.5 & .5 & 1.0 & 1.0 \\ 0 & 1.0 & 1.0 & 3.0 & 2.0 \end{bmatrix}\begin{bmatrix} 500 & .2 & 0 & 129 \\ 0 & .2 & 0 & 0 \\ 1560 & .32 & 1.7 & 6 \\ 0 & 0 & 0 & 0 \\ 460 & 0 & 0 & 0 \end{bmatrix} = \begin{bmatrix} 1490 & .56 & .85 & 67.5 \\ 2480 & .52 & 1.70 & 6.0 \end{bmatrix}$ (columns A, B_1, B_2, C; rows *I*, *II*)

Section 2–6, page 136

1. $25^{-1} = .04, \left(\frac{2}{3}\right)^{-1} = \frac{3}{2}, (-5)^{-1} = -\frac{1}{5}, .75^{-1} = 1.333, 11^{-1} = \frac{1}{11}$ **3.** $\begin{bmatrix} -5 & 2 \\ 3 & -1 \end{bmatrix}$ **5.** $\begin{bmatrix} 3 & -2 \\ -4 & 3 \end{bmatrix}$

7. $\begin{bmatrix} -\frac{5}{3} & 1 \\ 2 & -1 \end{bmatrix}$ **9.** $\begin{bmatrix} \frac{7}{3} & -3 & -\frac{1}{3} \\ -\frac{8}{3} & 3 & \frac{2}{3} \\ \frac{4}{3} & -1 & -\frac{1}{3} \end{bmatrix}$ **11.** $\begin{bmatrix} 0 & 2 & -1 \\ 1 & 1 & -1 \\ -\frac{2}{3} & -1 & 1 \end{bmatrix}$ **13.** No inverse. **15.** No inverse.

17. $\begin{bmatrix} \frac{3}{2} & -\frac{1}{2} \\ -2 & 1 \end{bmatrix}$ **19.** $\begin{bmatrix} -\frac{3}{7} & \frac{5}{7} & -\frac{11}{7} \\ \frac{2}{7} & -\frac{1}{7} & -\frac{2}{7} \\ \frac{2}{7} & -\frac{1}{7} & \frac{5}{7} \end{bmatrix}$

21. (a) $\left[\begin{array}{ccc|c} 3 & 4 & -5 & 4 \\ 2 & -1 & 3 & -1 \\ 1 & 1 & -1 & 2 \end{array}\right]$ **(b)** $\begin{bmatrix} 3 & 4 & -5 \\ 2 & -1 & 3 \\ 1 & 1 & -1 \end{bmatrix}$ **(c)** $\begin{bmatrix} 3 & 4 & -5 \\ 2 & -1 & 3 \\ 1 & 1 & -1 \end{bmatrix}\begin{bmatrix} x_1 \\ x_2 \\ x_3 \end{bmatrix} = \begin{bmatrix} 4 \\ -1 \\ 2 \end{bmatrix}$

23. (a) $\left[\begin{array}{cc|c} 4 & 5 & 2 \\ 3 & -2 & 7 \end{array}\right]$ **(b)** $\begin{bmatrix} 4 & 5 \\ 3 & -2 \end{bmatrix}$ **(c)** $\begin{bmatrix} 4 & 5 \\ 3 & -2 \end{bmatrix}\begin{bmatrix} x \\ y \end{bmatrix} = \begin{bmatrix} 2 \\ 7 \end{bmatrix}$ **25.** $\begin{bmatrix} 1 & 3 \\ 2 & -1 \end{bmatrix}\begin{bmatrix} x_1 \\ x_2 \end{bmatrix} = \begin{bmatrix} 5 \\ 6 \end{bmatrix}$

27. $\begin{bmatrix} 1 & 2 & -3 & 4 \\ 1 & 1 & 0 & 1 \\ 3 & 2 & 1 & 2 \end{bmatrix}\begin{bmatrix} x_1 \\ x_2 \\ x_3 \\ x_4 \end{bmatrix} = \begin{bmatrix} 0 \\ 5 \\ 4 \end{bmatrix}$ **29.** No inverse. **31.** $\begin{bmatrix} \frac{1}{9} & \frac{1}{3} & \frac{5}{9} \\ \frac{1}{3} & 0 & -\frac{1}{3} \\ -\frac{2}{9} & \frac{1}{3} & -\frac{1}{9} \end{bmatrix}\begin{bmatrix} 2 \\ 0 \\ 1 \end{bmatrix} = \begin{bmatrix} \frac{7}{9} \\ \frac{1}{3} \\ -\frac{5}{9} \end{bmatrix}$

33. $\begin{bmatrix} -\frac{14}{17} & \frac{8}{17} & \frac{4}{17} & \frac{5}{17} \\ \frac{3}{17} & -\frac{9}{17} & \frac{4}{17} & \frac{5}{17} \\ \frac{5}{17} & \frac{2}{17} & \frac{1}{17} & -\frac{3}{17} \\ \frac{18}{17} & -\frac{3}{17} & -\frac{10}{17} & -\frac{4}{17} \end{bmatrix}\begin{bmatrix} 5 \\ 6 \\ 1 \\ 7 \end{bmatrix} = \begin{bmatrix} 1 \\ 0 \\ 1 \\ 2 \end{bmatrix}$

35. $\begin{bmatrix} -5 & 2 \\ 3 & -1 \end{bmatrix}\begin{bmatrix} 3 \\ 8 \end{bmatrix} = \begin{bmatrix} 1 \\ 1 \end{bmatrix}$ $\begin{bmatrix} -5 & 2 \\ 3 & -1 \end{bmatrix}\begin{bmatrix} 4 \\ 9 \end{bmatrix} = \begin{bmatrix} -2 \\ 3 \end{bmatrix}$ $\begin{bmatrix} -5 & 2 \\ 3 & -1 \end{bmatrix}\begin{bmatrix} 3 \\ 7 \end{bmatrix} = \begin{bmatrix} -1 \\ 2 \end{bmatrix}$

37. $\begin{bmatrix} 0 & 3 & -1 \\ -2 & 0 & 1 \\ -1 & -1 & 1 \end{bmatrix}\begin{bmatrix} 6 \\ 5 \\ 14 \end{bmatrix} = \begin{bmatrix} 1 \\ 2 \\ 3 \end{bmatrix}$ $\begin{bmatrix} 0 & 3 & -1 \\ -2 & 0 & 1 \\ -1 & -1 & 1 \end{bmatrix}\begin{bmatrix} -5 \\ -3 \\ -8 \end{bmatrix} = \begin{bmatrix} -1 \\ 2 \\ 0 \end{bmatrix}$ $\begin{bmatrix} 0 & 3 & -1 \\ -2 & 0 & 1 \\ -1 & -1 & 1 \end{bmatrix}\begin{bmatrix} 4 \\ 3 \\ 9 \end{bmatrix} = \begin{bmatrix} 0 \\ 1 \\ 2 \end{bmatrix}$

39. $\begin{bmatrix} 2 & -5 & 2 \\ -1 & -5 & 3 \\ 0 & 2 & -1 \end{bmatrix}\begin{bmatrix} 2 \\ 5 \\ 10 \end{bmatrix} = \begin{bmatrix} -1 \\ 3 \\ 0 \end{bmatrix}$ $\begin{bmatrix} 2 & -5 & 2 \\ -1 & -5 & 3 \\ 0 & 2 & -1 \end{bmatrix}\begin{bmatrix} 3 \\ 2 \\ 4 \end{bmatrix} = \begin{bmatrix} 4 \\ -1 \\ 0 \end{bmatrix}$

41. $\begin{bmatrix} 4 & -3 \\ 3 & -2 \end{bmatrix}\begin{bmatrix} 18 & 20 & 5 & 20 \\ 5 & 18 & 1 & 27 \end{bmatrix} = \begin{bmatrix} 57 & 26 & 17 & -1 \\ 44 & 24 & 13 & 6 \end{bmatrix}$
so the message is 57, 44, 26, 24, 17, 13, −1, 6.

43. $\begin{bmatrix} -2 & 3 \\ -3 & 4 \end{bmatrix}\begin{bmatrix} 49 & -5 & -61 \\ 38 & -3 & -39 \end{bmatrix} = \begin{bmatrix} 16 & 1 & 5 \\ 5 & 3 & 27 \end{bmatrix}$
so the message is PEACE.

Section 2–7, page 144

1. **(a)** \$.25 **(b)** \$.40 **(c)** electricity **(d)** steel **(e)** steel **3.** $\begin{bmatrix} 5.00 \\ 7.25 \end{bmatrix}$ **5.** $\begin{bmatrix} 3.1 \\ 3.4 \\ 5.5 \\ 4.5 \end{bmatrix}$ **7.** $\begin{bmatrix} 1.60 & .40 \\ .55 & 1.70 \end{bmatrix}$

9. $I - A = \begin{bmatrix} .80 & -.60 \\ -.40 & .90 \end{bmatrix}$ $(I - A)^{-1} = \begin{bmatrix} \frac{15}{8} & \frac{10}{8} \\ \frac{5}{6} & \frac{10}{6} \end{bmatrix}$

$D = \begin{bmatrix} 24 \\ 12 \end{bmatrix}$ total output $= \begin{bmatrix} \frac{15}{8} & \frac{10}{8} \\ \frac{5}{6} & \frac{10}{6} \end{bmatrix}\begin{bmatrix} 24 \\ 12 \end{bmatrix} = \begin{bmatrix} 60 \\ 40 \end{bmatrix}$

$D = \begin{bmatrix} 8 \\ 6 \end{bmatrix}$ total output $= \begin{bmatrix} 22.5 \\ 16.7 \end{bmatrix}$

11. $(I - A)^{-1} = \begin{bmatrix} \frac{15}{7} & \frac{10}{7} \\ \frac{5}{6} & \frac{10}{6} \end{bmatrix}$

$D = \begin{bmatrix} 42 \\ 84 \end{bmatrix}$ output $= \begin{bmatrix} 210 \\ 175 \end{bmatrix}$ $D = \begin{bmatrix} 0 \\ 10 \end{bmatrix}$ output $= \begin{bmatrix} 14.3 \\ 16.7 \end{bmatrix}$

$D = \begin{bmatrix} 14 \\ 7 \end{bmatrix}$ output $= \begin{bmatrix} 40 \\ 23.3 \end{bmatrix}$ $D = \begin{bmatrix} 42 \\ 42 \end{bmatrix}$ output $= \begin{bmatrix} 150 \\ 105 \end{bmatrix}$

13. $(I - A)^{-1}\begin{bmatrix} \frac{25}{12} & \frac{10}{12} & \frac{10}{12} \\ \frac{10}{3} & \frac{10}{3} & \frac{10}{3} \\ \frac{35}{18} & \frac{10}{9} & \frac{25}{9} \end{bmatrix}$

$D = \begin{bmatrix} 36 \\ 72 \\ 36 \end{bmatrix}$ output $= \begin{bmatrix} 165 \\ 480 \\ 250 \end{bmatrix}$ internal demand $\begin{bmatrix} 129 \\ 408 \\ 214 \end{bmatrix}$

$D = \begin{bmatrix} 36 \\ 0 \\ 18 \end{bmatrix}$ output $= \begin{bmatrix} 90 \\ 180 \\ 120 \end{bmatrix}$ internal demand $\begin{bmatrix} 54 \\ 180 \\ 102 \end{bmatrix}$

$D = \begin{bmatrix} 3 \\ 0 \\ 0 \end{bmatrix}$ output $= \begin{bmatrix} 6.25 \\ 10.00 \\ 5.83 \end{bmatrix}$ internal demand $\begin{bmatrix} 3.25 \\ 10.00 \\ 5.83 \end{bmatrix}$

$D = \begin{bmatrix} 0 \\ 18 \\ 18 \end{bmatrix}$ output $= \begin{bmatrix} 30 \\ 120 \\ 70 \end{bmatrix}$ internal demand $\begin{bmatrix} 30 \\ 102 \\ 52 \end{bmatrix}$

15. internal demand $= \begin{bmatrix} 9 \\ 9 \\ 13 \end{bmatrix}$ consumer demand $\begin{bmatrix} 1 \\ 1 \\ 7 \end{bmatrix}$

Review Exercises, Chapter 2, page 147

1. $\left(\frac{38}{16}, \frac{17}{16}\right)$ **2.** $\left(\frac{37}{11}, -\frac{3}{11}\right)$ **3.** $(6, -4)$ **4.** $(4, -3, 5)$ **5.** $\left(2, 0, \frac{1}{3}\right)$ **6.** $x = 4 - z,\ y = 1 - 2z$

7. $(-6, 2, -5)$ **8.** $\left(\frac{2}{3}, \frac{1}{2}\right)$ **9.** $x = \frac{3}{4}$ **10.** $\begin{bmatrix} -3 & -12 \\ 6 & -21 \end{bmatrix}$ **11.** $\begin{bmatrix} -3 & -2 \\ 6 & 7 \end{bmatrix}$ **12.** $\begin{bmatrix} 4 & 6 \\ -2 & 2 \end{bmatrix}$

13. $\begin{bmatrix} 11 & -3 \\ 7 & -1 \\ 3 & 0 \end{bmatrix}$ **14.** not possible **15.** [3] **16.** $\begin{bmatrix} 4 & 4 & 0 \\ 20 & 17 & -8 \end{bmatrix}$ **17.** not possible **18.** $\begin{bmatrix} -4 & -7 \\ -3 & -5 \end{bmatrix}$

19. $\begin{bmatrix} -\frac{5}{2} & \frac{6}{2} \\ \frac{7}{2} & -4 \end{bmatrix}$ **20.** not possible **21.** $\begin{bmatrix} -2 & -6 & 15 \\ 0 & -1 & 2 \\ 1 & 2 & -5 \end{bmatrix}$ **22.** not possible **23.** $\left[\begin{array}{ccc|c} 6 & 4 & -5 & 10 \\ 3 & -2 & 0 & 12 \\ 1 & 1 & -4 & -2 \end{array}\right]$

24. $\begin{bmatrix} 1 & 0 & 0 & 0 \\ 0 & 1 & 0 & 1 \\ 0 & 0 & 1 & -1 \end{bmatrix}$ **25.** $\begin{bmatrix} 1 & 0 & -\frac{3}{5} & 0 \\ 0 & 1 & \frac{9}{5} & 0 \\ 0 & 0 & 0 & 1 \end{bmatrix}$ **26.** $\begin{bmatrix} 1 & 0 & \frac{7}{13} \\ 0 & 1 & -\frac{5}{13} \\ 0 & 0 & 0 \\ 0 & 0 & 0 \end{bmatrix}$ **27.** 23 field goals, 13 free throws

28. **(a)** (57, 13) **(b)** (198, 158) **29.** $20,000 in bonds and $30,000 in stocks

CHAPTER 3

Section 3–1, page 157

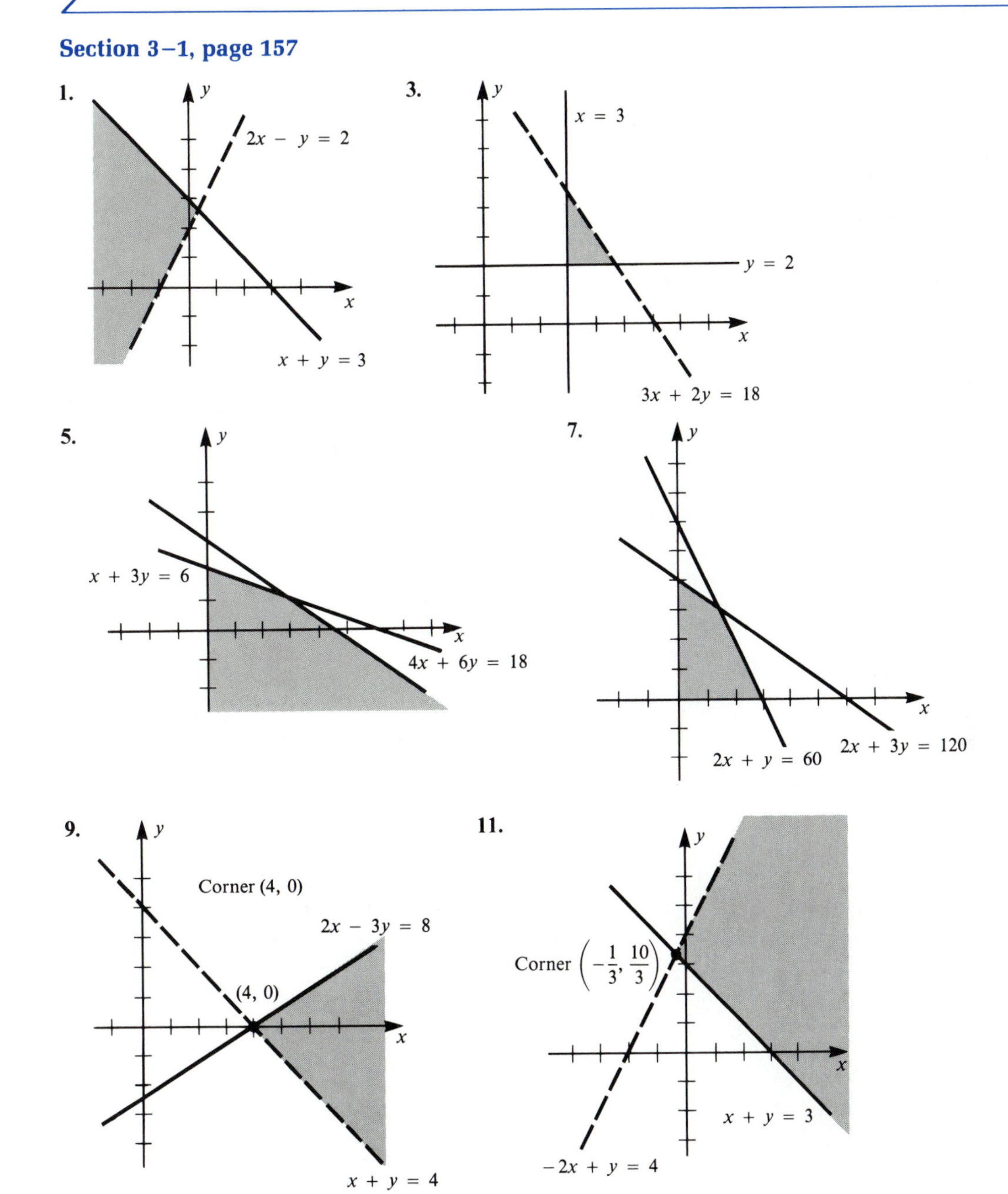

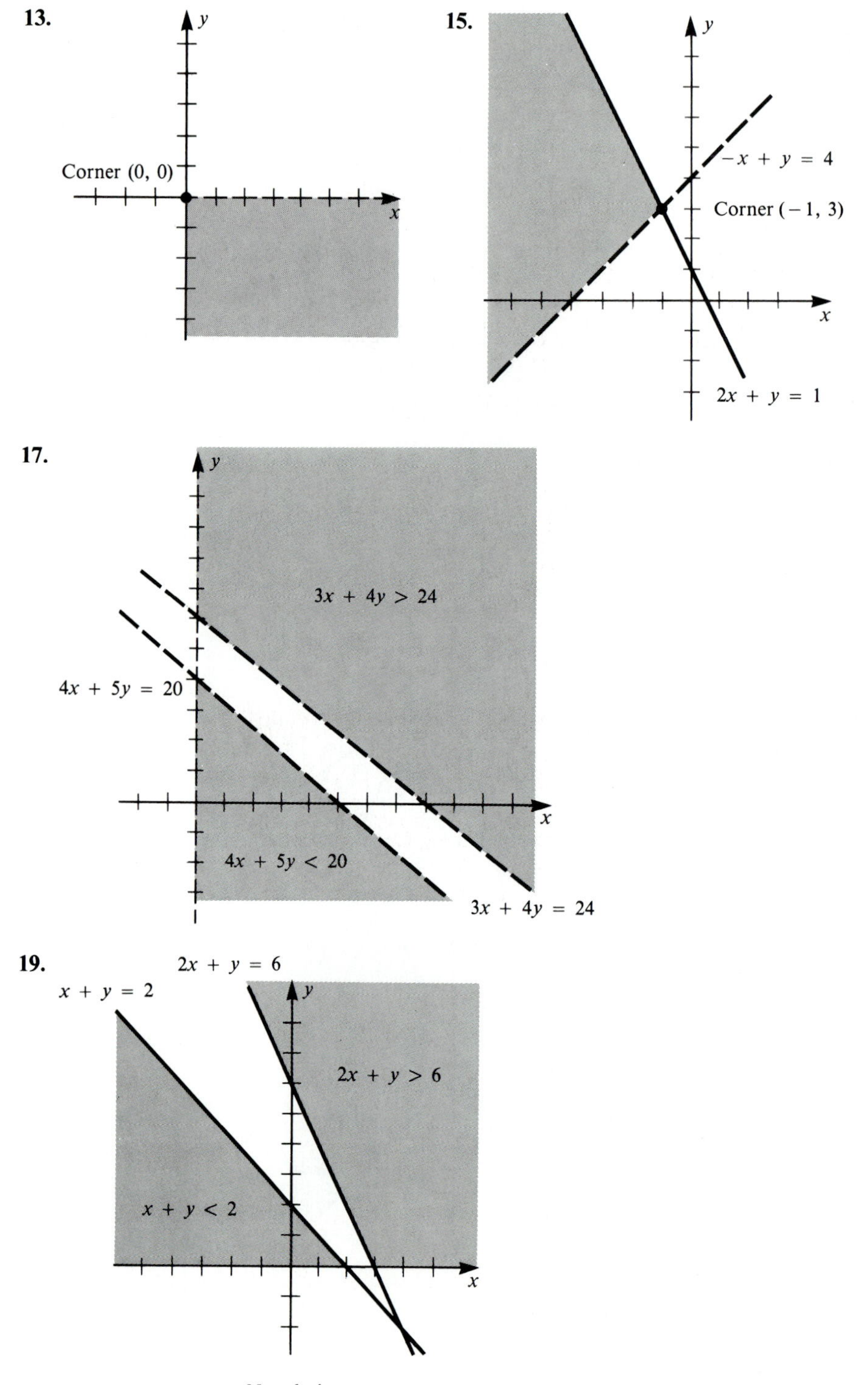
13.
y
Corner (0, 0)
x
15.
y
−x + y = 4
Corner (−1, 3)
x
2x + y = 1
17.
y
3x + 4y > 24
4x + 5y = 20
x
4x + 5y < 20
3x + 4y = 24
19.
2x + y = 6
x + y = 2
y
2x + y > 6
x + y < 2
x

No solution

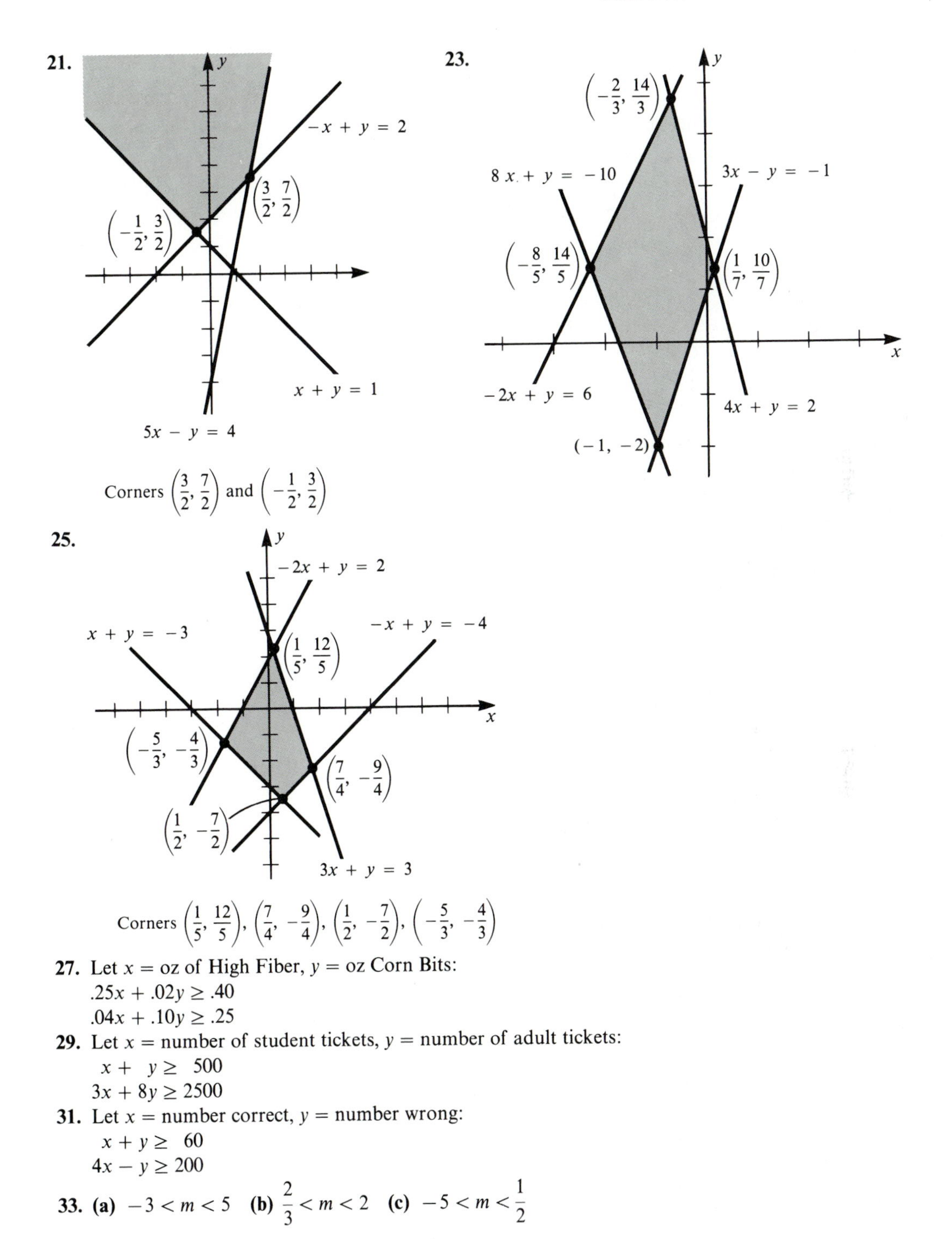

Corners $\left(\frac{3}{2}, \frac{7}{2}\right)$ and $\left(-\frac{1}{2}, \frac{3}{2}\right)$

Corners $\left(\frac{1}{5}, \frac{12}{5}\right)$, $\left(\frac{7}{4}, -\frac{9}{4}\right)$, $\left(\frac{1}{2}, -\frac{7}{2}\right)$, $\left(-\frac{5}{3}, -\frac{4}{3}\right)$

27. Let x = oz of High Fiber, y = oz Corn Bits:
$.25x + .02y \geq .40$
$.04x + .10y \geq .25$

29. Let x = number of student tickets, y = number of adult tickets:
$x + y \geq 500$
$3x + 8y \geq 2500$

31. Let x = number correct, y = number wrong:
$x + y \geq 60$
$4x - y \geq 200$

33. **(a)** $-3 < m < 5$ **(b)** $\frac{2}{3} < m < 2$ **(c)** $-5 < m < \frac{1}{2}$

Section 3–2, page 169

1. Let x = number of style A, y = number of style B:
$x + 2y \leq 110$ (labor restriction)
$x + \ \ y \leq \ 80$ (space restriction)
$x \geq 0, y \geq 0$
Maximize $z = 50x + 40y$

3.

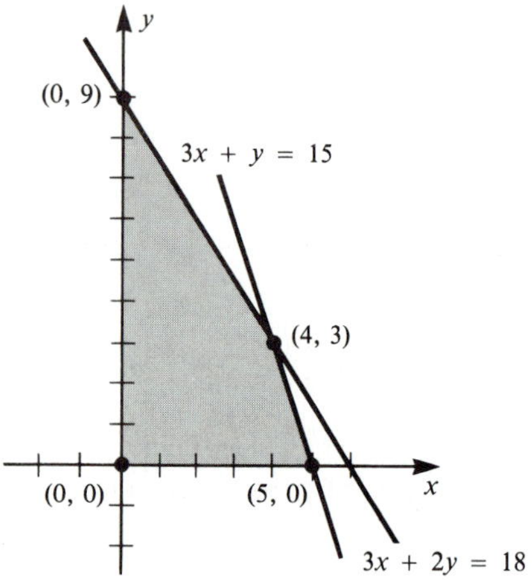

at $(0, 9)$ $z = 108$
$(4, 3)$ $z = 116$
$(5, 0)$ $z = 100$
$(0, 0)$ $z = 0$
so maximum z is 116 at $(4, 3)$

5. Maximum z is 6 at (2, 1). **7.** Maximum z is 26 at (8, 2).

9. Maximum z is 90 at (3, 5), (0, 10), and all points on that line segment. **11.** Maximum z is 50 at (0, 10).

13. Maximum z is 240 at $(-5, 20)$, $(-20, 8)$, and points on the line segment between.

15. Maximum z is 30.4 at $\left(\frac{14}{5}, \frac{2}{5}\right)$. **17.** No feasible solution.

19. Maximum profit is \$27,000 for 30 acres of strawberries and 10 acres tomatoes.

21. Maximum profit is \$18,000 with 500 premium and 1500 regular tires.

23. Maximum profit is \$2400 when 120 are shipped from Akron and none from Cincinatti, or 60 from Akron and 30 from Cincinatti (or any 2 values on the line segment between).

25. Maximum profit is \$5200 for 800 barrels of gas and 600 barrels heating oil.

27. Maximum efficiency number is 1780 for 50 B250's and 20 X100's.

29. Let x = number 20-passenger buses, y = number 30-passenger buses:
Maximize $z = 20x + 30y$ subject to
$18{,}000x + 22{,}000y \leq 572{,}000$
$x + y \leq 30$
$x \geq 0, y \geq 17$
Maximum number passengers is 780 with 26 30-passenger buses, 0 20-passenger buses.

31. **(a)** The maximum value of z is 100 at (8, 5) and (10, 0), so it is maximum at all points on the line segment between. The slope of that line is $m = -\frac{5}{2}$. The slope of the objective function is also $-\frac{5}{2}$.

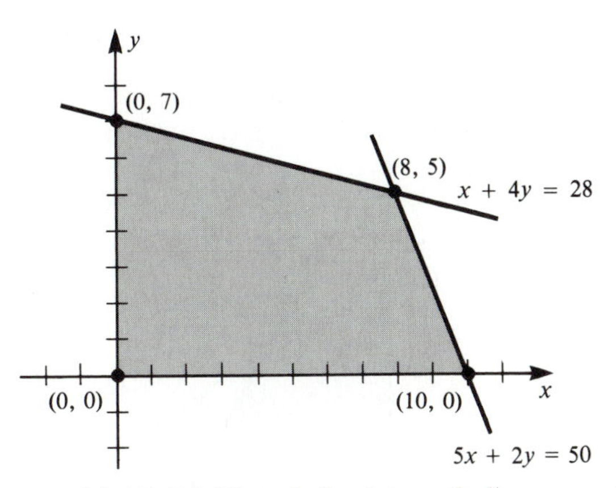

(b) The maximum value of z is 180 at (12, 20), (24, 10), and all points on the line segment joining them. The slope of that line is $-\frac{5}{6}$, the same as the objective function.

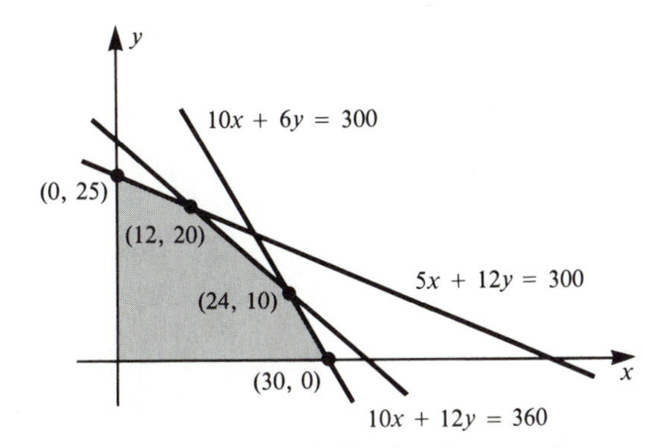

(c) The minimum value of z is 300 at (6, 16), (24, 4), and all points between that are on $2x + 3y = 60$ $\left(\text{its slope is } -\frac{2}{3}\right)$. The slope of $10x + 15y = 300$ is also $-\frac{2}{3}$.

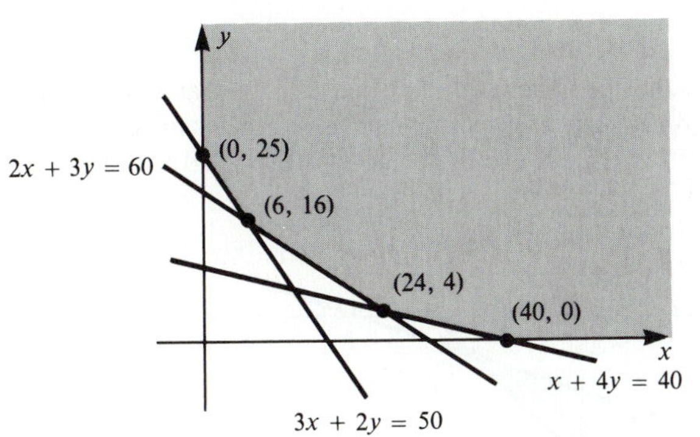

(d) The maximum value of z is 2400 at (0, 100), (60, 75), and all points between that are on the line $5x + 12y = 1200$. The slope of this line is $-\frac{5}{12}$, the same as the slope of $10x + 24y = 2400$.

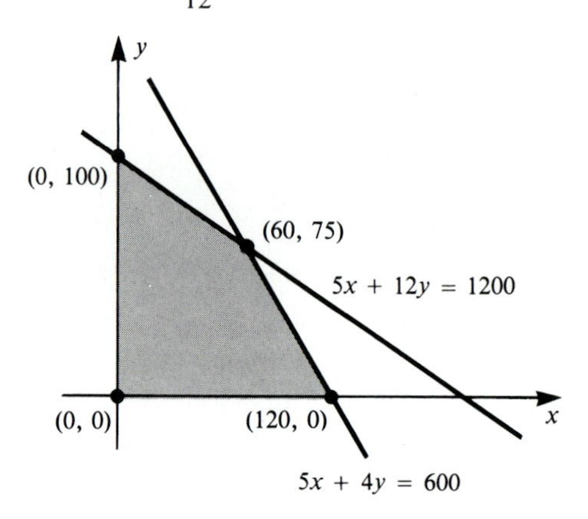

Review Exercises, Chapter 3, page 173

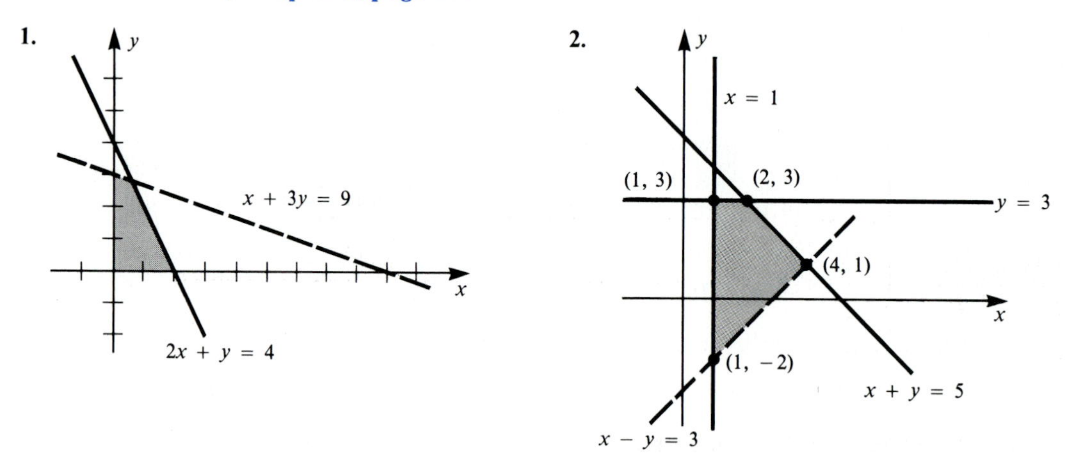

3. Corners are $(-9, -5)$, $(-1, -5)$, and $(3, -1)$.

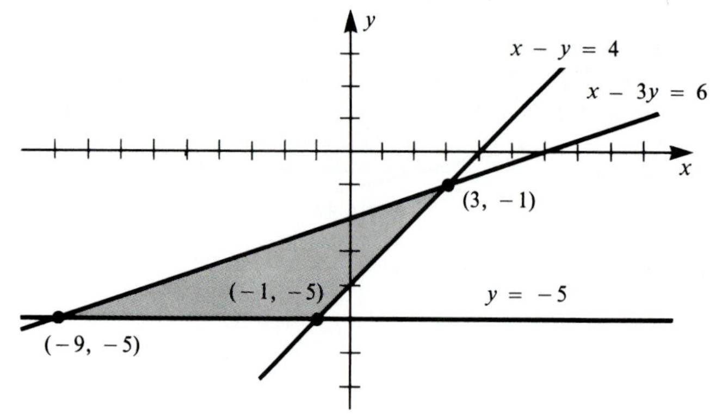

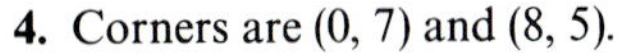

4. Corners are (0, 7) and (8, 5).

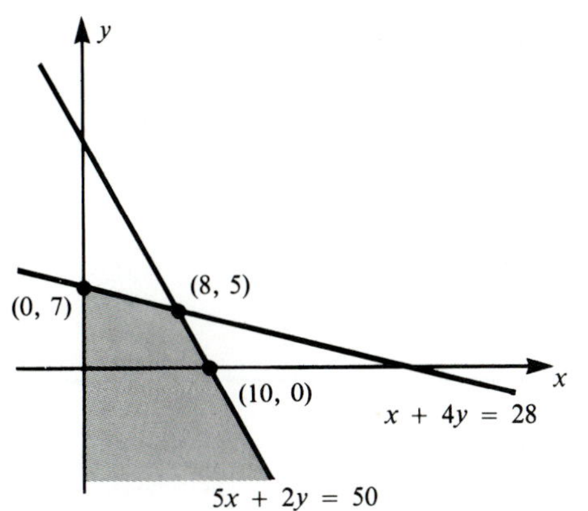

5. Corners are (−4, 2), (4, 8), (5, 0), and (−2, 0).

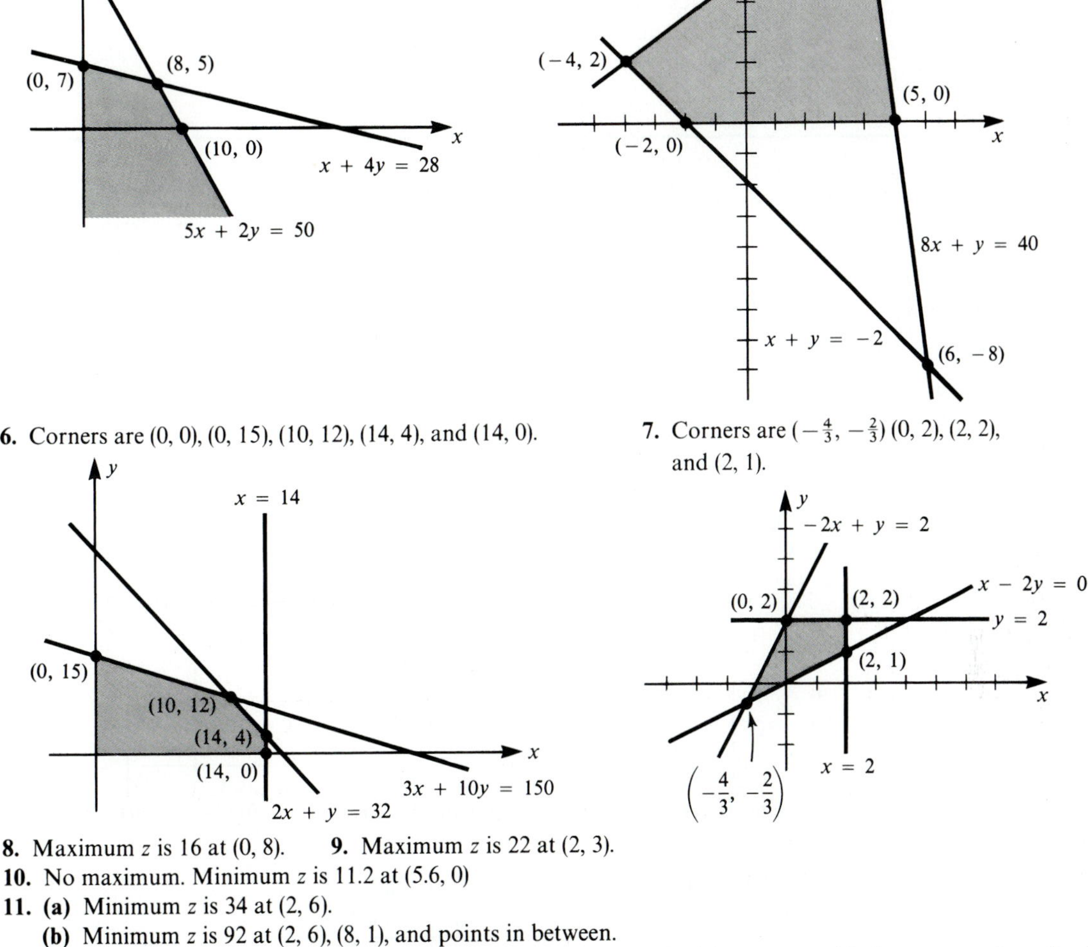

6. Corners are (0, 0), (0, 15), (10, 12), (14, 4), and (14, 0).

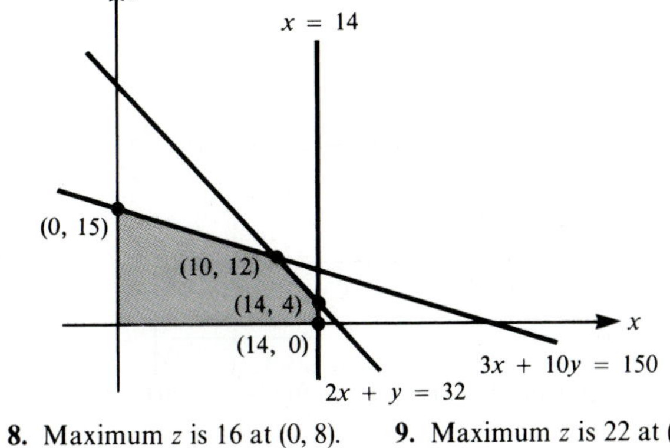

7. Corners are $\left(-\frac{4}{3}, -\frac{2}{3}\right)$ (0, 2), (2, 2), and (2, 1).

8. Maximum z is 16 at (0, 8). **9.** Maximum z is 22 at (2, 3).

10. No maximum. Minimum z is 11.2 at (5.6, 0)

11. **(a)** Minimum z is 34 at (2, 6).

(b) Minimum z is 92 at (2, 6), (8, 1), and points in between.

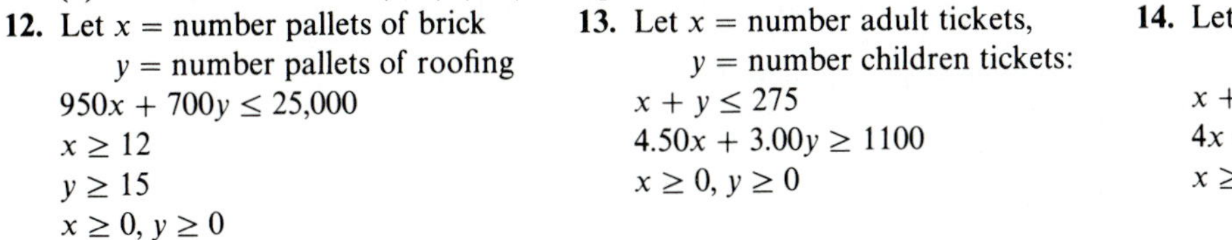

12. Let x = number pallets of brick
y = number pallets of roofing
$950x + 700y \leq 25{,}000$
$x \geq 12$
$y \geq 15$
$x \geq 0, y \geq 0$

13. Let x = number adult tickets,
y = number children tickets:
$x + y \leq 275$
$4.50x + 3.00y \geq 1100$
$x \geq 0, y \geq 0$

14. Let x = number of suits
y = number of dresses
$x + 2y \leq 80$
$4x + 2y \leq 150$
$x \geq 0, y \geq 0$

15. 640 bars of standard and 280 bars of premium yield a revenue of $85,600.

16. Let x = lbs peanuts
y = lbs cashews
$z = .75x + 1.40y$

(a) Constraints are
$x + y \geq 1200$
$y \geq \frac{1}{3}(x + y)$
$x \leq 600$
$x \geq 0, y \geq 0$
Minimum cost is $1290 using 600 lbs peanuts and 600 lbs cashews

(b) Constraints are
$x + y \geq 1200$
$y \geq \frac{1}{3}(x + y)$
$x \leq 900$
$x \geq 0, y \geq 0$
Minimum cost is $1160 using 900 lbs peanuts and 450 lbs cashews

CHAPTER 4

Section 4–1, page 183

1. Maximize $[50 \quad 80]\begin{bmatrix} x_1 \\ x_2 \end{bmatrix}$ subject to $\begin{bmatrix} 7 & 15 \\ 3 & 14 \end{bmatrix}\begin{bmatrix} x_1 \\ x_2 \end{bmatrix} \leq \begin{bmatrix} 30 \\ 56 \end{bmatrix}$ and $\begin{bmatrix} x_1 \\ x_2 \end{bmatrix} \geq \begin{bmatrix} 0 \\ 0 \end{bmatrix}$

3. Maximize $[42 \quad 26 \quad 5]\begin{bmatrix} x_1 \\ x_2 \\ x_3 \end{bmatrix}$ subject to $\begin{bmatrix} 13 & 4 & 23 \\ 4 & 15 & 7 \\ 3 & -2 & 32 \end{bmatrix}\begin{bmatrix} x_1 \\ x_2 \\ x_3 \end{bmatrix} \leq \begin{bmatrix} 88 \\ 92 \\ 155 \end{bmatrix}$ and $\begin{bmatrix} x_1 \\ x_2 \\ x_3 \end{bmatrix} \geq \begin{bmatrix} 0 \\ 0 \\ 0 \end{bmatrix}$

5.
$$\begin{aligned} 2x_1 + 3x_2 + s_1 \quad\quad &= 9 \\ x_1 + 5x_2 \quad\quad + s_2 &= 16 \end{aligned}$$

7.
$$\begin{aligned} x_1 + 7x_2 - 4x_3 + s_1 \quad\quad\quad &= 150 \\ 5x_1 + 9x_2 + 2x_3 \quad\quad + s_2 \quad\quad &= 435 \\ 8x_1 - 3x_2 + 16x_3 \quad\quad\quad + s_3 &= 345 \end{aligned}$$

9.
$$\begin{aligned} 2x_1 + 6x_2 + s_1 \quad\quad\quad &= 9 \\ x_1 - 5x_2 \quad\quad + s_2 \quad\quad &= 14 \\ -3x_1 + x_2 \quad\quad\quad + s_3 &= 8 \\ -3x_1 - 7x_2 &= z \end{aligned}$$

11.
$$\begin{aligned} 6x_1 + 7x_2 + 12x_3 + s_1 \quad\quad\quad &= 50 \\ 4x_1 + 18x_2 + 9x_3 \quad\quad + s_2 \quad\quad &= 85 \\ x_1 - 2x_2 + 14x_3 \quad\quad\quad + s_3 &= 66 \\ -420x_1 - 260x_2 - 50x_3 &= z \end{aligned}$$

13.
$$\left[\begin{array}{ccccc|c} 4 & 5 & 1 & 0 & 0 & 10 \\ 3 & 1 & 0 & 1 & 0 & 25 \\ \hline -3 & -17 & 0 & 0 & 1 & 0 \end{array}\right]$$

15.
$$\left[\begin{array}{ccccccc|c} 16 & -4 & 9 & 1 & 0 & 0 & 0 & 128 \\ 8 & 13 & 22 & 0 & 1 & 0 & 0 & 144 \\ 5 & 6 & -15 & 0 & 0 & 1 & 0 & 225 \\ \hline -20 & -45 & -40 & 0 & 0 & 0 & 1 & 0 \end{array}\right]$$

17. Let x_1 = number of cartons of screwdrivers, x_2 = number of cartons of chisels, x_3 = number of cartons of putty knives:

$$\left[\begin{array}{cccccc|c} 3 & 4 & 5 & 1 & 0 & 0 & 2200 \\ 15 & 12 & 11 & 0 & 1 & 0 & 8500 \\ \hline -5 & -6 & -5 & 0 & 0 & 1 & 0 \end{array}\right]$$

19. Let x_1 = number oz of steak, x_2 = number oz of potato, x_3 = number oz salad:

$$\left[\begin{array}{cccccc|c} 20 & 50 & 56 & 1 & 0 & 0 & 1000 \\ 1.5 & 3.0 & 2.0 & 0 & 1 & 0 & 35 \\ \hline -.5 & -1.0 & -9.0 & 0 & 0 & 1 & 0 \end{array}\right]$$

21. Let x_1 = number of desks, x_2 = number of cabinets, x_3 = number of chairs:

$$\left[\begin{array}{ccccccc|c} 3 & 9 & 1 & 1 & 0 & 0 & 0 & 810 \\ 4 & 1 & 2 & 0 & 1 & 0 & 0 & 400 \\ 2 & 1 & 2 & 0 & 0 & 1 & 0 & 100 \\ \hline -16 & -12 & -6 & 0 & 0 & 0 & 1 & 0 \end{array}\right]$$

23. Let x_1 = number of Alpine, x_2 = number of Cub, x_3 = number of Aspen:

$$\left[\begin{array}{cccccc|c} 30 & 15 & 60 & 1 & 0 & 0 & 8400 \\ 36 & 24 & 32 & 0 & 1 & 0 & 7680 \\ \hline -8 & -4 & -12 & 0 & 0 & 1 & 0 \end{array}\right]$$

Section 4–2, page 197

1. $x_1 = 6, x_2 = 12, s_1 = 0, s_2 = 0, z = 24$ **3.** $x_1 = 2, x_2 = 1, s_1 = 0, s_2 = 0, z = 14$
5. $x_1 = 160, x_2 = 0, z = 1600$ **7.** $x_1 = 0, x_2 = 100, x_3 = 50, z = 22{,}500$
9. $x_1 = 0, x_2 = 75, x_3 = 12.5, z = 475$ **11.** $x_1 = 24, x_2 = 0, z = 792$ **13.** $x_1 = 20, x_2 = 20, x_3 = 20, z = 1200$
15. $x_1 = 62.5, x_2 = 75, x_3 = 35, z = 1700$ **17. 17.** Unbounded, no solution **19.** Unbounded, no solution
21. $x_1 = 0, x_2 = 15, x_3 = 5, x_4 = 0, z = 50$
23. 316.67 screwdrivers, 312.5 chisels, 0 putty knives; maximum profit \$3,458.33.
25. 30 note pads, 0 loose leaf paper, 120 spiral notebooks; profit \$1740.
27. 550 machines from *A*, 175 from *B*. and none from *C*; profit \$10,100.
29. 36 of table *X* and none of *Y*; profit \$288.

Section 4–3, page 207

1. (a) 17, 3, −33, 6, 0 **(b)** (0, 0), (2, 2), (2, 1), (2, 3) **(c)** (2, 3) **3.** 40, 26, 0, 13

5.

Point	s_1	s_2	s_3	Point on Boundary	Point in Feasible Region
(5, 10)	15	32	18	no	yes
(8, 10)	12	14	−3	no	no
(5, 13)	0	29	0	yes	yes
(11, 13)	−6	−7	−42	no	no
(10, 12)	0	0	−29	no	no
(15, 11)	0	−29	−58	no	no

7. **(i) (a)** (0, 0, 900, 2800, 0) **(b)** $x_1 = 0, x_2 = 0$
(ii) (a) (180, 0, 0, 1360, 540) **(b)** $x_2 = 0$
$5x_1 + 2x_2 = 900$
(iii) (a) (100, 200, 0, 0, 700) **(b)** $5x_1 + 2x_2 = 900$
$8x_1 + 10x_2 = 2800$

9. (a) $\frac{36}{7}$ **(b)** 6

Section 4–4, page 218

1. $\begin{bmatrix} 2 & 4 \\ 1 & 0 \\ 3 & 2 \end{bmatrix}$ **3.** $\begin{bmatrix} 4 & 1 & 6 & 2 \\ 3 & 8 & -7 & 4 \\ 2 & -2 & 1 & 6 \end{bmatrix}$

5. **(a)** $\left[\begin{array}{cc|c} 6 & 5 & 30 \\ 8 & 3 & 42 \\ \hline 25 & 30 & 1 \end{array}\right]$ **(b)** $\left[\begin{array}{cc|c} 6 & 8 & 25 \\ 5 & 3 & 30 \\ \hline 30 & 42 & 1 \end{array}\right]$ **(c)** $\left[\begin{array}{ccccc|c} 6 & 8 & 1 & 0 & 0 & 25 \\ 5 & 3 & 0 & 1 & 0 & 30 \\ \hline -30 & -42 & 0 & 0 & 1 & 0 \end{array}\right]$

7. **(a)** $\left[\begin{array}{cc|c} 22 & 30 & 110 \\ 15 & 40 & 95 \\ 20 & 35 & 68 \\ \hline 500 & 700 & 1 \end{array}\right]$ **(b)** $\left[\begin{array}{ccc|c} 22 & 15 & 20 & 500 \\ 30 & 40 & 35 & 700 \\ \hline 110 & 95 & 68 & 1 \end{array}\right]$ **(c)** $\left[\begin{array}{cccccc|c} 22 & 15 & 20 & 1 & 0 & 0 & 500 \\ 30 & 40 & 35 & 0 & 1 & 0 & 700 \\ \hline -110 & -95 & -68 & 0 & 0 & 1 & 0 \end{array}\right]$

9. $x_1 = 2, x_2 = 0, z = 6$ **11.** $x_1 = \frac{1}{4}, x_2 = \frac{3}{8}, x_3 = 0, z = 16$ **13.** $x_1 = 0, x_2 = 0, x_3 = 0, x_4 = 2, z = 8$

15. Initial tableau:

$$\begin{array}{ccccc|c} y_1 & y_2 & x_1 & x_2 & w & \\ \hline -18{,}000 & & & & & \end{array}$$

$$\begin{array}{ccccc} y_1 & y_2 & x_1 & x_2 & w \end{array}$$
$$\left[\begin{array}{ccccc|c} 800 & 280 & 1 & 0 & 0 & 22{,}000 \\ 500 & 150 & 0 & 1 & 0 & 12{,}000 \\ \hline -18{,}000 & -9{,}000 & 0 & 0 & 1 & 0 \end{array}\right]$$

Dallas 15 days; New Orleans 32 days; Cost \$714,000

Section 4–5, page 230

1. $x_1 = 70, x_2 = 0, z = 1050$ **3.** $x_1 = 4, x_2 = 14, z = 300$

5. Constraints:

$$\begin{aligned} 5x_1 + 3x_2 + s_1 \quad &= 220 \\ 4x_1 + 10x_2 - s_2 + a_1 &= 80 \\ -3x_1 + 7x_2 \quad + a_2 &= 42 \end{aligned}$$

Objective function:

$z = 20x_1 + 17x_2 - Ma_1 - Ma_2$

7. No solution **9.** $x_1 = 10, x_2 = 15, z = 120$

11. Constraints:

$$\begin{aligned} 2x_1 + 4x_2 + 7x_3 + s_1 \quad &= 42 \\ -x_1 + 3x_2 - x_3 - s_2 + a &= 10 \end{aligned}$$

Objective function:

$z = 3x_1 + 7x_2 + x_3 - Ma$

13. $x_1 = 24, x_2 = 0, z = 360$ **15.** $x_1 = 2.5, x_2 = .5, z = 4.5$

17. $x_1 = 2, x_2 = 6, z = 8$ **19.** $x_1 = 3, x_2 = 1, z = 5$ **21.** $x_1 = 3, x_2 = 0, z = 9$ **23.** No solution

25. 40 Custom, 60 Executive, minimum cost of \$7600

Review Exercises, Chapter 4, page 233

1.
$$\begin{aligned} 6x_1 + 4x_2 + 3x_3 + s_1 \qquad\qquad &= 220 \\ x_1 + 5x_2 + x_3 \quad + s_2 \qquad &= 162 \\ 7x_1 + 2x_2 + 5x_3 \qquad\quad + s_3 &= 139 \end{aligned}$$

2.
$$\begin{aligned} 5x_1 + 3x_2 + s_1 \qquad\qquad &= 40 \\ 7x_1 + 2x_2 \quad + s_2 \qquad &= 19 \\ 6x_1 + 5x_2 \qquad\quad + s_3 &= 23 \end{aligned}$$

3.
$$\begin{aligned} 6x_1 + 5x_2 + 3x_3 + 3x_4 + s_1 \quad &= 89 \\ 7x_1 + 4x_2 + 6x_3 + 2x_4 \quad + s_2 &= 72 \end{aligned}$$

4.
$$\begin{aligned} 7x_1 + 5x_2 + s_1 \quad &= 14 \\ 3x_1 + 6x_2 \quad + s_2 \quad &= 25 \\ 4x_1 + 3x_2 \quad + s_3 \quad &= 29 \\ -3x_1 - 7x_2 \quad + z &= 0 \end{aligned}$$

5.
$$\begin{aligned} 10x_1 + 12x_2 + 8x_3 + s_1 \quad &= 24 \\ 7x_1 + 13x_2 + 5x_3 \quad + s_2 \quad &= 35 \\ -20x_1 - 36x_2 - 19x_3 \quad + z &= 0 \end{aligned}$$

6.
$$\begin{aligned} 9x_1 + 7x_2 + x_3 + x_4 + s_1 \quad &= 84 \\ x_1 + 3x_2 + 5x_3 + x_4 \quad + s_2 \quad &= 76 \\ 2x_1 + x_2 + 6x_3 + 3x_4 \quad + s_3 \quad &= 59 \\ -5x_1 - 12x_2 - 8x_3 - 2x_4 \quad + z &= 0 \end{aligned}$$

7.
$$\begin{aligned} 3x_1 + 7x_2 + s_1 \quad &= 14 \\ 9x_1 + 5x_2 \quad + s_2 \quad &= 18 \\ x_1 - x_2 \quad + s_3 \quad &= 21 \\ -9x_1 - 2x_2 \quad + z &= 0 \end{aligned}$$

8.
$$\begin{aligned} x_1 + x_2 + x_3 + s_1 \quad &= 20 \\ 4x_1 + 5x_2 + x_3 \quad + s_2 \quad &= 48 \\ 2x_1 - 6x_2 + 5x_3 \quad + s_3 \quad &= 38 \\ -x_1 - 5x_2 - 4x_3 \quad + z &= 0 \end{aligned}$$

9.
$$\begin{aligned} x_1 + x_2 + x_3 + s_1 \quad &= 15 \\ 2x_1 + 4x_2 + x_3 \quad + s_2 \quad &= 44 \\ -6x_1 - 8x_2 - 4x_3 \quad + z &= 0 \end{aligned}$$

10. $x_1 = 1, x_2 = \frac{10}{3}, z = \frac{65}{3}$ **11.** $x_1 = 3, x_2 = 0, z = 32$

12. $x_1 = 14, x_2 = 0, x_3 = 3, z = 48$ **13.** $x_1 = 0, x_2 = 8, z = 32$ **14.** Unbounded feasible region, no solution

15. $x_1 = 0, x_2 = 100, x_3 = 50, z = 650$ **16.** $x_1 = 60, x_2 = 132, z = 456$ **17.** For (3, 2, 1), $s_1 = 44$ and $s_2 = 78$. For (5, 10, 3), $s_1 = 0$ and $s_2 = 50$.

18. (a) x_3 **(b)** x_2 **19.** $\frac{56}{17}$ **20.** $\begin{bmatrix} 3 & 4 & 5 \\ 1 & 0 & 7 \\ -2 & 6 & 8 \end{bmatrix} \begin{bmatrix} 4 & -5 \\ 3 & 6 \\ 2 & 12 \\ 1 & 9 \end{bmatrix}$

21. (a) Augmented matrix:

$$\left[\begin{array}{cc|c} 4 & 5 & 52 \\ 7 & 14 & 39 \\ \hline 30 & 17 & 1 \end{array}\right]$$

Dual problem tableau:

$$\left[\begin{array}{ccccc|c} 4 & 7 & 1 & 0 & 0 & 30 \\ 5 & 14 & 0 & 1 & 0 & 17 \\ \hline -52 & -39 & 0 & 0 & 1 & 0 \end{array}\right]$$

(b) Augmented matrix:

$$\left[\begin{array}{ccc|c} 20 & 35 & 15 & 130 \\ 40 & 10 & 6 & 220 \\ 35 & 22 & 18 & 176 \\ \hline 100 & 225 & 145 & 1 \end{array}\right]$$

Dual problem initial tableau:

$$\left[\begin{array}{ccccccc|c} 20 & 40 & 35 & 1 & 0 & 0 & 0 & 100 \\ 35 & 10 & 22 & 0 & 1 & 0 & 0 & 225 \\ 15 & 6 & 18 & 0 & 0 & 1 & 0 & 145 \\ \hline -130 & -220 & -176 & 0 & 0 & 0 & 1 & 0 \end{array}\right]$$

22. $x_1 = 6, x_2 = 4, x_3 = 6, z = 252$ **23.** $x_1 = 0, x_1 = 200, z = 3{,}000$

24. $x_1 = 34.62, x_2 = 61.54, x_3 = 0, z = 1096.15$ **25.** $x_1 = 0, x_2 = 10, z = -20$

26. $x_1 = .75, x_2 = 2.75, z = 4.25$ **27.** $\left[\begin{array}{cccccc|c} 3 & 2.5 & 3.5 & 1 & 0 & 0 & 3{,}200 \\ 26 & 20 & 22 & 0 & 1 & 0 & 18{,}000 \\ \hline -7.50 & -9 & -11 & 0 & 0 & 1 & 0 \end{array}\right]$

28. $\left[\begin{array}{ccccc|c} 500 & 100 & 1 & 0 & 0 & 18{,}000 \\ 250 & 90 & 0 & 1 & 0 & 12{,}000 \\ \hline -20{,}000 & -5{,}000 & 0 & 0 & 1 & 0 \end{array}\right]$

CHAPTER 5

Section 5–1, page 242

1. **(a)** True **(b)** False **(c)** False **(d)** False **(e)** True **(f)** True **(g)** False **(h)** True **(i)** False **(j)** True
3. {M, i, s, p} **5.** {16, 18, 20, ...} **7.** $A \neq B$ **9.** $A = B$ **11.** $A \neq B$ **13.** Not a subset
15. Is a subset **17.** Is a subset
19. **(a)** $\varnothing, \{-1\}, \{2\}, \{4\}, \{-1, 2\}, \{-1, 4\}, \{2, 4\}, \{-1, 2, 4\}$
(b) $\varnothing, \{4\}$
(c) $\varnothing, \{-3\}, \{5\}, \{6\}, \{8\}, \{-3, 5\}, \{-3, 6\}, \{-3, 8\}, \{5, 6\}, \{5, 8\}, \{6, 8\}, \{-3, 5, 6\}, \{-3, 5, 8\}, \{-3, 6, 8\}, \{5, 6, 8\}, \{-3, 5, 6, 8\}$

21. Empty **23.** Not empty **25.** Empty **27.** Empty **29.** Empty **31.** {a, b, c, x, y, z, d}
33. {5, 2, 9, 4, 3, 0, 8} **35.** {1, 2, 3} **37.** {1, 2, 3, 6, 9} **39.** {2, 3} **41.** $\varnothing$ **43.** {1, 2, 3, 6}
45. {1, 2, 3, 6, 9} **47.** 1, 4, 9 **49.** −2, 3, 4 **51.** **(a)** $\subset$ **(b)** neither **(c)** = **(d)** neither **(e)** $\subset$
53. $A \cap B = \{x \mid x$ is an integer that is a multiple of $35\}$ **55.** disjoint **57.** not disjoint
59. $A \cap B$ = set of students at Miami Bay U who are taking both finite math and American history.

Section 5–2, page 251

1. $A' = \{1, 4, 6\}$ **3.** $A' = \{3, 24, 25\}$
5. **(a)** {−3, 4, 12} **(b)** $B' = \{4, 14, 21\}$ **(c)** {4} **(d)** {−3, 4, 12, 14, 21} **7.** **(a)** 10 **(b)** 5 **(c)** 4 **(d)** 11
9. 7 **11.** 19 **13.** **(a)** 3 **(b)** 1 **(c)** 0 **(d)** 6 **(e)** 7 **15.** **(a)** 4 **(b)** 6 **(c)** 0 **(d)** 8 **(e)** 0 **17.** 35
19. **(a)** 21 **(b)** 9 **(c)** 23 **(d)** 28 **(e)** 3 **(f)** 3 **(g)** 19
21.

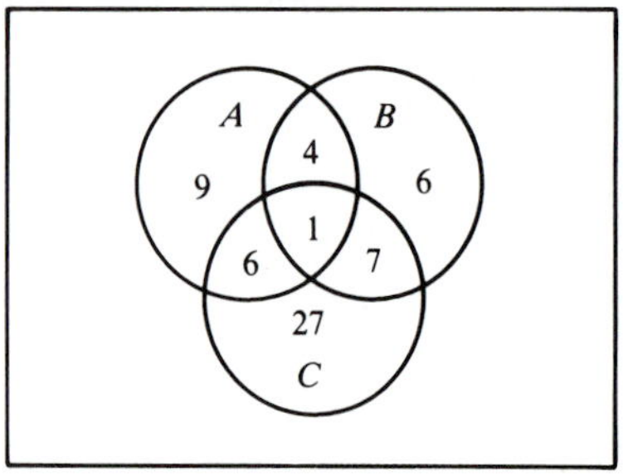

23. 225 **25.** **(a)** 8 **(b)** 27 **(c)** 4 **(d)** 31 **27.** **(a)** 21 **(b)** 10 **(c)** 8 **(d)** 56 **(e)** 29 **(f)** 62
29. **(a)** 42 **(b)** 9 **(c)** 44 **(d)** 24 **31.** **(a)** 210 **(b)** 60 **(c)** 35 **(d)** 115 **33.** 68
35. $n(A \cup B)$ must be at least as large as $n(A)$. **37.** **(a)** 30 **(b)** 80 **39.** **(a)** 3 **(b)** 7 **(c)** 34
41. This information accounts for only 133 students.

43.

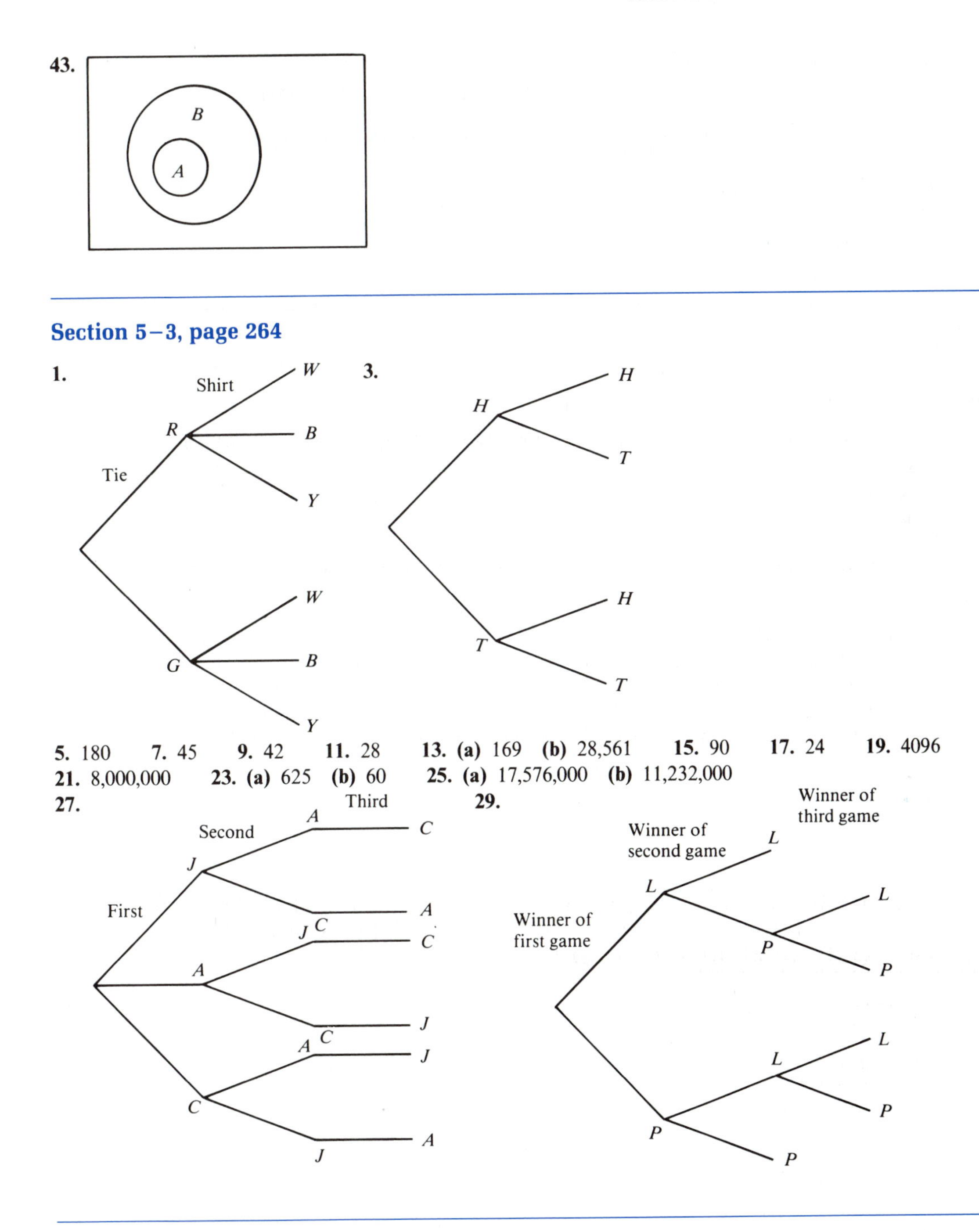

Section 5–3, page 264

5. 180 **7.** 45 **9.** 42 **11.** 28 **13.** **(a)** 169 **(b)** 28,561 **15.** 90 **17.** 24 **19.** 4096
21. 8,000,000 **23.** **(a)** 625 **(b)** 60 **25.** **(a)** 17,576,000 **(b)** 11,232,000

Section 5–4, page 277

1. 6 **3.** 120 **5.** 720 **7.** 840 **9.** 95,040 **11.** 360 **13.** 970,200 **15.** 840 **17.** 15 **19.** 286
21. 126 **23.** 1 **25.** 120 **27.** 2520 **29.** 336 **31.** **(a)** 5040 **(b)** 210 **33.** 6

35. **(a)** 24 **(b)** 840 **37.** 9! **39.** **(a)** 24 **(b)** 64 **41.** {a, b}, {a, c}, {a, d}, {b, c}, {b, d}, {c, d} **43.** 15 **45.** 455 **47.** 210 **49.** 5600 **51.** 4200 **53.** 55 **55.** 24 **57.** 5040 **59.** $C(52, 13)$ **61.** 35 **63.** $C(35, 30)$ **65.** **(a)** $n(n-1)$ **(b)** $n(n-1)(n-2)$ **(c)** n **(d)** $n(n-1)(n-2)(n-3)(n-4)$ **67.** 17,576,000 **69.** **(a)** 64 **(b)** 20 **(c)** 4 **71.** 24 **73.** 826 **75.** **(a)** 20,475 **(b)** 37,128 **77.** **(a)** 11,760 **(b)** 6885 **79.** 116,280 **81.** **(a)** 3003 **(b)** 1176

Section 5–5, page 284

1. 369,600 **3.** 35 **5.** 1260 **7.** 15 **9.** 1680 **11.** 17,153,136 **13.** 630,630 **15.** 3150

Review Exercises, Chapter 5, page 285

1. **(a)** True **(b)** False **(c)** False **(d)** False **(e)** True **(f)** False **(g)** True **(h)** False **(i)** True **(j)** False **(k)** True **(l)** False **(m)** True **(n)** False **(o)** False

2. $A' = \{-1, 1, 3\}$ $B' = \{0, 3, 4\}$
$(A \cap B)' = \{-1, 0, 1, 3, 4\}$ $A' \cap B' = \{3\}$
$A' \cup B' = \{-1, 0, 1, 3, 4\}$
$A \cup A' = \{-2, -1, 0, 1, 2, 3, 4\}$

3. 49 **4.** 14

5.

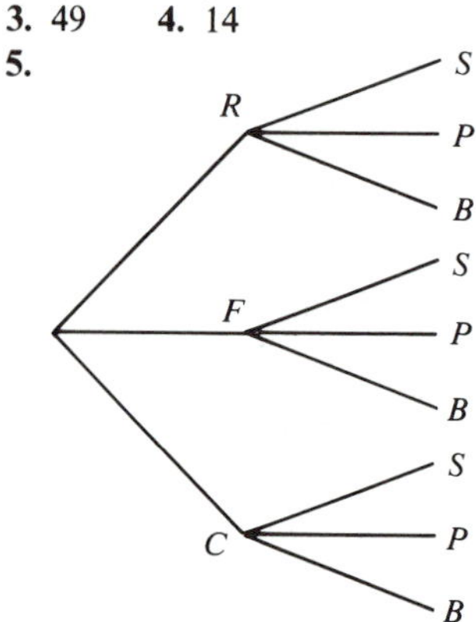

6. 15,120 **7.** **(a)** 6,760,000 **(b)** 3,276,000 **8.** 1680 **9.** 3003 **10.** 756,756 **11.** 3,121,200 **12.** 864 **13.** 6300 **14.** 13,860 **15.** **(a)** 2370 **(b)** 455 **(c)** 10,100 **(d)** 175,560 **(e)** 6 **(f)** 120 **(g)** 560 **(h)** 69,300 **16.** **(a)** 1472 **(b)** 400 **(c)** 1272 **(d)** 22 **(e)** 420 **17.** This information yields only 58 people. **18.** **(a)** 3 **(b)** 11 **(c)** 9 **(d)** 3 **19.** **(a)** 20 **(b)** 6 **(c)** 23 **20.** 96 **21.** 2730 **22.** 10 **23.** 15 **24.** **(a)** 8000 **(b)** 6840 **25.** 180 **26.** $C(150, 6) \times C(60, 3) \times C(15, 2)$ **27.** 17,280 **28.** 15,120 **29.** 20 **30.** 15,504 **31.** 2,598,960 **32.** **(a)** 1680 **(b)** 126 **(c)** 5040 **(d)** 1 **(e)** 24 **(f)** 35 **(g)** 70 **33.** 14 **34.** **(a)** 216 **(b)** 120 **(c)** 108 **35.** 210 **36.** 10,920 **37.** $\varnothing$, {red}, {white}, {blue}, {red, white}, {red, blue}, {white, blue}, {red, white, blue} **38.** 1680 **39.** 5040

40.

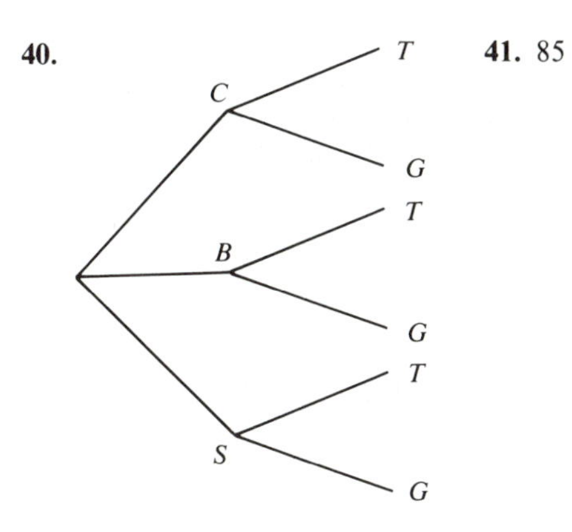

41. 85

CHAPTER 6

Section 6–1, page 298

1. **(a)** {T, F}
(b) all letters of alphabet
(c) {1, 2, 3, 4, 5, 6}
(d) {HHH, HHT, HTH, THH, HTT, THT, TTH, TTT}
(e) {Mon, Tue, Wed, Thurs, Fri, Sat, Sun}
(f) {GC win, GC lose, GC tie} or {BC win, BC lose, BC tie}
(g) {A, B, C, D, F} or the grade 0–100
(h) {Susan Leann, Susan Dana, Susan Julie, Leann Dana, Leann Julie, Dana Julie}
(i) {1, 2, . . . , 30, 31} or (Mon., Tue., Wed., . . . , Sun.)
(j) List of students or {male, female}

3. **(a)** valid **(b)** valid **(c)** not valid **(d)** not valid **(e)** valid **(f)** not valid **(g)** valid **(h)** not valid **(i)** not valid

5. **(a)** .47 **(b)** .53 **7.** **(a)** .46 **(b)** .54 **(c)** .55 **9.** $P(A) = .2$ $P(B) = .3$ $P(C) = .4$ $P(D) = .1$

11. $P(A) = .55$ $P(B) = .20$ $P(C) = .25$ **13.** **(a)** $\frac{4}{52}$ **(b)** $\frac{8}{52}$ **(c)** $\frac{1}{4}$ **(d)** $\frac{1}{2}$

15. $P(\text{Miniburger}) = \frac{140}{800}$, $P(\text{Burger}) = \frac{345}{800}$, $P(\text{Big Burger}) = \frac{315}{800}$

17. $P(\text{Below } 40) = \frac{16}{180}$, $P(40\text{–}49) = \frac{3}{20}$, $P(50\text{–}55) = \frac{1025}{1800}$ $P(\text{over } 55) = \frac{345}{1800}$.

Section 6–2, page 306

1. **(a)** 1 **(b)** 3 **(c)** 4 **(d)** 5 **(e)** 6 **(f)** 4 **(g)** 2 **(h)** 1 **3.** $\frac{5}{8}$ **5.** $\frac{1}{6}$ **7.** $\frac{45}{136}$ **9.** $\frac{1}{221}$

11. **(a)** $\frac{5}{9}$ **(b)** 0 **(c)** 1 **13.** $\frac{3}{4}$ **15.** $\frac{594}{1330}$ **17.** **(a)** $\frac{1}{1365}$ **19.** .1 **21.** $\frac{5}{9}$ **23.** $\frac{48}{143}$

25. **(a)** $\frac{6}{28}$ **(b)** $\frac{3}{28}$ **(c)** $\frac{10}{28}$ **27.** **(a)** $\frac{5}{14}$ **(b)** $\frac{9}{14}$ **29.** **(a)** $\frac{4}{5}$ **(b)** $\frac{1}{5}$ **31.** $\frac{324}{570}$ **33.** **(a)** $\frac{1}{4}$ **(b)** $\frac{1}{16}$
35. **(a)** $\frac{1}{6}$ **(b)** $\frac{1}{2}$ **37.** 50 **39.** $\frac{1}{15}$

Section 6–3, page 317

1. $E \cup F = \{1, 2, 3, 4, 5, 7, 9\}$; $E \cap F = \{1, 3\}$; $E' = \{2, 4, 6, 8, 10\}$
3. $E \cup F$ is the set of students passing English or not passing chemistry.
$E \cap F$ is the set of students passing English and not passing chemistry.
E' is the set of students at State U not passing English.
5. $\frac{2}{5}$ **7.** 0.3 **9.** 0.7 **11.** yes **13.** no **15.** yes **17.** **(a)** $\frac{1}{270{,}725}$ **(b)** $\frac{2}{270{,}725}$
19. **(a)** $\frac{1}{2}$ **(b)** $\frac{1}{3}$ **(c)** $\frac{1}{2}$ **(d)** $\frac{2}{3}$ **21.** **(a)** $\frac{5}{6}$ **(b)** $\frac{1}{6}$ **(c)** $\frac{1}{2}$ **(d)** $\frac{1}{3}$ **(e)** $\frac{1}{3}$
23. **(a)** $\frac{1}{10}$ **(b)** $\frac{1}{10}$ **(c)** $\frac{1}{2}$ **(d)** $\frac{1}{2}$ **(e)** $\frac{1}{2}$ **(f)** $\frac{8}{10}$ **(g)** $\frac{9}{10}$ **25.** $\frac{4}{10}$ **27.** mutually exclusive
29. mutually exclusive **31.** **(a)** $\frac{21}{22}$ **(b)** $\frac{37}{44}$ **33.** $1 - \frac{[C(46, 3)]}{[C(50, 3)]} = \frac{221}{980}$
35. **(a)** $\frac{120}{400}$ **(b)** $\frac{220}{400}$ **(c)** $\frac{55}{400}$ **(d)** $\frac{285}{400}$ **37.** **(a)** $\frac{4}{35}$ **(b)** $\frac{20}{35}$
39. **(a)** $\frac{7}{58}$ **(b)** $\frac{11}{58}$ **(c)** $\frac{17}{58}$ **(d)** $\frac{8}{58}$ **(e)** $\frac{13}{58}$ **(f)** $\frac{22}{58}$ **41.** **(a)** $\frac{4}{11}$ **(b)** $\frac{2}{11}$ **(c)** $\frac{4}{11}$ **(d)** 0 **(e)** $\frac{1}{11}$ **(f)** $\frac{10}{11}$
43. $\frac{1}{6}$ **45.** $\frac{1}{56}$ **47.** **(a)** $\frac{1}{12}$ **(b)** $\frac{1}{144}$ **(c)** $\frac{1}{12}$
49.

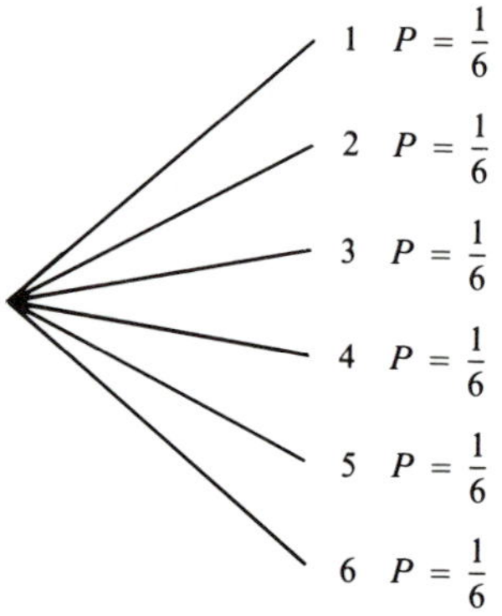

1 $P = \frac{1}{6}$
2 $P = \frac{1}{6}$
3 $P = \frac{1}{6}$
4 $P = \frac{1}{6}$
5 $P = \frac{1}{6}$
6 $P = \frac{1}{6}$

P (even number < 6) $= P(4) + P(6) = \frac{1}{3}$

(a) $\frac{2}{6}$ **(b)** $\frac{2}{6}$
51. **(a)** $\frac{1}{8}$ **(b)** $\frac{7}{8}$ **53.** $\frac{1}{7}$

Section 6–4, page 332

1. **(a)** $\frac{3}{4}$ **(b)** $\frac{1}{2}$ **3.** **(a)** $\frac{6}{10}$ **(b)** $\frac{3}{7}$ **5.** $P(K \mid F) = \frac{1}{10}$, $P(F \mid K) = 1$ **7.** $\frac{2}{11}$ **9.** $\left(\frac{4}{52}\right)\left(\frac{4}{51}\right)$

11. **(a)** $\left(\frac{4}{52}\right)\left(\frac{4}{52}\right)$ **(b)** $\frac{1}{16}$ **(c)** $\frac{1}{4}$ **13.** $\frac{1}{6}$ **15.** **(a)** $\frac{1}{3}$ **(b)** $\frac{1}{3}$ **(c)** $\frac{2}{3}$ **17.** $\frac{1}{6}$

19. **(a)** $\frac{1}{60}$ **(b)** $\frac{1}{10}$ **(c)** 0 **(d)** $\frac{1}{10}$ **(e)** $\frac{1}{10}$ **(f)** $\frac{6}{10}$ **21.** **(a)** .65 **(b)** .68 **(c)** .76

23. **(a)** .06 **(b)** .56 **(c)** .38 **(d)** .94 **25.** **(a)** .24 **(b)** .04 **(c)** .96 **(d)** .76 **27.** **(a)** $\frac{5}{12}$ **(b)** $\frac{1}{2}$

29. **(a)** No. A ring may have both a diamond and a ruby.
(b) Yes. $P(E \cap F) = \frac{1}{4}$, $P(E)P(F)\left(\frac{1}{2}\right)\left(\frac{1}{2}\right) = \frac{1}{4}$

31. **(a)** $P(E \cup F) = .7$, $P(E \mid F) = .5$ **(b)** No, $P(E \mid F) \neq P(E)$ **33.** **(a)** $\frac{9}{10}$ **(b)** $\frac{2}{3}$ **35.** $\frac{C(40, 4)C(60, 8)}{C(100, 12)}$

37. $\frac{2700}{168{,}113}$ **39.** $P(F \cap G) = \frac{6}{52}$, $P(F)P(G) = \left(\frac{1}{2}\right)\left(\frac{12}{52}\right) = \frac{6}{52}$, so they are independent

41. **(a)**

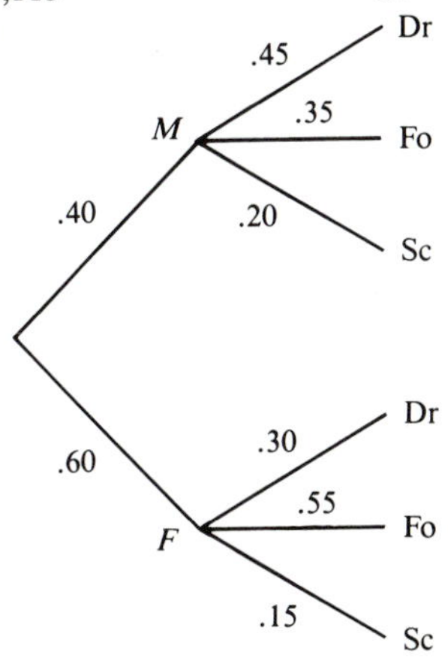

M (.40): Dr .45, Fo .35, Sc .20

F (.60): Dr .30, Fo .55, Sc .15

(b) .47
(c) .18
(d) No, $P(E \cap F) = .33$ and $P(E)P(F) = .282$

43. No. $P(F \cap G) = \frac{834}{3519}$, $P(F)P(G) = \left(\frac{887}{3519}\right)\left(\frac{1768}{3519}\right)$ **45.** **(a)** $\frac{1}{10}$ **(b)** $\frac{1}{90}$ **(c)** $\frac{9}{10}$

47. $\frac{C(15, 2)C(20, 2)C(10, 2)C(5, 2)}{C(50, 8)}$ **49.** $\frac{1}{8}$ **51.** **(a)** $\frac{C(4, 1)C(48, 12)}{C(52, 13)}$ **(b)** $\frac{C(4, 4)C(48, 9)}{C(52, 13)}$ **(c)** $\frac{C(48, 13)}{C(52, 13)}$

53. **(a)** $\frac{1}{11}$ **(b)** $\frac{6}{11}$ **(c)** $\frac{1}{11}$ **55.** **(a)** $\frac{1}{25}$ **(b)** $\frac{2}{5}$ **(c)** $\frac{4}{25}$ **(d)** $\frac{1}{5}$ **(e)** $\frac{11}{25}$ **57.** $\frac{8}{19}$

59. **(a)** $\frac{1}{810{,}000}$ **(b)** $\frac{1}{27{,}000}$ **(c)** $\frac{203}{250}$ **61.** $P(E \cap E) = P(E)$, $P(E \mid E) = \frac{P(E \cap E)}{P(E)} = \frac{P(E)}{P(E)} = 1$

63. If $E \subset F$, then $F \cap E = E$. Then, $P(F \mid E) = \frac{P(F \cap E)}{P(E)} = \frac{P(E)}{P(E)} = 1$

Section 6–5, page 347

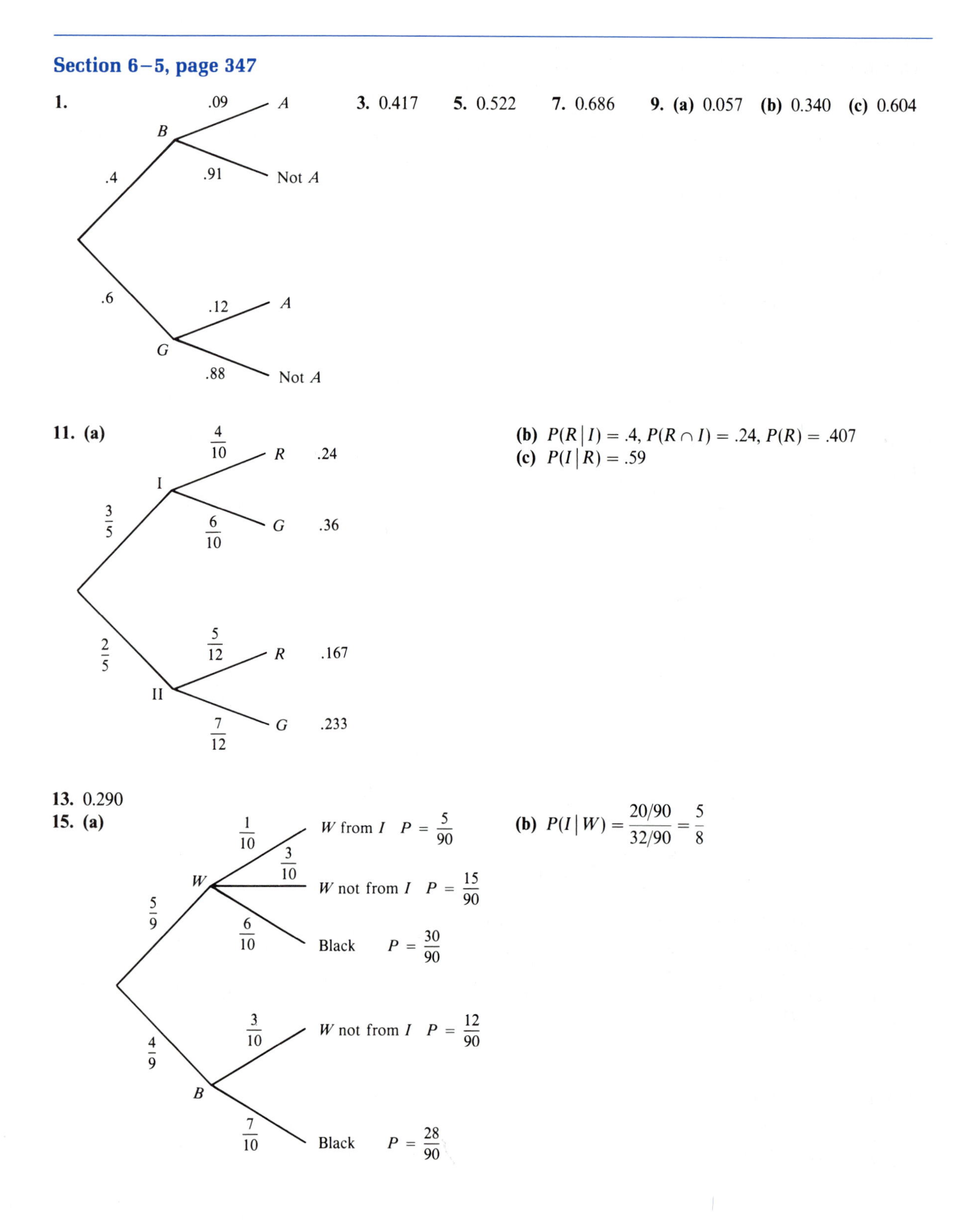

3. 0.417 **5.** 0.522 **7.** 0.686 **9.** **(a)** 0.057 **(b)** 0.340 **(c)** 0.604

11. **(b)** $P(R|I) = .4$, $P(R \cap I) = .24$, $P(R) = .407$
(c) $P(I|R) = .59$

13. 0.290

15. **(b)** $P(I|W) = \dfrac{20/90}{32/90} = \dfrac{5}{8}$

17. 0.991

19. **(a)** $P(\text{Under 25} \mid \text{Accident}) = 0.18$ **(b)** $P(\text{Over 50} \mid \text{Accident}) = 0.30$

(c) $P(\text{Accident} \mid \text{Under 25}) = \dfrac{P(\text{Accident})P(\text{Under 25} \mid \text{Accident})}{P(\text{Under 25})} = 0.225$

21. **(a)** $P(\text{Fraud}) = P(\text{Exceeds})P(\text{Fraud} \mid \text{Exceeds}) + P(\text{Not Exceeds})P(\text{Fraud} \mid \text{Not Exceeds}) = .11(.20) + .89(.03) = .0487$

(b) $P(\text{Exceeds} \mid \text{Fraud}) = \dfrac{.022}{.0487} = .452$

23. $P(A \mid \text{Defective}) = .171$
$P(B \mid \text{Defective}) = .086$
$P(C \mid \text{Defective}) = .628$
$P(D \mid \text{Defective}) = .114$

25. 0.324 **27.** 0.294

Section 6–6, page 358

1. 0.0154 **3.** 0.0486 **5.** 0.00122 **7.** 0.0701 **9.** 0.2344 **11.** 0.00045 **13.** 0.0879

15.

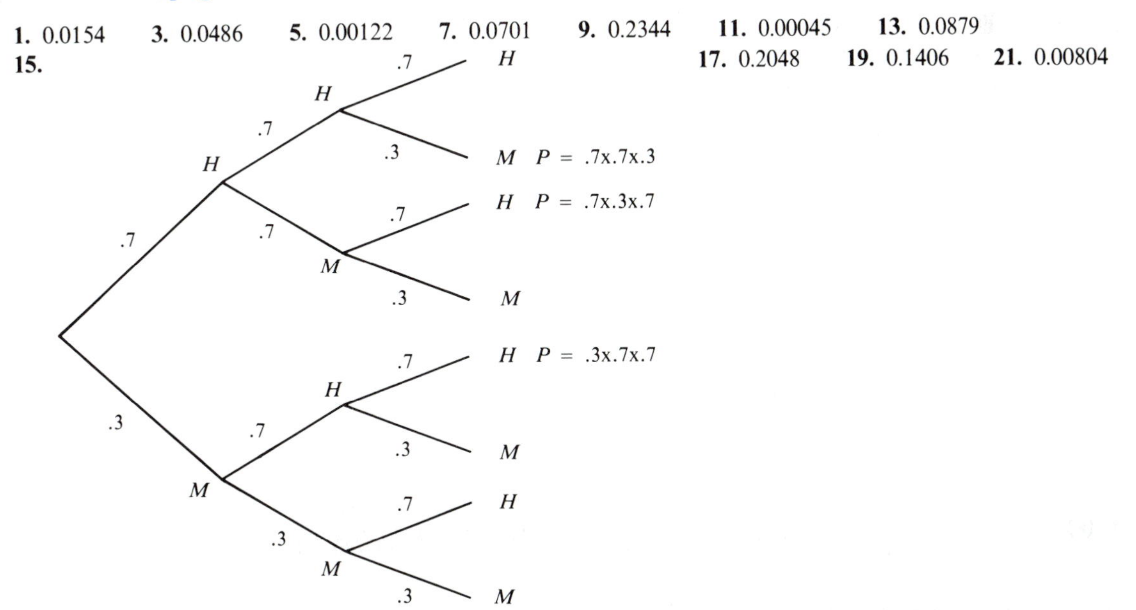

The probability of 2 hits and 1 miss is 3 (.7x.7x.3) = .441

17. 0.2048 **19.** 0.1406 **21.** 0.00804

23. no sixes, 0.4823; 1 six, 0.3858; 2 sixes, 0.1157; 3 sixes, 0.0154; 4 sixes, 0.00077. **25.** 0.2344

27. 0.2458 **29.** 0.2765 **31.** 0.2150 **33.** 0.1209 **35.** 0.0355 **37.** 0.3633

39. **(a)** 0.0439 **(b)** 0.0547 **(c)** 0.0547 **41.** **(a)** 0.1975 **(b)** 0.3951 **(c)** 0.2963 **(d)** 0.0988 **(e)** 0.0123

43. 0.0080 **45.** 0.5906 **47.** 0.6330

Section 6–7, page 371

1. **(a)** transition **(b)** not transition **(c)** transition **(d)** transition **3.** **(a)** 10% **(b)** 0.6 **(c)** 0.9 **(d)** 40%

5. **(a)** 0.4 **(b)** 0.8 **(c)** 0.3 **7.** [.675 .325] **9.** $M_1 = [.282 \quad .718]$, $M_2 = [.32616 \quad .67384]$

11. $M_1 = [.175 \quad .375 \quad .450]$, $M_2 = [.1800 \quad .4075 \quad .4125]$ **13.** $[\frac{2}{3} \quad \frac{1}{3}]\begin{bmatrix} .6 & .4 \\ .8 & .2 \end{bmatrix} = [\frac{2}{3} \quad \frac{1}{3}]$

15. $[\frac{13}{28} \quad \frac{15}{28}]\begin{bmatrix} .25 & .75 \\ .65 & .35 \end{bmatrix} = [\frac{13}{28} \quad \frac{15}{28}]$

17. $MT = [.25 \quad .33 \quad .42]$,
$(MT)T = [.259 \quad .323 \quad .418]$,
$((MT)T)T = [.2577 \quad .3233 \quad .419]$,

$$T^3 = \begin{bmatrix} .249 & .345 & .406 \\ .242 & .386 & .372 \\ .276 & .260 & .464 \end{bmatrix}$$

$[.3 \quad .3 \quad .4]T^3 = [.2577 \quad .3233 \quad .419]$

19. $x = .4$ **21.** Next year 42% contribute, 58% do not. In two years 40.2% contribute.

23. [.6 .4], [.38 .62], [.314 .686], [.2942 .7058], [.28826 .71174], [.286478 .713522], [.2859434 .7140566]
It appears to be approaching [.285 .714]. The steady state matrix is actually $[\frac{2}{7} \quad \frac{5}{7}]$.

25. $[\frac{4}{17} \quad \frac{10}{17} \quad \frac{3}{17}]$ **27.** $[\frac{18}{57} \quad \frac{29}{57} \quad \frac{10}{57}]$ **29.** $[\frac{7}{29} \quad \frac{16}{29} \quad \frac{6}{29}]$ **31.** $[\frac{10}{11} \quad \frac{1}{11}]$ **33.** $[\frac{3}{11} \quad \frac{4}{11} \quad \frac{4}{11}]$

35.

		To				
		1	*2*	*3*	*4*	*5*
From	*1*	0	$\frac{1}{2}$	0	0	$\frac{1}{2}$
	2	$\frac{1}{2}$	0	$\frac{1}{2}$	0	0
	3	0	$\frac{1}{2}$	0	$\frac{1}{2}$	0
	4	0	0	$\frac{1}{2}$	0	$\frac{1}{2}$
	5	$\frac{1}{2}$	0	0	$\frac{1}{2}$	0

37. One-fourth red, one-half pink, one-fourth white **39.** **(a)** 0.06 **(b)** 0.05 **(c)** 0.80 **(d)** 0.04

Review Exercises, Chapter 6, page 378

1. No **2.** $P(a) = 0.1$, $P(b) = 0.1$, $P(c) = 0.2$, $P(d) = 0.6$ **3.** $\frac{146}{360}$ **4.** **(a)** $\frac{2}{7}$ **(b)** $\frac{4}{7}$ **(c)** 0 **5.** $\frac{17}{30}$

6. **(a)** $\frac{1}{2}$ **(b)** $\frac{17}{30}$ **(c)** $\frac{2}{5}$ **(d)** 0 **(e)** $\frac{7}{30}$ **(f)** $\frac{1}{30}$ **7.** $\frac{1}{22}$ **8.** $\frac{2}{3}$ **9.** $\frac{8}{52}$ **10.** **(a)** $\frac{1}{6}$ **(b)** $\frac{1}{4}$ **(c)** $\frac{1}{12}$

11. $\frac{1}{32}$ **12.** **(a)** $\frac{28}{52}$ **(b)** $\frac{22}{52}$ **(c)** $\frac{16}{52}$ **13.** **(a)** $\frac{3}{4}$ **(b)** $\frac{13}{66}$ **14.** **(a)** $\frac{48}{52}$ **(b)** $\frac{40}{52}$ **15.** **(a)** $\frac{3}{13}$ **(b)** $\frac{10}{13}$

16. **(a)** $P(F) = \frac{4}{9}$, $P(G) = \frac{5}{9}$, $P(H) = \frac{8}{90}$

(b) $P(F \cap H) = \frac{5}{90}$ (warped birch)

(c) $P(F \cup H) = \frac{43}{90}$ (birch or warped)

(d) $P(F' \cup H) = \frac{55}{90}$ (pine or warped)

(e) $P(F \cup H)' = \frac{47}{90}$ (not birch nor warped)

17. 0.3456 **18.** 0.00154 **19.** No. $P(E)P(F) = \left(\frac{20}{52}\right)\left(\frac{4}{52}\right)$, $P(E \cap F) = \frac{4}{52}$

20. Yes. $P(R)P(Q) = \left(\frac{1}{2}\right)\left(\frac{4}{52}\right)$, $P(R \cap Q) = \frac{2}{52}$ **21. (a)** $\frac{1}{1296}$ **(b)** $\frac{24}{1296}$ **(c)** $\frac{1}{72}$ **22. (a)** $\frac{5}{36}$ **(b)** $\frac{1}{6}$

23. (a) 0.036 **(b)** 0.196 **(c)** 0.084

24. (a) 0.40 **(b)** 0.70 **(c)** $\frac{7}{9}$ **(d)** No. $P(LI)P(BH) = .27$, $P(LI \cap BH) = .35$ **(e)** no **25. (a)** 0.379 **(b)** 0.170

26. (a)

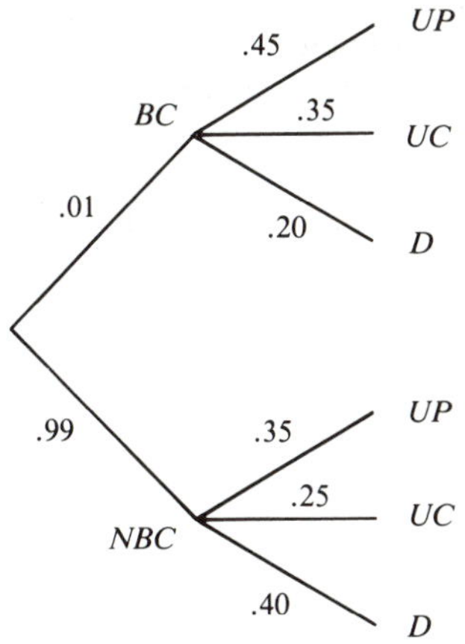

BC (.01): UP .45, UC .35, D .20

NBC (.99): UP .35, UC .25, D .40

(b) 0.045

27. (a) 0 **(b)** $\frac{1}{30}$ **(c)** $\frac{1}{30}$ **(d)** $\frac{1}{2}$ **(e)** $\frac{7}{30}$ **28.** 0.913 **29.** 0.266 **30.** 0.670 **31.** 0.0171

CHAPTER 7

Section 7–1, page 394

1. (a) 2 **(b)** 3 **(c)** 4 **(d)** 2 **(e)** 1

3.

Category	Frequency
85–99	7
100–114	8
115–129	13
130–144	7
145–159	5

5. (a) 67 **(b)** 19 **(c)** 22 **(d)** cannot determine **(e)** 40 **(f)** cannot determine

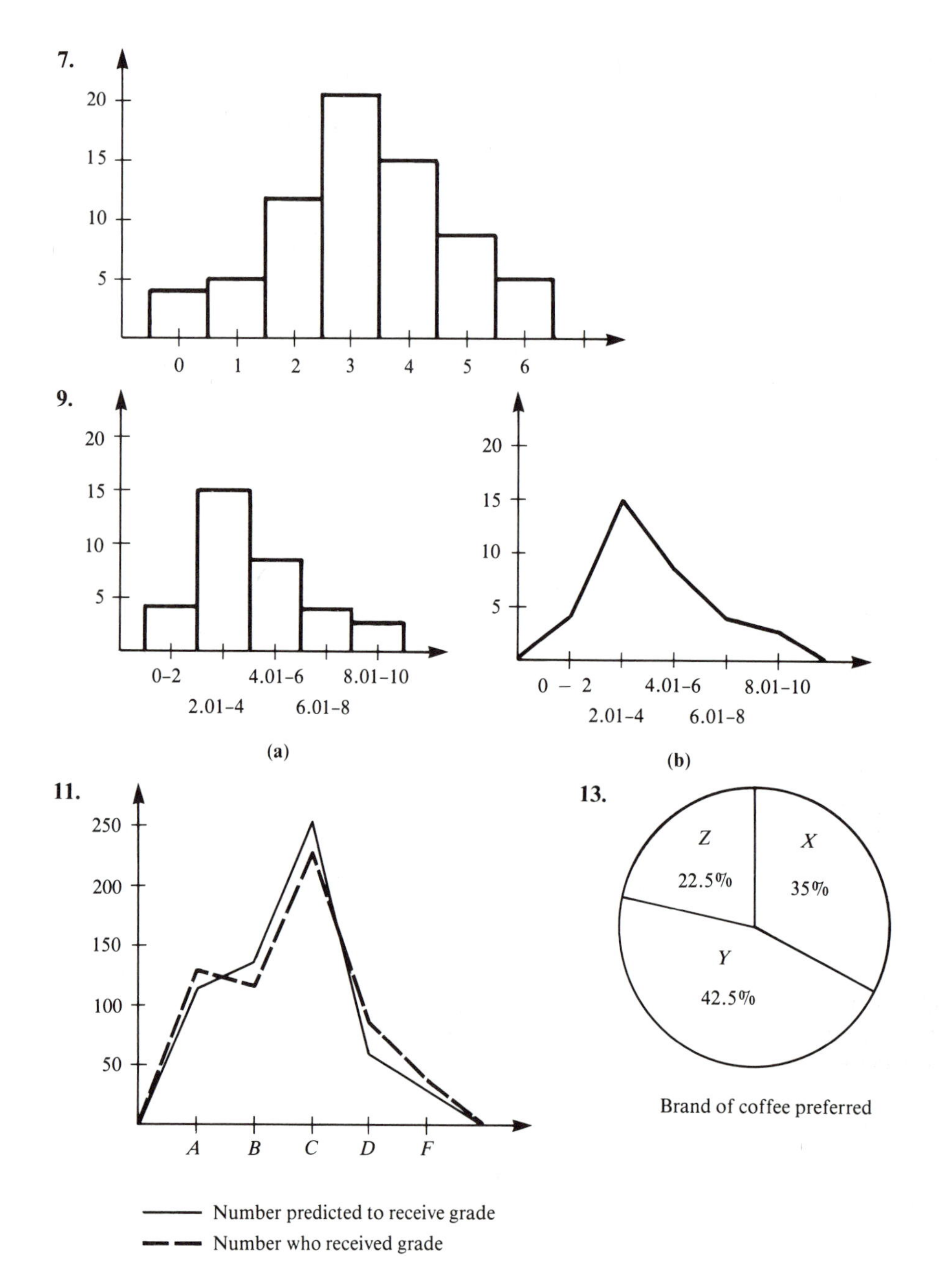
7.
20
15
10
5
0
1
2
3
4
5
6
9.
20
15
10
5
0–2
2.01–4
4.01–6
6.01–8
8.01–10
(a)
20
15
10
5
0 – 2
2.01–4
4.01–6
6.01–8
8.01–10
(b)
11.
250
200
150
100
50
A
B
C
D
F
Number predicted to receive grade
Number who received grade
13.
Z
22.5%
X
35%
Y
42.5%
Brand of coffee preferred

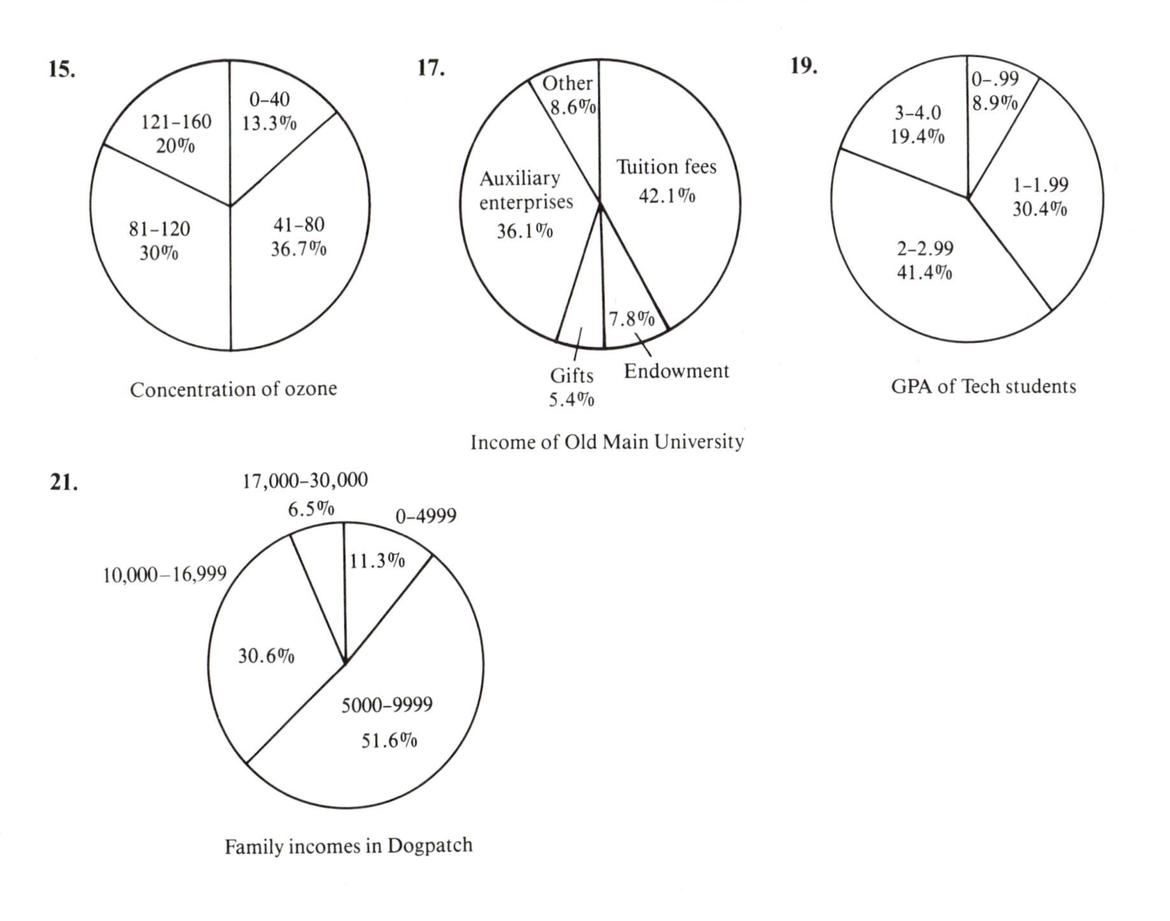

Section 7–2, page 408

1. 6 **3.** 3.9 **5.** 0.667 **7.** 4.825 **9.** 77.5 **11.** 185.25 **13.** 75.9 **15.** 76 **17.** 9 **19.** 23
21. 72 **23.** 5 **25.** 1 and 5 **27.** no mode **29.** \$138.40 **31.** Mean = \$3.97, median = \$3.95
33. 82 **35.** \$13,355 **37.** 160.7 miles **39.** It is 82 or larger **41.** \$1.30

43. **(a)** It is 7.4, 3 more than 4.4.
(b) It is 10 more than 4.4.
(c) It is $2 \times 4.4 = 8.8$.
(d) It is 6×4.4.
(e) When a constant is added to each score, it increases the mean by that constant. When each score is multiplied by a constant, the mean is multiplied by that constant.

Section 7–3, page 420

1. Mean = 16, variance = 15.5, $\sigma = \sqrt{15.5} = 3.94$ **3.** Mean = 9, variance = 10.4, $\sigma = \sqrt{10.4} = 3.22$
5. Mean = 35, variance = 266.2, $\sigma = 16.32$ **7.** Mean = -2, variance = 11.67, $\sigma = 3.42$
9. Mean = 5.5, variance = 130.92, $\sigma = 11.44$ **11.** Mean = 16.6, variance = 50.072, $\sigma = 7.076$
13. Mean = 8.8, variance = 13.47, $\sigma = 3.67$ **15.** Mean = 75.1, variance = 126.29, $\sigma = 11.24$

17. (a) The standard deviation would help us to determine if the rainfall is about the same each month or if most of it occurs in one or two months.
(b) The standard deviation could help us to determine if most of the grades were clustered near 82.
(c) It would help us to know if the temperature was near 80 degrees all day or if it became very hot during the day.
(d) Were there times when the speed was well over 50 mph?
19. (a) For lathe I, $\sigma = .0036$. For lathe II, $\sigma = .0047$.
(b) Lathe I is more consistent because σ is smaller.
21. (a) Six 3's, one 2, and one 4.
(b) Four 2's, no 3's, and four 4's
(c) No, because the largest possible value of the sum of squared deviations is 8 for which $\sigma = 1$.
23. 33rd out of 110 **25.** 173
27. The Z score for Joe was $-.75$, and the Z score for Josephine was -1.43. Joe scored better because he had the larger Z score.

Section 7–4, page 428

1.

Outcome	X
HHH	3
HHT	2
HTH	2
THH	2
TTH	1
THT	1
HTT	1
TTT	0

3. 0, 1, 2, 3 **5. (a)** 0, 1, 2, 3, 4 **(b)** 0, 1, 2, 3

7. (a) 1, 2, 3, . . . , to length of longest word **(b)** discrete **9.** discrete **11.** discrete **13.** discrete **15.** yes

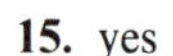

17.

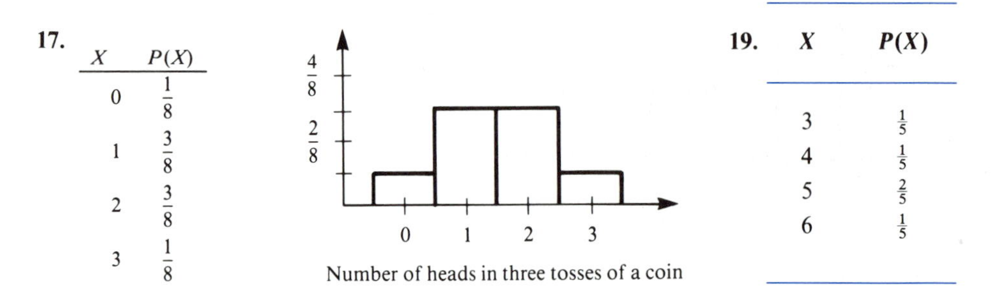

X	$P(X)$
0	$\frac{1}{8}$
1	$\frac{3}{8}$
2	$\frac{3}{8}$
3	$\frac{1}{8}$

19.

X	$P(X)$
3	$\frac{1}{5}$
4	$\frac{1}{5}$
5	$\frac{2}{5}$
6	$\frac{1}{5}$

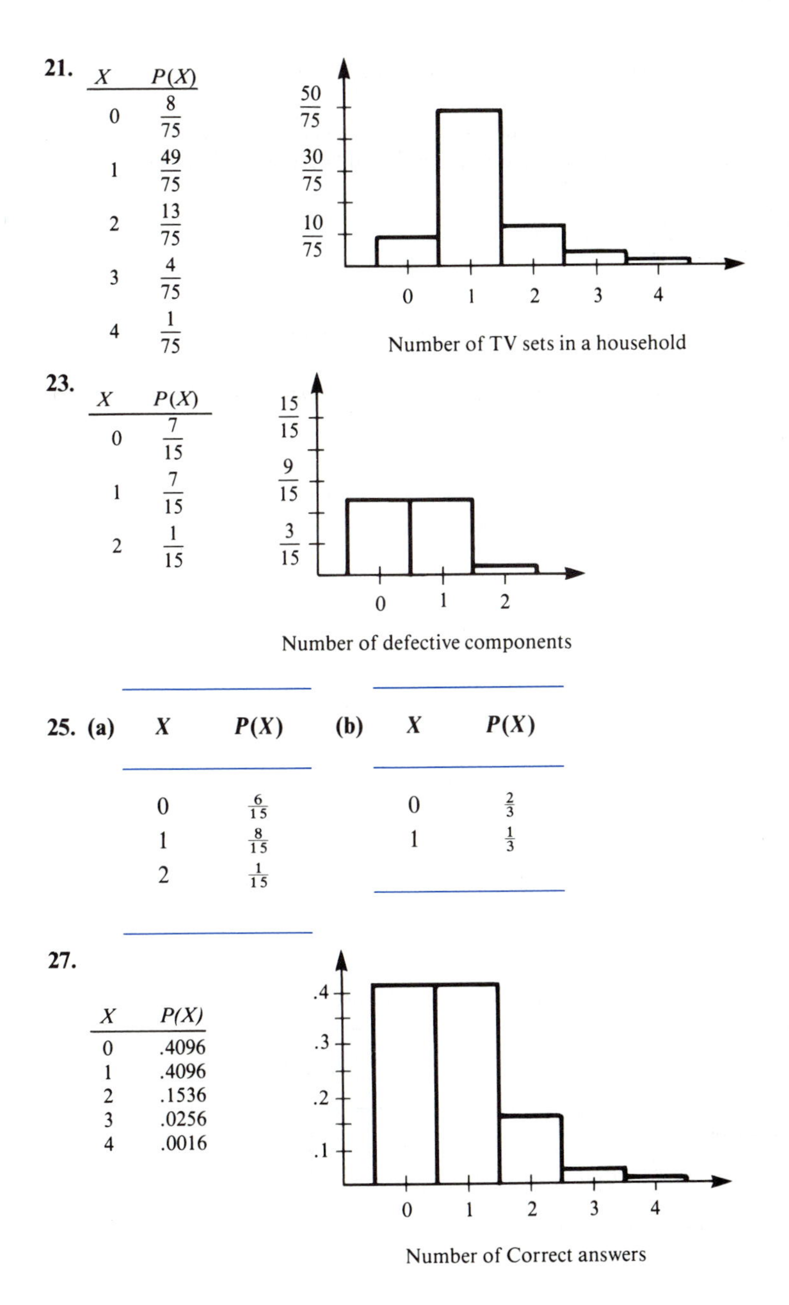

21.

X	$P(X)$
0	$\frac{8}{75}$
1	$\frac{49}{75}$
2	$\frac{13}{75}$
3	$\frac{4}{75}$
4	$\frac{1}{75}$

23.

X	$P(X)$
0	$\frac{7}{15}$
1	$\frac{7}{15}$
2	$\frac{1}{15}$

25. (a)

X	$P(X)$
0	$\frac{6}{15}$
1	$\frac{8}{15}$
2	$\frac{1}{15}$

(b)

X	$P(X)$
0	$\frac{2}{3}$
1	$\frac{1}{3}$

27.

X	$P(X)$
0	.4096
1	.4096
2	.1536
3	.0256
4	.0016

Section 7–5, page 436

1. 12.8 **3.** 325 **5.** \$5.00 **7. (a)** \$.50 **(b)** \$1.50 **9. (a)** \$227.50 **(b)** \$28,437.50
11. $\text{var}(X) = 1049$, $\sigma(x) = 32.39$ **13.** $\mu = 34$, $\sigma(x) = 24.98$
15. Venture *A* has an expected value of \$23,000 and venture *B* \$28,000, so *B* is the better risk.
17. 3.55 **19. (a)** \$15.95 **(b)** \$2,392,500

Section 7–6, page 452

1. 19.15% **3.** 9.87% **5.** 36.43% **7.** 49.81% **9.** 27.34% **11.** 0.1772 **13.** 0.3413 **15.** 48.44% **17.** 29.60% **19.** −3.0 **21.** 0.05 **23.** 2.65 **25.** 38.98% **27.** 13.36% **29.** 0.1360 **31.** 0.4207 **33.** 0.3174 **35.** 44.30% **37.** 65.54% **39.** 30.85% **41.** 1.41 **43.** 1.70 **45.** **(a)** 0.0977 **(b)** 0.2033 **(c)** 0.3256 **47.** **(a)** 478 **(b)** 16 **(c)** 64 **49.** **(a)** 0.5468 **(b)** 0.1555 **(c)** 0.5000, 0.7422 **(d)** 352 **(e)** 0.1079 **51.** 0.621 **53.** 16.98% **55.** 0.0968 **57.** 74.86%

Section 7–7, page 462

1.

X	$P(x)$
0	.1681
1	.3602
2	.3087
3	.1323
4	.0284
5	.0024

3.

X	$P(x)$
0	.0778
1	.2592
2	.3456
3	.2304
4	.0768
5	.0102

5.

X	$P(x)$
0	.4823
1	.3858
2	.1157
3	.0154
4	.0008

7.

X	$P(X)$
0	.0625
1	.2500
2	.3750
3	.2500
4	.0625

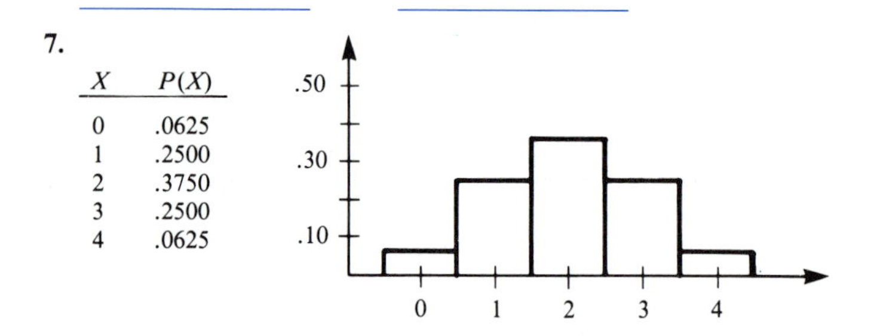

9.

X	$P(X)$
0	.0256
1	.1536
2	.3456
3	.3456
4	.1296

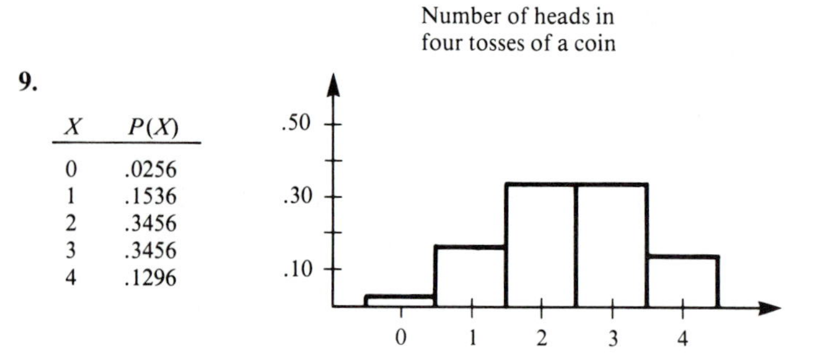

11. $C(90, 40)(.4)^{40}(.6)^{50}$ **13.** $C(75, 35)(.25)^{35}(.75)^{40}$ **15.** $\mu = 20$, $\text{var}(x) = 12$, $\sigma(x) = 3.46$ **17.** $\mu = 312$, $\text{var}(x) = 149.76$, $\sigma(x) = 12.24$ **19.** $\mu = 37.6$, $\text{var}(x) = 34.592$, $\sigma(x) = 5.88$

21. $np = 35$, $nq = 15$, so normal distribution is a good estimate.
23. $np = 36$, $nq = 4$, so normal distribution is not a good estimate.
25. $np = 12.5$, $nq = 12.5$, so normal distribution is a good estimate. **27.** **(a)** 0.0377 **(b)** 0.9450 **(c)** 0.6971
29. **(a)** 0.9830 **(b)** 0.0008 **(c)** 0.0162 **31.** 0.5167 **33.** 0.4041 **35.** 0.1493 **37.** 0.9357 **39.** 0.9265
41. **(a)** 47 **(b)** 0.3745

Section 7–8, page 470

1. 0.561, 0.719 **3.** 0.193, 0.407 **5.** 0.149, 0.451 **7.** 0.451, 0.549 **9.** **(a)** 0.403, 0.497 **(b)** 0.394, 0.506
11. **(a)** 0.218, 0.342 **(b)** 0.236, 0.324 **13.** 0.757 to 0.863 **15.** 0.434 to 0.646 **17.** **(a)** 9604 **(b)** 16,641
19. 156 **21.** 0.504 to 0.696 **23.** 40.3% to 45.7%

Review Exercises, Chapter 7, page 473

1.

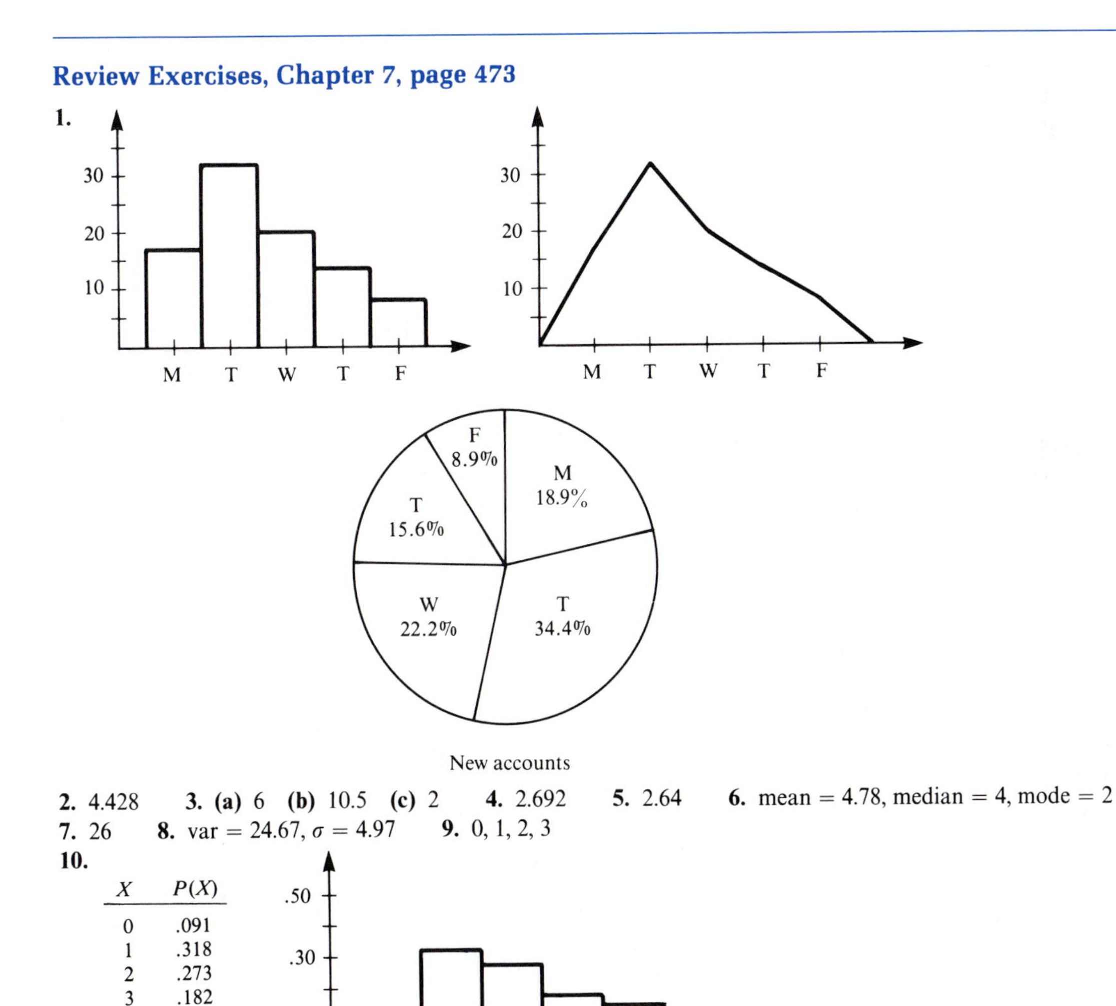

New accounts

2. 4.428 **3.** **(a)** 6 **(b)** 10.5 **(c)** 2 **4.** 2.692 **5.** 2.64 **6.** mean = 4.78, median = 4, mode = 2
7. 26 **8.** var = 24.67, $\sigma = 4.97$ **9.** 0, 1, 2, 3
10.

X	$P(X)$
0	.091
1	.318
2	.273
3	.182
4	.136

11. 2.85 **12.** mean = 3.74, variance = 2.292 **13.** 2.72 **14. (a)** 0.4082 **(b)** 0.7011 **(c)** 0.0475
15. 0.0227 **16. (a)** 0.0918 **(b)** 0.2981 **(c)** 0.3779

17.

X	$P(x)$
0	.3164
1	.4219
2	.2109
3	.0469
4	.0039

18. $\mu = 7.7, \sigma = 2.237$ **19. (a)** 0.6368 **(b)** 0.0009 **(c)** 0.1669

20. (a) 0.0704 **(b)** 0.0466 **(c)** 0.02 **21. (a)** 0.312 to 0.588 **(b)** 0.193 to 0.407 **22.** 0.575 to 0.892
23. 1068 **24. (a)** 1.20 **(b)** -1.00 **(c)** 217 **25.** 80th percentile
26. His Z score was .78 and hers was 1.08, so she had the better score.

CHAPTER 8

Section 8–1, page 485

1. \$66.00 **3.** \$90.00 **5.** \$31.66 **7.** $I =$ \$18.00, $A =$ \$318.00 **9.** \$120.00 **11.** \$52.50 **13.** \$116.38
15. \$660.00 **17.** \$860 **19.** 6.5% **21.** 8.5% **23.** \$3024 **25.** \$7478.25 **27.** \$1607.14
29. discount = \$138.75, PR = \$1711.25 **31.** discount = \$94.58, PR = \$390.42 **33.** \$1050
35. \$3750 annually, \$22,500 total **37. (a)** \$750, **(b)** \$54,000 **39.** 3 years **41.** 3 months **43.** \$650
45. interest = \$150.00, Principal = \$29.95 **47.** 9.3% **49.** \$6576.73 **51.** \$39,175.26
53. simple interest = \$8.67 = simple discount **55.** 10.53%
57. 10.1% discount is \$104.52, 10.3% simple interest is \$103.00, so simple interest is better. **59.** \$49.5 billion
61. \$225,000 **63.** \$16,000 payments, \$320,000 total **65.** \$982,500 **67.** 7.2% **69.** \$1,988,125

Section 8–2, page 497

1. (a) 1.225043 **(b)** 1.1040807 **(c)** 1.0613634 **(d)** 1.071225 **(e)** 1.1925185 **(f)** 1.1314081
3. (a) 7.5% **(b)** 3.75% **(c)** 1.25% **5. (a)** \$15,386.24 **(b)** \$5,386.24 **7. (a)** \$29,386.56 **(b)** \$9,386.56
9. (a) \$2,382.04 **(b)** \$382.04 **11.** \$1004.08 **13.** \$1965.65
15. (a) \$15,972.00 **(b)** \$16,081.20 **(c)** \$16,138.68 **17.** \$18,061.10 **19.** \$483.48 **21.** \$720.74
23. \$34.08 **25.** \$10,299.68 **27.** \$2020.87 **29.** 6.09% **31.** 12 years **33.** 6 years **35.** 10%
37. 11.5 years
39. 8.8% compounded semiannually has an effective rate of 8.9936%, while 8.6% compounded quarterly has an effective rate of 8.8813%, so 8.8% is better.
41. 9.2% compounded quarterly is better because its effective rate is 9.522%.
43. 10% compounded quarterly is better because its effective rate is 10.38%. **45.** 42,665
47. \$26,115.30 **49.** 8% **51.** 6% **53.** \$16,000 lump sum. The other will be worth \$15,919.80 **55.** 8%
57. about 14 years **59.** \$162,135.86

Section 8–3, page 505

1.

First Deposit	798.60
Second Deposit	726.00
Third Deposit	660.00
Fourth Deposit	600.00
Total	2784.60

3. **(a)** 18.59891 **(b)** 29.77808 **(c)** 14.19203 **(d)** 19.59863 **(e)** 12.68250 **5.** \$58,666.00 **7.** \$17,800.80 **9.** \$15,210.93 **11.** \$153,872.46 **13.** \$60,470.64 **15.** \$1,244.21 **17.** \$4,341.11 **19.** \$65.74 **21.** \$3,869.93 **23.** \$764.73 **25.** \$13,435.19 **27.** \$4,802.44 **29.** \$306.72 **31.** \$51,299.00 **33.** \$20,969.37 **35.** \$34,365.00 **37.** about 9% **39.** **(a)** \$14,486.56 **(b)** \$1,158.92

Section 8–4, page 516

1. 12.84926 **3.** 10.83777 **5.** \$33,550.40 **7.** \$9,456.96 **9.** \$3,782.79 **11.** \$336,637.00 **13.** \$2,400.00 **15.** \$15,000 **17.** \$1,476.14 **19.** \$3,207.36 **21.** \$13,077.68 **23.** \$194.97 **25.** \$167.29 **27.** **(a)** \$297,270.00 **(b)** \$222,270.00 **29.** **(a)** \$850.00 interest, \$20.96 interest **(b)** \$261,288.00 **31.** \$46,576.21 **33.** \$7600 **35.** \$574.49 **37.** \$9,818.15 **39.** \$65,405.72 **41.** \$83.68, \$669.44 **43.** \$33,698.88 **45.** \$2,847.71

Review Exercises, Chapter 8, page 519

1. \$90.00 **2.** \$55.00 **3.** \$960.00 **4.** \$840.00 **5.** \$1,350.00 **6.** \$7,231.25 **7.** discount = \$1530, proceeds = \$6970 **8.** \$1,077.84 **9.** \$1,125.22 **10.** interest = \$107.76, amount = \$607.76 **11.** 9 years **12.** \$1,215.51 **13.** 6.09% **14.** 5.0625% **15.** 26.824% **16.** The effective rate of 12% compounded quarterly is 12.551%, so 13% is better. **17.** 18% compounded quarterly is better because its effective rate is 19.25%. **18.** \$5832 **19.** **(a)** \$10,099.82 **(b)** \$2,099.84 **20.** **(a)** \$3,307.50 **(b)** \$307.50 **21.** **(a)** \$9,043.65 **(b)** \$4,043.65 **22.** \$161,270.40 **23.** \$17,496 **24.** \$6030.23 **25.** \$33,172.45 **26.** 11.5 years **27.** 19 quarters or 4.75 years **28.** **(a)** 24.29737 **(b)** 29.77808 **(c)** 21.57856 **29.** \$2,189.94 **30.** \$4,774.56 **31.** \$5,637.09 **32.** \$7,663.40 **33.** \$7,546.73 **34.** \$7,335.93 **35.** \$7,605.46 **36.** \$5,637.09 **37.** \$69,556.27 **38.** \$1,447.55 **39.** \$21,061.80 **40.** \$34,058.17 **41.** \$98,108.58 **42.** \$10,837.77 **43.** \$33,648.51 **44.** 5 years **45.** 5 years **46.** \$10,018.26 **47.** \$301.39 **48.** \$627.98 **49.** \$1,207.68 **50.** \$647.52, \$6,475.20 **51.** \$3,855.88, \$96,397 **52.** \$885.72, \$21,257.28 **53.** \$29,991.02 **54.** \$14,794.91 **55.** \$2,434.91 **56.** interest = \$650.00, principal = \$18.59, total paid in 30 years = \$240,692.40

CHAPTER 9

Section 9–1, page 533

1.

	H	T
H	$-\$1$	$-50¢$
T	$-50¢$	$\$2$

3.

R	*1*	*2*
1	2	−3
2	−3	4

5. **(a)** R gets 10 **(b)** 10 **(c)** 10 **(d)** 8

7. value = 0; saddle point (1, 1); strategy r_1 and c_1 **9.** not strictly determined
11. value = 0, saddle point (2, 2); strategy r_2 and c_2
13. **(a)** value = 0; saddle point (2, 2); solution is strategies r_2 and c_2
(b) value = −1; saddle point (1, 3); solution is r_1 and c_3
(c) not strictly determined
(d) value = 0; saddle point (2, 2); solution is r_2 and c_2
15. R should adopt r_1 and C should adopt c_1 because (1, 1) is the saddle point. R will then gain 120 units from C.
17. Since (2, 2) is the saddle point, A should adopt strategy a_2 and B should adopt b_2, giving A a gain of 10 units of land.
19. Each should adopt their third strategy because the saddle point is (3, 3). The payoff is a purchase of 300 acres of land by R at the expense of C.

Section 9–2, page 544

1. $\frac{49}{8}$ **3.** 0 **5.** **(a)** −4 **(b)** $\frac{13}{2}$ **(c)** $\frac{17}{2}$ **(d)** 15 **(e)** 5.62 **(f)** 24.3

7. R: $\begin{bmatrix} \frac{11}{13} & \frac{2}{13} \end{bmatrix}$, C: $\begin{bmatrix} \frac{4}{13} & \frac{9}{13} \end{bmatrix}$, value $= \frac{60}{13}$ **9.** R: $\begin{bmatrix} \frac{1}{2} & \frac{2}{3} \end{bmatrix}$, C: $\begin{bmatrix} \frac{5}{6} & \frac{1}{6} \end{bmatrix}$, value $= \frac{19}{3}$

11. offense: $\begin{bmatrix} 0 & \frac{14}{17} & \frac{3}{17} \end{bmatrix}$, defense: $\begin{bmatrix} \frac{7}{17} & 0 & \frac{10}{17} \end{bmatrix}$, value $= \frac{72}{17} = 4.24$ **13.** **(a)** $\begin{bmatrix} 0 & 4 \\ 2 & 1 \end{bmatrix}$ **(b)** $\begin{bmatrix} 6 & 2 & -2 & 1 \\ 4 & 1 & 9 & 2 \\ 2 & 4 & 6 & 8 \end{bmatrix}$

15. **(a)** not fair **(b)** fair **(c)** strictly determined, so not fair **17.** 25% tourists, 75% locals
19. 90% industry, 10% tourism **21.** 83% standard, 17% economy

Review Exercises, Chapter 9, page 548

1. **(a)** not strictly determined
(b) strictly determined with saddle point (2, 2) and value 4
(c) strictly determined with saddle point (2, 2) and value 275
(d) strictly determined with saddle point (3, 2) and value 2
2. The game is strictly determined with saddle point (1, 2), so R should adopt r_1 and C should adopt c_2.
3. **(a)** The payoff matrix is not strictly determined, so there is no solution.
(b) The saddle point is (2, 1), so the offense should adopt r_2 and the defense c_1. The payoff is a gain of 4 yards by the offense.
4. The saddle point is (3, 2), so X should use r_3 and Y should use c_2. **5.** **(a)** 7.16 **(b)** 6.80 **(c)** 2.40

6. **(a)** R: [.3 .7], C: [.6 .4], value = 4.8

(b) R: $\begin{bmatrix} \frac{4}{11} & \frac{7}{11} \end{bmatrix}$, C: $\begin{bmatrix} \frac{1}{11} & \frac{10}{11} \end{bmatrix}$, value $= \frac{48}{11}$

(c) row 2 and column 2 are dominated by row 1 and column 1, so r_2 and c_2 are deleted, giving

$$\begin{bmatrix} 3 & 7 \\ 6 & 5 \end{bmatrix}$$

For the matrix the strategies are R: [.2 .8], C: [.4 .6], and value = 5.4.

7. 24% field, 76% greenhouse

Index

M

N

O

P

Q

R

S

T

U

V

W

X

Y

Z